Compendium of Soybean Diseases and Pests

FIFTH EDITION

Edited by

G. L. Hartman
U.S. Department of Agriculture–Agricultural Research Service
Department of Crop Sciences, National Soybean Research Center
University of Illinois, Urbana

J. C. Rupe
Department of Plant Pathology
University of Arkansas, Fayetteville

E. J. Sikora
Department of Entomology and Plant Pathology
Auburn University, Auburn, Alabama

L. L. Domier
U.S. Department of Agriculture–Agricultural Research Service
Department of Crop Sciences
University of Illinois, Urbana

J. A. Davis
Department of Entomology
Louisiana State University Agricultural Center, Baton Rogue

K. L. Steffey
Dow AgroSciences LLC, Indianapolis, Indiana

APS PRESS
The American Phytopathological Society
St. Paul, Minnesota, U.S.A.

Photo Credits
Front cover: Courtesy Chungking/Shutterstock.
Back cover: Top left and bottom right: Courtesy G. L. Hartman; Top right and bottom left: Courtesy T. K. Herman.
Interior images: All credits appear in place with the images.

Library of Congress Card Catalog Number: 2015941927
International Standard Book Number: 978-0-89054-473-0

First edition published 1975
Second edition published 1982
Third edition published 1989
Fourth edition published 1999
Fifth edition published 2015

Printed in the United States of America on acid-free paper

The American Phytopathological Society
3340 Pilot Knob Road
St. Paul, Minnesota 55121 U.S.A.

Preface

This fifth edition of *Compendium of Soybean Diseases and Pests* builds on previous editions to provide a thorough, authoritative, and practical reference for plant pathologists and others working with soybean diseases in the field, in diagnostic laboratories, in plant clinics, and in academic institutions. This revised edition is both more international in scope and more comprehensive than previous editions. It will be useful to advisors in government departments of agriculture and regulatory agencies, agribusiness representatives, crop consultants and scouts, county agents, educators and students, producers, and researchers worldwide.

Each section of the compendium has been revised and updated by scientists who are authorities in specific areas of soybean research. (Those who prepared sections on individual diseases, pests, and other topics are acknowledged at the ends of their respective sections and also listed in the Contributors section that follows this Preface.) New sections in the fifth edition describe arthropod pests and their management, summarize the molecular detection of pathogens, and provide an international perspective on the importance of soybean diseases in the world's leading soybean-producing countries. Two hundred twenty-eight new images have been added to this edition, as well, for a total of 260 illustrations.

In addition, the lists of Selected References provided throughout the book have been revised and include contemporary scientific papers along with classic papers, books, monographs, and proceedings. For additional literature on soybean diseases and their management, readers are encouraged to consult the first four editions of *Compendium of Soybean Diseases* (dating back to 1975) and the specific publications identified for each section.

The fifth edition of *Compendium of Soybean Diseases and Pests* is organized into five main sections:

- The Introduction provides background information about the history and growth of the soybean plant along with a contemporary discussion of the importance of soybean pathogens and pests in the world's chief soybean-producing countries.
- Part I addresses infectious diseases, including those caused by bacteria and mollicutes, fungi and oomycetes, nematodes, and viruses; a section on seed and postharvest pathology is provided, as well.
- Part II, which is new to this edition, covers arthropod pests and provides strategies for their management.
- Part III discusses noninfectious and stress disorders, including those caused by weather-related factors, nutrient deficiencies and toxicities, and various chemicals and pollutants.
- Part IV provides strategies for managing soybean diseases and pests. The discussion focuses on biological enhancements, cultural practices, host plant resistance, and pesticide use.

This new edition also includes several appendices that identify common names of diseases of soybean (along with their pathogens) and arthropod pests associated with soybean. A comprehensive glossary and index are provided, as well.

Throughout the book, the common names of diseases and pests have mostly been retained from the previous edition. Within each disease/pest section, the most recent scientific name of the pathogen/pest is provided, as well. Authorities for the scientific names are included in the appendices. The primary sources used for scientific names and authorities of pathogens/pests include the List of Prokaryotic Names with Standing in Nomenclature (LPSN) database, the U.S. Department of Agriculture (USDA)–Agricultural Research Services (ARS) Fungal Database, the International Committee on Taxonomy of Viruses (ICTV) database, the Nematode Species Descriptions database of the Nematology Lab at the University of Nebraska–Lincoln, and the Common Names of Insects Database of the Entomological Society of America. Within the text, plants are mostly referred to by their common names; scientific names are also provided for those plants whose common names may be unfamiliar to readers. The primary resource for scientific names of plants is the USDA Natural Resources Conservation Service PLANTS Database.

The management strategies recommended at the ends of the disease/pest sections are purposely general and reflect the principles and cultural practices applicable in most soybean-growing areas of the world. Within each Management section, resistance (if known) is discussed along with other management strategies. More information on resistance can be found in specific publications; a general list of sources is provided by the USDA–ARS National Plant Germplasm System (www.ars-grin.gov). A discussion of strategies for managing soybean disease is presented in Part IV of the book. For current information on chemical management and cultural practices, readers should consult recent literature, competent area specialists, extension or advisory plant pathologists, or informed agricultural suppliers.

Numerous databases are available for obtaining up-to-date information about soybean. The USDA National Agricultural Statistics Service (www.nass.usda.gov) provides timely, accurate, and useful data in service to U.S. agriculture, including the total soybean areas planted and harvested and production in volume and dollar values over many decades. In addition, the USDA–ARS and Iowa State University website SoyBase and the Soybean Breeder's Toolbox (www.soybase.org) provides up-to-date information on integrating genetics and molecular biology. Another database that is useful for providing a global view of hectares planted and produced is that of the United Nations Food and Agriculture Organization (www.fao.org).

The website of the United Soybean Board (USB) (www.unitedsoybean.org) also makes available a variety of helpful resources. On the site, the section Farmer Resources provides information on the role of soybeans in animal agriculture, biotechnology, exports, global opportunities, and soy foods, along with many other interesting aspects of soybeans. Additional resources include information on research projects, weather forecasts, commodity price information, and trade and export information that can help soybean producers and industry leaders expand their knowledge and stay informed of important developments.

Acknowledgments

Special thanks are given to Theresa Herman and Jessica Stewart of the University of Illinois for their editing and organizational contributions. Additional thanks go to C. A. Clark, D. E. Hershman, C. R. Grau, and D. L. Kerns for their review of this fifth edition of *Compendium of Soybean Diseases and Pests*.

The editors wish to thank the authors identified in the Contributors list on pages v–viii for preparing new sections and updating previous sections for this edition. Finally, the editors wish to thank the following individuals, who provided new images for this edition:

S. Beniwal, Virginia Polytechnic Institute and State University, Blacksburg, U.S.A.
S. Bundy, New Mexico State University, Las Cruces, U.S.A.
E. C. Corrigan, Brandt Consolidated, Inc., Bloomington, Illinois, U.S.A.
M. L. Ellis, Iowa State University, Ames, U.S.A.
K. Fothergill, University of Tennessee at Martin, U.S.A.
R. B. Hammond, OARDC, The Ohio State University, Wooster, U.S.A.
T. K. Herman, University of Illinois, Urbana, U.S.A.
M. Iwaki, Virginia Polytechnic Institute and State University, Blacksburg, U.S.A.
C. C. Marett, Iowa State University, Ames, U.S.A.
J. Méndez-Lozano, Instituto Politécnico Nacional, CIIDIR, Unidad Sinaloa, Departamento de Biotecnología Agrícola, Maestría en Recursos Naturales y Medio Ambiente, Guasave, Sinaloa, México
M. L. Pawlowski, University of Illinois, Urbana, U.S.A.
R. Scott, USDA–ARS, Beltsville, Maryland, U.S.A.
J. H. Taylor, University of Arkansas, Fayetteville, U.S.A.
G. L. Tylka, Iowa State University, Ames, U.S.A.
L. Wen, University of Illinois, Urbana, U.S.A.

Contributors

P. Agudelo
Department of Entomology, Soils, and Plant Sciences
Clemson University
Clemson, South Carolina, U.S.A.

A. M. R. Almeida
Embrapa Soja
Londrina, Paraná, Brazil

M. V. Avanzato
Department of Plant Pathology
University of Arkansas
Fayetteville, Arkansas, U.S.A.

L. L. Buschman
Department of Entomology
Kansas State University
Manhattan, Kansas, U.S.A.

G. Cai
Department of Plant Pathology and Crop Physiology
Louisiana State University Agricultural Center
Baton Rouge, Louisiana, U.S.A.

A. Catchot
Department of Entomology
Mississippi State University
Mississippi State, Mississippi, U.S.A.

A. Chanda
Department of Plant Pathology and Crop Physiology
Louisiana State University Agricultural Center
Baton Rouge, Louisiana, U.S.A.

P. Chen
Department of Agronomy
University of Arkansas
Fayetteville, Arkansas, U.S.A.

M. I. Chilvers
Department of Plant, Soil, and Microbial Sciences
Michigan State University
East Lansing, Michigan, U.S.A.

L. M. Costamilan
Embrapa Trigo
Passo Fundo, Rio Grande do Sul, Brazil

V. Damsteegt
USDA–ARS Foreign Disease/Weed Science Research Unit
Fort Detrick, Maryland, U.S.A.

J. A. Davis
Department of Entomology
Louisiana State University Agricultural Center
Baton Rouge, Louisiana, U.S.A.

W. P. Dias
Embrapa Soja
Londrina, Paraná, Brazil

L. L. Domier
USDA–ARS and Department of Crop Sciences
University of Illinois
Urbana, Illinois, U.S.A.

A. E. Dorrance
Department of Plant Pathology
The Ohio State University
Columbus, Ohio, U.S.A.

N. Dufault
Plant Pathology Department
University of Florida
Gainesville, Florida, U.S.A.

C. V. Godoy
Embrapa Soja
Londrina, Paraná, Brazil

C. R. Grau
Department of Plant Pathology
University of Wisconsin
Madison, Wisconsin, U.S.A.

L. E. Gray
USDA–ARS and Department of Crop Sciences
University of Illinois
Urbana, Illinois, U.S.A.

J. K. Greene
Department of Entomology, Soils, and Plant Sciences
Clemson University
Clemson, South Carolina, U.S.A.

G. K. Gupta
Parmar University of Horticulture and Forestry
Regional Fruit Research Station
Mashobra, Shimla, India

A. Hager
Department of Crop Sciences
University of Illinois
Urbana, Illinois, U.S.A.

S. Hardaningsih
Indonesian Legumes and Tuber Crops Research Institute
Malang, Indonesia

G. L. Hartman
USDA–ARS and Department of Crop Sciences
National Soybean Research Center
University of Illinois
Urbana, Illinois, U.S.A.

R. M. Harveson
Institute of Agriculture and Natural Resources
University of Nebraska
Lincoln, Nebraska, U.S.A.

J. S. Haudenshield
USDA–ARS and Department of Crop Sciences
University of Illinois
Urbana, Illinois, U.S.A.

A. A. Henning
Embrapa Soja
Londrina, Paraná, Brazil

A. Herbert
Tidewater Agricultural Research and Extension Center
Virginia Polytechnic Institute and State University
Suffolk, Virginia, U.S.A.

C. B. Hill
Department of Crop Sciences
University of Illinois
Urbana, Illinois, U.S.A.

J. H. Hill
Department of Plant Pathology
Iowa State University
Ames, Iowa, U.S.A.

T. Hymowitz
Department of Crop Sciences
University of Illinois
Urbana, Illinois, U.S.A.

M. Kato
Agriculture, Forestry, and Fisheries Research Council
Tsukuba, Japan

S. R. Koenning
Department of Plant Pathology
North Carolina State University
Raleigh, North Carolina, U.S.A.

P. U. Kuruppu
Department of Plant Pathology and Crop Physiology
Louisiana State University Agricultural Center
Baton Rouge, Louisiana, U.S.A.

L. F. Leandro
Department of Plant Pathology and Microbiology
Iowa State University
Ames, Iowa, U.S.A.

B. R. Leonard
Macon Ridge Research Station
Louisiana State University Agricultural Center
Winnsboro, Louisiana, U.S.A.

S. Li
USDA–ARS
James Whitten Delta States Research Center
Stoneville, Mississippi, U.S.A.

Y. Li
National Center for Soybean Improvement
Nanjing Agricultural University
Nanjing, China

G. Lorenz
Department of Entomology
University of Arkansas
Lonoke, Arkansas, U.S.A.

J. G. Lundgren
USDA–ARS
North Central Agricultural Research Laboratory
Brookings, South Dakota, U.S.A.

D. K. Malvick
Department of Plant Pathology
University of Minnesota
St. Paul, Minnesota, U.S.A.

E. C. McGawley
Department of Plant Pathology and Crop Physiology
Louisiana State University Agricultural Center
Baton Rouge, Louisiana, U.S.A.

A. Mengistu
USDA–ARS Crop Genetics Research Unit
West Tennessee AgResearch and Education Center
Jackson, Tennessee, U.S.A.

M. C. Meyer
Embrapa Soja
Balsas, Maranhão, Brazil

D. S. Mueller
Department of Plant Pathology and Microbiology
Iowa State University
Ames, Iowa, U.S.A.

J. D. Mueller
Department of Plant Pathology
Clemson University
Clemson, South Carolina, U.S.A.

F. R. Musser
Department of Entomology
Mississippi State University
Mississippi State, Mississippi, U.S.A.

R. M. Nair
AVRDC–The World Vegetable Center Regional Center
for South Asia
Hyderabad, Andhra Pradesh, India

B. D. Nelson
Department of Plant Pathology
North Dakota State University
Fargo, North Dakota, U.S.A.

M. E. Newman
Department of Entomology and Plant Pathology
University of Tennessee
Jackson, Tennessee, U.S.A.

T. L. Niblack
Department of Plant Pathology
The Ohio State University
Columbus, Ohio, U.S.A.

G. R. Noel
USDA–ARS and Department of Crop Sciences
University of Illinois
Urbana, Illinois, U.S.A.

M. E. O'Neal
Department of Entomology
Iowa State University
Ames, Iowa, U.S.A

C. Overstreet
Department of Plant Pathology and Crop Physiology
Louisiana State University Agricultural Center
Baton Rouge, Louisiana, U.S.A.

G. B. Padgett
Macon Ridge Research Station
Louisiana State University Agricultural Center
Winnsboro, Louisiana, U.S.A.

L. D. Ploper
Estación Experimental Agroindustrial Obispo Colombres
Tucumán, Argentina

P. P. Price, III
Macon Ridge Research Station
Louisiana State University Agricultural Center
Winnsboro, Louisiana, U.S.A.

R. D. Riggs
Department of Plant Pathology
University of Arkansas
Fayetteville, Arkansas, U.S.A.

C. S. Rothrock
Department of Plant Pathology
University of Arkansas
Fayetteville, Arkansas, U.S.A.

J. C. Rupe
Department of Plant Pathology
University of Arkansas
Fayetteville, Arkansas, U.S.A.

J. S. Russin
Department of Plant Pathology and Crop Physiology
Louisiana State University
Baton Rouge, Louisiana, U.S.A.

S. Sabanadzovic
Department of Biochemistry, Molecular Biology, Entomology, and Plant Pathology
Mississippi State University
Mississippi State, Mississippi, U.S.A.

R. W. Schneider
Department of Plant Pathology and Crop Physiology
Louisiana State University Agricultural Center
Baton Rouge, Louisiana, U.S.A.

N. E. Schroeder
Department of Crop Sciences
University of Illinois
Urbana, Illinois, U.S.A.

C. D. S. Seixas
Embrapa Soja
Londrina, Paraná, Brazil

N. D. Sharma
Department of Plant Pathology
J. N. Agricultural University
Jabalpur, India

E. J. Sikora
Department of Entomology and Plant Pathology
Auburn University
Auburn, Alabama, U.S.A.

E. C. Silva
Plant Pathology and Crop Physiology Department
Louisiana State University Agricultural Center
Baton Rouge, Louisiana, U.S.A.

J. B. Sinclair
Department of Crop Sciences
University of Illinois
Urbana, Illinois, U.S.A.

J. F. Smith
Department of Entomology and Plant Pathology
Mississippi State University
Mississippi State, Mississippi, U.S.A.

R. M. Soares
Embrapa Soja
Londrina, Paraná, Brazil

T. N. Spurlock
Southeast Research and Extension Center (SEREC)
Department of Plant Pathology
University of Arkansas
Monticello, Arkansas, U.S.A.

K. L. Steffey
Dow AgroSciences LLC
Indianapolis, Indiana, U.S.A.

S. D. Stewart
Department of Entomology and Plant Pathology
University of Tennessee
Jackson, Tennessee, U.S.A.

A. U. Tenuta
Ontario Ministry of Agriculture and Food
Ridgetown, Ontario, Canada

K. J. Tilmon
Plant Science Department
South Dakota State University
Brookings, South Dakota, U.S.A.

K. V. Tindall
DuPont Pioneer
Union City, Tennessee, U.S.A.

S. A. Tolin
Department of Plant Pathology
Virginia Polytechnic Institute and State University
Blacksburg, Virginia, U.S.A.

G. L. Tylka
Department of Plant Pathology and Microbiology
Iowa State University
Ames, Iowa, U.S.A.

I. E. Tzanetakis
Department of Plant Pathology
University of Arkansas
Fayetteville, Arkansas, U.S.A.

N. A. Ward-Gauthier
Department of Plant Pathology
University of Kentucky
Lexington, Kentucky, U.S.A.

J. Whalen
Department of Entomology and Wildlife Ecology
University of Delaware
Newark, Delaware, U.S.A.

K. A. Wise
Department of Botany and Plant Pathology
Purdue University
West Lafayette, Indiana, U.S.A.

A. Wrather
Plant Sciences
University of Missouri
Delta Center, Missouri, U.S.A.

H.-C. Yang
Department of Crop Sciences
University of Illinois
Urbana, Illinois, U.S.A.

X. B. Yang
Department of Plant Pathology and Microbiology
Iowa State University
Ames, Iowa, U.S.A.

Y. F. Zhao
Department of Crop Sciences
University of Illinois
Urbana, Illinois, U.S.A.

J. Zhou
Department of Biochemistry and Biophysics and Institute for Plant Genomics and Biotechnology
Texas A&M University
College Station, Texas, U.S.A.

Contents

Part II. Arthropod Pests

Part III. Noninfectious and Stress Disorders

Part IV. Soybean Disease and Pest Management Strategies

Compendium of Soybean Diseases and Pests

FIFTH EDITION

Introduction

History and Growth of the Soybean Plant

Soybean (*Glycine max* [L.] Merr. [Fabaceae, Phaseoleae]) emerged as a domesticate during the Zhou Dynasty (1046–256 BCE) in the eastern half of northern China. Domestication is a process of trial and error, not a time-specific event. In the case of soybean, this process likely took place during the Shang Dynasty (ca. 1500–1100 BCE) or perhaps earlier. Thus, for several thousand years, people in eastern Asia have been using soybean seeds for animal feed, for human food, and as a medicine to treat a number of human disorders.

Soybean was introduced to North America by Samuel Bowen in 1765. Approximately 175 years later, it had become a widely planted, major crop in the United States. Several reviews are available on the origin and early history of soybean (see Selected References: Qui and Chang, 2010; Shurtleff and Akiko, 2013, 2014).

Soybean is the domesticated member of the 26 species that comprise the genus *Glycine* (Table 1). The geographic distribution of the wild perennial species is mostly limited to Australia—the home of 24 wild perennial *Glycine* species. Only a few successful crosses have been made between these perennial species and soybean, and few if any potentially valuable traits have been transferred to commercial soybean cultivars.

Germplasm Collections

There are perhaps more than 100,000 accessions of *G. max,* probably fewer than 10,000 accessions of *G. soja,* and approximately 3,500 accessions of wild perennial *Glycine* species in germplasm collections throughout the world. The exact number of accessions is unknown, and some of them may be duplicates. Major collections can be found in Australia, Brazil, China, Germany, India, Indonesia, Japan, Russia, South Korea, and the United States. Many of the smaller collections are found throughout Africa, Asia, and Europe. Active collection programs are underway, and thus, the number of accessions will continue to grow.

The U.S. Department of Agriculture (USDA) *Glycine* Germplasm Collection, located on the campus of the University of Illinois at Urbana–Champaign, is one of the largest in the world (Table 2). Only China maintains a larger collection. The USDA distributes *Glycine* seed free of charge to scientists around the world; unfortunately, many countries have restrictive policies concerning *Glycine* seed exchange. The search for genetic resistance to pathogens, pests, and abiotic stress problems is currently and historically the largest use of germplasm.

Sources of resistance have been identified for a number of major soybean pathogens and pests. Commercial cultivars are available with acceptable resistance and/or tolerance to some economically significant pathogens and pests. However, as pathogens and pests evolve, germplasm collections will serve as invaluable resources for screening for additional sources of resistance. The availability of high-quality and diverse germplasm is essential, because producing genetic resistance continues to be a major option in reducing yield losses to plant pathogens and pests.

Utilization

The two primary products of soybean are oil and meal. Most modern-day cultivars have an oil content of 18–23% and a protein content of 30–40%. The refined fraction of crude soybean oil is converted to margarine, shortening, mayonnaise, salad oils and dressings, and sandwich spreads. The lecithin fraction of crude soybean oil is used as an emulsifying, surface-active, antispattering, and stabilizing agent. A small percentage of soybean oil is used in industrial products, such as paint, ink, varnish, soap, linoleum, rubber fabrics, cosmetics, and biodiesel fuel. The recent release of soybean cultivars containing high-oleic or low-linolenic fatty acids offers the possibility of increasing the uses of soybean both for human consumption and for new industrial products.

Most soybean meal is used as a source of high protein in animal feeds for the production of beef, butter, eggs, fish, lamb, milk, pork, and poultry. Industrial soybean protein is used in adhesives, wallboard, and paper coatings. Utilization of soybean protein in the form of concentrates, isolates, and textured protein for human consumption is increasing.

In parts of Asia, soybean is used in liquid, powder, and curd forms to make *miso* (fermented soy paste), *shoyu* (soy sauce), *tofu* (soy curd), *natto* (fermented soy cheese), *tempeh, yuba, kinako, hammanatto, kochu chang,* and soymilk. These soy products are also being manufactured and consumed in the Western Hemisphere because of the growing popularity of the vegetarian diet.

World Production

Soybean is the leading oilseed crop produced and consumed in the world. In 2012, world oilseed production was 467 million metric tons (MMT). The top crop producers were soybean (57%), rapeseed (13%), cottonseed (10%), peanut (8%), and sunflower seed (8%).

In 2013, soybean was grown in at least 70 countries, and the annual production was 268 MMT. The leading producers were the United States (31%), Brazil (31%), and Argentina (19%) (Fig. 1), followed by China (5%), India (4%), Paraguay (3%), and Canada (2%). Soybean production on the continent of Africa was less than 1% of total world production, but several studies have indicated the potential for a growth in production there as the demands for oil and for meal continue to increase.

Plant Improvement

The soybean plant has undergone dramatic changes since the 1980s, and further development will no doubt continue at a rapid pace as scientific discoveries and innovations are applied. In the years before soybean breeding, pure-line selections from plant introductions were used for commercial production. Public breeding programs initially developed soybean cultivars through plant-breeding efforts. Private breeding programs

TABLE 1. Species in the Subgenera *Glycine* and *Soja*, along with the Three-Letter Code Associated with Each Species, the Chromosome Number (2n), the Genome Type, and the Geographic Distribution[a]

Subgenus *Glycine* Willd.[b]	Code	2n	Genome Type[c]	Geographic Distribution
G. albicans Tind. & Craven	ALB	40	I	Australia
G. aphyonota B. E. Pfeil	APH	40	?	Australia
G. arenarea Tind.	ARE	40	H	Australia
G. argyria Tind.	ARG	40	A	Australia
G. canescens F. J. Herman	CAN	40	A	Australia
G. clandestine Wendl.	CLA	40	A	Australia
G. curvata Tind.	CUR	40	C	Australia
G. cyrtoloba Tind.	CYR	40	C	Australia
G. falcata Benth.	FAL	40	F	Australia
G. gracei B. E. Pfeil & Craven	GRA	40	A	Australia
G. hirticaulis Tind. & Craven	HIR	40	H	Australia
		80	?	Australia
G. lactovirens Tind. & Craven	LAC	40	I	Australia
G. latifolia (Benth.) Newell & Hymowitz	LAT	40	B	Australia
G. latrobeana (Meissn.) Benth.	LTR	40	A	Australia
G. microphylla (Benth.) Tind.	MIC	40	B	Australia
G. montis-douglas B. E. Pfeil & Craven	MON	40	?	Australia
G. peratosa B. E. Pfeil & Tind.	PER	40	A	Australia
G. pindanica Tind. & Craven	PIN	40	H	Australia
G. pullenii B. E. Pfeil, Tind., & Craven	PUL	?	?	Australia
G. rubiginosa Tind. & B. E. Pfeil	RUB	40	A	Australia
G. stenophita B. E. Pfeil & Tind.	STE	40	B	Australia
G. syndetika B. E. Pfeil & Craven	SYN	40	A	Australia
G. tabacina (Labill.) Benth.	TAB	40	B	Australia
		80	Complex	Australia, South Pacific Islands
G. tomentella Hayata	TOM	38	E	Australia
		40	D	Australia, Papua New Guinea
		78	Complex	Australia, Papua New Guinea
		80	Complex	Australia, Papua New Guinea, Indonesia, Philippines, Taiwan
Subgenus *Soja* (Moench) F. J. Herman				
G. max (L.) Merr.	MAX	40	G	Worldwide
G. soja Sieb. & Zucc.	SOJ	40	G	China, Japan, Korea, Russia, Taiwan (wild soybean)

[a] Adapted from Hymowitz (2004) and Pfeil et al. (2006).
[b] Two species designations have not been resolved and do not appear in this table: *G. dolichocarpa* Tateishi & Ohashi (DOL), which was found in Taiwan and appears to be a member of the *G. tomentella* complex, and *G. pescadrensis* Hayata (PES), which was found in Australia, Japan, and Taiwan and appears to be a member of the *G. tabacina* complex.
[c] Species carrying the same letter have more genomic similarities than species carrying different letters.

started in the 1970s, and by 2015, private cultivars dominated the market in the leading soybean-producing countries.

Much of the rapid improvement in breeding has been accomplished through marker-assisted selection, which began in the 1990s. The first soybean cultivars that were genetically modified to be insensitive to glyphosate were released at that time, as well.

Steady gains in yields have been recorded in soybean for more than 85 years—specifically, a rate of 20 kg per ha per year, based on experimental plots of northern U.S. cultivars. However, larger increases in yield may come from advances in biotechnology, which will produce improvements in multiple traits. These traits will certainly involve improved disease and pest resistance, along with improved photosynthetic and respiratory efficiency to allow for greater allocation of nutrients to developing pods (promoting greater pod survival) and to optimize the capacity of beneficial microbes (maximizing soybean growth).

Plant Growth and Development

The soybean plant is highly responsive to its environment. Preparing the seedbed properly, planting seeds to the correct depth, and limiting weed competition can all improve the overall growth of soybean plants. The productivity of a soybean crop can be altered by changes in cultivar selection, planting date, row spacing, and seeding rates.

The relationship between nodulating bacteria and soybean may also affect crop productivity. When infected by nodulating bacteria, soybean roots form an association with the pathogen in which the plant provides carbohydrates and mineral nutrients to the pathogen, whereas the pathogen provides nitrogen to the host soybean plant. This symbiotic relationship is beneficial to both the plant and the pathogen. Thus, it is important to establish nodulating bacteria via inoculation in a field where soybean has never been grown.

Another microbial association that benefits plants is that with mycorrhizal fungi. These fungi form a network of filaments that associate with plant roots and draw nutrients from the soil, benefiting the plant. This alliance stimulates plant growth, may improve drought tolerance, and optimizes the intake of phosphorus and other nutrients for the plant's benefit.

The soil moisture and temperature requirements of soybean are quite similar to those of maize. The minimum germination temperature for soybean seeds is 12°C, whereas that for maize is 10°C, and the rates of germination for both species increase with warmer soils. Thus, the primary areas of soybean production, like those of maize, are similar and broad in scope—from temperate regions to subtropical and even tropical regions. Soybean seeds require a moisture content of 50% (dry weight basis) for germination. Hence, soil moisture at planting is critical to achieving uniform germination.

Soybean plants grow vegetatively for a few weeks before floral initiation, which is associated with a decreasing photoperiod

TABLE 2. U.S. Department of Agriculture *Glycine* Germplasm Collection[a]

Glycine spp.	Entries	*Glycine* spp.	Entries
G. arenaria	5	*G. microphylla*	33
G. argyrea	14	*G. peratosa*	8
G. canescens	123	*G. pescadrensis*	68
G. clandestina	90	*G. pindanica*	4
G. curvata	9	*G. rubiginosa*	39
G. cyrtoloba	48	*G. soja*	1,181
G. dolichocarpa	13	*G. stenophita*	27
G. falcata	29	*G. syndetika*	6
G. latifolia	44	*G. tabacina*	142
G. latrobeana	6	*G. tomentella*	299
G. max[b]	19,541		

[a] USDA–ARS, National Genetic Resources Program. *Germplasm Resources Information Network (GRIN)* [Online Database]. National Germplasm Resources Laboratory, Beltsville, Maryland. Available online at www.ars-grin.gov/cgi-bin/npgs/html/site_holding.pl?SOY (summary statistics generated 4-12-15).

[b] Holdings for *G. max* (soybean) include plant introductions, germplasm releases, genetic types, and modern, old, and private cultivars.

TABLE 3. Vegetative and Reproductive Growth Stages of Soybean[a,b]

Stage	Vegetative
VE	Emergence: Cotyledons above soil surface.
VC	Cotyledon: Unifoliolate leaves unrolled sufficiently so leaf edges are not touching.
V1	First node: Fully developed leaves at unifoliate node.
V2	Second node: Fully developed trifoliate leaf at first node above unifoliate node.
V3	Third node: Three nodes on main stem with fully developed leaves, beginning with unifoliate node.
VN	*N:* Number of nodes on main stem with fully developed leaves, beginning with unifoliate nodes.
Stage	**Reproductive**
R1	One flower at any node.
R2	Flower at node immediately below uppermost node with completely unrolled leaf.
R3	Pod 0.5 cm long at one of four uppermost nodes with completely unrolled leaf.
R4	Pod 2.0 cm long at one of four uppermost nodes with completely unrolled leaf.
R5	Beans beginning to develop (can be felt when pod is squeezed) at one of four uppermost nodes with completely unrolled leaf.
R6	Pod contains full-sized green beans at one of four uppermost nodes with completely unrolled leaf.
R7	Pods yellowing; 50% leaves are yellow (physiological maturity).
R8	95% of pods are brown (harvest maturity).

[a] Adapted from Fehr et al. (1971).

[b] To use this method to report a disease occurrence at a specific stage of soybean growth, first select a random sample of soybean plants for disease assessment. The sample may consist of individual leaves or plants, groups of plants, or all plants in a plot. Next, calculate the average infection for the units in the sample. Finally, determine the vegetative stage by counting the number of nodes on the main stem (beginning with the unifoliate node) that have a completely unrolled leaf and finding the corresponding entry in the table.

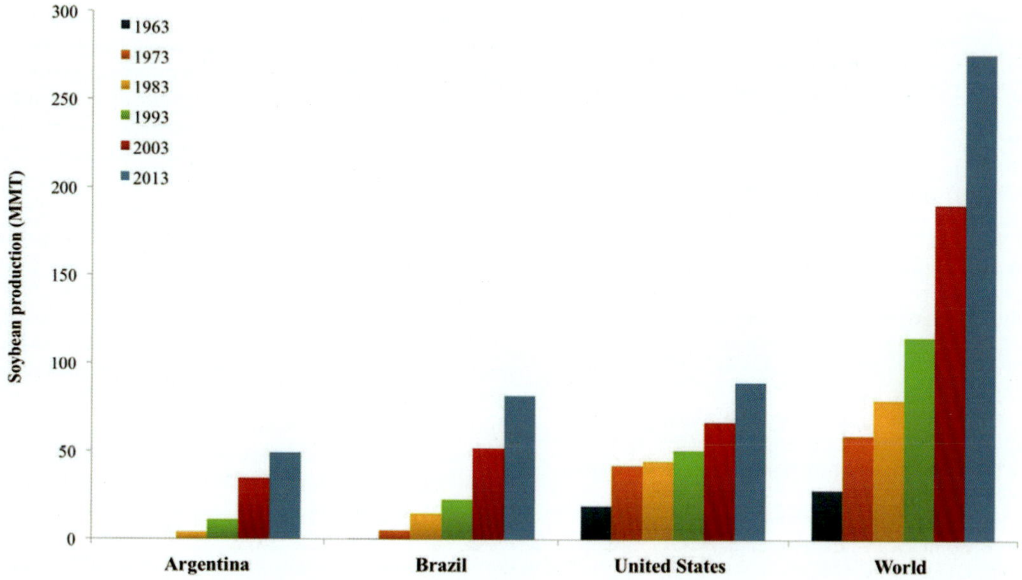

Fig. 1. Soybean production in million metric tons (MMT) in the top three producing countries and total world production from 1963 to 2013. (Data from Food and Agriculture Organization of the United Nations, Statistics Division [www.faostat.fao.org]. Graph courtesy G. L. Hartman)

and influenced by temperature. Soybean plants flower over a 2- to 4-week period and can resist short periods of moisture stress by slowing growth. However, for seeds to fully develop, adequate moisture is needed during the pod-filling growth stages. Soybean grows well in a range of soil types, having a pH between 5.8 and 7.0. The optimum temperature for plant growth and development is 25–30°C.

In tropical regions, soybean completes its growth cycle within 5 months, whereas in temperate regions, the cycle may be completed in as little as 3 months. A standardized method of reporting soybean growth stages is used to record progress through the vegetative and reproductive stages (Table 3). The terminology associated with this method is used in reporting data associated with growth stages for multiple purposes, including field scouting and writing directions for pesticide labels.

Because soybean is a photoperiodic plant, it is classified based on completion of its life cycle from seed to seed. In North America, a relative maturity group (MG) system is used to indicate the region of cultivar adaptation, ranging from MG 000 to MG X. Cultivars designated MG 000 mature the earliest and are adapted to Canada, whereas cultivars designated MG IX and X are adapted to semitropical and tropical production areas.

Selected References

Ainsworth, E. A., and Ort, D. R. 2010. How do we improve crop production in a warming world? Plant Physiol. 154:526-530.

Ainsworth, E. A., Yendrek, C. R., Skoneczka, J. A., and Long, S. P. 2012. Accelerating yield potential in soybean: Potential targets for biotechnological improvement. Plant Cell Environ. 35:38-52.

American Soybean Association (ASA). 2013. Soy Stats. ASA, St. Louis, MO.

Fehr, W. R., Caviness, C. E., Burmood, D. T., and Pennington, J. S. 1971. Stage of development descriptions for soybeans, *Glycine max* (L.) Merrill. Crop Sci. 11:929-931.

Food and Agriculture Organization (FAO). 2006. FAO Statistical Yearbook 2005–2006. FAO, Rome.

Fox, C., Cary, T., Colgrove, A., Nafziger, E., Haudenshield, J. S., Hartman, G. L., Specht, J., and Diers, B. W. 2013. Estimating soybean genetic gain for yield in the northern United States—Influence of cropping history. Crop Sci. 53:2473-2482.

Hoeft, R. G., Nafziger, E. D., Johnson, R. R., and Aldrich, S. R. 2000. Modern Corn and Soybean Production. McSo, Champaign, IL.

Hymowitz, T. 2004. Speciation and cytogenetics. Pages 97-136 in: Soybeans: Improvement, Production, and Uses. H. R. Boerma and J. E. Specht, eds. Agron. Monogr. 16. 3rd ed. American Society of Agronomy–Crop Science Society of America–Social Science Society of America (ASA–CSSA–SSSA), Madison, WI.

Kollipara, K. P., Singh, R. J., and Hymowitz, T. 1995. Genomic relationships to the genus *Glycine* (Fabaceae: Phaseoleae): Use of monoclonal antibody to the soybean Bowman-Birk inhibitor as a genome marker. Am. J. Bot. 82:1104-1111.

Kollipara, K. P., Singh, R. J., and Hymowitz, T. 1997. Phylogenetic and genomic relationships in the genus *Glycine* Willd. based on sequences from the ITS region of nuclear rDNA. Genome 40:57-68.

National Agricultural Statistics Service (NASS). 2013. Quick Stats. U.S. Department of Agriculture, NASS. Available online: http://nass.usda.gov

Orf, J. H. 2008. Breeding, genetics and production of soybeans. Pages 33-65 in: Soybeans: Chemistry, Production Processing, and Utilization. L. A. Johnson, P. J. White, and R. Galloway, eds. American Oil Chemists' Society (AOCS) Press, Urbana, IL.

Pfeil, B. E., Craven, L. A., Brown, A. H. D., Murray, B. G., and Doyle, J. J. 2006. Three new species of northern Australian *Glycine* (Fabaceae, Phaseoleae): *G. gracei*, *G. monti-douglas* and *G. syndetika*. Aust. Syst. Bot. 19:245-258.

Qui, L.-J., and Chang, R.-Z. 2010. The origin and history of soybean. Pages 1-23 in: Soybean: Botany, Production, and Uses. G. Singh, ed. CABI, Wallingford, Oxfordshire, U.K.

Ray, D. K., Mueller, N. D., West, P. C., and Foley, J. A. 2013. Yield trends are insufficient to double global crop production by 2050. PLoS One 8(6):e66428.

Shurtleff, W., and Akiko, A., 2013. History of Whole Dry Soybeans, Used as Beans, or Ground, Mashed or Flaked (240 BCE to 2013). SoyInfo Center, Lafayette, CA.

Shurtleff, W., and Akiko, A. 2014. History of Soybeans and Soyfoods in China and Taiwan, and in Chinese Cookbooks, Restaurants, and in Chinese Work with Soybeans Outside China (1024 BCE to 2014). SoyInfo Center, Lafayette, CA.

U.S. Soybean Export Council (USSEC). 2006. Quality standards for U.S. soybeans and soy products. Pages 2.1-2.11 in: U.S. Soy: International Buyers' Guide. USSEC, Saint Louis, MO.

(Prepared by T. Hymowitz, R. L. Nelson,
J. B. Sinclair, and G. L. Hartman;
revised by T. Hymowitz and G. L. Hartman)

Worldwide Importance of Soybean Pathogens and Pests

Soybean diseases may be classified as biotic (infectious) or abiotic (noninfectious). Various unfavorable environmental and nutritional conditions cause abiotic diseases. Diseases caused by biotic agents can be transmitted from an infected to a healthy plant and cause disease when conditions are favorable for infection and colonization. Bacteria (including mycoplasmalike organisms), fungi, nematodes, parasitic plants, and viruses all cause infectious diseases. (Latin binomials and common names for the causal organisms of biotic pathogens are provided in Appendix I, and scientific and common names of arthropods and other vectors that transmit some viruses are provided in Appendix II.) Feeding by arthropods and nematodes can predispose plants to infection from root-infecting or wilt-inducing agents and from seed-decaying organisms by providing wounds for the pathogens' entry. These factors must all be present and in balance for extensive disease development to occur.

More than 200 pathogens are known to affect soybean, and approximately 35 are believed to be economically important. In addition, there have been many reports of associations between soybean and microorganisms in which the pathology is poorly understood. Generally, multiple pathogens and pests can be found in plants wherever the crop is grown, and all plant parts are susceptible to a number of pathogens that reduce the quality and quantity of seed yields. The extent of loss depends on these key factors: (1) the pathogen or condition involved; (2) the state of plant development and health when infection occurs; (3) the severity of the disease on individual plants; and (4) the number of plants infected or affected. In many diseases, a latent period elapses between infection and symptom expression.

As the value of soybean has increased, so has the importance of soybean diseases and pests. Although most diseases and pests on soybean are well known, with some descriptions dating back more than 100 years, others have been identified relatively recently. As soybean production continues to expand—especially in South America and Africa—new pathogens and variants of known pathogens and pests will inevitably be identified.

The economic importance of any single disease or pest often varies from one geographic area to another. Given this, it is not possible to accurately predict the impact of a disease or a pest globally or even regionally, although estimates can be used as guides to determine which disease or pest is most important. Loss estimates are continuously in a state of flux, and changes in pathogen–pest genetics, host susceptibility, and the environment are all factors in this dynamic interaction, which occurs over time. Nonetheless, predictions about the importance of diseases and pests are needed to allocate research resources and to understand the long-term dynamics of how diseases and pests increase or decrease over time in relation to environmental conditions.

TABLE 4. Estimated Reductions in Soybean Yields Caused by Diseases for World's Three Largest Soybean-Producing Countries, plus Total Additive Loss and Loss Attributed to Each Disease[a]

Disease(s)	Reduction in Yield (MMT)[b]			Total Additive Loss (MMT)	Loss Attributed to Disease (%)[c]
	Argentina	Brazil	United States		
Soybean rust	0.05	4.72	0.02	4.79	2.6
Soybean cyst nematode	0.02	0.52	3.37	3.91	2.2
Brown spot	1.18	0.34	0.54	2.05	1.1
Charcoal rot	0.91	0.36	0.70	1.96	1.1
Purple seed stain and Cercospora leaf blight	1.09	0.72	0.09	1.89	1.0
Sudden death syndrome	0.77	0.32	0.74	1.83	1.0
Phytophthora root and stem rot	0.07	0.00	1.46	1.53	0.8
Seedling diseases	0.05	0.08	1.09	1.21	0.7
Anthracnose	0.05	0.22	0.49	0.76	0.4
Sclerotinia stem rot	0.14	0.20	0.36	0.70	0.4
Brown stem rot	0.02	0.04	0.50	0.56	0.3
Diseases caused by root-knot and other nematodes	0.02	0.26	0.22	0.50	0.3
Frogeye leaf spot	0.02	0.04	0.35	0.41	0.2
Pod and stem blight	0.18	0.00	0.21	0.39	0.2
Virus diseases	0.05	0.10	0.20	0.35	0.2
Rhizoctonia aerial blight	0.00	0.30	0.01	0.31	0.2
Phomopsis seed decay	0.05	0.13	0.12	0.30	0.2
Bacterial diseases	0.02	0.13	0.10	0.25	0.1
Stem canker	0.00	0.00	0.21	0.21	0.1
Fusarium root rot	0.00	0.04	0.17	0.21	0.1
Downy mildew	0.00	0.05	0.15	0.20	0.1
Seed diseases	0.05	0.07	0.02	0.13	0.1
Sclerotium blight	0.00	0.06	0.01	0.06	0.0
Other diseases	0.05	0.00	0.00	0.05	0.0
Powdery mildew	0.00	0.00	0.00	0.00	0.0
Total	**4.75**	**8.70**	**11.12**	**24.57**	**13.6**

[a] Adapted from Wrather et al. (2010) (2006 data).
[b] MMT = million metric tons.
[c] Based on total production of 220.4 MMT produced in these three countries, which was 82% of total world production in 2006.

As an example of the importance of biotic diseases in soybean production, consider that in 2006, an estimated loss of 59.9 million metric tons (MMT) was reported from the top eight soybean-producing countries in the world. Those eight countries produced 213.1 MMT of soybeans, which means 28% of the crop was lost because of diseases. When recalculated for the three largest soybean-producing countries (Argentina, Brazil, and the United States), the reduction in yield was 13.6% (Table 4). More conservative estimates put the annual loss at approximately 11%.

Similar estimates of global losses are not available for soybean arthropod pests. Yet one estimate for one pest—the soybean aphid—indicates that the loss could be as high as 11 MMT a year in North America alone. In addition to the soybean aphid, a number of other soybean pests cause annual yield losses (see Part II, Arthropod Pests).

Selected References

Eastburn, D., Degennaro, M., DeLucia, E., Dermody, O., and McElrone, A. 2010. Elevated atmospheric carbon dioxide and ozone alter soybean diseases at SoyFACE. GCB Bioenergy 16:320-330.

Hartman, G. L., West, E., and Herman, T. 2011. Crops that feed the world 2. Soybean-worldwide production, use, and constraints caused by pathogens and pests. Food Secur. 3:5-17.

Kim, C., Schaible, G., Garrett, L., Lubowski, R., and Lee, D. 2008. Economic impacts of the U.S. soybean aphid infestation: A multi-regional competitive dynamic analysis. Agric. Res. Econ. Rev. 37:227-242.

McGee, D. C. 1992. Soybean Diseases: A Reference Source for Seed Technologists. American Phytopathological Society, St. Paul, MN.

Walla, W. J., ed. 1988. Soybean Disease Atlas. Southern Soybean Disease Workers. Texas A&M University, College Station.

Wrather, A., Shannon, G., Balardin, R., Carregal, L., Escobar, R., Gupta, G. K., Ma, Z., Morel, W., Ploper, D., and Tenuta, A. 2010. Effect of diseases on soybean yield in the top eight producing countries in 2006. Plant Health Progress. doi:10.1094/PHP-2010-0125-01-RS.

Wyllie, T. D., and Scott, D. H. 1988. Soybean Diseases of the North Central Region. American Phytopathological Society, St. Paul, MN.

(Prepared by G. L. Hartman)

Soybean Diseases in Argentina

Soybean Production

Soybean cultivation was attempted in Argentina in 1862 and again in the 1910s and 1920s, but it did not become widespread until the early 1960s. Since the 1970–1971 growing season, there has been a continuous and significant increase in soybean production. In 2013, 19.4 million ha produced 49.3 million metric tons of harvested soybean seeds, or 17.8% of the world's production (Table 5). In 2012, Argentina was the leading world exporter of soybean oil (45% of the world market) and soybean meal (44% of the world market), which represented 20% of total exports for the nation.

Approximately 80% of soybean is produced in the Pampean region (provinces of Buenos Aires, Córdoba, La Pampa, San Luis, and Santa Fe), whereas 9% is produced in the northwestern region (provinces of Catamarca, Jujuy, Salta, Santiago del Estero, and Tucumán) and 11% from the eastern region (provinces of Chaco, Corrientes, Entre Ríos, Formosa, and Misiones). Soybean is planted either as a single annual crop or as a second crop following wheat.

Soybean Diseases

The number and severity of biotic diseases in Argentine-grown soybean have increased steadily over the years, particularly since the early 1990s. Several factors have contributed to the escalation of disease in this country. Some of the problems have been related to seed production: at first, poor seed production techniques, but in later years, the failure to use certified seed. Other problems have been related to the crop management practices that became popular during the 1990s, which favored the survival of pathogens and the development of diseases. Continued monocropping and no-tillage systems—both practices widely used in modern soybean production—have been associated with many disease problems.

Approximately 40 diseases caused by microbes occur on soybean grown in Argentina. The impact of these diseases is variable. Some of the diseases are endemic and do not usually result in yield losses, whereas others have caused significant losses under favorable conditions.

Diseases of Foliage, Upper Stems, and Seeds Caused by Fungi and Oomycetes

Foliar diseases found on soybean in Argentina include downy mildew, frogeye leaf spot, rust, and target spot. Soybean rust is one of the most recently reported diseases. It was first found in March 2002 in the province of Misiones in the northeastern part of the country near Paraguay and Brazil, but it was not detected in the major soybean production regions until the end of the 2003–2004 growing season. Over subsequent growing seasons, levels of rust severity and yield losses were highly variable throughout the country. As of 2015, however, rust had not caused major damage except in a few instances, in part because of effective fungicide use. In addition, because the pathogen does not survive from year to year in the main soybean production regions, the arrival of inoculum from sources outside these regions is required to initiate an epidemic.

TABLE 5. Soybean Production, Area Harvested, Yield, and Percentage of World Production for 10 Largest Soybean-Producing Countries[a]

Country	Production (million metric tons)	Area Harvested (million ha)	Yield (metric tons/ha)	World Production (%)
Argentina	49.3	19.4	2.5	17.8
Bolivia	2.3	1.2	1.9	0.8
Brazil	81.7	27.9	2.9	29.6
Canada	5.2	1.8	2.9	1.9
China	12.5	6.6	1.9	4.5
India	11.9	12.2	1.0	4.3
Paraguay	9.1	3.1	3.0	3.3
Ukraine	2.8	1.4	2.1	1.0
United States	89.5	30.7	2.9	32.4
Uruguay	3.2	1.2	2.7	1.2
World	**276.4**	**111.3**	**2.5**	—

[a] Food and Agriculture Organization of the United Nations, Statistics Division (FAOSTAT) (2013 data).

Among diseases that affect stems, Sclerotinia stem rot was historically the most significant, with yield losses in some fields reaching 55%. However, since the mid-1990s, effective management techniques based primarily on using early maturing cultivars and early planting dates have reduced the impact of this disease.

Similarly, southern stem canker was an important disease in the middle to late 1990s. Severe epidemics occurred in the central and northern regions of Argentina during the 1996–1997 and 1997–1998 growing seasons, resulting in 100% yield losses in many fields because only one-third of the cultivars available at that time had genetic resistance to stem canker. The use of resistant cultivars in the following seasons rapidly reduced the significance of this disease. The causal agent of northern stem canker was identified in 1999 in the provinces of Buenos Aires, Córdoba, and Santa Fe. This development has created the possibility of a second stem canker disease occurring in the country, because the genes that confer resistance to the southern stem canker pathogen do not necessarily protect against the northern stem canker pathogen (see Part I, the section Stem Canker).

The late-season disease complex—a combination of various soybean diseases that affect foliage, stems, pods, and seeds during the middle to late reproductive growth stages—has caused significant yield losses in Argentina, especially in the northwestern and eastern regions. The main diseases included in this complex are Alternaria leaf spot, anthracnose, brown spot, Cercospora blight, frogeye leaf spot, pod and stem blight, and target spot. The occurrence of this disease complex has increased gradually since the 1990s, and under favorable environmental conditions, it can cause yield losses as great as 40%.

During the 1999–2000 growing season, a severe epidemic of frogeye leaf spot occurred in northwestern Argentina, producing yield losses of up to 48% in susceptible cultivars and prompting growers to make major changes in cultivar selection. Other important outbreaks of frogeye leaf spot occurred in the Pampean region during the 2008–2009 and 2009–2010 growing seasons and caused heavy losses in fields planted with widely used susceptible cultivars.

Diseases of Lower Stems and Roots Caused by Fungi and Oomycetes

Seedling diseases of soybean include Phytophthora root and stem rot, which was first reported in Argentina in 1979. Races 1, 2, 4, 9, 13, and 17 of the pathogen have been identified. However, this disease is not considered a major disease in Argentina's main soybean-growing regions. Other soybean seedling diseases are produced by the pathogens that cause charcoal rot, Fusarium root rot, Pythium damping-off and root rot, Rhizoctonia root rot, and Sclerotium blight. In addition, brown stem rot and Rhizoctonia root rot are both of concern late in the growing season.

As of 2015, the most important disease caused by root-infecting pathogens was sudden death syndrome, which is caused by several species of *Fusarium*. This disease was first identified in the northern part of the Pampean region during the 1991–1992 growing season and then again in the northwestern region during the 1992–1993 growing season. The predominant pathogen of this disease in all soybean-growing regions of Argentina was found to be *F. tucumaniae;* secondary pathogens included *F. virguliforme, F. crassistipitatum,* and *F. brasiliense.* Yield losses of up to 15% and 90% have been reported in the central and northern regions, respectively. In addition to sudden death syndrome, other root and lower-stem rots are caused by various species of *Fusarium.*

Another prevalent disease in Argentina is charcoal rot. During the 2000s, it has caused significant losses in regions in which warm and dry conditions prevail during the growing season. Yield losses close to 100% have been reported in some locations in central and northern Argentina in fields with poor soils and little water retention capacity. The most severe epidemics have occurred during the 2000–2001, 2002–2003, 2008–2009, and 2011–2012 growing seasons.

Diseases Caused by Nematodes

The most important nematodes affecting soybean production in Argentina are the soybean cyst nematode, the southern nematode, and the tropical root-knot nematode. The prevalence of the soybean cyst nematode in the Pampean region has remained consistent since the late 1990s, but a decrease in the population has been detected since the early 2000s. In the northwestern region of Argentina, field surveys have indicated the occurrence of two types of the soybean cyst nematode. Root-knot nematodes have been detected in several provinces of central and northwestern Argentina.

Diseases Caused by Viruses

Soybean mosaic is the most prevalent virus disease of soybean in Argentina. Less frequently detected diseases include alfalfa mosaic, bean common mosaic, Brazilian bud blight, cucumber mosaic, and peanut mottle. In the northwestern region, whitefly-transmitted begomoviruses are frequently reported. Molecular characterization has indicated the presence of at least three geminiviruses: *Euphorbia mosaic virus, Soybean blistering mosaic virus,* and *Tomato yellow spot virus.*

Disease Management

Because of the epidemics that have occurred in Argentina since the mid-1990s, special attention is given to strategies of disease management. Integrated management programs—which combine biological, chemical, and cultural methods—are recommended, and a particular focus is applied to the economic aspects of crop production.

Cultivars are available with high levels of resistance to frogeye leaf spot, Phytophthora root and stem rot, stem canker, and the soybean cyst nematode and with moderate levels of resistance to downy mildew, Sclerotinia stem rot, sudden death syndrome, and target spot. The use of resistant cultivars was the key to overcoming epidemics of frogeye leaf spot and stem canker during the 1990s in northwestern Argentina.

Crop rotation—one of the most efficient cultural methods for managing many soybean diseases—is practiced only in a small proportion of the total soybean production area in Argentina. Over the years, the increased demand for soybean and the low costs of producing this crop have resulted in more land being devoted to soybean production and consequently less rotation with other crops. Successful management of foliage diseases has been observed in fields rotated with maize.

Although destroying crop residues via tillage is effective in lowering the survival of most pathogens that infect the aerial parts of soybean plants, this practice is not generally recommended for soybean production in Argentina. Conservation tillage systems (e.g., no-tillage) are preferred to manage soil erosion, to increase physical and chemical soil fertility, and to favor the water balance of the system. Several million hectares could not have been converted to soybean production in Argentina if these conservation tillage systems had not been adopted. Nevertheless, these systems enhance overseasoning of the inocula of many pathogens.

Another important strategy recommended for managing soybean diseases is the use of high-quality, pathogen-free seed. Even so, most Argentine soybean growers plant seed that they produce and do not always use the proper technology. These practices have contributed to the spread of diseases into new areas and fields.

Other cultural practices recommended by management programs are oriented to prevent or reduce soil compaction and improve soil drainage (for charcoal rot, Phytophthora root and stem rot, sudden death syndrome, and other diseases caused by *Fusarium* spp.) and to maintain good soil fertility (for charcoal rot, seed diseases, and the soybean cyst nematode). Additional strategies that should be considered are planting at earlier dates (for charcoal rot, late-season disease complex, and soybean mosaic); using early maturing cultivars (for Sclerotinia stem rot, sudden death syndrome, and soybean rust); applying appropriate row spacing and seeding rates (for charcoal rot and Sclerotinia stem rot); controlling weeds (for Sclerotinia stem rot); irrigating (for charcoal rot); and timely harvesting (for seed diseases).

Recommendations for chemical control include both seed-treatment fungicides and foliar-applied fungicides. Seed treatments are used extensively to increase seed germination when poor-quality seed is used and when planting conditions are not favorable for germination and emergence. Foliar applications have increased steadily since 2000. Data from field experiments in Argentina's different soybean-growing regions have shown the benefits of these applications in terms of reducing late-season diseases, increasing yields, and improving seed quality. Yield increases resulting from fungicide use generally have been higher in the northern region of the country than in the northern area of the Pampean region. The fungicides used, whether alone or in combination, belong to various chemical groups (e.g., benzimidazoles, carboxamides, strobilurins, and triazoles).

Most disease problems in Argentina have been solved using effective management techniques. Even so, the implementation of these techniques has been incomplete, and soybean production continues to suffer significant losses because of diseases. Argentine growers face the constant threat of new epidemics occurring and of new diseases or new races or stains of known pathogens emerging.

Selected References

Aoki, T., Scandiani, M. M., and O'Donnell, K. 2011. Phenotypic, molecular phylogenetic, and pathogenetic characterization of *Fusarium crassistipitatum* sp. nov., a novel soybean sudden death syndrome pathogen from Argentina and Brazil. Mycoscience 53:167-186.

Carmona, M., Sautua, F., Perelman, S., Reis, E. M., and Gally, M. 2011. Relationship between late soybean diseases complex and rain in determining grain yield responses to fungicide applications. J. Phytopathol. 159:687-693.

O'Donnell, K., Sink, S., Scandiani, M. M., Luque, A., Colletto, A., Biasoli, M., Lenzi, L., Salas, G., González, V., Ploper, L. D., Formento, N., Pioli, R. N., Aoki, T., Yang, X. B., and Sarver, B. A. J. 2010. Soybean sudden death syndrome species diversity within North and South America revealed by multilocus genotyping. Phytopathology 100:58-71.

Pioli, R. N., Morandi, E. N., and Bisaro, V. 2001. First report of soybean stem canker caused by *Diaporthe phaseolorum* var. *caulivora* in Argentina. Plant Dis. 85:95.

Ploper, L. D. 2004. Economic importance of and control strategies for the major soybean diseases in Argentina. Pages 606-614 in: Proc. World Soybean Res. Conf., 7th. F. Moscardi, C. B. Hoffman-Campo, O. F. Saraiva, and P. R. Galerani, eds. Embrapa Soybean, Londrina, Brazil.

Ploper, L. D., González, V., Gálvez, M. R., Devani, M. R., Ledesma, F., and Zamorano, M. A. 2001. Frogeye leaf spot caused by *Cercospora sojina* in northwestern Argentina. Plant Dis. 85:801.

Ploper, L. D., González, V., Gálvez, M. R., de Ramallo, N. V., Zamorano, M. A., García, G., and Castagnaro, A. P. 2005. Detection of soybean rust caused by *Phakopsora pachyrhizi* in northwestern Argentina. Plant Dis. 89:774.

Rodriguez-Pardina, P. E., Hanada, K., Laguna, I. G., Zerbini, F. M., and Ducasse, D. A. 2010. Molecular characterization and relative incidence of bean- and soybean-infecting begomoviruses in northwestern Argentina. Ann. Appl. Biol. 158:69-78.

Rossi, R. L. 2003. First report of *Phakopsora pachyrhizi,* the causal organism of soybean rust in the province of Misiones, Argentina. Plant Dis. 87:102.

(Prepared by L. D. Ploper)

Soybean Diseases in Brazil

Soybean Production

Brazil is one of the largest producers of soybean in the world, producing 81.7 million metric tons (MMT) in 2013 (Table 5). The area of cultivated soybean production in this country doubled from the 1998–1999 growing season to the 2012–2013 growing season, increasing from 12.9 million ha to 27.9 million ha. One of the main factors in the successful expansion of soybean production in Brazil has been the development of cultivars adapted to low latitudes south and north of the equator. This development has enabled the expansion of production into the Cerrado region (central Brazil), which produces more than 50% of Brazil's soybean crop (as of 2015).

Soybean is planted from October to December in many parts of Brazil. Under irrigation in central Brazil, soybean can be grown year-round.

Soybean Diseases

Disease problems vary according to region because of the large variation in weather conditions across Brazil. Even so, with the dramatic increase in production area has come a dramatic increase in diseases—in particular, soybean rust.

In addition to soybean rust, diseases such as frogeye leaf spot and southern stem canker have been important in the past but have become better controlled with resistance. For example, southern stem canker, which was identified during the 1988–1989 growing season in the state of Paraná (southern Brazil), caused yield losses of up to 100% in the mid-1990s. Other diseases that have more regional or local importance include brown stem rot, Phytophthora root and stem rot, Rhizoctonia aerial blight, and Sclerotinia stem rot, along with some diseases caused by nematodes and viruses. In addition, northern stem canker was identified during the 2005–2006 growing season in the state of Rio Grande do Sul (southern Brazil), but significant losses in commercial fields were neither observed nor reported.

Diseases Caused by Fungi and Oomycetes

A number of foliar diseases affect soybean in Brazil, but soybean rust has had the largest impact on production and management practices. Soybean rust caused by *Phakopsora pachyrhizi* was identified in May 2001 in the state of Paraná. Soybean rust caused by *P. meibomiae,* a less virulent species, was identified in Brazil in 1979, when it caused sporadic epidemics in regions with moderate temperatures, and it has remained a minor pathogen of soybean.

After the initial detection of *P. pachyrhizi,* soybean rust spread quickly into important soybean production regions in Brazil. Conditions in much of the country are conducive for year-round survival of the pathogen, affecting soybean grown during the winter in the central Cerrado region, as well as alternative hosts. From the first growing season in 2001–2002 after detection through 2012, annual losses were estimated at 15.5 MMT of soybean, valued at $3.8 billion.

Sclerotinia stem rot, which can be found in central and southern Brazil, is problematic in regions above 700 m altitude, with moderate air temperatures and frequent rain events during the flowering stage. Rhizoctonia aerial blight is an important disease in the northern region, where warm, wet weather conditions prevail. In the southern region, diseases such as brown stem rot and Phytophthora root and stem rot are significant. Charcoal rot is a serious problem in the southwestern, southern, and northern regions of Brazil, where environmental stresses (e.g., low rainfall, high temperatures) frequently occur during the growing season.

Diseases Caused by Nematodes

A number of nematodes affect soybean production in Brazil. The most destructive are the soybean cyst nematode, the lesion nematode, the reniform nematode, the southern root-knot nematode, and the tropical root-knot nematode.

The soybean cyst nematode was first detected in 1991–1992, and by 2015, it had infested approximately 3 million ha. Losses resulting from this parasite have decreased, however, because of crop rotation with nonhost species (mainly corn) and the use of genetically resistant soybean cultivars. The genetic variability of the nematode continues to be problematic. A number of types of soybean cyst nematode have been identified, and those that overcome the Hartwig source of resistance are of the greatest concern.

The lesion nematode has increased in importance in Brazil, especially in areas with sandy soils (less than 15% clay). The reniform nematode has increased in importance in the state of Mato Grosso do Sul (central-western Brazil), especially when soybean crops have followed cotton in rotation.

Diseases Caused by Viruses

Plant virologists have been concerned about the expansion of the Brazilian soybean crop because of its proximity to other botanical species, some of which are hosts of viruses that infect soybean. In the early 2000s, *Cowpea mild mottle virus,* transmitted by whiteflies, caused severe losses among susceptible soybean cultivars. Other viruses found on soybean in Brazil include *Bean golden mosaic virus, Bean pod mottle virus, Soybean mosaic virus,* and *Tobacco streak virus.*

Disease Management

Resistance is a key element in the successful management of most important soybean diseases in Brazil, including frogeye leaf spot, Phytophthora root and stem rot, southern stem canker, and soybean rust, along with diseases caused by the root-knot nematode and the soybean cyst nematode. Soybean-breeding programs in this country have focused on incorporating disease resistance into elite genetic material whenever possible, and as a result, several diseases have been eliminated as economic threats. For example, frogeye leaf spot was first identified in Brazil in 1971 and was considered a major disease until the incorporation of resistance gene *Rcs3* from cultivar Davis. Since the mid-1990s, the disease has practically been absent from the fields, although some races of the pathogen found in northern Brazil have overcome this resistance. Another example is southern stem canker, for which four resistance genes—*Rdc1, Rdc2, Rdc3,* and *Rdc4*—have been incorporated into commercial cultivars, resulting in nearly complete control of the disease. Resistance gene deployment using *Rpp2, Rpp4,* and *Rpp5* has also been effective, at least in part, in controlling soybean rust.

Rotating soybean with nonhost species is recommended for controlling the lesion, root-knot, and reniform nematodes, as well as diseases such as Sclerotinia stem rot. For instance, rotating soybean with species of *Crotalaria* or pearl millet (*Pennisetum glaucum*) was found to suppress the lesion nematode population. The use of tillage practices is also recommended for lesion nematode control; disking has been used to expose nematodes to sunlight and thus desiccation.

In some cases, changing planting times is recommended to help prevent disease. Growers are advised to plant early maturing cultivars in the beginning of the season and to reduce the length of the planting period to help the host avoid infection by the soybean rust pathogen. In 2013, 12 states in Brazil adopted a period of 60–90 days from July to September that is host free, during which farmers are restricted from planting soybean except under strictly controlled conditions. This practice was first adopted in 2007 to break the continuous cycle of the fungus (*P. pachyrhizi*) and to delay the onset of the epidemic in the regular season. Following this practice has been effective in delaying rust epidemics.

A major control strategy used by Brazilian growers is the use of fungicides. Recommendations for applying fungicides (mainly sulfur and benzimidazoles) were first made during the 1996–1997 growing season to control powdery mildew and then some years later to control late-season diseases, such as brown spot and Cercospora leaf blight. With the introduction of *P. pachyrhizi* in 2001, fungicide use greatly increased at first and then stabilized after some years to an average of two or three applications per field per season. The use of triazoles alone was recommended in the first growing season after detection of the pathogen, but since 2007, following observations of reduced efficiency, only mixtures of triazoles and strobilurins have been recommended for rust control. The weakened efficiency of triazoles observed 6 years after the introduction of rust in Brazil has been associated with selection of less sensitive populations of *P. pachyrhizi*.

Increases have also been noted in some diseases that are not readily controlled by triazoles and strobilurins, including anthracnose and target spot. When these diseases occur, an additional application of benzimidazoles is required. However, isolates of the target spot pathogen have developed resistance to benzimidazoles in several regions of Brazil. Other diseases, such as Sclerotinia stem rot, can be controlled by benzimidazoles, which can also be used to treat seeds if they are suspected of harboring the pathogen.

New compounds from the carboxamide group have been shown useful in controlling some diseases and may be applied more frequently in the future. New compounds from the pyridinamine and dicarboximide groups, which may also have efficacy in controlling Sclerotinia stem rot and other diseases, are being tested along with other fungicides in a nationwide network of uniform field trials.

Selected References

Almeida, A. M. R., Piuga, F. F., Marin, S. R. R., Kitajima, E. W., Gaspar, J. O., Oliveira, T. G., and Moraes, T. G. 2005. Detection and partial characterization of a carlavirus causing stem necrosis of soybean in Brazil. Fitopatol. Bras. 30:191-194.

Costamilan, L. M., Yorinori, J. T., Almeida, A. M. R., Seixas, C. D. S., Binneck, E., Araújo, M. R., and Carbonari, J. A. 2008. First report of *Diaporthe phaseolorum* var. *caulivora* infecting soybean plants in Brazil. Trop. Plant Pathol. 33:381-385.

Dias, W. P., Campos, V. P., Kiihl, R. A. S., Arias, C. A. A., and Toledo, J. F. F. 2005. Genetic control in soybean of resistance to soybean cyst nematode race 4+. Euphytica 145:321-329.

Scherm, H., Christiano, R. S. C., Esker, P. D., Del Ponte, E. M., and Godoy, C. V. 2009. Quantitative review of fungicide efficacy trials for managing soybean rust in Brazil. Crop Prot. 28:774-782.

Yorinori, J. T., Paiva, W. M., Frederick, R. D., Costamilan, L. M., Bertagnolli, P. F., Hartman, G. L., Godoy, C. V., and Nunes, J., Jr. 2005. Epidemics of soybean rust (*Phakopsora pachyrhizi*) in Brazil and Paraguay from 2001 to 2003. Plant Dis. 89:675-677.

(Prepared by C. V. Godoy, A. M. R. Almeida, L. M. Costamilan, W. P. Dias, A. A. Henning, M. C. Meyer, C. D. S. Seixas, and R. M. Soares)

Soybean Diseases in Canada

Soybean Production

Soybean has been grown in Canada for more than 100 years. Initially, it was grown as an alternative forage crop. In the 1940s, it became an important cash crop in southern Ontario when it was grown to meet the increasing need for edible oil, meal, and food-grade soybeans. Commercial production was concentrated in southern Ontario until the mid-1970s, when the development of earlier-maturing soybean cultivars allowed a significant expansion of production: from 190,000 ha in 1973 to 1.83 million ha in 2013, for a total annual soybean production of 5.2 million metric tons (Table 5).

Although Ontario is still the site of 60% of Canadian soybean production (as of 2015), Quebec and Manitoba have seen major expansions. Potential areas of future production expansion in Canada include northern Ontario; the Atlantic provinces of New Brunswick, Nova Scotia, and Prince Edward Island; and western Manitoba, southeastern Saskatchewan, and southern Alberta in the Canadian Prairies. In 2014, approximately 75% of the soybeans produced in Canada were crushed for oil and meal, and the remaining soybeans were exported.

Soybean Diseases

Approximately 40 soybean diseases occur in Canada, and they occur with the highest frequency in Ontario. The most significant yield-limiting soybean diseases are soybean cyst, early season seed and seedling blights, and diseases in the late-season complex, including sudden death syndrome, Sclerotinia stem rot, pod and stem blight, and Phomopsis seed decay. Sclerotinia stem rot and Phytophthora root and stem rot have been problematic since the early years of soybean production. Their occurrence each year depends on many factors, including environmental conditions, cultivar susceptibility, and tillage practices.

In the 1990s, the frequency and severity of soybean diseases increased (especially in Ontario) because of changing agronomic practices and pathogen dynamics. For instance, the adoption of a minimum-tillage system to combat soil erosion resulted in increased root rots and diseases caused by pathogens that overwinter in crop residue. Moreover, as new diseases developed—such as Fusarium root rot, soybean cyst, and sudden death syndrome—so did new races and pathotypes (i.e., species) of pathogens that cause Phytophthora root and stem rot, Pythium root rot, and soybean cyst. Other factors that have contributed to the increase in diseases include the greater demand for soybeans, resulting in shorter crop rotation; adverse weather, causing stressful growing conditions; the lack of cultivars resistant to emerging and endemic diseases; production practices aimed at generating maximum yields, such as earlier planting dates; and higher plant populations.

Diseases Caused by Bacteria

Bacterial diseases—including bacterial blight, bacterial pustule, and wildfire—are found in most soybean production areas, especially Ontario. These diseases do not cause significant economic losses, however.

Diseases Caused by Fungi and Oomycetes

Foliage and upper-stem diseases—including anthracnose, brown spot, downy mildew, frogeye leaf spot, powdery mildew, and purple stain—are observed on soybean in most seasons but rarely reduce yields. Brown spot can be problematic, however, especially if plants are under moisture or nutritional stress early in the season.

Sclerotinia stem rot has historically been a major problem in Manitoba, Ontario, and Quebec. Disease severity fluctuates from year to year and is greatest when cool, wet environmental conditions coincide with pod fill through maturity. Northern stem canker has become more common, and the incidence of this disease is often greatest in minimum-tillage fields. Phomopsis seed rot and pod and stem blight are weather dependent and most problematic in Ontario when warm, wet conditions occur at crop maturity, resulting in a delayed harvest and causing lower soybean seed quality and farm-gate value. This is especially true of soybeans grown for the identity-preserved market. Soybean rust infection was first confirmed very late in the 2007 growing season in Ridgetown, Ontario. Fortunately, the crop had reached physiological maturity, so no impact on yield was observed.

Seed rots, seedling blights, and/or root diseases are annual problems and cause early season emergence issues that affect plant density. In Ontario soybean crops, yield losses to these diseases are surpassed only by losses caused by the soybean cyst nematode. Late-season Fusarium and Rhizoctonia root rots and Fusarium wilt have increased in severity since the early 2000s in all growing regions, caused primarily by the increased frequency of soybean cropping and insufficient genetic resistance in commercial cultivars. Yield reductions to these diseases have been as high as 20% in some fields.

Phytophthora root and stem rot has been a problem since the 1960s and remains one of the most destructive diseases of soybean in Canada. Although commercial soybean cultivars with race-specific or partial resistance are available, isolates of the causal fungus with novel patterns of virulence continue to develop. The fungus is found in all soil types, but most losses occur on crops planted in poorly drained, fine-textured clay soils.

Sudden death syndrome has increased in frequency since symptoms of the disease were first noted in 1993. This disease is found primarily on crops grown in sandy loam soils in association with infestation of the soybean cyst nematode. Sudden death syndrome has become permanently established in southwestern Ontario, and losses occur each year.

Other diseases, such as charcoal rot and brown stem rot, are increasing in importance in all regions of Canada. They have become particularly predominant in southwestern Ontario, however, especially when adverse conditions occur in August.

Diseases Caused by Nematodes

Soybean cyst continues to be the most economically important soybean disease in Ontario, and since the soybean cyst nematode was initially confirmed near Chatham in southwestern Ontario in 1987, it has been detected in the other major soybean production regions in the province. Yield losses as high as 100% have been observed in double-cropped soybeans and parts of fields, and losses of 20–40% are not uncommon when a susceptible cultivar is grown. Fortunately, cultivars resistant to the soybean cyst nematode are available, and yield losses have declined in areas with a history of soybean cyst. However, as observed in other northern soybean production regions, new types of cyst nematodes have developed that are able to infect cultivars that derive their resistance from sources PI 88788 and Peking. Foliar symptoms may or may not be evident in infested fields.

In 2013, soybean cyst was detected in St. Anicet, Quebec, which is adjacent to the eastern Ontario border. Monitoring efforts in Canada are ongoing and focused on both early soybean cyst detection and education to minimize the spread of the nematode into noninfested areas and provinces.

Diseases Caused by Viruses

Alfalfa mosaic, bean pod mottle, soybean mosaic, soybean vein necrosis, and tobacco ringspot are virus diseases found in all soybean-growing regions of Canada, but they seldom cause significant yield losses in production fields. Virus infections impact seed quality, in part, by causing discoloration; this is a concern for food-grade soybean growers, because it can cause grade discounts and lost premiums. Because of increases in the populations of insect vectors (e.g., soybean aphids, bean leaf beetles, thrips), virus-like symptoms are likely to increase.

Disease Management

The most effective strategy for managing diseases is to grow resistant cultivars. Soybean cultivars are recommended with the genes *Rps1c, Rps1k,* and *Rps8,* which provide resistance to Phytophthora root and stem rot. In fields with a history of soybean cyst, cultivars with resistance sources PI 88788, Peking, and PI 437654 in their backgrounds should be considered.

For most foliar and root diseases, cultural practices for management of disease include rotation with corn or wheat. To manage soybean cyst specifically, soybean should be rotated with vegetables or cover crops, and good sanitation practices should be followed, including weed control. Improving soil structure and drainage will help reduce the impact of Phytophthora root and stem rot.

Pesticide applications, especially seed treatments, have become increasingly common. The use of pesticides is recommended for early planting, for fields with histories of seed and seedling diseases, and for seed lots that may have reduced vigor or obvious symptoms of Phomopsis seed decay or purple seed stain. Foliar-applied fungicides are not common and mostly not recommended.

Selected References

Agriculture: Crops and horticulture. n.d. Statistics Canada. Available online at www5.statcan.gc.ca/subject-sujet/subtheme-soustheme.action?pid=920&id=2024&lang=eng&more=0

Anderson, T. R., and Tenuta A. U. 1998. First report of *Fusarium solani* f. sp. *glycines* causing sudden death syndrome of soybean in Canada. Plant Dis. 82:448.

Anderson, T. R., and Tenuta, A. U. 2003. Soybean diseases in Canada. Pages 155-168 in: Diseases of Field Crops in Canada. K. L. Bailey, B. D. Gossen, R. K. Gugel, and R. A. A. Morrall, eds. Canadian Phytopathological Society, University Extension Press, University of Saskatchewan, Saskatoon, Canada.

Anderson, T. R., Welacky, T. W., Olechowski, T. H., Ablett, G., and Ebsary, B. A. 1988. First report of *Heterodera glycines* in Ontario, Can. Plant Dis. 72:453.

Canadian Soybean Council (CSC). 2013. Canadian Soybean Dispatch. 7th ed. CSC, Guelph, Ontario. Available online at www.soybeancouncil.ca/LinkClick.aspx?fileticket=bkkpxHzAnLQ%3D&tabid=182&language=en-US

Isard, S. A., Barnes, C. W., Hambleton, S., Ariatti, A., Russo, J. M., Tenuta, A., Gay, D. A., and Szabo, L. J. 2011. Predicting soybean rust incursions into the North American continental interior using crop monitoring, spore trapping, and aerobiological modeling. Plant Dis. 95:1346-1357.

Mimee, B., Peng, H., Popovic, V., Yu, G., Duceppe, M.-O., Tetreault, M.-P., and Belair, G. 2014. First report of soybean cyst nematode (*Heterodera glycines* Inchinohe) on soybean in the province of Quebec, Can. Plant Dis. 98:429.

Wrather, A., Shannon, G., Balardin, R., Carregal, L., Escobar, R., Gupta, G. K., Ma, Z., Morel, W., Ploper, D., and Tenuta, A. 2010. Effect of diseases on soybean yield in the top eight producing countries in 2006. Plant Health Progress. doi:10.1094/PHP-2010-0125-01-RS.

(Prepared by A. U. Tenuta)

Soybean Diseases and Pests in China

Soybeans were first produced as a crop in China several thousand years ago. In 2013, Chinese soybean production accounted for 4.5% of the world total, or 12.5 million metric tons (Table 5). That year, China was the world's fourth-largest soybean producer, as well as the largest soybean importer. The estimated total area of soybeans harvested in China in 2013 was 6.6 million ha, with an average yield of 1.9 metric tons per ha.

China has three main soybean-growing regions: northeast China, the Huang–Huai–Hai River valleys, and southern China. Generally, soybeans are planted in April or May and harvested from September through October.

Soybean Diseases and Pests

More than 30 soybean diseases caused by bacteria, fungi, viruses, and various pests have been reported in China. The most important diseases, in terms of causing significant economic losses, include bacterial blight, bean pod mottle, brown spot, brown stem rot, downy mildew, frogeye leaf spot, Phytophthora root and stem rot, pod and stem blight, Rhizoctonia root rot, Sclerotinia stem rot, soybean cyst, soybean mosaic, and soybean rust, as well as diseases caused by the common cutworm, the globular stink bug, the soybean aphid, the bean pyralid, and the soybean pod borer. The relative importance of each disease and pest varies by year and region.

Diseases Caused by Fungi and Oomycetes

Fungi and oomycetes cause a number of diseases in China, but not all of these diseases cause economic losses. Anthracnose, brown spot, downy mildew, seedling diseases, and soybean rust are the most significant. The distribution and occurrence of soybean rust has been well documented, with yield losses of up to 50% in areas during high rainfall in some years. Phytophthora root and stem rot has been reported more recently, with the first confirmed isolation of the pathogen in 1989 from a field in northeast China. The pathogen has since been found in most soybean-growing regions, from northern to southern China.

Diseases Caused by Nematodes

The soybean cyst nematode is a devastating pest in China. Eight types of soybean cyst nematode have been identified. In Heilongjiang and Jilin Provinces, a type of soybean cyst nematode was found to be widespread, whereas two other types were predominant in the Huang and Huai River valleys.

Thousands of soybean accessions have been screened for resistance since the 1970s, and many sources of resistance have been identified. Most of the resistant cultivars have black or brown seed coats, rather than yellow seed coats.

Diseases Caused by Viruses

Soybean mosaic is the major virus disease in China, and it occurs in all soybean-growing regions. The symptoms of soybean mosaic vary with the plant genotype and the age at which the plant first becomes infected, as well as the virus strain and the environmental conditions. In China, the virus has evolved into many different strains. The National Center for Soybean Improvement at Nanjing Agricultural University investigated the distribution of viral strains among 24 soybean-growing regions in China and identified 21 strains.

In 2011, a novel strain named "Recombinant soybean mosaic virus" was identified in Chongqing and later found to be prevalent in the central and southern provinces of China. Sequence and phylogenetic analyses suggest that this new strain represents a recombination event between two different viruses.

Soybean Insects

The soybean aphid was first reported in China in 1962 and later found in most of the country's soybean-growing regions; major outbreaks have occurred in northeast China and in the Huang–Huai–Hai River valleys, especially in dry years. In some cases, the yield loss caused by the soybean aphid has exceeded 50%.

The soybean pod borer is another widely distributed insect in the Huang–Huai–Hai River valleys. It feeds mainly on young soybean seeds inside the pods, causing both yield loss and poor seed quality.

The common cutworm has become one of the primary leaf-feeding insects of soybean plants in southern China. It can feed on plants of several taxa, and at the adult stage, it is hard to control using insecticides.

Disease and Pest Management

Soybean resistance is utilized whenever possible to manage diseases such as Phytophthora root and stem rot, soybean cyst, soybean mosaic, and soybean rust and to manage insect pests such as the soybean pod borer and the soybean aphid. Additional sources of resistance to different strains of the pathogen are needed because of the diversity in virulence found in collected isolates. Three genes—*Rsc4* (*Rsv4*), *Rsc8*, and *Rsc14*—have been fine-mapped and are used in marker-assisted breeding programs to develop virus-resistant cultivars.

Hundreds of soybean cultivars have been screened for resistance to the soybean aphid, and several resistant sources have been found. In 2009, soybean cultivars P203 and P746 were identified as sources of resistance and confirmed to have single dominant resistance genes to the soybean aphid, and in 2011, cultivar Zhongdou 27 was found to have two major resistance genes. For the common cutworm, soybean cultivars with both antibiosis and antixenosis types of resistance have been identified, and molecular markers associated with resistance have been mapped and used in breeding programs.

In addition to planting resistant cultivars, other biological and cultural practices are used in China to manage specific diseases on soybean. For soybean cyst, recommended practices include rotating crops, maintaining soil fertility, and coating soybean seeds with biocontrol bacteria and/or fungi. For soybean rust, altering the planting season is recommended in certain regions, so that growers plant during the spring instead of the fall, and furrow irrigation is recommended instead of overhead irrigation. For management of soybean mosaic, planting virus-free seed, planting early in the growing season, and reducing the aphid population in the field are all recommended practices. For the common cutworm, trap lamps have been used and shown to be effective in controlling mounted adults.

Recommended uses of pesticides include seed coating with nematicides to control soybean cyst and applying foliar fungicides to control soybean rust. Insecticides have been used on the common cutworm at the larval stage, but these treatments have been ineffective at the adult stage. In an effort to control soybean aphids, insecticides have been used over the years in many regions, resulting in soybean aphid populations developing resistance to commonly used chemicals. Moreover, soybean aphids from different regions show varying sensitivities to insecticides.

Selected References

Kim, H., Xing, G. N., Wang, Y. F., Zhao, T. J., Yu, D. Y., Yang, S. P., Li, Y., Chen, S. Y., Palmer, R. G., and Gai, J. Y. 2014. Constitution of resistance to common cutworm in terms of antibiosis and antixenosis in soybean RIL populations. Euphytica 196:137-154.

Kong, X. C., Li, H. M., Geng, T., Huang, W. K., and Peng, D. L. 2012. Resistance evaluation of soybean varieties and germplasms to the races no. 3 and no. 4 of soybean cyst nematode *Heterodera glycines*. Plant Prot. 38:146-150.

Li, K., Yang, Q. H., Zhi, H. J., and Gai, J. Y. 2010. Identification and distribution of *Soybean mosaic virus* strains in southern China. Plant Dis. 94:351-357.

Lu, W. G., Gai, J. Y., and Li, W. D. 2006. Sampling survey and identification of races of soybean cyst nematode (*Heterodera glycines* Ichinohe) in Huang-Huai Valleys. Sci. Agric. Sin. 39:306-312.

Ma, Y., Wang, D. G., Li, H. C., Zheng, G. J., Yang, Y. Q., Li, H. W., and Zhi, H. J. 2011. Fine mapping of the R locus for resistance to soybean SC14Q mosaic virus in soybean. Euphytica 181:127-135.

Meng, F., Han, Y., Teng, W., Li, Y., and Li, W. 2011. QTL underlying the resistance to soybean aphid (*Aphis glycines* Matsumura) through isoflavone-mediated antibiosis in soybean cultivar 'Zhongdou 27.' Theor. Appl. Genet. 123:1459-1465.

Sun, S., Zhao, J., Wu, X., Guo, N., Wang, Y., Gai, J. Y., and Xing, H. 2008. Resistance of soybean germplasm to *Phytophthora* in Huanhuai Valley. Soybean Sci. 27:465-470.

Tang, Q., Cui, L., Miao, M., Li, D., Yin, W., Zheng, X., and Wang, Y. 2011. Genetic structure analysis of *Phytophthora sojae* population from several provinces in China. J. Nanjing Agric. Univ. 34:73-77.

Wang, D. G., Ma, Y., Liu, N., Yang, Z. L., Zheng, G. J., and Zhi, H. J. 2011. Fine mapping and identification of the soybean RSC4 resistance candidate gene to *Soybean mosaic virus*. Plant Breed. 130:653-659.

Wang, D. G., Ma, Y., Yang, Y. Q., Liu, N., Li, C. Y., Song, Y. P., and Zhi, H. J. 2011. Fine mapping and analyses of RSC8 resistance candidate genes to *Soybean mosaic virus* in soybean. Theor. Appl. Genet. 122:555-565.

Xiao, L., Hu, Y., Wang, B., and Wu, T. 2013. Genetic mapping of a novel gene for soybean aphid resistance in soybean (*Glycine max* [L.] Merr.) line P203 from China. Theor. Appl. Genet. 126:2279-2287.

Xiao, L., Zhong, Y., Zhang, J., Wang, B., and Wu, T. 2012. Inheritance of resistance to *Aphis glycines* in soybean P746 from China. J. Econ. Entomol. 105:2167-2171.

Yan, R. H., Yang, Z. Y., Liu, B. Q., Chen, J., and Dong, L. C. 2008. Selection and breeding of a new soybean variety Jiyu no. 91 with high-yield and pod borer-resistance. Soybean Bull. 4:42-43.

Yang, Y. Q., Lin, J., Zheng, G. J., Zhang, M. C., and Zhi, H. J. 2014. Recombinant soybean mosaic virus is prevalent in Chinese soybean fields. Arch Virol. doi:10.1007/s00705-014-1980-z.

Zhang, W., Song, S. Y., Jin, Q. M., Li, H., and Sha, H. L. 2005. Identification of resistance of new soybean varieties (lines) to soybean pod borer in Jilin province. J. Jilin Agric. Sci. 30:40-41.

(Prepared by Y. Li)

Soybean Diseases in India

Soybean Production

In India, soybean has been the largest oilseed crop, in terms of both area cultivated and volume produced, since 2005. The crop has shown unprecedented growth since 1970, when the area in cultivation was 30,000 ha and the volume produced was 14,000 metric tons. By 2013, these levels had increased to 12.2 million ha and 11.9 million metric tons (Table 5). That year, soybean comprised 42% of the total oilseed and 25% of the edible oil production in India, and the crop earned approximately $127 million in foreign exchange through exports of defatted oil cake. Because of the rapid expansion in cultivated area and production volume of soybean, India was ranked fourth worldwide in 2013 in area harvested.

In India, soybean is grown mainly in the states of Andhra Pradesh, Chattisgarh, Gujarat, Karnataka, Madhya Pradesh, Maharashtra, and Rajasthan as a rain-fed crop during the rainy season. It has long been a popular crop in the northeastern states, particularly Nagaland. Over the years, cultivation of soybean has been instrumental in improving the socioeconomic condition of growers with small farms and marginal (i.e., less-fertile) land in the rain-fed agroecosystems of central and peninsular India.

Edible soybean is becoming increasingly popular in the northeast state of Jharkhand. This protein-rich, large-seeded soybean offers both functional and nutraceutical benefits for humans. Consuming this type of soybean can improve families' diets; moreover, it provides an additional food in programs that serve meals to children in schools and otherwise focus on alleviating poverty.

Soybean can serve as an oilseed, as a fresh vegetable, and in other culinary, confectionary, and ready-to-eat purposes. Given these diverse uses, soybean will certainly have a place in India's agricultural future.

Soybean Diseases

In the early years of Indian soybean production, diseases were not of concern, but continuous cultivation of this crop has led to increased disease incidence. Although the disease scenario varies with the location, 35 soybean diseases commonly occur in India. The most important diseases are anthracnose (on stems and pods), bacterial pustule, bud blight, Cercospora leaf blight, purple seed stain, Rhizoctonia aerial blight, Sclerotium blight, soybean mosaic, and yellow mosaic. These diseases occur regularly and are widely distributed throughout soybean-growing areas of the country. Other diseases are regionally important, such as bacterial blight, brown spot, charcoal rot, frogeye leaf spot, Fusarium blight or wilt, Fusarium pod and collar rot, Myrothecium leaf spot, rust, and target leaf spot. In addition, Aristastoma leaf spot, Choanephora leaf blight, Phoma leaf blight, and phyllody associated with no-podding syndrome (mollicute), Phyllosticta leaf spot, powdery mildew, red crown rot, and Sclerotinia stem rot also occur on soybean.

Yield losses caused by soybean diseases can be considerable. In 1994, anthracnose, Fusarium root rot, pod and stem blight, Rhizoctonia aerial blight, and bacterial and viral diseases caused an estimated loss of 420,000 metric tons (10.7% of total production). In 2006, anthracnose, bacterial blight, bacterial pustule, brown spot, Cercospora leaf blight, charcoal rot, frogeye leaf spot, Fusarium root rot, pod and stem blight, purple seed stain, Rhizoctonia aerial blight, Sclerotium blight and other seedling diseases, rust, and virus diseases caused an estimated loss of 820,000 metric tons (9.3% of total production).

Diseases Caused by Bacteria

Bacterial pustule appears after a succession of rainy days, usually after plants are 35 days old. Under these conditions, bacterial pustule can cause yield losses of 27–60%. The pathogen survives up to 20 months in soybean seed and for 7 months in infected leaves. The disease is prevalent wherever susceptible cultivars of soybean are grown.

Diseases Caused by Fungi and Oomycetes of Foliage and Upper Stems, Pods, and Seeds

Alternaria leaf spot occurs frequently in the Indian states of Delhi, Karnataka, Madhya Pradesh, and Uttarakhand. The disease becomes severe under high temperatures and humidity. Normally, it appears on mature plants at flowering and during the podding stage, causing drying and defoliation of leaves and pod blight. Alternaria leaf spot generally causes a 10–20% yield loss under favorable conditions, but in 1997, a 50% yield loss was observed at Indore in Madhya Pradesh.

Anthracnose on stems and pods (causing pod blight) occurs predominantly and regularly in these states: Assam, Chhattisgarh, Delhi, Himachal Pradesh, Karnataka, Madhya Pradesh, Meghalaya, Nagaland, Rajasthan, and Uttarakhand. The pathogen produces symptoms from the early seedling stage to maturity. Severe seedling mortality may occur and require that the crop be resown.

Brown spot occurs regularly in Himachal Pradesh and occasionally in Madhya Pradesh, Maharashtra, Uttarakhand, and the northeastern states. Brown spot causes substantial yield losses, because only a few soybean genotypes exhibit a fair level of resistance.

Frogeye leaf spot occurs under high humidity and moderate temperatures. This disease has been reported from Himachal Pradesh, Karnataka, Madhya Pradesh, Rajasthan, Uttar Pradesh, Uttarakhand, and the northeastern states

Myrothecium leaf spot occurs regularly in some parts of Karnataka, Madhya Pradesh, Maharashtra, Rajasthan, and Uttarakhand. The disease can appear on crops 25–30 days old and cause yield losses as high as 18–23%.

Pod blight, which is prevalent in almost all soybean-growing areas of India, is caused by a group of pathogens that produce a variety of symptoms. In various parts of the country, different pathogens are associated with pod blight, including those pathogens that cause Alternaria leaf spot, anthracnose, charcoal rot, Myrothecium leaf spot, pod and stem blight, and purple seed stain.

Purple seed stain has become more prevalent since the first outbreak was reported in 1971. It occurs in Delhi, Karnataka, Madhya Pradesh, and Maharashtra. Yield losses of 20% have been reported, and an approximately 30% loss in seed viability has been observed, as well. A loss in germination has been observed when more than 25% of the seed surface is stained, whereas postemergence seedling mortality has been observed when more than 50% of the seed surface is stained.

Rhizoctonia aerial blight appears regularly in areas of high rainfall—predominantly, the states of Chattisgarh, Madhya Pradesh, and Uttarakhand—and causes substantial yield losses.

Plant mortality of up to 65% has been reported from Madhya Pradesh.

Rust is a major disease and has the potential to cause yield losses of 20–100%, depending on the time of infection, the cultivar, and the climatic conditions. Rust was previously considered to be of minor importance in India; before 1977, the disease was restricted to the northeastern states and reported in the low hills of Uttar Pradesh and West Bengal. After 1993, the disease became more prevalent, and since then, it has been reported across the country. It occurs regularly in Karnataka, Maharashtra, and the northeastern states from the third week of July to the last week of September. In southern India, self-sown and winter-sown soybean as sole crops or crops intercropped with sugarcane in irrigated areas are known to harbor the rust pathogen in the off-season, thereby providing sources of primary inoculum for rainy-season crops. The onset and spread of rust are favored by a temperature range of 19–23°C (minimum) to 26–28°C (maximum), a humidity level of 83–90%, a preponderance of cloudy weather, and an intermittent pattern of rainfall. The presence of dew on plants' leaves (i.e., leaf wetness) and favorable temperatures are the main factors in the onset and spread of rust.

Diseases Caused by Fungi and Oomycetes of Lower Stems and Roots

Charcoal rot was considered a minor disease of soybean in India until 2004, when rainfall was erratic. Since then, the disease has occasionally been severe in the major soybean-growing states. Charcoal rot has the potential to cause yield losses of up to 80%. Dry conditions, relatively low levels of moisture and nutrients, and temperatures of 25–35°C all create favorable conditions for development of the disease. In 1997, an epiphytotic of charcoal rot occurred in the Guna district of Madhya Pradesh during the seedling stage.

Sclerotium blight occurs mostly in the states of Assam, Karnataka, Madhya Pradesh, Meghalaya, Nagaland, Rajasthan, and Uttarakhand. This disease causes yield losses of 30–40% under favorable conditions; seedling mortality of 2–65% has also been reported, appearing as early as 15 days after emergence under high soil moisture conditions.

Diseases Caused by Viruses

Soybean mosaic is found in almost all soybean-growing areas in India, and virus seed transmission rates range from 7 to 52%. Yellow mosaic disease is most serious in Delhi, Haryana, Karnataka, Madhya Pradesh, Punjab, Rajasthan, Uttar Pradesh, and Uttarakhand. Nucleotide analysis of the virus isolate from Delhi revealed more than 89% identity with the virus that causes Mungbean yellow mosaic.

Indian bud blight is caused by Peanut bud necrosis virus in India, but in other countries, it is caused by *Tobacco ringspot virus*. This disease occurs predominantly in Chhattisgarh, Delhi, Karnataka, Madhya Pradesh, Maharashtra, and Rajasthan, and yield losses ranging from 12 to 80% have been reported in the country. To date, no genotype has been found to be resistant or moderately resistant to this disease.

Diseases Caused by Nematodes

Nematode problems in India have received limited attention, and no systematic work has been done on nematode diseases. Even so, populations ranging from 300 to 500 nematodes per 250 g of soil of the southern root-knot nematode and the reniform nematode have been recorded in Uttar Pradesh.

Disease Management

Planting resistant and/or moderately resistant cultivars is recommended for control of Alternaria leaf spot, anthracnose, bacterial pustule, frogeye leaf spot, Myrothecium leaf spot, and Rhizoctonia aerial blight. For soybean rust, the resistance conferred by the genes *Rpp1, Rpp2, Rpp3,* and *Rpp4* is considered ineffective, but planting rust-tolerant cultivars is recommended. Similarly, for soybean mosaic and yellow mosaic, selecting moderately resistant cultivars is recommended.

Cultural management practices include using clean, vigorous, certified seed, if available. Rogueing of infected plants and burning of infected plant parts is recommended for the management of most foliar and stem pathogens. Other practices used singly or in combination include deep summer plowing, crop rotation or mixed cropping with cotton or cereals, and timely applications of optimum doses of fertilizers. Maintaining low plant populations to promote vigorous growth is recommended for most diseases but especially those caused by soilborne pathogens. An additional recommendation for charcoal rot is to flood the field 3–4 weeks before sowing or to maintain high soil moisture via irrigation during crop growth. For soybean rust, planting soybean plants in the off-season is discouraged and rogueing of self-sown plants is recommended. Another management technique used for Sclerotium blight includes applying neem cake to the soil in disease hotspots. Cultural management techniques for soybean mosaic and yellow mosaic include cultivating two or more cultivars per season, planting different cultivars at least after every other year, rogueing infected plants and plant parts, and practicing weed-free cultivation.

For the chemical control of bacterial pustule, recommendations include seed treatments of streptocycline and foliar applications of copper oxychloride plus streptocycline or kasugamycin/validamycin plus copper oxychloride. Many of the foliar diseases caused by fungi (including purple seed stain, charcoal rot, and Sclerotium blight) can be managed using seed treatments with captan, carbendazim, thiram, or some combination of them or with the biocontrol agents *Trichoderma harzianum* and/or *T. viride*. Foliar application of fungicides such as carbendazim, mancozeb, and thiophanate-methyl is also recommended for control of most foliar fungal pathogens. For soybean rust, specifically, the recommendation is two or three applications of hexaconazole, oxycarboxin, propiconazole, or triadimefon 35–40 days after sowing in rust hotspots.

Treatment to control insect vectors is recommended for the management of viruses. Strategies include presowing a soil application of phorate, seed treatment with thiamethoxam for yellow mosaic, and foliar application of thiamethoxam or methyl demeton for both insect vectors and yellow mosaic or ethofenprox for yellow mosaic.

Selected References

Chalam, V. C., Khetarpal, R. K., Mishra, A., Jain, A., and Gupta, G. K. 2004. Seed transmission of *Soybean mosaic virus* in soybean cultivars. J. Mycol. Plant Pathol. 34:86-87.

Gawade, D. B., Suryawanshi, A. P., Patil, V. B., Zagade, S. N., and Wadje, A. G. 2009. Screening of soybean cultivars against anthracnose caused by *Colletotrichum truncatum*. J. Plant Dis. Sci. 4:124-125.

Gupta, G. K. 2004. Soybean diseases and their management. Pages 145-168 in: Soybean production and improvement in India. N. B. Singh, G. S. Chauhan, A. K. Vyas, and O. P. Joshi, eds. National Research Centre for Soybean, Indore, India.

Gupta, G. K., and Sharma, S. K. 2009. Influence, optimization, energy budgeting and monetary considerations of different time intervals between fungicidal seed treatment and sowing on the Sclerotium blight disease of soybean. Crop Prot. 30:1-5.

Gupta, G. K., Kurundkar, B. P., Khot, G. G., Raut, V. M., and Sharma, S. K. 2007. Epidemiology and race scenario of soybean rust in India. Sopa Dig. 4:12-16.

Gupta, G. K., Sharma, S. K., and Ramteke, R. 2012. Biology, epidemiology and management of the pathogenic fungus *Macrophomina phaseolina* (Tassi) Goid with special reference to charcoal rot of soybean (*Glycine max* [L.] Merrill). J. Phytopathol. 160:167-180.

Ramteke, R., Gupta, G. K., and Joshi, O. P. 2004. Evaluation of rust resistance of soybean (*Glycine max* [L.] Merrill) under field condition. Indian J. Agr. Sci. 74:623-624.

Ramteke, R., Gupta, G. K., Gill, B. S., Varma, R. K., and Lal, S. K. 2007. Development of soybean lines resistant to Yellow mosaic virus. Soybean Res. 5:71-74.

Sharma, S. K., Gupta, G. K., and Khare, M. N. 2007. Purple seed stain (*Cercospora kikuchii* Mat. & Tomo.) disease of soybean—A review. JNKVV Res. J. 41:151-166.

Sharma, S. K., Gupta, G. K., and Ramteke, R. 2011. Distribution, symptoms, epidemiology and management of *Colletotrichum dematium* var. *truncatum* ([Schw.] Andrus & W. D. Moore) with special reference to anthracnose of soybean (*Glycine max* [L.] Merrill). Soybean Res. 9:31-52.

Shukla, A. K. 1994. Pilot estimation of soybean (*Glycine max*) yield losses by various levels of bacterial pustule (*Xanthomonas campestris* pv. *glycines*) infection. Integr. J. Pest Manag. 40:249-251.

Usharani, K. S., Surendranath, B., Haq, Q. M. R., and Malathi, V. G. 2004. Yellow mosaic virus infecting soybean in northern India is distinct from the species infecting soybean in southern and western India. Curr. Sci. India 86:845-850.

(Prepared by G. K. Gupta and R. M. Nair)

Soybean Diseases in Japan

Soybean Production

In 2013, Japan produced 199,990 metric tons of soybean. The soybean production area is approximately 131,000 ha, and 80% is planted in upland fields converted from paddy fields. Planting begins at the end of May in Hokkaido, the northernmost island of Japan, and in mid-June in the central and western regions of the country. Harvest occurs at the end of September in Hokkaido and in November and December in central and western Japan.

All harvested soybeans are used for food production: as grain for cooking and for making soy curd (*tofu*), fermented soy cheese (*natto*), fermented soy paste (*miso*), and soy sauce (*shoyu*). High-quality soybean grain is in demand from Japanese consumers and processing companies.

Soybean Diseases

The soybean diseases known to cause damage in Japan include bacterial pustule, downy mildew, Phytophthora stem and root rot, purple seed stain, red crown rot, Sclerotium blight, soybean mosaic, soybean dwarf, and soybean cyst. The relative importance of each disease varies by region and by year.

Other diseases occur to lesser degrees. Brown spot occurs, but its significance is not known. Sclerotinia stem rot and brown rot occasionally cause damage in Japan's northern regions. Soybean rust usually occurs just before harvest and has little impact on yield, but premature defoliation can occur if an epidemic breaks out early in the season.

Diseases Caused by Bacteria

Bacterial pustule is an important disease in central and western Japan and becomes severe after a heavy rainfall. Premature defoliation can cause a yield loss of up to 30%.

Diseases Caused by Fungi and Oomycetes

Downy mildew does not cause severe yield losses in Japanese soybean crops, but it reduces the commercial value of soybeans intended for food production, especially black soybeans. Similarly, purple seed stain (which occurs in all soybean production areas) does not cause severe yield losses but reduces the commercial value of soybeans for food production. Purple seed stain also reduces the availability of healthy seeds, which are necessary to obtain good plant stands.

Phytophthora root and stem rot and Pythium root rot cause preemergence decay and postemergence rot when soybean fields are flooded or wet; sometimes, farmers have to replant seeds. Red crown rot (also known as "black root rot") is one of the most important soilborne diseases in all regions of Japan. Symptoms usually appear on leaves after flowering, and diseased plants have reduced numbers of pods and decreased seed weight. Severely diseased plants sometimes die prematurely, reducing yield. This disease is particularly severe in fields with high moisture. Sclerotium blight is an important disease in the central and western regions of Japan. It causes preemergence decay, damping off, and take-all and continues to occur even after dry weather ensues. This disease usually occurs after intertillage and ridging have been performed.

Diseases Caused by Nematodes

The soybean cyst nematode is more problematic in upland fields than in fields converted from paddy fields. Three nematode groups are recognized—namely, those formerly known as race 1, race 3, and race 5. The former two groups are more common than the latter group.

Diseases Caused by Viruses

In Japan, soybean mosaic is the most important disease caused by a virus. Five strains of the pathogen, *Soybean mosaic virus,* have been identified (A, B, C, D, and E), and strains A and B are most common. The soybean aphid can transmit the virus.

Soybean dwarf is another significant virus disease, mostly in the northern regions of Japan. Four strains of the pathogen, *Soybean dwarf virus* (YS, YP, DS, and DP), are recognized by the symptoms and species of vectors that transmit them.

In 2005, a soilborne viral disease referred to as "soybean leaf rugose mosaic" was identified in Japan. The proposed name of the virus (which has yet to be officially recognized) is "Soybean leaf rugose mosaic virus."

Disease Management

A cultivar with resistance to bacterial pustule is available, and cultivars with reduced susceptibility are available for red crown rot. Some cultivars also are resistant to the soybean cyst nematode (races 1 and 3) and to *Soybean mosaic virus* strains A and B.

Sowing healthy, vigorous seed is important for producing healthy plant stands, and so is providing good soil drainage, which helps reduce seed and seedling diseases. For management of Phytophthora root and stem rot, tilling before sowing, ridge culture planting, and field drainage systems are recommended. Red crown rot can be controlled using long rotations (up to 3 years) with rice paddy fields, whereas Sclerotium blight can be reduced by flooding fields for 1 year. Crop rotation is recommended for managing soybean cyst, and delayed planting of an early cultivar can reduce soybean aphid infestation and thereby reduce the incidence of soybean dwarf.

Pesticides are used to control some soybean diseases in Japan but not on a regular basis. Copper fungicides are registered for bacterial pustule (but not widely used), and fungicides are also recommended for control of downy mildew (but rarely used), Phytophthora root and stem rot, and red crown rot (for which only one fungicide is registered as a seed treatment). For management of purple seed stain, azoxystrobin, thiophanate-methyl, and triazole fungicides are sprayed once or twice 2–6 weeks after flowering. Strains of the fungus that causes purple seed stain that have resistance to thiophanate-methyl have been detected in some areas. For management of Sclerotiorum blight, soil fumigants and soil-drenching fungicides are registered but not commonly used. Soil application or seed treatment with insecticides is used to control aphids, which can transmit virus diseases.

Selected References

Fujita, Y. 1990. Ecology and control of purple seed stain of soybean caused by *Cercospora kikuchii.* Bull. Tohoku Natl. Agric. Exp. Stn. 81:51-109.

Kato, M., Minamida, K., Tojo, M., Kokuryu, T., Hamaguchi, H., and Shimada, S. 2013. Association of *Pythium* and *Phytophthora* with pre-emergence seedling damping-off of soybean grown in a field converted from a paddy field in Japan. Plant Prod. Sci. 16:95-104.

Kuroda, T., Nabata, K., Hori, T., Ishikawa, K., and Natsuaki, T. 2010. Soybean leaf rugose mosaic virus, a new soilborne virus in the family *Potyviridae*, isolated from soybean in Japan. J. Gen. Plant Pathol. 76:382-388.

Mukobata, H., Iwata, T., Morikawa, T., Yamazaki, Y., and Sekihara, J. 2006. Occurrence and chemical control of tiophanete-methyl-resistant strains of soybean purple seed stain fungus, *Cercospora kikuchii,* in Toyama prefecture, Japan. Bull. Toyama Agric. Res. Ctr. 23:61-71.

Nakagawa, A., and Shimada, S. 1990. Survey on the occurrence of soybean dumping-off diseases in the upland field converted from paddy field in Kinki-Chugoku Districts. Bull. Chugoku Natl. Agric. Exp. Stn. 15:19-27.

Nishi, K., Sato, F., Karasawa, T., Sato, T., Fukuda, T., and Takahasi, H. 1999. Ecology and control of root necrosis of soybean caused by *Calonectria crotalariae*. Bull. Natl. Agric. Res. Cent. 30:11-109.

Sugimoto, T., Kato, M., Yoshida, S., Matsumoto, I., Kobayashi, T., Kaga, A., Hajika, H., Yamamoto, R., Watanabe, K., Aino, M., Matoh, T., Walker, D. R., Biggs, A. R., and Ishimoto, M. 2012. Pathogenic diversity of *Phytophthora sojae* and breeding strategies to develop Phytophthora-resistant soybeans. Breed. Sci. 61:511-522.

Takahasi K., Tanaka, T., Iida, W., and Tsuda, Y. 1980. Studies on virus diseases and causal viruses of soybean in Japan. Bull. Tohoku Natl. Agric. Exp. Stn. 62:1-130.

Terauchi, H., Kanematsu, S., Honda, K., Mikoshiba, Y., Ishiguro, K., and Hidaka, S. 2001. Comparison of complete nucleotide sequences of genomic RNAs of four *Soybean dwarf virus* strains that differ in their vector specificity and symptom production. Arch. Virol. 146:1885-1898.

(Prepared by M. Kato)

Soybean Diseases of Importance in the United States

Soybean Production

In 2013, the United States was the leading soybean producer in the world, producing nearly one-third (32.4%) of the total volume (Table 5). That year, U.S. growers harvested soybeans on 30.7 million ha, and their crops yielded a total of 89.5 million metric tons (MMT). The top 10 states for soybean production in 2013 (in decreasing order) were Illinois, Iowa, Minnesota, Missouri, Indiana, Nebraska, North Dakota, South Dakota, Ohio, and Kansas. The area harvested in the top three producing states—Illinois, Iowa, and Minnesota—was 10.2 million ha, or 33% of the total area harvested. Approximately 80% of the harvested area in any given year is from the north-central states.

Soybean Diseases

Soybean diseases occur throughout the 37 states that produce soybean. Although disease importance varies by state and even by regions within a state, some diseases are likely to occur independently of specific soil conditions, rainfall patterns, and cultivar planted.

Annual surveys of soybean diseases throughout all soybean production areas in the United States have not systematically been completed, but disease loss estimates have been published over the years in most states based on the expertise of plant pathologists. Although the yield loss varies from year to year, it has been estimated at approximately 11% of total production.

Based on the average of annual estimates from 2006 to 2009, the loss caused by all diseases was 11.2 MMT, or 13% of the average total production during those 4 years. Soybean cyst was ranked as the top disease (30.9% of total loss), followed by seedling diseases (11.2%), Phytophthora root and stem rot (10.4%), sudden death syndrome (6.3%), Sclerotinia stem rot (5.4%), charcoal rot (5.3%), brown spot (4.7%), brown stem rot (3.8%), anthracnose (3.5%), and pod and stem blight (2.3%). Other diseases are regionally important (such as soybean rust), and estimates of economic losses from other recently reported diseases have not been determined. (For instance, soybean vein necrosis was first reported in 2008.)

Diseases Caused by Bacteria

Two bacterial diseases common in U.S. soybean crops are bacterial blight and bacterial pustule. In general, bacterial blight develops early in the season under cool conditions, whereas bacterial pustule develops later in the season under warm conditions. Losses from both diseases may range from 4 to 40% under extreme conditions, but mostly, neither disease is problematic.

Some systemic bacterial diseases occur sporadically, such as wilt and tan spot. These diseases may be of local importance.

Diseases Caused by Fungi and Oomycetes

Among the soybean diseases caused by fungi and oomycetes, the most important are the seedling diseases, which have many causal pathogens and include those that cause Fusarium root rot, Phytophthora root and stem rot, and Pythium damping-off and root rot. The mid- to late-season diseases of greatest importance are anthracnose, brown spot, brown stem rot, charcoal rot, pod and stem blight, Sclerotinia stem rot, and sudden death syndrome.

Many of these diseases have greater regional than national importance, based on their occurrence from south to north. For example, Rhizoctonia aerial blight, Sclerotium blight, and soybean rust are more problematic in the southern growing region, whereas Sclerotinia stem rot and brown stem rot are more problematic in the northern growing region. The same is true of southern stem canker and northern stem canker, although these differences are not always entirely regional; epidemics can be driven by weather patterns and inoculum sources, which vary from year to year.

Diseases Caused by Nematodes

Soybean cyst was first reported in the United States in 1954 and is now widespread within all soybean-producing states. Populations of the soybean cyst nematode differ in their ability to parasitize soybean cultivars. The virulence of a specific population can be distinguished using a set of indicator lines representing seven sources of resistance; this method has been used in breeding programs across the United States.

After the soybean cyst nematode, the root-knot nematode is the next most important in terms of causing disease. The importance of other nematodes—such as the lance, lesion, reniform, and sting nematodes—has not been well documented.

Diseases Caused by Viruses

Many soybean viruses have been reported, but collectively, they represent only 1–2% of the total yield losses caused by soybean diseases. The virus diseases most commonly reported are bean pod mottle, cucumber mosaic, soybean mosaic, and tobacco ringspot. The effects of viruses documented in the early to mid-2000s, such as soybean dwarf and soybean vein necrosis, have yet to be determined.

Disease Management

Soybean growers in the United States purchase their seed primarily from commercial companies, and most of the soybean cultivars grown are genetically modified with herbicide resistance. Nearly all commercial cultivars have been bred and selected to have resistance to one or more pathogens. For example, most cultivars have resistance to Phytophthora root rot and stem rot and to the soybean cyst nematode. These cultivars are often marginally effective, however, if they are improperly matched with the fungal and nematode populations, respectively, that exist in individual fields.

For soybean growers, cultivar selection is an important tool in disease management, and some states offer programs with unbiased, state-specific disease evaluations of cultivars. In specific regions of the country, cultivars are advertised that have resistance to frogeye leaf spot and stem canker. Many cultivars have resistance to bacterial blight and bacterial pustule, although these qualities are not necessarily advertised by commercial seed companies. Even though diseases such as charcoal rot, Sclerotinia stem rot, and sudden death syndrome have significant effects on U.S. soybean production, relatively few cultivars on the market are advertised to have resistance to them.

In addition to the use of resistant cultivars, crop rotation is an effective strategy for managing disease on soybean. Crop rotation practices have not changed much since the 1970s. Most soybeans are grown after corn, but in some cases, they are grown after wheat as a single or a double crop within one season.

Seed and foliar fungicide applications have increased dramatically since the mid-2000s, to the extent that nearly 100% of commercial seed is fungicide coated. In some states, especially those in which soybean rust is problematic, most fields are treated with at least one application of a foliar fungicide.

Future soybean production in the United States is expected to increase proportionally to increased demand. To provide the solutions needed to combat both current and anticipated diseases and thereby meet this demand, more efficacious fungicides must be developed (either as seed treatments or foliar applications) and genetic resources must be further explored (through traditional breeding and molecular transformation).

Selected References

Chawla, S., Bowen, C. R., Slaminko, T. L., Hobbs, H. A., and Hartman, G. L. 2013. A public program to evaluate commercial soybean cultivars for pathogen and pest resistance. Plant Dis. 97:568-578.

Hartman, G. L., and Hill, C. B. 2010. Diseases of soybean and their management. Pages 276-279 in: Soybean: Botany, Production, and Uses. G. Singh, ed. CABI, Wallingford, Oxfordshire, U.K..

Roy, K. W., Baird, R. E., and Abney, T. S. 2000. A review of soybean (*Glycine max*) seed, pod, and flower mycofloras in North America, with methods and a key for identification of selected fungi. Mycopathologia 150:15-27.

Wrather, J. A., and Koenning, S. R. 2009. Effects of diseases on soybean yields in the United States 1996 to 2007. Plant Health Progress. doi:10.1094/PHP-2009-04401-01-RS.

Wrather, J. A., Anderson, T. R., Arsyad, D. M., Gai, J., Ploper, L. D., Porta-Puglia, A., Ram, H. H., and Yorinori, J. T. 1997. Soybean disease loss estimates for the top 10 soybean producing countries in 1994. Plant Dis. 81:107-110.

Wrather, J. A., Anderson, T. R., Arsyad, D. M., Tan, Y., Ploper, L. D., Porta-Puglia, A., Ram, H. H., and Yorinori, J. T. 2001. Soybean disease loss estimates for the top ten soybean-producing countries in 1998. Can. J. Plant Pathol. 23:115-121.

(Prepared by G. L. Hartman)

Part I. Infectious Diseases

Diseases Caused by Bacteria

Bacteria are small, prokaryotic microorganisms that can be seen with the naked eye only en masse. Prokaryotic cells usually do not have nuclei, mitochondria, and chloroplasts. Bacteria can be found in diverse environments and have different types of associations with plants. Bacteria that persist on the outsides of plants are referred to as "epiphytes," and those that persist on the insides of plants are referred to as "endophytes."

Most epiphytic bacteria are saprophytes, which simply grow or survive on the plant surface to obtain nutrients. Some epiphytic bacteria are parasites and cause diseases; they are called "pathogens." Some bacteria live within the plant either by causing disease or by establishing a mutually beneficial relationship termed "symbiosis." This section focuses on those bacteria that cause diseases in soybean plants and those nitrogen-fixing bacteria that form symbiotic relationships with soybean.

Soybean plants infected with various bacterial pathogens and symbionts can show different types of symptoms. These symptoms include yellowing (chlorosis), water soaking, cell death (necrosis), abnormal growth (tumors), blockage of water-conducting tissue (wilting), and breakdown of tissue structure (dry and soft rots). These symptoms are not unique to bacterial pathogens, however, and can be caused by other organisms, including fungi, nematodes, and viruses. Accurate diagnosis is essential to determine the cause of disease symptoms.

Bacteria lack active mechanisms for the penetration of plants. Thus, most plant-pathogenic bacteria infect plants through wounds produced by wind, hail, abrasion by surrounding branches, animal and insect feeding, mechanical injury associated with cultural practices, and lesions and wounds caused by other pathogens, as well as through natural openings in plants, such as the hila of seeds, hydathodes, lenticels, nectaries, stomata, and natural junctions (e.g., leaf scars). Splashing water, windblown rain, infected plant parts, and movement via soil can disseminate bacterial plant pathogens locally. Dissemination can also occur via other living organisms, such as insects and nematodes, which can transmit bacterial cells to infection courts. Many bacteria infect the insides of seeds or infest the outsides of seeds and remain viable in or on them for one or more seasons. When bacteria are seedborne, dissemination may occur by seed transportation or by trade locally, nationally, or internationally.

Many plant-pathogenic bacteria survive poorly in soil but well in and on plant debris. Bacterial diseases are typically severe when environmental conditions favor the growth and multiplication of bacterial cells that often require high temperature and high humidity. Temperature and moisture are major factors influencing the occurrence and severity of diseases caused by plant-pathogenic bacteria. Plants with water-soaked tissues are often more susceptible to invasion by bacteria. The status of plant growth and development also can affect the occurrence and severity of bacterial infection.

Successful management of bacterial diseases on plants requires the application of several practices. Major preventive measures include avoiding the introduction and dissemination of bacteria and using cultural practices to remove infected plant debris. Probably the most effective means of controlling bacterial diseases are to use resistant cultivars and to plant pathogen-free seed.

Selected References

Brenner, D. J., Krieg, N. R., Staley, J. T., and Garrity, G. M., eds. 2005. Bergey's Manual of Systematic Bacteriology. Vol. 2, The Proteobacteria. 2nd ed. Williams & Wilkins/Springer–Verlag, New York.

Bull, C. T., De Boer, S. H., Denny, T. P., Firrao, G., Saux, M. F., Saddler, G. S., Scortichin, M., Stead, D. E., and Takikawa, Y. 2008. Demystifying the nomenclature of bacterial plant pathogens. J. Plant Pathol. 90:403-417.

Dworkin, M., Falkow, S., Rosenberg, E., Schleifer, K. H., and Stackebrandt, E., eds. 2006. The Prokaryotes: A Handbook on the Biology of Bacteria. 3rd ed. Springer–Verlag, New York.

Goto, M. 1992. Fundamentals of Bacterial Plant Pathology. Academic Press, New York.

Jackson, R. W., ed. 2009. Plant Pathogenic Bacteria: Genomics and Molecular Biology. Caister Academic Press, Norfolk, U.K.

Janse, J. D. 2005. Phytobacteriology, Principles and Practice. CABI, Wallingford, Oxfordshire, U.K.

Lelliott, R. A., and Stead, D. E. 1987. Methods for the Diagnosis of Bacterial Diseases of Plants. Blackwell Scientific, Oxford, U.K.

Schaad, N. W., Jones, J. B, and Chun, W., eds. 2001. Laboratory Guide for Identification of Plant Pathogenic Bacteria. 3rd ed. American Phytopathological Society, St. Paul, MN.

Young, J. M., Bradbury, J. F., Davis, R. E., Dickey, R. S., Ercolani, G. L., Hayward, A. C., and Vidaver, A. K. 1991. Nomenclatural revisions of plant pathogenic bacteria and list of names, 1980–1988. Rev. Plant Pathol. 70:211-221.

Young, J. M., Saddler, G. S., Takikawa, Y., De Boer, S. H., Vauterin, L., Garden, L., Gvozdyak, R. I., and Stead, D. E. 1996. Names of plant pathogenic bacteria 1864–1995. Rev. Plant Pathol. 75:721-763.

(Prepared by Y. F. Zhao)

Bacterial Blight

Bacterial blight of soybean occurs worldwide and is one of the most common bacterial diseases of this crop. Bacterial blight occurs throughout the United States but is most commonly observed during cool, wet conditions. The disease is most prevalent in fields during the early to the mid-season under moist conditions. Outbreaks often follow thunderstorms.

Estimates of yield losses under extreme conditions range from 5 to 18% in South Korea and from 4 to 40% in the United States, but usually, the disease causes no measurable losses. Bacterial blight is of greater concern when soybean is grown for seed.

Symptoms and Signs

Seedborne inoculum may produce seedlings that express brown lesions on the margins of cotyledons early in the growing season. The lesions enlarge and turn dark brown, and then the affected tissue collapses. Young seedlings may have stunted growth. If the infection reaches the shoot apical meristem, the seedling may die.

Young leaves are most susceptible to bacterial infection, and disease symptoms are most noticeable on leaves in the upper part of the canopy. The initial symptom is the appearance of small, angular, water-soaked lesions on leaves (Fig. 2). The lesion centers soon dry out and turn brown to black; they also develop water-soaked margins and yellow halos. The presence of lesions with yellow halos is one of the characteristics of bacterial blight that distinguish it from other diseases, such as brown spot. The angular lesions may enlarge and coalesce, resulting in large, irregular blighted areas. Affected tissue often drops out or tears away, causing the leaves to appear ragged. Infected young leaves may be distorted, stunted, and chlorotic. Premature defoliation may occur on heavily infected plants.

Blight lesions can also occur on stems, petioles, and pods. Petiole and stem lesions are often large and black. Pod lesions, however, begin as small, water-soaked spots that enlarge and later merge, turning dark brown to black with age. If pod infection occurs, bacterial blight becomes seedborne. Infected seeds may eventually be covered with a layer of bacterial cells (slimy growth). Stored seeds may appear healthy or may shrivel, develop sunken or raised lesions, or become slightly discolored (Fig. 3).

Causal Organism

Pseudomonas savastanoi pv. *glycinea,* the causal bacterium, is a motile, gram-negative rod (1.2–1.5 × 2.3–3.3 μm) with one to several polar flagella. It weakly produces a yellow-green, water-soluble, diffusible fluorescent pigment on King's B medium. Colonies on rich media are circular, smooth, and glistening and have entire margins; they are white and raised. Sometimes, colonies become mucoid. The optimal temperature for growth of the bacterium is 24–28°C.

The bacterium produces levan and induces a hypersensitive response on tobacco. Like many gram-negative bacteria, *P. savastanoi* pv. *glycinea* employs a type III secretion system, and its secreted effector proteins cause disease on soybean. The bacterium also produces coronatine: a nonhost-specific phytotoxin that is responsible for inducing the yellow halos and may function in suppressing host defenses.

Strains of *P. savastanoi* pv. *glycinea* can be organized into at least nine physiological races based on their reactions to nine host cultivars (i.e., Acme, Centennial, Chippewa, Flambeau, Harosoy, Lindarin, Merit, Norchief, and Peking). Most of the races are known to occur in the United States, but several studies have shown that the predominant race is race 4, which can cause disease on all nine cultivars. The first avirulence gene (*avrA1*) was cloned from race 6 of *P. savastanoi* pv. *glycinea* in 1984, and subsequently, several more avirulence genes (*avrB1, avrB2, avrD1*) have been identified and characterized from other races. Whole-genome sequence analysis of *P. savastanoi* pv. *glycinea* has indicated that the pathogen is most closely related to *P. savastanoi* pv. *phaseolicola,* which causes bean halo blight.

Disease Cycle and Epidemiology

P. savastanoi pv. *glycinea* survives in infested crop debris and in seeds. During the growing season, the presence of infected pods results in bacterium-carrying seeds. Because bacterial blight is typically an early season disease, the major sources of inoculum are contaminated seeds and plant debris. Primary infections on cotyledons often result in secondary lesions on seedlings. The pathogen may be disseminated during the growing season by splashed or wind-driven water droplets, especially during windy rainstorms and cultivation when foliage is wet.

The bacterium can survive epiphytically on leaf surfaces and buds, and under favorable conditions, it can enter leaves through natural openings, such as stomata and wounds. The bacterium multiplies mainly in the intercellular spaces of the mesophyll and reaches high populations in these tissues. Re-

Fig. 2. Small, angular, yellow to light-brown leaf lesions, characteristic of bacterial blight (*Pseudomonas savastanoi* pv. *glycinea*). (Courtesy J. B. Sinclair)

Fig. 3. Seed discoloration, caused by *Pseudomonas savastanoi* pv. *glycinea,* the bacterial blight pathogen. (Courtesy J. B. Sinclair)

search has shown that the toxin coronatine, which is produced by the pathogen, promotes the opening of stomata for infection and also inhibits photosynthesis.

After the pathogen enters the plant, water-soaked lesions usually develop within 3–7 days. The optimum temperature for infection is 24–27°C, which is also the optimal temperature for the pathogen to produce coronatine. Disease development is also favored by cool, wet conditions. Bacterial blight is frequently more prevalent in areas with heavy rainfall and windy conditions. Early infection provides new inoculum for a secondary cycle to occur at any time during the growing season, and this condition persists until hot, dry weather limits the likelihood of disease development.

Management

In areas in which bacterial blight is a potential problem, growers should not plant highly susceptible cultivars but rather disease-tolerant cultivars. Currently, four resistance genes (*Rpglb, Rpg2, Rpg3,* and *Rpg4*) have been identified, which recognize four avirulence genes (*avrB1, avrA, avrB2,* and *avrD1*), respectively. Soybean germplasm line LL 489-605 (derived from plant introduction [PI] 437821) is also resistant.

Because bacterial blight is seedborne, planting pathogen-free seed will reduce its occurrence. Cultural practices are also usually effective in controlling bacterial blight, including rotating crops, managing residue, and avoiding cultivation and other field activities (e.g., spraying) when foliage is wet. Completely burying soybean residue can greatly reduce the risk of bacterial blight (although doing so is not recommended in some areas).

Selected References

Bender, C. L., Alarcón-Chaidez, F., and Gross, D. C. 1999. *Pseudomonas syringae* phytotoxins: Mode of action, regulation and biosynthesis by peptide and polyketide synthetases. Microbiol. Mol. Biol. Rev. 63:266-292.

Cross, J. E., Kennedy, B. W., Lambert, J. W., and Cooper, R. L. 1966. Pathogenic races of the bacterial blight pathogen of soybeans, *Pseudomonas glycinea*. Plant Dis. Rep. 50:557-560.

Fett, W. F., and Sequeira, L. 1981. Further characterization of the physiologic races of *Pseudomonas glycinea*. Can. J. Bot. 59:283-287.

Gardan, L., Shafik, H., Belouin, S., Broch, R., Grimont, F., and Grimont, P. A. D. 1999. DNA relatedness among the pathovars of *Pseudomonas syringae* and description of *Pseudomonas tremae* sp. nov. and *Pseudomonas cannabina* sp. nov. (ex Sutic and Dowson 1959). Int. J. Syst. Bacteriol. 49:469-478.

Huynh, T. V., Dahlbeck, D., and Staskawicz, B. J. 1989. Bacterial blight of soybean: Regulation of a pathogen gene determining host cultivar specificity. Science 245:1374-1380.

Hwang, I., and Lim, S. M. 1992. Effects of individual and multiple infections with three bacterial pathogens on disease severity and yield of soybeans. Plant Dis. 76:195-198.

Melotto, M., Underwood, W., and He, S. Y. 2008. Role of stomata in plant innate immunity and foliar bacterial diseases. Annu. Rev. Phytopathol. 46:101-122.

Nickell, C. D., Lim, S. M., and Eathington, S. 1994. Registration of soybean germplasm line LL 89-605, resistant to brown stem rot and bacterial blight. Crop Sci. 34:1134.

Park, E. W., and Lim, S. M. 1986. Effects of bacterial blight on soybean yield. Plant Dis. 70:214-217.

Prom, L. K., and Venette, J. R. 1997. Races of *Pseudomonas syringae* pv. *glycinea* on commercial soybean in eastern North Dakota. Plant Dis. 81:541-544.

Qi, M., Wang, D., Bradley, C., and Zhao, Y. F. 2011. Genome sequence analyses of *Pseudomonas savastanoi* pv. *glycinea* and *in silico* subtractive hybridization-based comparative genomics with nine phytopathogenic pseudomonads. PLoS One 6:e16451.

Staskawicz, B. J., Dahlbeck, D., and Keen, N. T. 1984. Cloned avirulence gene of *Pseudomonas syringae* pv. *glycinea* determines race-specific incompatibility on *Glycine max*. Merr. Proc. Natl. Acad. Sci. U.S.A. 81:6024-6028.

Staskawicz, B. J., Dahlbeck, D., Keen, N., and Napoli, C. 1987. Molecular characterization of cloned avirulence genes from race 0 and race 1 of *Pseudomonas syringae* pv. *glycinea*. J. Bacteriol. 169:5789-5794.

Tamaki, S., Dahlbeck, D., Staskawicz, S., and Keen, N. T. 1988. Characterization and expression of two avirulence genes cloned from *Pseudomonas syringae* pv. *glycinea*. J. Bacteriol. 170:4846-4854.

Zhao, Y., Thilmony, R., Bender, C. B., Schaller, A., He, S. Y., and Howe, G. A. 2003. Virulence systems of *Pseudomonas syringae* pv. *tomato* promote bacterial speck disease in tomato by targeting the jasmonate signaling pathway. Plant J. 36:485-499.

Zhou, J., Rodriguez-Zas, S., Aldea, M., Li, M., Zhu, J., Gonzalez, D. O., Vodkin, L. O., DeLucia, E., and Clough, S. J. 2005. Expression profiling soybean response to *Pseudomonas syringae* reveals new defense-related genes and rapid HR-specific downregulatation of photosynthesis. Mol. Plant-Microbe Interact. 18:1161-1174.

(Prepared by Y. F. Zhao)

Bacterial Pustule

Bacterial pustule has been reported in most soybean-growing areas of the world in which warm temperatures and frequent showers prevail during the growing season. The disease may cause premature defoliation, which may decrease yield by reducing seed size and number.

Symptoms and Signs

The initial symptom of bacterial pustule is the development of minute, pale-green spots with elevated centers on either or both leaf surfaces. Later, small, raised, light-colored pustules form in the centers of lesions, usually in those on the lower leaf surfaces (Fig. 4). The pustules form through hypertrophy and hyperplasia. These symptoms are sometimes confused with those of soybean rust.

The lesions vary from specks to large, irregularly shaped, mottled-brown areas, which develop when the lesions coalesce. Leaves become ragged when dead areas are torn away by wind, and a severe level of disease often results in some defoliation. Leaf spots sometimes form without developing pustules. Small, reddish-brown, slightly raised spots may develop on pods of susceptible cultivars.

Symptoms of pustule sometimes resemble those of blight. However, unlike blight lesions, pustule lesions are not water

Fig. 4. Small, reddish-brown lesions on a small leaflet and petiole, characteristic of bacterial pustule (*Xanthomonas axonopodis* pv. *glycines*). (Courtesy G. L. Hartman)

soaked in the early stages of development, and they usually have minute, raised pustules in the centers, especially if viewed from the lower surface.

Causal Organism

The causal organism of bacterial pustule, *Xanthomonas axonopodis* pv. *glycines,* is a motile, gram-negative rod (0.5–0.9 × 1.4–2.3 µm) with a single polar flagellum. Colonies on beef infusion agar are small, circular, and smooth and have entire margins; they are initially pale yellow and become deep yellow with age.

The bacterium is slow growing in culture and produces auxins, bacteriocin, and exopolysaccharides. It also liquefies gelatin, produces acid but not gas from sucrose, and rapidly hydrolyzes starch. The optimal temperature for growth is 30–33°C. Races of the pathogen may exist.

Disease Cycle and Epidemiology

X. axonopodis pv. *glycines* overseasons in seeds, in surface crop residue, and in the rhizosphere of wheat roots. Strains of the bacterium infect common bean and cowpea.

The bacterium enters the plant through natural openings and wounds and multiplies intercellularly. It spreads via splashing water and windblown rain and during cultivation when foliage is wet. New infections may occur throughout the growing season when wet or rainy conditions prevail. In contrast to bacterial blight, bacterial pustule is not inhibited by high temperatures. The disease can be severe in tropical regions.

Management

X. axonopodis pv. *glycines* is seedborne; therefore, planting pathogen-free seed will reduce the occurrence of bacterial pustule. Employing cultural practices such as rotating with a nonhost crop (e.g., maize) and avoiding field activities when foliage is wet may also reduce the occurrence of pustule. Growers should plant resistant cultivars that derive their resistance from a recessive gene (*rxp*), originally derived from the cultivar CNS.

Selected References

Athinuwat, D., Prathuangwong, S., Cursino, L., and Burr, T. 2009. *Xanthomonas axonopodis* pv. *glycines* soybean cultivar virulence specificity is determined by *avrBs3* homolog *avrXg1*. Phytopathology 99:996-1004.

Bernard, R. L., and Lindahl, D. A. 1972. Registration of Williams soybean. Crop Sci. 12:716.

Goradia, L., Hartman, G. L., and Daniel, S. L. 2009. Evaluation of glyphosate-resistant soybean cultivars for resistance to bacterial pustule. Eur. J. Plant Path. 124:331-335.

Groth, D. E., and Braun, E. J. 1989. Survival, seed transmission, and epiphytic development of *Xanthomonas campestris* pv. *glycines* in the north-central United States. Plant Dis. 73:326-330.

Hartwig, E. E., and Lehman, S. G. 1951. Inheritance of resistance to the bacterial pustule disease in soybeans. Agron. J. 43:226-229.

Kaewnum, S., Prathuangwong, S., and Burr, T. J. 2005. Aggressiveness of *Xanthomonas axonopodis* pv. *glycines* isolates to soybean and hypersensitivity responses by other plants. Plant Pathol. 54:409-415.

Manjaya, J. G., and Pawar, S. E. 1999. New genes for resistance to *Xanthomonas campestris* pv. *glycines* in soybean (*Glycine max* [L.] Merr.) and their inheritance. Euphytica 106:205-208.

Narvel, J. M., Jakkula, L. R., Phillips, D. V., Wang, T., Lee, S.-H., and Boerma, H. R. 2001. Molecular mapping of *Rxp* conditioning reaction to bacterial pustule in soybean. J. Hered. 92:267-270.

Thowthampitak, J., Shaffer, B. T., Prathuangwong, S., and Loper, J. E. 2008. Role of *rpfF* in virulence and exoenzyme production of *Xanthomonas axonopodis* pv. *glycines,* the causal agent of bacterial pustule of soybean. Phytopathology 98:1252-1260.

Vauterin, L., Rademaker, J., and Swings, J. 2000. Synopsis on the taxonomy of the genus *Xanthomonas*. Phytopathology 90:677-682.

(Prepared by J. B. Sinclair; revised by G. L. Hartman)

Wildfire

Wildfire (also known as "blackfire") occurs on a number of crop plants, including cowpea, eggplant, soybean, tobacco, and tomato, but it is best known on tobacco. Wildfire is a minor problem on soybean in Brazil and the United States.

Symptoms and Signs

The wildfire bacterium causes brown, necrotic spots on leaves. The spots vary in size and shape and are nearly always surrounded by broad, yellow halos (Fig. 5). Some spots may be restricted and develop without sharply delineated chlorotic zones; these spots are dark brown to black, in contrast to the usual light-brown, more expansive type of lesion. Under damp conditions, the lesions enlarge and coalesce, forming large dead areas; when these areas become dry and tear away, the leaves become tattered. In severe cases of wildfire, plants may defoliate prematurely.

Pustule lesions are natural infection courts for the wildfire bacterium. A pustule formed by *Xanthomonas axonopodis* pv. *glycines,* the pustule pathogen, can usually be found in the center of a wildfire lesion.

Causal Organism

Wildfire is caused by *Pseudomonas syringae* pv. *tabaci:* a motile, gram-negative rod (0.5–1.2 × 1.4–3.3 µm) with a polar flagellum. The bacterium produces a green, water-soluble, fluorescent pigment in culture, and it liquefies gelatin and produces acid but not gas from sucrose. Colonies on nutrient agar are white, circular, and slightly raised and have translucent margins. The optimal temperature for growth is 24–28°C.

Isolates from soybean are morphologically, physiologically, serologically, and pathologically similar to those from tobacco. Tobacco isolates are generally more virulent on tobacco than on soybean, but soybean isolates are equally virulent on both.

Disease Cycle and Epidemiology

P. syringae pv. *tabaci* may overseason in infected plant debris and seeds. It grows on root surfaces of many crop plants and is spread by splashing water and windblown rain.

Water congestion of plant tissues, caused by intense rainfall, is often required for invasion and infection. Seasonal temperature variations kill the bacterium and may influence its geographic distribution.

Management

The general control measures for bacterial blight should be followed to manage wildfire. Because wildfire is seedborne, planting pathogen-free seed will reduce its occurrence. Cul-

Fig. 5. Light-brown, necrotic spots surrounded by yellow halos—symptoms of bacterial wildfire (*Pseudomonas syringae* pv. *tabaci*). (Courtesy J. B. Sinclair)

tural practices are also usually effective in controlling this disease: namely, rotating crops, managing residue, and avoiding cultivation and other field activities (e.g., spraying) when foliage is wet. Completely burying soybean residue can greatly reduce the risk of wildfire (although doing so is not recommended in some areas).

Selected References

Hong, J. K., Sung, C. H., Kim, D. K., Yun, H.-T., Jung, W., and Kim, K. D. 2012. Differential effect of delayed planting on soybean cultivars varying in susceptibility to bacterial pustule and wildfire in Korea. Crop Prot. 42:244-249.

Horst, R. K. 2013. Bacterial diseases. Pages 69-90 in: Westcott's Plant Disease Handbook. R. K. Horst, ed. Springer Science, Dordrecht, Netherlands.

Ribeiro, R. de L. D., Hagedorn, D. J., Durbin, R. D., and Uchytil, T. F. 1979. Characterization of the bacterium inciting bean wildfire in Brazil. Phytopathology 69:208-212.

(Prepared by G. L. Hartman)

Bacterial Wilt and Bacterial Tan Spot

Three systemic bacterial diseases of soybean have been reported sporadically from several locations in the world. However, they are extremely uncommon in soybean production, unlike the two most prominent bacterial diseases found worldwide: bacterial pustule and bacterial blight. These uncommon diseases include two wilt-type diseases caused by the seedborne pathogens *Curtobacterium flaccumfaciens* pv. *flaccumfaciens* and *Ralstonia solanacearum* and a third disease referred to as "bacterial tan spot," also caused by *C. flaccumfaciens* pv. *flaccumfaciens*.

C. flaccumfaciens pv. *flaccumfaciens* was identified in the early 1920s as the cause of a destructive vascular wilt disease of common bean in South Dakota. The pathogen has since been documented to have an extensive host range within the Leguminosae, affecting cowpea, pea, soybean, and several other species of bean (e.g., azuki bean [*Vigna angularis*], hyacinth bean [*Lablab purpureus*], lima bean [*Phaseolus lunatus*], mung bean [*V. radiata*], and scarlet runner bean [*P. coccineus*]). In 1963, a similar wilt disease of soybean caused by the same pathogen was reported in Iowa. Tan spot disease was first described in Iowa in 1975 as distinct from wilt based on different symptoms but also attributed to *C. flaccumfaciens* pv. *flaccumfaciens*. This pathogen has been detected on soybean only in Brazil, Canada, Russia, and the United States.

Fig. 6. Yellowing and wilting, both symptoms of bacterial wilt of soybean, caused by *Curtobacterium flaccumfaciens* pv. *flaccumfaciens*. (Courtesy R. M. Harveson)

Ralstonia solanacearum (prev. *Pseudomonas solanacearum*) is a serious and universally recognized wilt pathogen on a wide range of plant hosts, including soybean. Affecting 200 species in 33 families, *R. solanacearum* has one of the largest host ranges of any plant-pathogenic bacterium. The common names for the various diseases this bacterium induces vary with the hosts that are affected. For example, the disease is referred to as "Granville wilt" on tobacco and as "southern bacterial wilt" on solanaceous plants such as potato and tomato. Furthermore, several races and/or biovars of *R. solanacearum* have been identified, although it is not clear which race or biovar incites wilt disease of soybean.

Symptoms and Signs

Symptoms of the wilt disease caused by *C. flaccumfaciens* pv. *flaccumfaciens* include leaf yellowing and progressive wilting (Fig. 6), as well as the formation of necrotic leaf lesions with yellow borders (Fig. 7) followed by stunting and eventual death of the plant. Seedlings may wilt during the day but recover when the temperature drops during the evening. Lower leaves (and later, younger leaves) may also show marginal necrosis but without water soaking (Fig. 8).

Plants infected with *R. solanacearum* develop foliar chlorosis and dark-brown, necrotic spots that elongate and form dark borders (Fig. 9). The necrotic tissues dry and may drop out, giving affected leaves a ragged appearance (Fig. 10). Some plants may continue to wilt severely and die, whereas others may be only slightly stunted. Some young plants may develop a rapid, severe wilt; others may wilt only slightly.

Fig. 7. Necrotic leaf lesions with yellow borders, characteristic of bacterial wilt of soybean (*Curtobacterium flaccumfaciens* pv. *flaccumfaciens*). (Courtesy R. M. Harveson)

Fig. 8. Marginal necrosis of leaves on soybean infected by the bacterial wilt pathogen, *Curtobacterium flaccumfaciens* pv. *flaccumfaciens*. (Courtesy R. M. Harveson)

Bacterial tan spot, also caused by *C. flaccumfaciens* pv. *flaccumfaciens,* produces foliar symptoms similar to those caused by *R. solanacearum*. The initial symptom of tan spot disease is the formation of small lesions, which enlarge until the entire leaflet becomes enveloped. Chlorosis forms in an oval or elongated pattern, frequently along the leaf margins and then progressing inward toward the midrib (Fig. 11). Affected tissues die, dry out, and turn brown (Fig. 12). Tan areas may also fall out under windy conditions, as occurs with disease caused by *R. solanacearum*. Further blighting and plant death have been documented, as well.

During periods of high moisture, opportunistic fungi may invade and colonize the dead tissues. The fruiting structures of many of these saprophytic fungi sometimes form within tan spot lesions, and the lesions are similar to those caused by Phyllosticta leaf spot (see later in Part I, Diseases Caused by Fungi and Oomycetes, the section Phyllosticta Leaf Spot). Given this, bacterial tan spot may be confused with Phyllosticta leaf spot; the two can be readily distinguished, however, through bacterial streaming and/or after isolation of the pathogen.

Causal Organisms

Curtobacterium flaccumfaciens pv. *flaccumfaciens* is an aerobic, motile bacterium. It produces short, gram-positive, coryneform-shaped rods (0.3–0.5 × 0.6–3.0 μm) that characteristically bend or snap, and it has one to three lateral or polar flagella. Colonies on beef agar are yellow, circular, smooth, flat or slightly convex, opaque, and slightly viscid, and they have entire margins. Colony growth on nutrient broth yeast (NBY) extract medium is slow and fluidal. Different color-variant strains of the pathogen have been reported that infect bean, including strains that develop yellow-, orange-, and pink-pigmented colonies and unique yellow strains that produce water-soluble, purple-to-blue pigments that diffuse into media or stain seeds (Fig. 13). Strains isolated from soybean have consisted only of the yellow and orange variants; however, the genetic differences between the pathogens that cause wilt and tan spot are uncertain because of how rarely they occur.

It was previously believed that bean isolates could infect only bean, but researchers have shown that isolates from both bean and soybean can infect either plant. All isolates are more virulent on bean than soybean, however, regardless of origin, suggesting that some resistance to the pathogen is naturally present in soybean. Furthermore, studies from the U.S. state of Nebraska have indicated that the pathogen is extremely diverse both phenotypically and genotypically. Cluster analysis of 67 isolates of *C. flaccumfaciens* pv. *flaccumfaciens* from multiple crops and sources in a composite data set based on amplified fragment length polymorphism (AFLP), pulsed-field gel electrophoresis (PFGE), and repetitive sequence-based polymerase chain reaction (rep-PCR) yielded two distinct groups; soybean *C. flaccumfaciens* pv. *flaccumfaciens* isolates were matched into both cluster groups.

Fig. 9. Chlorotic and necrotic foliar symptoms, characteristic of bacterial wilt (*Ralstonia solanacearum*). (Courtesy R. M. Harveson)

Fig. 10. Oblong lesions with necrotic centers that have dried and fallen out, characteristic of both bacterial wilt (*Ralstonia solanacearum*) and tan spot (*Curtobacterium flaccumfaciens* pv. *flaccumfaciens*). (Courtesy R. M. Harveson)

Fig. 11. Symptoms of bacterial tan spot (*Curtobacterium flaccumfaciens* pv. *flaccumfaciens*): oblong, chlorotic spots that turn necrotic and progress from the leaf margin to the midrib. (Courtesy R. M. Harveson)

Fig. 12. Foliar necrotic lesions, characteristic of tan spot (*Curtobacterium flaccumfaciens* pv. *flaccumfaciens*). (Courtesy R. M. Harveson)

Ralstonia solanacearum is a motile, gram-negative rod (0.5 × 1.5 μm) with a polar flagellum. Colonies on agar are small, opalescent (becoming brown), irregularly shaped, smooth, and mucoid. On differential media, such as peptone–glucose plus 2,3,5-triphenyltetrazolium chloride (TTC), the bacterium produces different pigments, including pink, red, blue, and brown. However, on standard media, such as NBY, pathogen colonies are cream to white colored and nonfluorescent (Fig. 14).

Disease Cycle and Epidemiology

The development of wilt disease caused by *C. flaccumfaciens* pv. *flaccumfaciens* is favored by high temperatures (greater than 30°C); optimal pathogen growth in culture occurs at 27–28°C. Infection occurs when the pathogen enters the vascular system through infected seed, the roots of developing seedlings, or wounds on leaves or stems. In contrast, wilt disease generally develops more rapidly and causes greater damage during periods of severe plant stress, perhaps caused by elevated temperature, moisture deprivation (Fig. 15), or mechanical damage of some kind.

C. flaccumfaciens pv. *flaccumfaciens* is seedborne, surviving both internally and externally on seed coats. Seed represents the major source of inoculum and means of pathogen dispersal, both long and short distances. In fact, the pathogen has been shown to remain viable up to 24 years on bean after being stored in a laboratory. The pathogen can also be soilborne, overwintering on infected plant residues and seeds from previous crops (Fig. 16), but its survival in soil alone without a substrate is poor.

When infected seeds are planted, tan spot infection of the unifoliolate and the first trifoliolate leaves may occur via the vascular system at temperatures of 25–30°C. Spread of the pathogen among seedlings is slow (2.2 m per week from a single infected plant), because it depends primarily on wounds being produced by leaf-to-leaf contact during storms or high-wind events. For example, bruising of leaves by hail was found to result in rapid spread of the disease, which reduced the seed yield by 17%; in Iowa, the disease reduced yield by 13% one year but caused no reduction the next year. The degree of crop damage is based on the number of seedlings that become diseased in a field and is determined by the percentage of infected seeds planted. In some cultivar seed lots, infection rates of up to 30% have been recorded.

R. solanacearum is occasionally seedborne but survives well in soil as a saprophyte. The pathogen becomes more virulent when conditions are increasingly favorable for infection (e.g., low temperature and abundant rainfall). The host is also more susceptible under conditions of severe plant stress and poor soil conditions.

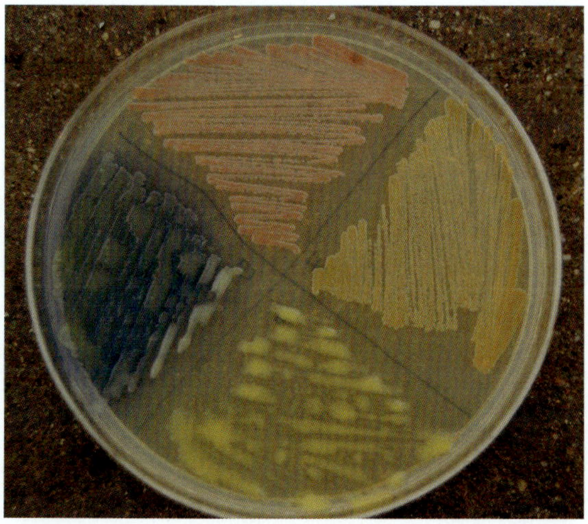

Fig. 13. Color variants of the bacterial wilt pathogen *Curtobacterium flaccumfaciens* pv. *flaccumfaciens* in culture on nutrient broth yeast extract medium. Reading clockwise from bottom: yellow, purple, pink, and orange. (Courtesy R. M. Harveson)

Fig. 15. Field distribution of bacterial wilt. Note the border effect of wilting and the mortality of plants on the field perimeter, where overhead irrigation was not consistently or evenly applied. (Courtesy R. M. Harveson)

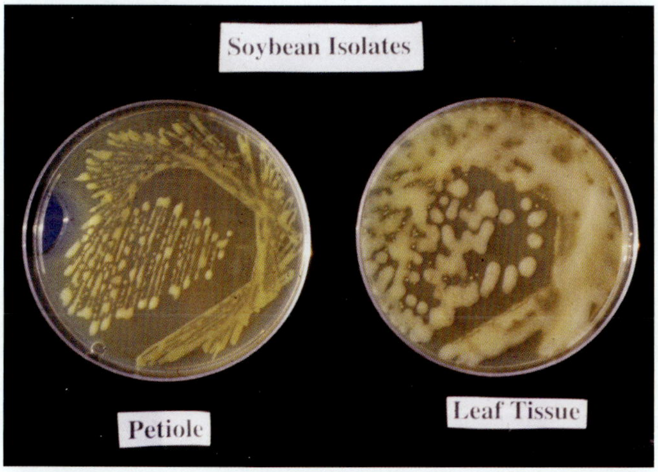

Fig. 14. Bacterial wilt pathogen *Ralstonia solanacearum*, displaying cream-colored colonies growing on nutrient broth yeast extract medium. (Courtesy R. M. Harveson)

Fig. 16. Volunteer dry bean plants from the previous season (center) infected with bacterial wilt and serving as the locus for infection in current-season soybeans (right and left). (Courtesy R. M. Harveson)

Management

Growers are highly encouraged to plant certified, pathogen-free seed; to rotate soybean with corn or small grains; and to avoid sequential plantings of common bean and soybean. Seed grown in dry climates is usually free of pathogen infection. If seeds are infected, planting early—when the soil temperature is cool (below 25°C)—will generally reduce the systemic infection of leaves. Tolerant and resistant cultivars should be planted when they are available.

C. flaccumfaciens pv. *flaccumfaciens* is considered an A2 quarantine pest in Europe and is subject to quarantine or phytosanitary regulations in some countries. Thus, restrictions on seed movement between countries and even within the United States can make it difficult for the dry bean and soybean industries to distribute affected seeds.

Selected References

Agarkova, I. V., Lambrecht, P. A., Vidaver, A. K., and Harveson, R. M. 2012. Genetic diversity among *Curtobacterium flaccumfaciens* pv. *flaccumfaciens* populations in the American High Plains. Can. J. Microbiol. 58:788-801.

Dunleavy, J. M. 1963. A vascular disease of soybeans caused by *Corynebacterium* sp. Plant Dis. Rep. 47:612-613.

Dunleavy, J. M. 1983. Bacterial tan spot, a new foliar disease of soybeans. Crop Sci. 23:473-476.

Dunleavy, J. M. 1985. Spread of bacterial tan spot of soybean in the field. Plant Dis. 69:1036-1039.

Harveson, R. M., and Schwartz, H. F. 2007. Bacterial diseases of dry edible beans in the central high plains. Plant Health Progress. Plant Management Network. doi:10.1094/PHP-2007-0125-01-DG.

Harveson, R. M., and Vidaver, A. K. 2007. First report of the natural occurrence of soybean bacterial wilt isolates pathogenic to dry beans in Nebraska. Plant Health Progress. Plant Management Network. doi:10.1094/PHP-2007- 0822-01-BR.

Harveson, R. M., and Vidaver, A. K. 2008. A new color variant of the dry bean bacterial wilt pathogen (*Curtobacterium flaccumfaciens* pv. *flaccumfaciens*) found in western Nebraska. Plant Health Progress. doi:10.1094/PHP-2008- 0815-01-BR.

Harveson, R. M., Schwartz, H. F., Vidaver, A. K., Lambrecht, P. A., and Otto, K. L. 2006. New outbreaks of bacterial wilt of dry bean in Nebraska observed from field infections. Plant Dis. 90:681.

Hayward, A. C. 1991. Biology and epidemiology of bacterial wilt caused by *Pseudomonas solanacearum*. Annu. Rev. Phytopathol. 29:65-87.

Hayward, A. C., and Hartman, G. L., eds. 1994. Bacterial Wilt: The Disease and Its Causative Agent *Pseudomonas solanacearum*. CABI, Wallingford, Oxfordshire, U.K.

McDonald, J. G., and Wong, E. 2000. High diversity in *Curtobacterium flaccumfaciens* pv. *flaccumfaciens* characterized by serology and rep-PCR genomic fingerprinting. Can. J. Plant Pathol. 22:17-22.

Soares, R. M., Fantinato, G. G. P., Darben, L. M., Marcelino-Guimaraes, F. C., Seixas, C. D. S., and Carneiro, G. E. de S. 2013. First report of *Curtobacterium flaccumfaciens* pv. *flaccumfaciens* on soybean in Brazil. Trop. Plant Pathol. 38:452-454.

Yabuuchi, E., Kosako, Y., Yano, I., Hotta, H., and Nishiuchi, Y. 1995. Transfer of two *Burkholderia* and an *Alcaligenes* species to *Ralstonia* gen. nov.: Proposal of *Ralstonia pickettii* (Ralston, Palleroni and Doudoroff 1973) comb. nov., *Ralstonia solanacearum* (Smith 1896) comb. nov. and *Ralstonia eutropha* (Davis 1969) comb. nov. Microbiol. Immunol. 39:897-904.

(Prepared by R. M. Harveson)

Bacillus Seed Decay

Bacillus seed decay is ubiquitous. It causes losses under hot (25–35°C), moist conditions in storage, in the field, and in experimental situations. Field losses of nearly 100% were reported in Nigeria when soybean seeds were planted in wet soil at temperatures greater than 30°C. The disease frequently appears on seeds being tested under warm-germination conditions (e.g., accelerated aging tests) but can occur in typical germination tests, as well.

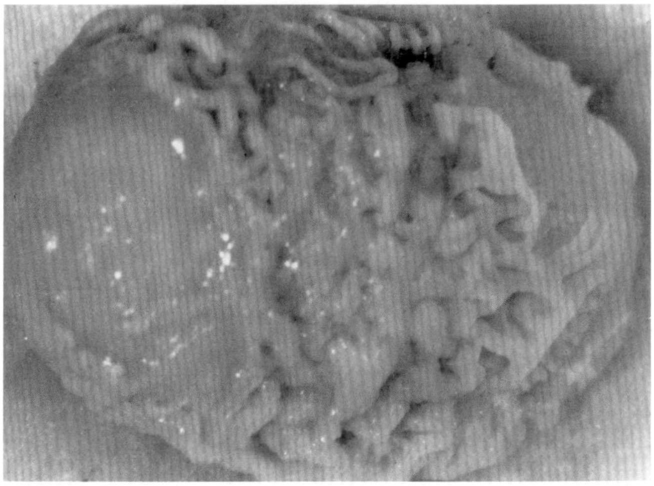

Fig. 17. *Bacillus subtilis* on a soybean seed. (Courtesy F. D. Tenne)

Symptoms and Signs

Bacillus seed decay is characterized by a soft decay of seeds, which increases in severity with high moisture and temperature levels (Fig. 17). In the laboratory at temperatures of 30°C or higher and relative humidity approaching 100%, decay occurs within 5 days. Seeds are initially colonized in the seed coat, without conspicuous symptoms. The bacterium produces rough or smooth, slimy, glistening colonies on potato dextrose agar (PDA) and soybean seeds.

Causal Organism

Bacillus subtilis is a soil resident and an epiphyte on soybean leaves and other plant tissues. The bacterium is a motile, gram-positive, aerobic rod (0.7–0.8 × 2.0–3.0 µm) with peritrichous flagella, and it forms endospores. Colonies on nutrient agar are white to cream colored and wrinkled or smooth and folded. The minimum and maximum temperatures for growth are 5° and 55°C, respectively.

Culture filtrates of several isolates of *B. subtilis* have been shown to be antagonistic to seven soybean pathogens in culture. Some strains have been shown to enhance stands when used as a seed treatment and thus show promise for the control of root rots and Sclerotinia stem rots (see later in Part I, Diseases Caused by Fungi and Oomycetes, the section Sclerotinia Stem Rot).

Disease Cycle and Epidemiology

B. subtilis is ubiquitous in the environment, and its associations with soybean and soybean seeds are well known. The conditions that promote the incidence of the bacterium on seed have not been determined, but the inoculum may come from its epiphytic existence on leaves or from soil and debris dust generated during harvest. Regardless, higher moisture levels and temperatures are both associated with seed decay.

Management

In soils with temperatures greater than 30°C, growers should consider using mulch or another means to reduce the soil temperature at planting time.

Selected References

Cubeta, M. A., Hartman, G. L., and Sinclair, J. B. 1985. Interaction between *Bacillus subtilis* and fungi associated with soybean seeds. Plant Dis. 69:506-509.

Scortichini, M., Rossi, M. P., Ricci, B., and Ndzoamba, B. 1989. Soybean (*Glycine max* [L.] Merr.) seed decay associated with *Bacillus subtilis* (Ehrenberg) Cohn in Gabon. FAO Plant Prot. Bull. 37:87-91.

Zhang, J. X., Xue, A. G., and Tambong, J. T. 2009. Evaluation of seed and soil treatments with novel *Bacillus subtilis* strains for control of soybean root rot caused by *Fusarium oxysporum* and *F. graminearum*. Plant Dis. 93:1317-1323.

Zhang, J. X., Xue, A. G., Morrison, M. J., and Meng, Y. 2011. Impact of time between field application of *Bacillus subtilis* strains SB01 and SB24 and inoculation with *Sclerotinia sclerotiorum* on the suppression of Sclerotinia stem rot in soybean. Eur. J. Plant Pathol. 131:95-102.

(Prepared by J. B. Sinclair; revised by G. L. Hartman)

Other Bacteria

Soybean has been reported to be susceptible to isolates of the following phytobacteria: *Burkholderia andropogonis, Curtobacterium plantarum, Erwinia amylovora, E. carotovora* subsp. *atroseptica, Pseudomonas savastanoi* pv. *phaseolicola, P. syringae* pv. *syringae,* and *Xanthomonas campestris* pv. *cannabis.*

In addition, strains of *Agrobacterium rhizogenes* and *A. tumefaciens* cause crown gall and are used for biotechnological research on many plants, including soybean. Induction of crown gall tumors and hairy root have been achieved under inoculated, laboratory conditions. Tumors have also been induced by *A. tumefaciens* in wild soybean and in many soybean cultivars.

Selected References

Dunleavy, J. M. 1989. *Curtobacterium plantarum* sp. nov. is ubiquitous in plant leaves and is seed transmitted in soybean and corn. Int. J. Syst. Bacteriol. 39:240-249.

Kovács, L. G., and Pueppke, S. G. 1993. The chromosomal background of *Agrobacterium tumefaciens* Chry5 conditions high virulence on soybean. Mol. Plant-Microbe Interact. 6:601-608.

Nikitina, K. V., and Korsakov, N. I. 1978. Bacterial diseases of soybean in the Soviet Far East and in southern regions of the U.S.S.R. Search for sources of resistance to them. (In Russian) Tr. Prikl. Bot. Genet. Sel. 62:13-18.

Pedersen, H., Christiansen, J., and Wyndaele, R. 1983. Induction and in vitro culture of soybean crown gall tumors. Plant Cell Rep. 2:201-204.

Wang, L. Z., Yin, G. C., Luo, J. F., Lei, B. J., Wang, J., Yao, Z. C., Li, X. L., Shae, Q. Q., Jiang, X. C., and Zhou, Z. Q. 1984. A study on tumor formation of soybean and gene transfer. Sci. Sin. B27:391-397.

(Prepared by J. B. Sinclair; revised by G. L. Hartman)

Ineffective Biological Nitrogen Fixation

Soybean is highly dependent on a symbiotic relationship with nitrogen-fixing bacteria, in which atmospheric nitrogen is fixed in the nodules and symbiotic nitrogen fixation is used for plant growth. This symbiotic relationship with nitrogen-fixing bacteria allows the two organisms to convert dinitrogen gas in the atmosphere to a plant-available form. The bacteria infect the plant root, forming a nodule at which biological nitrogen fixation occurs. This association between the two organisms supplies, on average, 50–60% of the nitrogen required for plant growth; the level of efficiency depends on the interaction of the bacterial strain and the soybean genotype.

Nitrogen fixation commences approximately 3 weeks after colonization. Large, irregularly shaped nodules with pink to red interiors (caused by leghemoglobin) develop early in the growing season, clustered near the root crown. Later, the nodules located on secondary roots become more important in nitrogen-fixing activity. Not all nodules are effective; thus, the number of nodules on the soybean root may not be a good indicator of adequate nitrogen-fixing ability.

Molecular and serological techniques are often used to further classify various strains of effective nitrogen fixers. Strain–cultivar interactions occur in which specific strains fix more nitrogen on some cultivars than others. Quantitative polymerase chain reaction (PCR) is one technique used to quantify the amount of soybean-associated rhizobia in soil using *nodZ* as the target gene.

Symptoms and Signs

Several situations can reduce nitrogen fixation and induce chlorosis. Nodules may not form in fields in which soybean has not been grown for several years if the seeds have not been inoculated with a viable, effective strain.

In soils amended with crop residue with a high carbon/nitrogen ratio, the resulting low level of soil inorganic nitrogen can cause plants to become chlorotic 2–3 weeks after plant emergence. Chlorosis occurs because nitrogen fixation has not begun and soil inorganic nitrogen levels are inadequate to meet plant requirements. Two situations in which soil inorganic nitrogen levels may be low are no-tillage production and double-cropping of soybean following wheat. Chlorosis can also occur late in the growing season in soils amended with a high level of nitrogen fertilizer or organic material (e.g., manure or biosolids [sewage sludge]).

Anaerobic soil conditions can also reduce nitrogen fixation. Flooding for 7 days at growth stage V1 has been shown to inhibit nodulation completely and at growth stage R2, to significantly reduce nodule activity and yields.

Chlorosis is also caused by the amino compound rhizobitoxine, which is produced in the nodules under certain conditions with some strains of nitrogen-fixing bacteria. The phytotoxin causes a temporary chlorosis of newly developed leaves in susceptible genotypes, along with a residual green color along the leaf midrib and to a lesser extent along the secondary veins (Fig. 18). The residual green color is especially prominent toward the base of the leaflet.

Causal Organisms

Soybean nodule-forming bacteria are gram negative, aerobic, non-spore-forming, and rod shaped. Motility occurs by one polar or subpolar flagellum, and growth is optimum at a pH of 6–7 and a temperature of 25–30°C. Bacterial cells are commonly pleomorphic under adverse growth conditions and exist in the soil for an extended period in the absence of the plant host. *Bradyrhizobium japonicum* strain USDA110 represents a

Fig. 18. Plant with chlorosis (left), resulting from an inefficient nodulating strain of bacteria, and a healthy plant (right). (Courtesy D. Wolf)

widely distributed type that was formerly known as *B. japonicum* Ia but has been renamed *B. diazoefficiens.*

Along with *B. japonicum/B. diazoefficiens, B. elkanii, B. liaoningense,* and *Sinorhizobium fredii* have been shown to form effective nodules on soybean and wild soybean, whereas a different reaction was produced by some strains of *B. yuanmingense,* which did not produce nodules on soybean. In addition, *B. huanghuaihaiense, B. daqingense,* and *S. sojae* were found to be effective microsymbionts of soybean. Some of the strains within these species can cause rhizobitoxine-induced chlorosis in soybean. *Bradyrhizobium* USDA strains 31, 46, 76, 94, and 130 have also been associated with producing rhizobitoxine symptoms on certain cultivars.

Epidemiology

The specific *Bradyrhizobium* strain–cultivar interaction plays a key role in the development of rhizobitoxine-induced chlorosis. In the field, the rapid onset of chlorosis generally occurs approximately 6 weeks after planting and lasts for 1–2 weeks, after which plants fully recover. Chlorosis is frequently spotty in the field and has not been consistently related to soil type, fertility level, or moisture status. Chlorosis has been shown to decrease with the increased application of nitrate; thus, nitrate uptake from the soil may counteract the development of chlorosis. Symptoms can be severe in greenhouse studies, but in most field situations, chlorosis is temporary and yield reductions are minimal.

Management

Selecting resistant cultivars is the most feasible strategy for managing rhizobitoxine problems. Dominant genes have unique features that restrict nodulation with specific strains (or serogroups) of *B. japonicum* (*Rj2, Rj4*), *B. elkanii* (*Rj3, Rj4*), and *S. fredii* (*Rfg1*). Recessive genes at three loci—*rj1, rj5,* and *rj6*—result in non-nodulation phenotypes, whereas *rj7* and *nts1* (i.e., nitrate-tolerant symbiosis 1) cause the formation of an unusually large number of nodules.

In the field, when symptoms do occur, the chlorotic cultivars grow out of the symptoms, causing a minimal impact on yield if the plants exhibit effective nodulation. Research efforts are focused on selecting and/or constructing strains of *B. japonicum* that are both highly effective and competitive with indigenous strains. In Brazil, researchers showed that the effectiveness of commercial inoculants could be increased by screening for higher N_2 fixation capacity within the commercial inoculant because of its population variability. Amendments of additional nitrogen through fertilizer may enhance production. Other options include deep placement of slow-release fertilizer below the nodulation zone and applying nitrogen during reproductive stages in high-yielding environments.

Selected References

Delamuta, J. R. M., Ribeiro, R. A., Ormeño-Orrillo, E., Melo, I. S., Martínez-Romero, E., and Hungria, M. 2013. Polyphasic evidence supporting the reclassification of *Bradyrhizobium japonicum* group Ia strains as *Bradyrhizobium diazoefficiens* sp. nov. Int. J. Syst. Evol. Microbiol. 63:3342-3351.

Erdman, L. W., Johnson, H. W., and Clark, F. A. 1957. Varietal responses of soybeans to a bacterial-induced chlorosis. Agron. J. 49:267-271.

Furseth, B. J., Conley, S. P., and Ané, J.-M. 2010. Enumeration of soybean-associated rhizobia with quantitative real-time polymerase chain reaction. Crop Sci. 50:2591-2596.

Hayashi, M., Saeki, Y., Haga, M., Harada, K., Kouchi, H., and Umehara, Y. 2012. *Rj* (*rj*) genes involved in nitrogen-fixing root nodule formation in soybean. Breed. Sci. 61:544-553.

LaFavre, J. S., and Eaglesham, A. R. J. 1986. Rhizobitoxine: A phytotoxin of unknown function which is commonly produced by bradyrhizobia. Plant Soil 92:443-452.

Owens, L. D., and Wright, D. A. 1965. Rhizobial-induced chlorosis in soybeans: Isolation, production in nodules, and varietal specificity of the toxin. Plant Physiol. 40:927-930.

Torres, A. R., Kaschuk, G., Saridakis, G. P., and Hungria, M. 2012. Genetic variability in *Bradyrhizobium* strains nodulating soybean (*Glycine max* [L.] Merrill). World J. Microbiol. Biotechnol. 28:1831-1835.

Vasilas, B. L., and Fuhrmann, J. J. 1993. Field response of soybean to nodulation by a rhizobitoxine-producing strain of *Bradyrhizobium.* Agron. J. 85:302-305.

Vijayan, R., Palaniappan, P., SA, T., Elavarasi, P., and Manoharan, N. 2013. Rhizobitoxine enhances nodulation by inhibiting ethylene synthesis of *Bradyrhizobium elkanii* from *Lespedeza* species: Validation by homology modeling and molecular docking study. World J. Pharm. Pharm. Sci. 2:4079-4094.

Zhang, X. X., Guo, H. J., Wang, R., Sui, X. H., Zhang, Y. M., Wang, E. T., Tian, C. F., and Chen, W. X. 2014. Genetic divergence of *Bradyrhizobium* strains nodulating soybeans as revealed by multilocus sequence analysis of genes inside and outside the symbiosis island. Appl. Environ. Microbiol. 80:3181-3190.

(Prepared by D. C. Wolf and L. C. Purcell;
revised by G. L. Hartman)

Diseases Associated with Mollicutes

A number of well-described and not-so-well described soybean diseases are caused by Mollicutes. The distinguishing feature of members of this class of bacteria is that they lack cell walls. They live on or in host cells, are usually 0.2–0.3 µm, and have very small genomes. Complete genome sequences for many of the Mollicutes reveal extensive horizontal gene transfer and genomic plasticity within some species. Their taxonomy is still in a state of flux.

Mollicutes vary in form, although most have sterols that make the cell membrane somewhat rigid. Spiroplasmas, which lack cell walls, are bounded by a trilaminar membrane and grow as mobile, helical, filamentous cells. Phytoplasmas commonly occur in the phloem of plants. Most of the diseases thought to be caused by these microbes are the so-called yellows types, for which characteristic symptoms include yellowing, shortening of internodes, phyllody, proliferation of axillary shoots (producing witches'-broom), sterility, virescence, and reduction in root growth. Phytoplasmas are suspected of causing several soybean diseases, including bud proliferation, machismo, phyllody, and witches'-broom.

Selected Reference

Mishra, S. 2014. Mollicutes and Plant Diseases. Discovery Publishing, New Delhi, India.

(Prepared by G. L. Hartman)

Bud Proliferation

Bud proliferation, which resembles machismo, has been reported in the U.S. states of Arkansas and Louisiana. The pathogen is seed transmitted and spread by the leafhopper *Scaphytopius acutus.* It is not mechanically transmitted.

Selected References

Dale, J. L., and Walters, H. J. 1985. Soybean bud proliferation of unknown etiology in Arkansas. Plant Dis. 69:811.

Derrick, K. S., and Newsom, L. D. 1984. Occurrence of a leafhopper-transmitted disease of soybeans in Louisiana. Plant Dis. 68:343-344.

(Prepared by J. B. Sinclair)

Machismo

Machismo was reported in Colombia in 1968 and in Mexico in 1984. This disease has been economically important in Colombia since 1974, with reported yield losses of up to 80%. In addition to soybean, machismo infects field bean, species of *Crotalaria* and *Desmodium, Galactia glaucescens,* least snoutbean (*Rhynchosia minima*), Madagascar periwinkle (*Catharanthus roseus*), and pigeon pea (*Cajanus cajan*).

Symptoms of machismo vary with the host, cultivar, and age of plant upon infection. The first symptoms appear at the time of flowering or after pod formation. Pods may be rigid, curved, flat, and thin; they are usually in an upright position, and no beans have formed. Alternatively, pods may be transformed into corrugated leaflike structures in a condition resembling phyllody. Once the floral organs have been transformed into leaflike structures, buds proliferate from leaf axils anywhere on the plant, causing witches'-broom (Fig. 19). The younger the plant when infected, the more severe the symptoms.

The sepals of infected plants may be longer, larger, and stronger than healthy sepals, giving the appearance of small leaves. The color intensity of the petals may be diminished. The flowers are smaller and remain closed, and the podlike or leaflike structures grow through the tips of the unopened flowers. Often, part of the plant displays symptoms, whereas the remainder appears healthy and has normal development of flowers, pods, and beans. When this occurs, the affected branch is usually close to the bottom of the plant.

The causal agent of machismo is transmitted by the leafhopper *Scaphytopius fuliginosus.* The disease has not been transmitted by mechanical inoculation, nor is there evidence of seed transmission.

Fig. 19. Plant displaying symptoms of machismo: a bud proliferation associated with phytoplasmas. (Courtesy M. E. Irwin)

Treating plants with tetracycline in the early stages of the disease prevents further symptom development; however, this treatment causes only a temporary remission of symptoms. Other methods of management include controlling weed hosts and rogueing infected plants.

Selected References

Fletcher, J., Irwin, M. E., Bradfute, O. E., and Granada, G. A. 1984. A machismo-like disease of soybeans in Mexico. Phytopathology (Abstr.) 74:857.

Granada, G. A. 1979. Machismo disease of soybeans: I. Symptomatology and transmission. Plant Dis. Rep. 63:47-50

Granada, G. A. 1979. Machismo disease of soybeans: II. Suppressive effects of tetracycline on symptom development. Plant Dis. Rep. 63:309-312.

(Prepared by J. B. Sinclair)

Phyllody and Witches'-Broom

Witches'-broom and phyllody resemble machismo. The pathogen causes a phylloid disorder of the floral organs, as well as shoot proliferation and reduced leaflets. Plants with these symptoms have been reported in Brazil, India, Indonesia, Japan, Malawi, Mozambique, Nigeria, and Tanzania. No comprehensive report about these diseases is available, although in Malawi and Mozambique, affected plants had 100% yield loss and the cause was associated with a 16SrII-C group of '*Candidatus* phytoplasma.'

The leafhopper vectors of witches'-broom are an *Orosius* sp. in Brazil, *O. argentatus* in India and Indonesia, and *O. orientalis* in Japan.

In India, symptoms of witches'-broom were suppressed when plants were treated with tetracycline.

Selected References

Dingra, K. L., and Chenulu, V. V. 1983. Symptomatology and transmission of witches' broom disease of soybean in India. Curr. Sci. 52:603-604.

Kumar, P. L., Sharma, K., Boahen, S., Tefera, H., and Tamò, M. 2011. First report of soybean witches'-broom disease caused by group 16SrII phytoplasma in soybean in Malawi and Mozambique. Plant Dis. 95:492.

Tarp, G., Lange, L., and Kongsdal, O. 1987. Seed-borne pathogens of major food crops in Mozambique. Seed Sci. Technol. 15:793-810.

(Prepared by J. B. Sinclair; revised by G. L. Hartman)

Clover Proliferation and Aster Yellows

Two distinct phytoplasmas belonging to the clover proliferation group (16SrVI) and the aster yellows group (16SrI), subgroup B, have been detected in the continental United States. Each was identified by restriction fragment length polymorphism (RFLP) of polymerase chain reaction- (PCR-) amplified 16S rDNA or sequence analysis, including the use of the *i*PhyClassifier using phytoplasma-specific primers. The clover proliferation relative has been detected in soybean grown in Alabama, Arkansas, Georgia, and Wisconsin, and the aster yellows subgroup B phytoplasma has been detected in Michigan and Wisconsin. Some symptomatic soybean plants have been infected with both phytoplasmas. Koch's postulates have not been fulfilled for either phytoplasma.

Fig. 20. Purple stem of a plant that has tested positive for the presence of phytoplasmas. (Courtesy C. R. Grau)

Fig. 23. Stem necrosis, associated with phytoplasmas. (Courtesy C. R. Grau)

Fig. 21. Epinasty and flower bud proliferation, associated with phytoplasmas. (Courtesy M. E. Lee and N. C. Kurtzweil)

Fig. 24. Stunting and bud proliferation, associated with phytoplasmas. (Courtesy M. I. Chilvers)

Fig. 22. Leaf vein discoloration, associated with phytoplasmas. (Courtesy C. R. Grau)

The symptoms putatively associated with plants infected with the phytoplasmas include bushy or stunted growth, purple streaks and patches on stems, yellow blotches or marginal necrosis on leaves, stems that remain green even after the leaves and pods have matured, bud proliferation, reduction in the number of pods per node, dull seed coat, and relatively dark, bluish foliage (Figs. 20–24). Specific symptoms have not been linked to a specific phytoplasma.

A novel phytoplasma infecting soybean has also been reported in Costa Rica, and the phytoplasma of group 16SrII, subgroup C, has been reported to infect soybean in Malawi and Mozambique.

Selected References

Derrick, K. S., and Newsom, L. D. 1984. Occurrence of a leafhopper-transmitted disease of soybeans in Louisiana. Plant Dis. 68:343-344.

Fletcher, J., Irwin, M. E., Bradfute, O. E., and Granada, G. A. 1984. Discovery of a mycoplasmalike organism associated with diseased soybeans in Mexico. Plant Dis. 68:994-996.

Kumar, P. L., Sharma, K., Boahen, S., Tefera, H., and Tamó, M. 2011. First report of soybean witches'-broom disease caused by group 16SrII phytoplasma in soybean in Malawi and Mozambique. Plant Dis. 95:492.

Lee, I.-M., Bottner-Parker, K. D., Zhao, Y., Villalobos, W., and Moreira, L. 2011. '*Candidatus* Phytoplasma costaricanum,' a novel phytoplasma associated with an emerging disease in soybean (*Glycine max*). Int. J. Syst. Evol. Microbiol. 61:2822-2826.

Lee, I.-M., Hammond, R. W., Davis, R. E., and Gundersen, D. E. 1993. Universal amplification and analysis of pathogen 16S rDNA for classification and identification of mycoplasmalike organisms. Phytopathology 83:834-842.

Lee, M. E., Grau, C. R., Lukaesko, L. A., and Lee, I.-M. 2002. Identification of aster yellows phytoplasmas in soybean in Wisconsin based on RFLP analysis of PCR-amplified products (16S rDNAs). Can. J. Plant Pathol. 24:125-130.

Maejima, K., Oshima, K., and Namba, S. 2014. Exploring the phytoplasmas, plant pathogenic bacteria. J. Gen. Plant Pathol. 80:210-221.

Mansfield, J., Genin, S., Magori, S., Citovsky, V., Sriariyanum, M., Ronald, P., Dow, M., Verdier, V., Beer, S., Machado, M., Toth, I., Salmond, G., and Foster, G. 2012. Top 10 plant pathogenic bacteria in molecular plant pathology. Mol. Plant Pathol. 13:614-629.

Namba, S. 2011. Phytoplasmas: A century of pioneering research. J. Gen. Plant Pathol. 77:345-349.

Sinclair, W. A., Iuli, R. J., Dyer, A. T., and Larsen, A. O. 1989. Sampling and histological procedures for diagnosis of ash yellows. Plant Dis. 73:432-435.

Tomlinsona, J. A., Boonhama, N., and Dickinson, M. 2010. Development and evaluation of a one-hour DNA extraction and loop-mediated isothermal amplification assay for rapid detection of phytoplasmas. Plant Pathol. 59:465-471.

Zhao, Y., Wei, W., Lee, I. M., Shao, J., Suo, X., and Davis, R. E. 2009. Construction of an interactive online phytoplasma classification tool, *i*PhyClassifier, and its application in analysis of the peach X-disease phytoplasma group (16SrIII). Int. J. Syst. Evol. Microbiol. 59:2582-2593.

(Prepared by C. R. Grau, I.-M. Lee, M. E. Lee, M. F. Heimann, and L. A. Lukaesko; revised by M. I. Chilvers)

Diseases Caused by Fungi and Oomycetes

Pathogens classified as fungi and oomycetes (commonly referred to "water molds") cause most of the known plant diseases, and numerous additional plant diseases caused by these microbes are discovered every year. Of the nearly 70,000 described species of fungi and oomycetes, more than 8,000 are plant pathogens.

The microbes that are fungi and oomycetes are a diverse group of organisms that lack chlorophyll and cannot carry out photosynthesis. Instead, they obtain sustenance from organic matter produced by other organisms. Fungi and oomycetes are adapted for survival in air, soil, and water, and they are typically in close association with other organisms. These microbes can live in both darkness and light and may require special lighting conditions for sporulation.

Fungi and oomycetes acquire nutrients in several ways. Saprophytic species utilize nonliving organic matter, such as dead leaves and stems, for food. Parasites require living hosts for obtaining nutrients. A large percentage of fungi can function either as saprophytes or as parasites. In general, obligate parasitic fungi and oomycetes tend to have narrower host ranges and to be more closely adapted to the biology of their hosts than the microbes in these groups that are capable of growing saprophytically.

Most fungi and oomycetes produce hyphae: microscopic, cellular, threadlike filaments that contain cytoplasm and organelles required to sustain life. Hyphae grow and branch to form a mycelium, which is the vegetative body of these microbes. Mycelial structures range in size from microscopic wefts of hyphae to large mushrooms and conks.

Typically, fungi and oomycetes reproduce and spread by means of spores. Some spores, as well as larger structures (e.g., sclerotia), are adapted for survival in the absence of a host. Asexual spores are generally produced in great numbers during the growing season and are often responsible for the development of epidemics. Sexual spores result from genetic recombination and therefore serve to maintain genetic diversity within populations of fungi and oomycetes. Air currents, splashing and flowing water, and human and animal activity can all disseminate spores. For some diseases, insects and animals are important not only as agents of dissemination but also as agents of inoculation.

Fungal taxonomy is based largely on the morphology of spores, the structures that produce them, and their development. Other characteristics, such as host specificity, are also used to classify fungi. Fungal taxonomy is changing rapidly, and in certain fungal groups, information on hyphal anastomosis groups, isozyme variation, DNA restriction fragment length polymorphisms (RFLPs), and other new kinds of taxonomic characters are being used to improve the system by which fungi are identified and classified. In the future, such advances will likely improve methods for diagnosing the causes of plant diseases.

Many fungi and oomycetes enter soybean plants indirectly through natural openings, such as stomata, hydathodes, nectaries, and lenticels. In other cases, these microbes infect plants through wounds made by wind, hail, blowing sand, people, insects, nematodes, equipment, and other agents. Hyphae and specialized structures such as penetration pegs generally penetrate plant tissue directly through a combination of enzyme action and pressure. Fungi and oomycetes can overseason on or in both living and dead plants, as well as seeds, soil, and even insects, on occasion. Many fungi and oomycetes have a broad host range that includes soybean, other crops, and weed hosts (in which they can overseason).

Physiological races and strains of fungi and oomycetes are morphologically indistinguishable. However, they differ in the species, varieties, and cultivars of host plants they can parasitize and the degree of virulence they exhibit.

Many fungi and oomycetes colonize soybean plants and seed tissues asymptomatically. Latently infected plants often develop symptoms if they are under stress or near the end of the growing season, when they are senescing. Among the genera known to have species that cause latent infections in soybean are *Cercospora, Colletotrichum, Diaporthe, Fusarium,* and *Macrophomina.* Research has shown that some of these species are present as endophytes (i.e., are present in symptomless plants but do not cause disease).

Many of the fungi and oomycetes that infect soybean infect multiple tissues. Some of these pathogens infect primarily upper-plant tissues, such as leaves, as is the case for *Phakopsora pachyrhizi,* which causes soybean rust. Others infect only lower-plant tissues, such as roots, as is the case for many root-rotting pathogens. However, many pathogens, such as species of *Colletotrichum* and others, are found on all parts of soybean plants.

The remainder of this discussion of fungi and oomycetes is divided into two sections. The first section addresses diseases of the foliage, upper stems, pods, and seeds, and the second section addresses diseases of the lower stems and roots. A disease that could be placed in either section has been placed in the section that is most commonly associated with the relevant pathogen.

Fungi and oomycetes known to infect soybean are also listed in Appendix I; some pathogens not addressed in the two following sections are included in that appendix, as well. Readers will note that in both Appendix I and the text, alternative names of fungi are identified exclusively as "synonyms," rather than as specific life cycle stages (e.g., "anamorph," "teleomorph"). This practice reflects the policy for naming fungi issued in 2011 by the International Botanical Congress: namely, that the use of multiple names for the same fungus be abandoned. (Readers interested in learning more about this policy are referred to the 2011 article by Hawksworth [see Selected References].) Going forward, accommodating this new policy may mean that the most familiar names of some fungi will change. Changes to the names of fungi, as well as other updates to taxonomy and nomenclature, will be made in the Common Names of Diseases list for soybean, which is available on the website of The American Phytopathological Society.

Selected References

Agrios, G. N. 2005. Plant Pathology. 5th ed. Academic Press, New York.

Alexopoulos, C. J., Mims, C. J. and Blackwell, M. 1996. Introductory Mycology. 4th ed. John Wiley & Sons, New York.

Carris, L., Little, C., and Stiles, C. 2012. Introduction to Fungi. Plant Health Instructor. American Phytopathological Society. doi:10.1094/PHI-I-2012-0426-1001.

Farr, D. F., and Rossman, A. Y. n.d. Fungal Databases, Systematic Mycology and Microbiology Laboratory. U.S. Department of Agriculture, Agricultural Research Service. Available online at http://nt.ars-grin.gov/fungaldatabases/

Hawksworth, D. L. 2011. A new dawn for the naming of fungi: Impacts of decisions made in Melbourne in July 2011 on the future publication and regulation of fungal names. IMA Fungus 2:155-162.

Heffer Link, V., Powelson, M. L., and Johnson, K. B. 2002 (updated 2012). Oomycetes. Plant Health Instructor. American Phytopathological Society. doi:10.1094/PHI-I-2002-0225-01.

Kirk, P. M., Cannon, P. F., Minter, D. W., and Stalper, J. A. 2011. Dictionary of the Fungi. 10th ed. Commonwealth Scientific and Industrial Research Organisation (CSIRO), Clayton South, Victoria, Australia.

Rossman, A. Y., Palm, M. E., and Spielman, L. J. 1987. A Literature Guide for the Identification of Plant Pathogenic Fungi. American Phytopathological Society, St. Paul, MN.

Sinclair, J. B., and Cerkauskas, R. F. 1996. Latent infection vs. endophytic colonization by fungi. Pages 3-29 in: Endophytic Fungi in Grasses and Woody Plants: Systematics, Ecology, and Evolution. S. C. Redlin and L. M. Carris, eds. American Phytopathological Society, St. Paul, MN.

(Prepared by G. L. Hartman)

Diseases of Foliage, Upper Stems, Pods, and Seeds

Alternaria Leaf Spot, Pod Necrosis, and Seed Decay

Alternaria leaf spot and pod necrosis have been reported in all soybean-growing areas of the world. Leaf spot appears occasionally on soybean at the seedling stage, but generally, symptoms of leaf spot and pod necrosis are more evident on plants approaching maturity. Because these diseases usually occur late in the growing season, yield losses are believed to be minimal.

Symptoms and Signs

Brown, necrotic leaf spots with concentric rings appear on infected foliage; the spots enlarge and may coalesce to form necrotic areas (Fig. 25). Infected leaves eventually desiccate and then drop prematurely. Similar symptoms appear on pods.

Seed decay has been associated with *Alternaria alternata* and *A. tenuissima:* opportunistic fungi that decay soybean pods and seeds after senescence, insect damage, and wounding. Seed symptoms include dark, irregularly shaped, sunken, spreading areas on the seed coat (Fig. 26). Seeds generally do not germinate, and if they germinate in soil, seedlings do not emerge. The seeds are not suitable for processing.

Causal Organisms

Alternaria spp. have been reported as the causal fungi of the leaf spot phases of these diseases. *A. atrans* is a weak parasite that invades the leaves through aphid punctures and sunburn injury. *A. tenuissima* has been reported to cause a leaf spot disease in India and a wilt disease in Kenya. *A. alternata* has been isolated consistently from necrotic areas of pods in Canada and is the prevalent species reported in seed surveys from Argentina.

Fig. 25. Brown, necrotic leaf spots and concentric rings, characteristic of Alternaria leaf spot (*Alternaria* spp.). (Courtesy J. B. Sinclair)

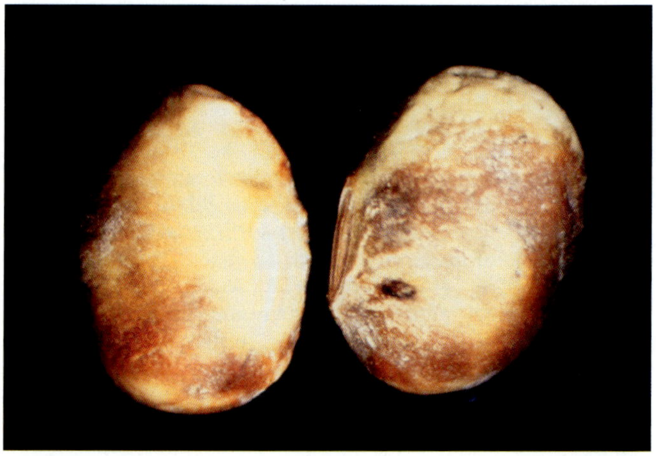

Fig. 26. Seed discoloration, caused by *Alternaria* spp. infection. (Courtesy J. B. Sinclair)

Fig. 27. Conidia of *Alternaria* sp. (Courtesy F. D. Tenne)

Alternaria spp. are easily identified by their characteristic conidia, which are an inverted club shape, mostly elongated at the tip and muriform at the bottom. The conidia are dark colored but usually lighter at the tips (Fig. 27), and they are borne in long, simple chains and mostly multiseptate.

Disease Cycle and Epidemiology

Soybean plants with chlorotic leaves and necrotic pods typically appear following periods of dry, warm weather in Ontario, Canada. The incidence of seed infection increases with delays in harvesting. The severity of decay is usually associated with beetle damage—particularly from *Cerotoma trifurcata*—and the greater the damage, the higher the incidence of Alternaria seed decay. Freezing also increases the incidence of seed infection.

Management

Diseases caused by *Alternaria* spp. rarely require specific intervention. However, the use of broad-spectrum dithiocarbamate fungicides for control has been reported from India.

Selected References

Broggi, L. E., Gonalez, H. H. L., Resnik, S. L., and Pacin, A. 2007. *Alternaria alternata* prevalence in cereal grains and soybean seeds from Entre Rios, Argentina. Rev. Iberoam. Micol. 24:47-51.

Clear, R., Nowicki, T. W., and Daun, J. K. 1989. Soybean seed discoloration by *Alternaria* spp. and *Fusarium* spp. effects on quality and production of fusariotoxins. Can. J. Plant Pathol. 11:308-312.

Kunwar, I. K., Manandhar, J. B., and Sinclair, J. B. 1986. Histopathology of soybean seeds infected with *Alternaria alternata*. Phytopathology 76:543-546.

Osorio, J. A., and McGee, D. C. 1992. Effects of freeze damage on soybean seed mycoflora and germination. Plant Dis. 76:879-882.

Shortt, B. J., Sinclair, J. B., Helms, C. G., Jeffords, M. R., and Kogan, M. 1982. Soybean seed quality losses associated with bean leaf beetles and *Alternaria tenuissima*. Phytopathology 72:615-618.

(Prepared by G. L. Hartman)

Anthracnose

Anthracnose, first reported in 1917 in Korea, occurs wherever soybean is grown. The disease reduces stands, seed quality, and yields by 16–26% or more in the United States, by 30–50% in Thailand, and by 100% in certain areas of Brazil and India. Losses increase as the percentage of infected pods rises.

Symptoms and Signs

Soybean is susceptible to anthracnose pathogens at all stages of development. Foliar symptoms, which develop after a prolonged period of high humidity, include necrosis of laminar veins, leaf rolling, petiole cankering, and premature defoliation (Fig. 28). Blighted plants may not grow as tall as healthy plants.

Fig. 28. Anthracnose (*Colletotrichum* spp.) on petioles and stems, with accompanying defoliation. (Courtesy J. B. Sinclair)

Fig. 29. Black acervuli of an anthracnose pathogen (*Colletotrichum* sp.) scattered on soybean petioles and stems with some charcoal rot on the lower main stem. (Courtesy G. L. Hartman)

During the early reproductive stages, symptoms typically appear on stems, petioles, and pods as irregularly shaped, brown areas and may resemble pod and stem blight (see the sections Phomopsis Seed Decay and Pod and Stem Blight). Leaves, pods, and stems may be infected without showing symptoms. However, symptoms and fruiting structures may be induced by treatment with paraquat or another desiccant. In the advanced stages of anthracnose (usually, the late reproductive stages), infected tissues are covered with black fruiting bodies (acervuli) (Figs. 29 and 30) that produce minute black spines (setae), which can be seen with the unaided eye. The setae are diagnostic for preliminary identification of the pathogen.

Preemergence and postemergence damping-off may occur when infected seeds are planted. Dark-brown, sunken cankers often develop on the cotyledons of emerging seedlings (Fig. 31). These cankers gradually extend up toward the epicotyl and down toward the radicle. During humid conditions, one or both cotyledons become water soaked, quickly wither, and fall off. The anthracnose fungi may grow from infected cotyledons into young stems, where numerous small, deep-seated cankers form and may kill the young plant.

Anthracnose of maturing plants causes serious losses, particularly during rainy periods, when shaded lower branches and leaves are killed. Premature defoliation may occur throughout the canopy when cankers girdle the leaf petioles. These cankers often occur where a leaflet joins a petiole, resulting in a shepherd's crook (Fig. 32). Affected plants senesce earlier than usual from the combined effects of stem and petiole lesions, and yields are reduced accordingly. When pods or pedicels are infected early, either no seeds form (a condition known as "pod blanking") or the seeds that develop are fewer or smaller than normal.

Seeds colonized by *Colletotrichum truncatum* may not show symptoms or may develop brown staining or small, irregularly shaped, gray areas with black specks. The fungus is confined at first to the seed coat. Infected seeds may die during germination, and if they do germinate, they may produce infected seedlings. When both *C. truncatum* and *Phomopsis longicolla* are present, they have an additive effect on the destruction of seed coat tissues (see the section Phomopsis Seed Decay).

Fig. 30. Lesions on soybean pods, characteristic of anthracnose (*Colletotrichum* spp.). (Courtesy H.-C. Yang)

Fig. 31. Anthracnose (*Colletotrichum* spp.) seed and seedling decay. (Courtesy T. K. Herman)

Causal Organisms

The most common pathogen associated with anthracnose is *Colletotrichum truncatum*. Other *Colletotrichum* spp. also reported on soybean are *C. chlorophyti, C. coccodes, C. gloeosporioides* (syn. *Glomerella cingulata*), *C. graminicola* (syn. *G. graminicola*), *C. incanum,* and *Glomerella glycines* (syn. *Colletotrichum* sp.). The host ranges for *C. chlorophyti* and *C. incanum* are not known, but all the other pathogens have wide host ranges. All the species are known or presumed to be seedborne in soybean.

C. truncatum is characterized by crowded, black acervuli borne on well-developed stromata (Fig. 33). Acervuli are oval to elongate, hemispherical to truncate conical, and erumpent, and they have numerous black, needlelike, intermixed long and short setae (3–8 × 60–300 µm) (Fig. 34). Conidia (2.0–4.5 × 17–31 µm), borne singly on conidiophores, are bluntly tapered, curved, unicellular, and hyaline; they usually produce one or two short germ tubes. Conidial masses are colorless to grayish white. Appressoria (5.7–11.3 × 11.7–30.2 µm) are elliptical or irregularly shaped and dark colored. *C. truncatum* grows on oatmeal agar and potato dextrose agar (PDA) under 12-h diurnal light at 25°C. It produces whitish colonies that eventually turn smoky black and have abundant acervuli.

Fig. 32. Leaf and stem collapse, characteristic of anthracnose (*Colletotrichum* spp.). (Courtesy H.-C. Yang)

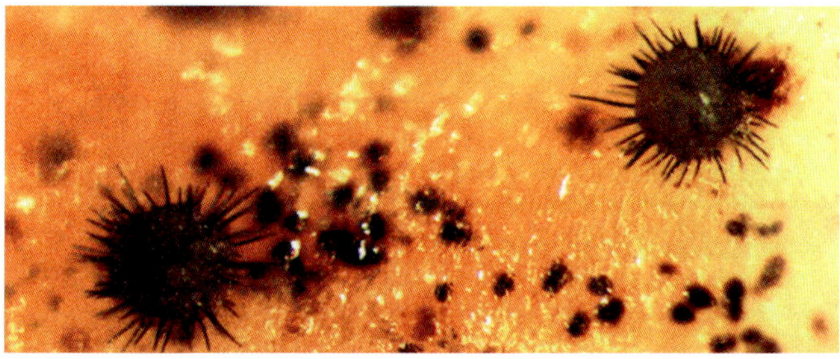

Fig. 33. Acervuli, setae, and masses of conidia on the acervuli of anthracnose pathogen *Colletotrichum truncatum.* (Courtesy H.-C. Yang)

C. chlorophyti forms black, irregularly shaped spots that are associated with fissures in the seed coat. These infected areas can expand, producing a black fungal mass that darkens the entire seed coat (Fig. 35). Acervuli produce setae that are straight, dark brown, and septate. Conidia (2.5–4.3 × 15.5–21.3 μm) are curved, have tapered ends and truncated bases, and are aseptate and hyaline. Conidial masses are light salmon in color. *C. chlorophyti* produces chlamydospores (5–12 μm in diameter) that are dark brown and clustered or chained together. No appressoria and perithecia are found. On PDA, *C. chlorophyti* initially produces white- to pink-colored, smooth-margined colonies that turn black with age.

C. incanum has acervuli that are subepidermal, are circular to elliptical, and have one to more than 10 setae. Setae (74–202 μm in length) are straight, black-brown, septate, and subacute. Conidia (2.3–3.7 × 16.9–21.9 μm) are more curved on one side than the other and have tapered ends with truncated bases; they are aseptate, unicellular, and hyaline. Conidial masses are colorless to pale salmon-orange. Appressoria (5.4–9.7 × 8.6–15.4 μm) are round, oval, or irregular in shape and dark colored. Sclerotia (60–280 μm in diameter) are black, globose, packed with masses of globules, subepidermal, and sometimes merged into irregular shapes. Colonies on PDA are flat and usually have regular margins, as well as dense mycelia. Colonies are initially white to pale olive-gray; they turn a mousey gray and finally become a blackish gray. Colonies on oatmeal agar are flat with smooth margins, have less dense mycelia, and have abundant surface acervuli and sclerotia.

G. glycines produces submerged perithecia that protrude through the epidermis from a stroma produced in the cortex. The perithecia (180–340 μm in diameter) form in groups of two or more and are short beaked, membranous, and globose (Fig. 36). Asci (7.0–13.5 × 30–106 μm) are oblong to bluntly clavate. Ascospores (2.7–5.5 × 15–35 μm) are smooth, hyaline, aseptate, straight to sigmoid, and mostly allantoid with rounded ends. Ascosporic masses are colorless to pale orange. Ascospores are ejected forcibly from the perithecia upon maturation (Fig. 37). Acervuli are black and have numerous brown setae. Conidia (4.0–6.6 × 15.7–22.5 μm) are straight, unicellular, bluntly tapered, and hyaline. Appressoria (6.9–10.3 × 7.1–13.0 μm) are rounded or oval and dark colored. On PDA, *G. glycines* colonies are yellowish white or gray; this pathogen forms acervuli and perithecia on PDA.

Isolates of these species vary considerably in colony characteristics, size of fruiting structures, and pathogenicity. Germ tubes arising from conidia of *C. truncatum* and *G. glycines* develop large, black appressoria when in contact with the surface of a soybean plant or another solid surface. A narrow infection peg extends from the underside of the appressorium and directly penetrates the cuticle and cell wall. After penetration, hyphae spread both between and within host cells.

Fig. 34. Acervuli of anthracnose pathogen *Colletotrichum truncatum* developing near a trichome, with setae and conidia. (Courtesy N. M. Kokalis)

Fig. 35. Black fungal mass on the decayed seed coat of a soybean seedling infected with anthracnose pathogen *Colletotrichum chlorophyti.* (Courtesy H.-C. Yang)

Disease Cycle and Epidemiology

The anthracnose pathogens overseason as mycelia, sclerotia, and acervuli in crop residue and in seeds. Seedborne inoculum may result in pre- or postemergence damping-off of seedlings or failed seed germination if the embryos are infected. Mycelium may also become established in infected seedlings without symptoms developing until the plants begin to mature. Stem

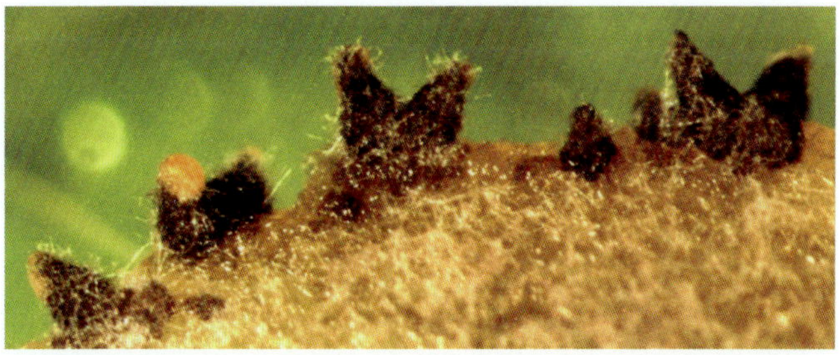

Fig. 36. Perithecia of anthracnose pathogen *Glomerella glycines* on soybean petioles. (Courtesy H.-C. Yang)

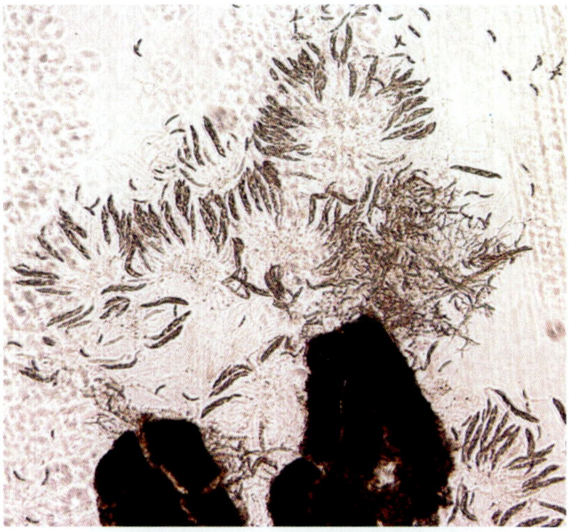

Fig. 37. Ascospores and perithecia of anthracnose pathogen *Glomerella glycines*. (Courtesy H.-C. Yang)

and pod infections usually occur in the reproductive stages of soybean development during warm, moist conditions.

The anthracnose pathogens presumably infect plants at all stages of development—particularly from bloom to pod fill. Disease-forecasting methods are used in the southeastern United States to predict the occurrence of infections and to prescribe fungicide treatment during these periods (see the section Management).

Detailed information about the environmental conditions that promote disease development is lacking. Conidia of *C. truncatum* and *G. glycines* germinate and form appressoria at temperatures below 35°C when the plant surface is wet. The conidia are short lived and sensitive to drying; 5 h of air drying can reduce germination by 98%. Infections typically occur if rain, dew, or fog provides free moisture for a period of 12 h or more.

Management

Soybean germplasm varies in susceptibility to anthracnose. Sources of resistance have been identified in some soybean germplasm (i.e., plant introductions and public soybean cultivars) but not in commercial, private cultivars. Using high-quality seeds that are free of pathogens and/or treated with a recommended fungicide may reduce the levels of inoculum associated with the pathogens that cause anthracnose. Crop residues should be plowed under in situations in which soil erosion is not a concern, and rotating crops may also reduce the inoculum load. Applying fungicides during the mid to late reproductive growth stages may also be effective, especially under conditions known to favor disease development and when needed for seed production fields.

Selected References

Bailey, J. A., and Jeger, M. J., eds. 1992. *Colletotrichum:* Biology, Pathology, and Control. CABI, Wallingford, Oxfordshire, U.K.

Chen, L. S., Chu, C., Liu, C. D., Chen, R. S., and Tsay, J. G. 2006. PCR-based detection and differentiation of anthracnose pathogens, *Colletotrichum gloeosporioides* and *C. truncatum*, from vegetable soybean in Taiwan. J. Phytopathol. 154:654-662.

Hartman, G. L., Manandhar, J. B., and Sinclair, J. B. 1986. Incidence of *Colletotrichum* spp. on soybeans and weeds in Illinois and pathogenicity of *Colletotrichum truncatum.* Plant Dis. 70:780-782.

Khan, M., and Sinclair, J. B. 1992. Pathogenicity of sclerotia- and nonsclerotia-forming isolates of *Colletotrichum truncatum* on soybean plants and roots. Phytopathology 82:314-319.

Manandhar, J. B., Hartman, G. L., and Sinclair, J. B. 1986. *Colletotrichum destructivum,* the anamorph of *Glomerella glycines.* Phytopathology 76:282-285.

Manandhar, J. B., Hartman, G. L., and Sinclair, J. B. 1988. Soybean germ plasm evaluation for resistance to *Colletotrichum truncatum.* Plant Dis. 72:56-59.

Manandhar, J. B., Kunwar, I. K., Singh, T., Hartman, G. L., and Sinclair, J. B. 1985. Penetration and infection of soybean leaf tissues by *Colletotrichum truncatum* and *Glomerella glycines.* Phytopathology 75:704-708.

Riccioni, L., Conca, G., and Hartman, G. L. 1998. First report of *Colletotrichum coccodes* on soybean in the United States. Plant Dis. 82:959.

Roy, K. W. 1982. Seedling diseases caused in soybean by species of *Colletotrichum* and *Glomerella.* Phytopathology 72:1093-1096.

Roy, K. W. 1996. Falcate-spored species of *Colletotrichum* on soybean. Mycologia 88:1003-1009.

Yang, H.-C., and Hartman, G. L. 2015. Methods and evaluation of soybean genotypes for resistance to *Colletotrichum truncatum.* Plant Dis. 99:143-148.

Yang, H.-C., Haudenshield, J. S., and Hartman, G. L. 2012. First report of *Colletotrichum chlorophyti* causing soybean anthracnose. Plant Dis. 96:1699.

Yang, H.-C., Haudenshield, J. S., and Hartman, G. L. 2014. *Colletotrichum incanum* sp. nov., a curved-conidial species causing soybean anthracnose in the USA. Mycologia 106:32-42.

Yang, H.-C., Stewart, J. M., and Hartman, G. L. 2013. First report of *Colletotrichum chlorophyti* infecting soybean seed in Arkansas, United States. Plant Dis. 97:1510.

(Prepared by H.-C. Yang and G. L. Hartman)

Aristastoma Leaf Spot and Yellowing

Aristastoma leaf spot and yellowing have been reported from India during the rainy season. A single leaf spot is 2.0–22.0 mm

Fig. 38. Diffuse chlorosis on lower leaves, caused by Aristastoma leaf spot (*Aristastoma* spp.). (Courtesy N. D. Sharma)

Fig. 39. Pycnidia and conidia of an *Aristastoma* sp. (Courtesy N. D. Sharma)

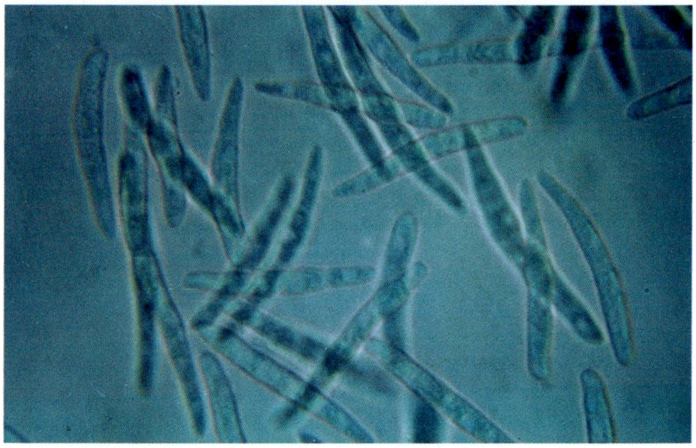

Fig. 40. Conidia of an *Aristastoma* sp. (Courtesy N. D. Sharma)

in diameter and irregularly shaped; it is often limited by a vein at the margin, has an indistinct edge, and has more or less alternating narrow, light- and dark-brown areas. A chlorotic halo is sometimes present, as well. Another distinct symptom is diffuse chlorosis on the lower leaves (Fig. 38), accompanied by stunting of the plant.

Aristastoma camarographioides and *A. guttulosum* are the two causal organisms of Aristastoma leaf spot and yellowing (Figs. 39 and 40).

Selected References

Dave, G. S., and Sharma, N. D. 1992. *Aristastoma guttulosum* Sutton, a new addition to the fungi of India. Acta Bot. Indica 20:305-306.

Maiti, S., and Dhar, V. 1982. A new zonate leaf spot of soybean caused by *Aristastoma camarographioides*. Trans. Brit. Mycol. Soc. 79:547-548.

(Prepared by N. D. Sharma)

Black Leaf Blight

Black leaf blight was first recorded in New South Wales, Australia, in 1982. In subsequent years, the disease has spread north and become established in the North Coast district of New South Wales. It has not been recorded in any other soybean-growing area of Australia. The initial outbreaks of black leaf blight were severe and estimated to have caused losses of 30–50%. In recent years, incidence of the disease and significant losses in yield have been sporadic.

Symptoms and Signs

Infected plants show severe spotting and blighting of leaves, stems, and pods (Fig. 41). Leaf spots are gray to dark gray to brown and have thin, distinct, very dark-brown margins separating them from healthy tissue. The spots are circular to oval and 1–10 mm in diameter. Larger, more irregularly shaped dead areas result from the joining of two or more spots and tear easily, giving heavily affected plants a tattered, blighted appearance. Leaf spots occur on both surfaces and penetrate the thickness of the leaf; on severely spotted leaves, the surrounding tissue is yellowed.

Fig. 41. Foliar spotting and blighting, characteristic of black leaf blight (*Arkoola nigra*). (Courtesy G. Stovold)

Lesions that develop on pods and stems are darker brown than leaf spots and oval to elongated, rather than circular. Heavily infected pods are shrunken, and the seeds directly below spots may be discolored. Infected plants develop a web of fine, dark-brown to black, glistening aerial mycelium. Infected leaves and pods may drop, and the fungus will colonize crop debris that remains after harvest (Fig. 42).

Causal Organism

Arkoola nigra causes black leaf blight. This fungus forms dark-brown to black aerial mycelium, which grows over living and dead leaves and stems and pods. Hyphae are dark brown to black, branched, septate, 14–20 µm wide, and smooth or finely verruculose; they are also closely branched at the apice and form raised, dark-brown appressoria up to 300 µm in diameter and 250–300 µm wide at the tip in contact with the plant. Appressoria consist of several short hyphal branches, under which penetration occurs through small, circular holes approximately 3.5 µm in diameter. Appressoria are more evident on the lower leaf surfaces than the upper leaf surfaces. Hyphae within host tissue are pale brown, intercellular, branched, and more variable in width than surface hyphae.

Pseudothecia develop from mycelium on dead and fallen leaves. They are superficial, sometimes erumpent, setose, black, and single or loosely clustered (clusters of up to six to 10). They are at first subglobose and then broadly ovoid to obpyriform, as well as uniloculate. Pseudothecia are 400–900 µm high, 300–450 µm wide in the lower half, and 200–350 µm wide in the upper half; each pseudothecium has a 40–70 µm wide ostiole. Pseudothecia contain 35–40 asci that are cylindrical, mainly eight spored, 24–30 × 230–300 µm, and bitunicate. Ascospores are uniseptate, 13–24 × 40–76 µm, ellipsoidal or fusiform, straight or slightly curved, and constricted at the septum; each ascopore has a pale-greenish wall, is smooth, and is covered by a thin, gelatinous sheath 2.0 µm thick. Pseudoparaphyses are abundant, hyaline, filiform, and branched with cells that are 10–26 µm long and 4–7 µm wide (Fig. 43).

Ascospores germinate at temperatures of 5–35°C (optimum is 20–30°C), and the fungus grows readily in culture. Colonies are black and have relatively open, radially oriented mycelia.

Disease Cycle and Epidemiology

A. nigra survives on air-dried host tissue for more than 2 years under laboratory conditions. In the field, it can be detected in heaps of residue after 6 months—sufficient time to enable its survival between successive soybean crops.

Infection of volunteer seedlings growing in accumulated debris and in seedlings growing in pots with infected debris

Fig. 42. Dead plant debris covered with mycelial strands of *Arkoola nigra,* the cause of black leaf blight. (Courtesy G. Stovold)

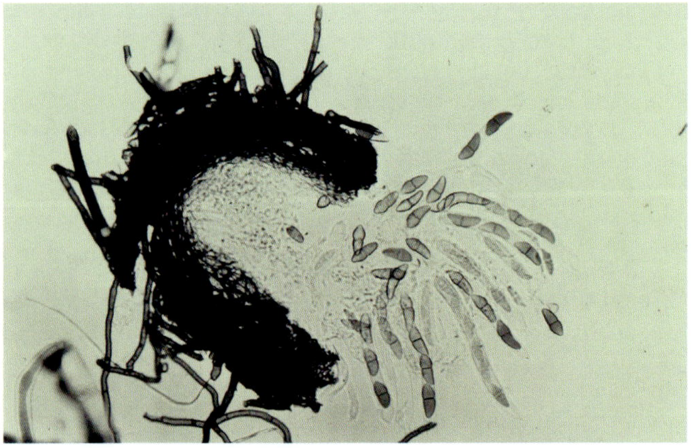

Fig. 43. Perithecia, asci, and ascopores of *Arkoola nigra,* which causes black leaf blight. (Courtesy G. Stovold)

has been observed. The lower stems and taproots of seedlings become infected with mycelium of *A. nigra;* this may be how the fungus becomes established and survives during the early stages of crop development. However, black leaf blight is not observed in commercial crops until after flowering, when there is a dense canopy of foliage. It is not known how the fungus survives from early to late season or what forms of inoculum are responsible for postflowering infection, and it is presumed that the fungus has some latent period from early infection to later in the growing season, when symptoms are observed.

Immature ascocarps have been found on debris in the field. However, the cycle of ascocarp development and ascospore release and dispersal is unknown.

Seed colonization of *A. nigra* occurs rarely, but it could be a means of introducing the pathogen into new areas. Infected seeds do not geminate.

Controlled inoculation on a range of other legumes found in the North Coast region of New South Wales indicated that some of these plants are alternative hosts of *A. nigra*. Even so, black leaf blight has not been observed on these legumes in the field.

The most severe symptoms of black leaf blight develop when the postflowering stages occur during persistent wet conditions. In the early years of the outbreak in New South Wales, severe losses were experienced. Since then, knowledge of the disease and the application of better management practices have reduced black leaf blight to minor significance.

Management

Growers should avoid planting successive crops of soybean, especially when black leaf blight has been observed in the current crop. In addition, growers should manage crop residue to encourage adequate breakdown between crops. High-quality seed that is relatively free of pathogens should be used, and it should be sown at the recommended density and time. Weeds should be controlled to avoid the development of a tall, dense canopy in the postflowering stages and thereby reduce humidity in the canopy.

In research trials, a single application of iprodione (postflowering) provided control of the pathogen and increased yields, compared with nontreated plots.

Selected References

Stovold, G. E., Walker, J., Desborough, P. J., and Francis, A. 1985. Page 44 in: Black leaf blight, a new disease of soybean. Proc. World Soybean Res. Conf., 3rd. Westview Press, Boulder, CO.

Walker, J., and Stovold, G. E. 1986. *Arkoola nigra* gen. et sp. nov. (*Venturiaceae*) causing black leaf blight of soybean in Australia. Trans. Brit. Mycol. Soc. 87:23-44.

Zhang, Y., Crous, P. W., Schoch, C. L., Bahkali, A. H., Guo, L. D., and Hyde, K. D. 2011. A molecular, morphological and ecological re-appraisal of *Venturiales*—A new order of *Dothideomycetes*. Fungal Divers. 51:249-277.

(Prepared by G. Stovold; revised by G. L. Hartman)

Brown Spot

Brown spot (also known as "Septoria leaf spot") was first reported in the United States in 1922 and now occurs in all production areas of the country. It has also been reported in the soybean-growing areas of Argentina, Brazil, Canada, China, Germany, Italy, Japan, Korea, Russia, Taiwan, and Yugoslavia.

Symptoms and Signs

Brown spot is primarily a leaf disease, although the seeds, pods, and stems of maturing plants may be infected. Irregularly shaped, dark-brown spots—varying in size from minute specks to lesions 4 mm in diameter—appear on both the upper and lower surfaces of unifoliolate leaves. These leaves quickly turn yellow and drop. Lesions frequently coalesce to form irregularly shaped blotches. Numerous irregularly shaped, light-brown lesions form on trifoliolate leaves (Fig. 44); these lesions gradually darken, becoming dark brown. During warm, wet periods, symptoms occur on the lower leaves; they progress to the upper leaves as the plant matures. Late in the growing season, leaves become rusty brown (often with chlorotic areas surrounding the lesions) and drop prematurely.

Irregularly shaped, brown lesions with indefinite margins form on the main stem, branches, petioles, and pods. These lesions range in size from small specks to areas of several square centimeters. Symptoms on these organs may not be sufficiently distinct from those of other diseases to be diagnostic.

A host-specific pathotoxin isolated from culture filtrate of *Septoria glycines* was found to cause typical symptoms on cotyledons and leaves.

Causal Organism

Septoria glycines, the cause of brown spot, forms brown pycnidia that are embedded in the tissues of mature lesions and open to the surface by large pores (ostioles). Pycnidia that form in leaf tissues are globose to conical–globose, whereas those that form in stems are considerably flatter because of their position. Pycnidia are 60–125 µm in diameter and can be either scattered or densely packed in host tissue. The pycnidial wall is thin and membranous. Conidia (pycnidiospores) are hyaline, filiform, curved, and 1.4–2.0 × 21–50 µm. They are indistinctly one to three septate, and the septa become noticeable at the time of germination. Conidia readily germinate in water on the surfaces of leaves. First, germ tubes extend from the ends of the spores, and later, hyphae develop from other conidial cells. Mature hyphae are densely branched, thick walled, and bead-like. Pathogenic variability among isolates of the fungus has not been detected.

Disease Cycle and Epidemiology

Plants infected with *S. glycines* often prematurely defoliate. Yield losses range from 8 to 34%, based on results from experimental field plots. Brown spot is most severe when soybean is grown continuously in the same field. When plants were inoculated in the field or the greenhouse, leaf symptoms developed on 29 legumes and on velvetleaf (*Abutilon theophrasti*).

Conidia and mycelia overseason on infected leaf and stem debris. At low frequencies, the fungus can be seedborne. Lesions that develop on infected cotyledons and unifoliolate leaves serve as sources of secondary inoculum.

Infection and disease development are favored by warm, moist conditions, which promote sporulation of the pathogen in primary lesions. Disease severity increases when periods of leaf

Fig. 44. Irregular, light-brown lesions on a distal trifoliolate leaf, characteristic of brown spot (*Septoria glycines*). (Courtesy G. L. Hartman)

wetness increase from 6 to 36 h. The optimum temperature for disease development is 25°C, but disease symptoms can develop from 15 to 30°C. Conidia are distributed by wind and splashing rain. Infection and defoliation proceed from the lower to the upper parts of the plant. Early senescence can occur even at low disease severity. Dry conditions halt the spread of the disease.

The fungus enters leaves through the stomata and grows intercellularly, killing cells next to the hyphae. The fungus penetrates pods through the stomata or by growing through placental and funicular tissue; it later invades the seed coat.

Management

Cultivars vary in susceptibility to brown spot. Some are considered to have partial resistance, but no genes are known to be associated with resistance in soybean.

Growers should rotate soybean with a nonlegume crop for at least 1 year. Tillage has had a variable effect on the severity of brown spot.

When economically feasible, a foliar fungicide should be applied from bloom to pod fill. One report has indicated that the disease can be managed with strobilurin fungicides applied at growth stage R3. In Argentina, applying fungicides during growth stages R3–R5 resulted in significant yield increases, especially during wet periods with high rainfall.

Selected References

Carmona, M., Sautua, F., Perelman, S., Reis, E. M., and Gally, M. 2011. Relationship between late soybean diseases complex and rain in determining grain yield responses to fungicide applications. J. Phytopathol. 156:687-693.

Cruz, C. D., Mills, D., Paul, P. A., and Dorrance, A. E. 2010. Impact of brown spot caused by *Septoria glycines* on soybean in Ohio. Plant Dis. 94:820-826.

Kamicker, T. A., and Lim, S. M. 1985. Field evaluation of pathogenic variability in isolates of *Septoria glycines*. Plant Dis. 69:744-746.

Lee, G. B., and Hartman, G. L. 1996. Reactions of *Glycine* species and other legumes to *Septoria glycines*. Plant Dis. 80:90-94.

Lee, G. B., Hartman, G. L., and Lim, S. M. 1996. Brown spot severity and yield of soybeans regenerated from calli resistant to a host-specific pathotoxin produced by *Septoria glycines*. Plant Dis. 80:408-413.

Schuh, W., and Adamowicz, A. 1993. Influence of assessment time and modeling approach on the relationship between temperature–leaf wetness periods and disease parameters of *Septoria glycines* on soybeans. Phytopathology 83:941-948.

Song, H. S., Lim, S. M., and Widholm, J. M. 1994. Selection and regeneration of soybeans resistant to the pathotoxic culture filtrates of *Septoria glycines*. Phytopathology 84:948-951.

Vernetti, F. J., Tomm, G. O., and Casela, C. R. 1985. Occurrence of *Mycosphaerella phaseolorum, Phyllosticta glycineum* and *Colletotrichum truncatum* in Mozambique. Fitopatol. Bras. 10:143-145.

(Prepared by G. L. Hartman)

Cercospora Leaf Blight and Purple Seed Stain

Cercospora leaf blight (CLB) and purple seed stain (PSS) are both caused by *Cercospora kikuchii*.

Yield losses attributed to CLB (also known as "Cercospora leaf spot") range from 30 to 50% in Brazil, Japan, Taiwan, Uganda, and Zambia. CLB is also present in most soybean-growing regions of the United States. In the Gulf Coast states, CLB is the most important disease of soybean in terms of management costs and yield losses, with losses of 15% estimated in some states. In addition, CLB has been associated with a disorder known as "green stem," in which the pods and seeds mature normally but the stems remain green; no direct relationship has been documented, however.

Cercospora leaf blight may cause premature defoliation, especially under hot, dry conditions. Leaf and petiole symptoms generally appear during the late reproductive growth stages (R5–R6), but this may vary from year to year. In managing CLB, the greatest concern is the difficulty controlling the disease with fungicides. Even though the pathogen has been shown to be highly sensitive to commonly used fungicides in in vitro tests, these products have been relatively ineffective under field conditions when applied at growth stages R3–R5.

Purple seed stain (also known as "purple blotch," "purple speck," "purple spot," and "lavender spot") does not reduce yield directly but can lead to poor grain quality, which may reduce the value of a harvested crop. Additionally, seed infection increases the possibility of delayed germination and lower oil content. Unlike CLB, PSS occurs worldwide.

PSS and CLB are distinct diseases, and the environmental factors that predispose plants to one disease do not necessarily cause a higher incidence or severity of the other. In addition, there is no correlation among cultivars for resistance to these two diseases.

Symptoms and Signs on Leaves

The first visual symptoms of Cercospora leaf blight may be observed on petioles during early reproductive growth (Fig. 45), but leaf symptoms generally do not appear until the later growth stages (R5–R6). The diagnostic symptom is that the plant's upper leaves, which are exposed to the sun, become purplish on their upper surfaces. The discoloration can deepen and extend over the entire upper surfaces of affected leaves, giving them a leathery, dark, reddish-purple or bronze appearance (Figs. 46 and 47). In less severe cases, these symptoms may be confined to small, discrete but irregularly shaped blotches (Fig. 48). (For a comparison of the effects of CLB with those of sunburn, compare Figure 48 with Figure 235 in Part III, Noninfectious and Stress Disorders.) A true blight symptom may also develop, in which necrotic patches develop on affected leaves, creating a patchy or tattered appearance (Fig. 49). In severe cases, these patches may coalesce, and the entire leaf may become necrotic (Fig. 50).

Fig. 45. Lesions of Cercospora leaf blight (*Cercospora kikuchii*) on a petiole. (Courtesy R. W. Schneider)

The purple-bronze and blight symptoms occur independently (i.e., severe coloration symptoms do not necessarily produce blighted leaves), and cultivars may differ in their propensity to display one symptom or the other. Blight often occurs without the purple-bronze symptoms. These changes in color are often mistaken for early senescence; however, green leaves usually appear below the symptomatic leaves, which have been exposed to direct sunlight. (The toxin, cercosporin, is light activated.)

Lesions on petioles and stems are slightly sunken, reddish-purple, and up to several millimeters long. These lesions may coalesce to the extent that the entire petiole is discolored (Fig. 51). Severe petiole infection usually results in defoliation, regardless of leaf symptom severity (Fig. 52).

On more susceptible cultivars, round, reddish-purple lesions (which later become purplish-black) occur on pods. A general dark red-purple discoloration (similar to that which occurs on the upper leaves) may also occur on upper stems, petioles, and pods exposed to the sun.

Symptoms and Signs on Seeds

Purple seed stain is conspicuous and easily distinguished on seeds. Seed discoloration varies from pink or pale purple to dark purple, and discolored areas range in size from specks to large, irregularly shaped blotches, which may cover the entire surface of the seed coat. Cotyledons are generally not discolored. Infected seeds may not show symptoms, but the pathogen can be recovered upon culturing on agar media.

Fig. 46. Purple discoloration on a sun-exposed leaf, characteristic of Cercospora leaf blight (*Cercospora kikuchii*). (Courtesy R. W. Schneider)

Fig. 48. Irregularly shaped blotches, symptomatic of Cercospora leaf blight (*Cercospora kikuchii*). (Courtesy R. W. Schneider)

Fig. 47. Advanced purple discoloration with bronze highlights, symptomatic of Cercospora leaf blight (*Cercospora kikuchii*). (Courtesy R. W. Schneider)

Fig. 49. Leaf yellowing and necrosis, characteristic of Cercospora leaf blight (*Cercospora kikuchii*). (Courtesy R. W. Schneider)

Heavily infected seeds can produce diseased seedlings and reduce stands. Surviving seedlings may display cotyledons that are shriveled, turn dark purple, and fall prematurely. A cotyledon infection may spread to the stem, producing necrotic areas that girdle the stem and kill a young plant. Less severely affected seedlings may survive but be stunted. If conditions are humid and warm during seedling emergence, a velvety, grayish-white growth of conidiophores and conidia may appear on infected portions of cotyledons and stems.

Causal Organism

Cercospora kikuchii sporulates abundantly under high humidity at 23–27°C within 3–5 days on infected plant tissues, including excised seed coats in the laboratory. Rapid sporulation also occurs when tissue is incubated during 12-h diurnal periods of light and darkness and when inoculated onto carrot leaf decoction agar, soybean plant tissue agar, and V8 juice agar. Conidiophores are produced in fascicles, with up to 20 stalks on a stroma. Conidiophores are yellowish-brown to dark brown at the bases, gradually lighter in color at the distal ends, and hyaline at the tips. Conidial scars are not found on newly formed conidiophores but are evident on old ones; the scars are 10–50 μm apart. Cultures generally lose their ability to sporulate upon subculturing.

A holoblastic conidium is produced at the tip of a conidiophore. As the conidium matures, a cross-wall forms at the attachment. Mature conidia detach easily, and conidiophores proliferate sympodially to form additional conidiogenous loci at the new apices.

Conidia are hyaline, acicular and truncate, straight or curved, and multiseptate with thickened hila. They may develop in an alternating manner on conidiophores, but this is not a consistent feature. Conidia germinate in distilled water, producing one to many germ tubes within 2–3 h. Conidiophores and conidia from Japan ranged from 85 to 200 μm and 70 to 164 μm, respectively, and had zero to 22 septations, whereas those from the U.S. state of North Carolina were 30–286 μm and 38.8–445.0 μm, respectively, and had two to 49 septations (Fig. 53). (Compare this description with that of *C. sojina;* see the section Frogeye Leaf Spot.)

Environmental factors—such as relative humidity and the composition and pH of the substrate—influence the development and morphology of *C. kikuchii.* The optimum temperature

Fig. 50. Necrosis of an entire leaf, symptomatic of Cercospora leaf blight (*Cercospora kikuchii*). (Courtesy R. W. Schneider)

Fig. 51. Severe petiole discoloration, characteristic of Cercospora leaf blight (*Cercospora kikuchii*). (Courtesy R. W. Schneider)

Fig. 52. Severe defoliation of leaves in the upper canopy, symptomatic of Cercospora leaf blight (*Cercospora kikuchii*). Note that the petioles remain attached to the stems. (Courtesy R. W. Schneider)

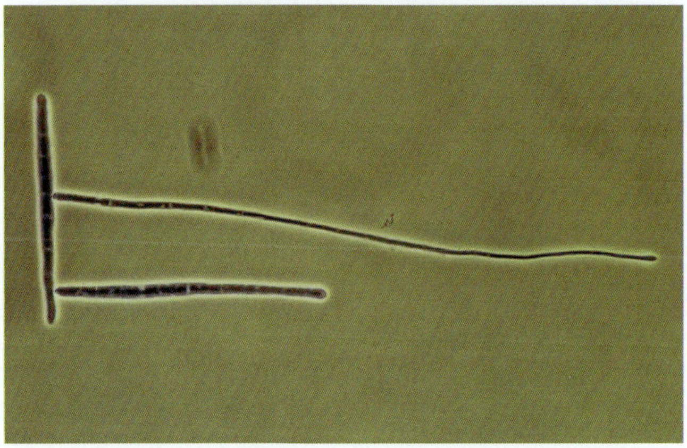

Fig. 53. Conidia of *Cercospora kikuchii* (top) and *C. sojina* (bottom). (Courtesy C. C. Yeh)

for sporulation on infected seeds plated on moist filter paper is 23–27°C. The minimum, maximum, and optimum temperatures for growth on potato dextrose agar (PDA) are 27, 35, and 12°C, respectively. The optimum pH for growth in culture is 5.9.

Growth on artificial media is uniformly dense, with deep folds radiating from the center. Colonies are white at the edges and light grayish-olive toward the center. The medium beneath the colony varies in color but is often dark purple with pink margins (Fig. 54). The pigment produced on agar looks similar to that produced by *Chaetomium cupreum* and some other fungi.

Young hyphae in culture are hyaline, septate, 2–4 µm thick, granular, and sometimes noduled. Older hyphae are pale brown, 3–5 µm thick, and closely septate. Orange-brown, thick-walled cells that are 6–15 µm in diameter (which resemble chlamydospores) form in old cultures.

Disease Cycle and Epidemiology

Primary inoculum may come from infected seeds or surface debris from previous crops. The incidence of infection generally increases with the increased incidence of dew periods. Germinated conidia can survive an extended period in the absence of leaf wetness if the level of relative humidity is high.

Germ tubes penetrate the plant through the stomata or directly through epidermal cells. Infections that arise from this first cycle of sporulation are mostly latent and colonize few cells. Evidence based on sequential sampling and molecular detection of *C. kikuchii* on field-grown leaves suggests that the leaves become infected during early vegetative growth, but these infections remain latent until the late reproductive stages (early R5–R6), at which time symptoms become evident. These early infections likely arise from infected seeds, which may not display the characteristic symptoms of purple seed stain.

A study of *C. kikuchii* isolates from the U.S. state of Louisiana demonstrated that this population was dominated by a new lineage; fewer than 5% of isolates clustered with the old lineage. The dominance of this new lineage was not caused by higher virulence. Instead, multiple lines of molecular and genetic evidence suggested the existence of a cryptic functional sexual stage in this fungus. Pathogen populations in South America and Japan were also found to be dominated by a single lineage, and whether the same lineage dominates populations in all three regions is unknown. The study using the Louisiana isolates also showed that pathogen populations from infected leaves were significantly different from those from infected seeds, suggesting that some isolates are more adapted to leaf infection, whereas others are more adapted to seed infection.

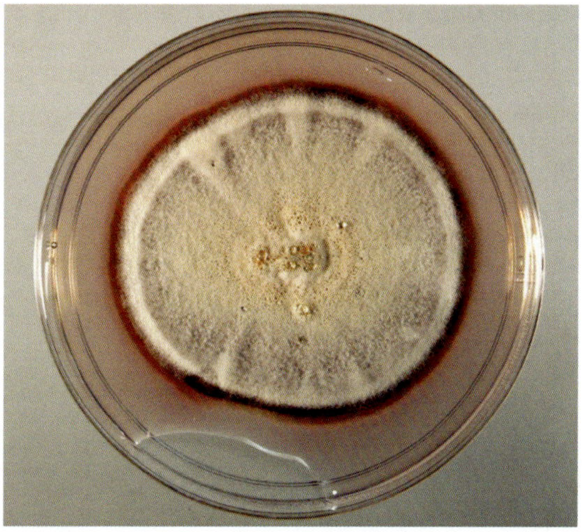

Fig. 54. Culture of *Cercospora kikuchii,* showing purple pigmentation in the agar medium. (Courtesy R. W. Schneider)

Management

Some soybean cultivars express resistance to Cercospora leaf blight and purple seed stain, but no cultivar is immune. The genetic mechanism of resistance to the two diseases may differ, and resistance to both diseases may vary by location and year, perhaps related to the high level of genetic variability in this pathogen. The use of clean seed is essential for disease management, although infected seeds (as assessed by plating on agar media) are often found in commercial seed lots in which PSS has not been observed.

Fungal isolates are known to differ in fungicide sensitivity, and some inconsistencies in field control have been reported for CLB. This may be attributed in part to the pathogen's development of resistance to thiophanate-methyl, which was first documented in Japan in the late 1980s. Resistance to thiophanate-methyl and to the strobilurin fungicides was later documented for isolates recovered in Louisiana in 2011. In general, despite late-season symptom development, fungicides should be applied during the late vegetative or early reproductive stages, and two fungicide applications may be required for efficacious disease control. The first application should commence at first flower (growth stages R1–R2), and the second has proven most effective when applied at early pod fill (growth stage R4). The efficacy of early fungicide applications is probably related to suppression of the rapid increase in latent fungal biomass beginning at growth stage R1. The triazoles have been found to perform better than the strobilurins, and using maximum label rates has been necessary. This situation is very different from that of frogeye leaf spot, for which the infection of leaves occurs by dispersed spores and the disease is readily controlled by fungicide applications at or shortly after symptom development (see the section Frogeye Leaf Spot). For control of PSS, seed treatments should be applied, along with in-season foliar fungicide applications between growth stages R3 and R5.

Management of CLB has also been achieved with foliar applications of iron. Test results have shown that an application at growth stage R5 suppresses disease development when leaf tissue concentrations surpass a threshold of approximately 280 ppm. These results are promising, especially since disease resistance and fungicide efficacy have been elusive in managing CLB.

Selected References

Almeida, A. M. R., Piuga, F. F., Marin, S. R. R., Binneck, E., Sartori, F., Costamilan, L. M., Teixeira, M. R. O., and Lopes, M. 2005. Pathogenicity, molecular characterization, and cercosporin content of Brazilian isolates of *Cercospora kikuchii*. Fitopatol. Bras. 30:594-602.

Cai, G., and Schneider, R. W. 2008. Population structure of *Cercospora kikuchii*, the causal agent of Cercospora leaf blight and purple seed stain in soybean. Phytopathology 98:823-829.

Cai, G., Schneider, R. W., and Padgett, G. B. 2009. Assessment of lineages of *Cercospora kikuchii* in Louisiana for aggressiveness and screening soybean cultivars for resistance to Cercospora leaf blight. Plant Dis. 93:868-874.

Chanda, A. K., Ward, N. A., Robertson, C. L., Chen, Z.-Y., and Schneider, R. W. 2014. Development of a quantitative polymerase chain reaction detection protocol for *Cercospora kikuchii* in soybean leaves and its use for documenting latent infection as affected by fungicide applications. Phytopathology 104:1118-1124.

Dakai, Y. 1984. Importance of fallen soybean leaves infected with *Cercospora kikuchii* Matsu. & Tomoyasu for disease spread. (In Japanese) Bull. Hiroshima Pref. Agric. Exp. Stn. 48:67-74.

Groenewald, J. Z., Nakashima, C., Nishikawa, J., Shin, H.-D., Park, J.-H., Jama, A. N., Groenewald, M., Braun, U., and Crous, P. W. 2013. Species concepts in *Cercospora:* Spotting the weeds among the roses. Stud. Mycol. 75:115-170.

Imazaki, I., Homma, Y., Kato, M., Vallone, S., Yorinori, J. T., Henning, A. A., Iizumi, H., and Koizumi, S. 2006. Genetic relationships between *Cercospora kikuchii* populations from South America and Japan. Phytopathology 96:1000-1008.

Orth, C. E., and Schuh, W. 1994. Resistance of 17 soybean cultivars to foliar, latent, and seed infection by *Cercospora kikuchii*. Plant Dis. 78:661-664.

Price, P. P., III, Purvis, M. A., Robertson, C. L., Schneider, R. W., and Padgett, G. B. 2014. Cercospora leaf blight pathogen resistant to fungicides. La. Agric. 57:28-29.

Schuh, W. 1993. Influence of interrupted dew periods, relative humidity, and light on disease severity and latent infections caused by *Cercospora kikuchii* on soybean. Phytopathology 83:109-113.

Silva, E. C., Chanda, A. K., Aroca, T. G., Robertson, C. L., Ward, B. M., and Schneider, R. W. 2014. Foliar applications of iron suppress Cercospora leaf blight and increase yields. Phytopathology (Poster Session) 233-P.

Walters, H. J. 1980. Soybean leaf blight caused by *Cercospora kikuchii*. Plant Dis. 64:961-962.

(Prepared by N. A. Ward-Gauthier, R. W. Schneider, A. Chanda, E. C. Silva, P. P. Price, III, and G. Cai)

Choanephora Leaf Blight

Choanephora leaf blight has been reported in tropical and subtropical areas of Indonesia, Thailand, and the United States. Yield losses of 5–10% and 20–40% have been reported in northern and central Thailand, respectively. Other reported hosts include cowpea, cucumber, mung bean (*Vigna radiata*), and tobacco, as well as species of *Abelmoschus, Gossypium, Hibiscus, Ipomoea, Jasminum, Petunia,* and *Pisum.*

Symptoms and Signs

Symptoms of Choanephora leaf blight occur in the field after 2–3 consecutive days of rain when the temperature is cool and the relative humidity is close to saturation. Infected areas of leaves first appear grayish, similar to leaves with hot-water scald injury, and then they turn dark. These symptoms first appear on mature leaves and occasionally on young leaves at the top of the plant; pods may also display symptoms.

If high humidity levels and cool temperatures prevail at night, then sporangiophores and sporangia will form; they can easily be seen by the unaided eye on infected parts of plants early in the morning (Fig. 55). Infected leaves may drop and adhere to the stems beneath them. Extensive defoliation and yield losses occur with continued favorable conditions for disease development.

Fig. 55. Leaf necrosis, associated with Choanephora leaf blight, and sporulation of *Choanephora infundibulifera* on the necrotic margin. (Courtesy S. Poonpolgul)

Causal Organisms

The fungus that causes Choanephora leaf blight is *Choanephora infundibulifera*. Sporangiophores bearing sporangiola arise from aerial hyphae, which are erect or ascending, and extend to 10 mm high; they are unbranched, hyaline, smooth walled, and up to 30 µm wide. They are apically dilated and form a clavate vesicle up to 45 µm in diameter; from it arise one to six unbranched or dichotomously branched, distally clavate secondary vesicles, which bear the sporangiola on apical dilations. As many as 12 or more sporangioliferous dilations are formed. Secondary vesicles usually remain attached at maturity.

Sporangiospores within sporangiola are broadly ellipsoidal to obovoid, brown or reddish-brown to pale brown, and smooth (or rarely indistinctly longitudinally striate). They remain inside the wall of the sporangiolum at maturity. Colonies grow rapidly on potato–carrot agar at 25°C.

C. cucurbitarum also causes Choanephora leaf blight on soybean. This fungus is commonly reported as causing Choanephora wet rot on cucurbits, which produces a rapid rot.

Disease Cycle and Epidemiology

The pathogen often attacks senescing plant parts, such as flowers. During extended periods of wet and warm conditions, the fungus will continue to grow, producing spores that can be disseminated to neighboring plants. The fungus may overseason on debris as spores.

Management

No reports are available on the resistance of commercial cultivars to Choanephora leaf blight. Similarly, no satisfactory chemical control has been reported, although triforine and copper oxychloride have been shown effective in laboratory studies.

Selected References

Roy, K. W. 1993. First report of Choanephora leaf blight of soybean in Mississippi. Plant Dis. 77:1262.

Roy, K. W., Baird, R. E., and Abney, T. S. 2000. A review of soybean (*Glycine max*) seed, pod, and flower mycofloras in North America, with methods and a key for identification of selected fungi. Mycopathologia 150:15-27.

Subba Rao, K. V., Padgett, G. B., Berner, K. K., Berggren, G. T., and Snow, J. P. 1990. Choanephora leaf blight of soybeans in Louisiana. Plant Dis. 74:614.

(Prepared by J. B. Sinclair; revised by G. L. Hartman)

Downy Mildew

Downy mildew was first reported in 1908 from Kashmir, India, and it was found in the United States in 1923. Since then, it has become one of the most common foliar diseases of soybean. Downy mildew is distributed worldwide but rarely produces significant damage. It may cause some defoliation and reduced seed quality and size, and it may lower yields by 9–18%.

Symptoms and Signs

Downy mildew appears on the upper surfaces of young leaves as discrete pale-green to light-yellow spots (2–8 mm in diameter), which enlarge to become pale- to bright-yellow lesions (Fig. 56). The sizes and shapes of the lesions vary, depending on the leaf age, the inoculum dose, and the race of the pathogen involved. Older lesions may turn grayish brown to dark brown and have yellowish-green margins; eventually, they may become entirely brown. On the lower surfaces of leaves (particularly under moist conditions), lesions are covered with tufts of grayish to pale-purple sporangiophores, which easily distinguish downy mildew from other foliar diseases of soybean (Fig. 57). The pathogen may not sporulate on old lesions. Severely infected leaves turn yellow and finally brown, curl at the edges, and drop prematurely.

A pod infection may occur without external symptoms. The interior of the pod and the seed coat may be encrusted with a whitish mass of mycelia and oospores. Seeds that are partly or completely encrusted with oospores often appear dull white and have cracks in their coats (Fig. 58). These seeds may be smaller or lighter in weight than normal seeds. Occasionally, mycelium can be found in the seed coat. Oospores on the seed may remain viable for up to 8 years.

Infected seeds can produce systemically infected seedlings. Symptoms first appear when the plants are approximately 2 weeks old. Light-green areas appear at the bases of young trifoliolate leaves and spread along the veins in a serrated or fanlike manner. Systemically infected plants may remain small and have stunted, mottled, gray-green leaves that curl down at the edges. Prolific sporangiophore production can occur on the lower leaf surfaces.

Fig. 56. Pale- to bright-yellow lesions on an upper leaf surface, caused by the downy mildew pathogen, *Peronospora manshurica.* (Courtesy J. B. Sinclair)

Fig. 57. Pale-brown to yellow lesions on a lower leaf surface, caused by the downy mildew pathogen, *Peronospora manshurica.* (Courtesy G. L. Hartman)

Fig. 58. Seeds encrusted with oospores (left) of *Peronospora manshurica,* which causes downy mildew, and healthy seeds (right). (Courtesy J. B. Sinclair)

Causal Organism

Peronospora manshurica is the causal agent of downy mildew. Hyphae of this oomycete are coenocytic and 7–10 μm wide, and they run intercellularly in the host tissues. Sporangiophores are slender (5–9 × 240–984 μm) and gray to pale violet; they branch dichotomously two to 10 times (Fig. 59). Sporangiophores emerge singly or in clusters from stomata on the lower surfaces of leaves. The terminal branchlets, or sterigmata (2–3 × 9–13 μm), are pointed and straight. Sporangia (usually 19–24 μm in diameter) are subhyaline and broadly elliptical to subglobose. They lack an apical papilla and germinate in water; a germ tube can emerge from any point on the sporangium. Zoospores are not produced.

Oospores are light brown or yellow, may be smooth or variously marked, and have reticulated walls (Fig. 60). They measure 20–45 μm in diameter, including the reticulations, which are more visible when stained with cotton blue. Oospores germinate via the formation of germ tubes. The viability of oospores can be determined by incubating them in 1.0 ml of

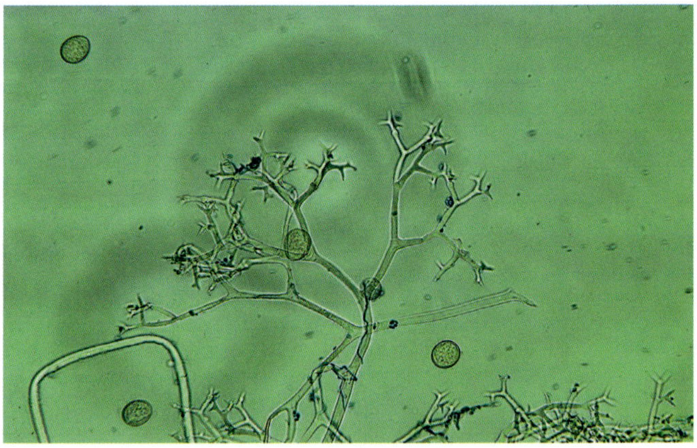

Fig. 59. Sporangiophores and sporangia of *Peronospora manshurica,* which causes downy mildew. (Courtesy F. D. Tenne)

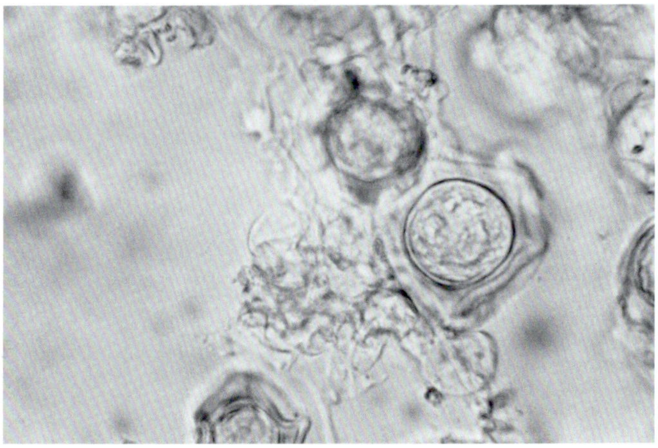

Fig. 60. Mature oospores of *Peronospora manshurica,* which causes downy mildew. (Courtesy H. C. Phatak)

distilled water for 48 h at 30°C and then adding one drop of a 1% solution of 2,3,5-triphenyltetrazolium chloride (TTC); after another 48 h, viable spores turn orange-red.

Haustoria are formed from intercellular mycelia and may branch up to five times within a cell. They vary in diameter from 1.0 to 3.5 µm and are generally long, cylindrical, tenuous, and curved.

Thirty-five physiological races of *P. manshurica* have been identified in the United States on the basis of disease reactions of differential cultivars.

Disease Cycle and Epidemiology

P. manshurica overseasons as oospores in leaves and on seeds. Oospore-encrusted seeds may produce systemically infected seedlings under cool conditions. Seedling infection probably occurs in the hypocotyl from oospores on the seed coat or leaf residues. Hyphae grow through the hypocotyl into the first two or three trifoliolate leaves and into some nodal buds. The first and second pairs of leaves to develop on systemically infected plants show large chlorotic areas. When dew is present, sporangia are produced on newly infected leaves and disseminated by air currents. Few or no sporangia form on old leaves. Under favorable conditions, sporangia germinate within 12 h.

The germ tube may enter a stoma or form an ovate appressorium (9–12 µm). Penetration pegs push between adjacent cell walls and into the intercellular spaces of the mesophyll. In susceptible soybean cultivars, the pathogen continues to ramify through the mesophyll and into the palisade layers. Haustoria frequently are formed within cells. Bundle sheath parenchyma cells of the larger veins stop hyphal advancement, with the result that the lesions acquire an angular shape. In resistant cultivars, hyphal growth stops abruptly and relatively few haustoria are formed.

The development of downy mildew is favored by high humidity and temperatures of 20–24°C. Sporulation occurs between 10 and 25°C; no sporulation occurs above 30°C or below 10°C.

Leaves become more resistant to downy mildew as they grow older. A soybean cultivar may vary in reaction to *P. manshurica* among trials, depending on leaf age at the time of infection. Lesions increase in number but decrease in size as leaves age. Exposure of plants to a high temperature before infection increases the likelihood of a resistant reaction.

Management

P. manshurica survives as oospores in crop residue and on the surfaces of soybean seeds. Crop rotation with a nonhost plant for 1 or more years and deep burial (if applicable) of infested crop residue can reduce primary inoculum. Planting pathogen-free seeds and fungicide-treated seeds is also an effective management practice.

Numerous sources of resistance to *P. manshurica* are present in soybean germplasm, although soybean cultivars are not generally characterized for downy mildew resistance. Cultivars with the resistance gene *Rpm* are resistant to races 1–32 but susceptible to race 33; few cultivars are resistant to race 33. Resistance has been reported in cultivars Kanrich (*Rpm*1) and plant introduction (PI) 88788 (*Rpm*2). Resistance also has been reported in cultivars Burlison, Columbia, Fayette, Harosoy, Tracy, PI 86150, and PI 90138.

Foliar application of potassium phosphite and soil amendments with calcium silicate may reduce the severity of downy mildew. Foliar-applied fungicides are not typically recommended for control of downy mildew, because the disease generally has a low level of overall severity and little impact on yield.

Selected References

Chowdhury, A. K., Srinives, P., Saksoong, P., and Tongpamnak, P. 2002. RAPD markers linked to resistance to downy mildew disease in soybean. Euphytica 128:55-60.

Dunleavy, J. M. 1987. Yield reduction in soybeans caused by downy mildew. Plant Dis. 71:1112-1114.

Lim, S. M. 1978. Disease severity gradient of soybean downy mildew from a small focus of infection. Phytopathology 68:1774-1778.

Lim, S. M. 1989. Inheritance of resistance to *Peronospora manshurica* races 2 and 33 in soybean. Phytopathology 79:877-879.

Lim, S. M., Bernard, R. L., Nickell, C. D., and Gray, L. E. 1984. New physiological race of *Peronospora manshurica* virulent to the gene *Rpm* in soybeans. Plant Dis. 68:71-72.

McKenzie, J. R., and Wyllie, T. D. 1971. The effect of temperature and lesion size on the sporulation of *Peronospora manshurica*. Phytopathology 71:321-326.

Nolla, A., Korndörfer, G. H., and Coelho, L. 2006. Efficiency of calcium silicate and carbonate in soybean disease control. J. Plant Nutr. 29:2049-2061.

Silva, O. C., Santos, H. A. A., Dalla Pria, M., and May-De Mio, L. L. 2011. Potassium phosphite for control of downy mildew of soybean. Crop Prot. 30:598-604.

Tai Roongruangsree, U., Kjerulf-Jensen, C., Olson, L. W., and Lange, L. 1987. Viability tests for the thick walled fungal spores (ex: oospores of *Peronospora manshurica*). J. Phytopathol. 123:244-252.

Tai Roongruangsree, U., Olson, L. W., and Lange, L. 1987. The seed-borne inoculum of *Peronospora manshurica,* causal agent of soybean downy mildew. J. Phytopathol. 123:233-243.

(Prepared by D. V. Phillips; revised by M. V. Avanzato)

Drechslera Blight

Drechslera blight was first reported from India in 1971 and then again in 1976. A report from Brazil describes a *Drechslera* sp. as an endophyte in soybean.

Symptoms include circular to angular, brown spots (0.5–2.0 cm), which form near leaf margins. The lesions enlarge and develop gray centers with dark-brown margins; occasionally, chlorotic halos form, as well. Lesions sometimes develop along the veins, resulting in leaf curling and drying. Drechslera blight may kill seedlings.

The pathogen, *Bipolaris glycines* (syn. *Drechslera glycines*), grows well on oat agar, producing brown, septate, profusely branched mycelium. Conidiophores are erect, 6–7 µm, simple or branched, septate, brown, and geniculate. Conidia are borne acrogenously and 9–13 × 31–94 µm, fusoid, straight or slightly curved, septate, and three to 12 celled with a basal hilum. They germinate by producing germ tubes from both ends of the conidium. The pathogen may be endophytic in soybean stems.

Drechslera blight is also caused by *Exserohilum rostratum* (syn. *Setosphaeria rostrata*).

Selected References

Kulkarni, C. S., Patil, B. C., and Patil, P. L. 1976. Leaf spot of soybean (*Glycine max* [L.] Merr.) caused by *Drechslera halodes* in India. J. Maharashtra Agric. Univ. 1:167.

Narayanasamy, P., and Durairaj, P. 1971. A new blight disease of soybean. Madras Agric. J. 58:711-712.

Pande, A., and Rao, V. G. 1998. A Compendium of Fungi on Legumes from India. Scientific Publishers, Jodhpur, India.

Pimentel, I. C., Glienke-Blanco, C., Gabardo, J., Stuart, R. M., and Azevedo, J. L. 2006. Identification and colonization of endophytic fungi from soybean (*Glycine max* [L.] Merrill) under different environmental conditions. Braz. Arch. Biol. Technol. 49:705-711.

(Prepared by J. B. Sinclair; revised by G. L. Hartman)

Frogeye Leaf Spot

Frogeye leaf spot (sometimes called "Cercospora leaf spot") was first reported on soybean in 1915 in Japan and later in 1924 in the United States. The disease occurs everywhere soybean is grown, but it is most destructive in warm regions under warm, humid conditions. The use of resistant cultivars, pathogen-free seed, and fungicide applications have reduced the incidence of the disease and prevented widespread economic losses. Generally, yields from susceptible cultivars may be reduced 30% if extensive leaf blighting occurs. However, losses of 40% and more have been reported from Nigeria.

Symptoms and Signs

Frogeye leaf spot is primarily a disease of foliage, but stems, pods, and seeds may also be affected. Lesions are circular to angular and vary in size from less than 1 to 5 mm in diameter (Fig. 61). Lesions begin as dark, water-soaked spots (with or without lighter centers) and develop into brown spots surrounded by narrow, dark reddish-brown margins. As lesions age, the central areas become ash gray to light brown. On the lower surfaces of leaves, the spots are darker and have light- to dark-gray centers while sporulating; they also have clusters of conidiophores and thin, reddish-brown margins (Fig. 62). The nonsporulating, older lesions are light to dark brown and translucent; their white centers contain minute, dark stromata. Several of these lesions may coalesce to form larger, irregularly shaped spots. When lesions are numerous enough to cover approximately 30% of the leaf area, a blighting phase often occurs, in which leaves wither quickly and fall prematurely.

Stem infections appear late in the season and are less common and distinctive than foliar infections. Stem lesions are two to four times as long as they are wide and may encompass up to half of the stem. Young lesions are often red-brown and have narrow, dark-brown to black margins. Stem lesions have flattened or slightly sunken centers. As these lesions age, their centers become brown and then a pale, smoky gray. Minute, dark stromata (often bearing clusters of conidiophores and numerous conidia) form in older lesions, giving them an almost black appearance.

Lesions on pods are circular to elongate, slightly sunken, and reddish brown (Fig. 63). Older lesions become brown to light gray and usually have narrow, dark-brown borders. The causal fungus frequently grows through the pod wall into maturing seeds.

Seeds infected with the causal fungus usually develop conspicuous light- to dark-gray or brown areas, which vary in size from specks to large blotches that cover the entire seed coat. Some lesions display alternating bands of light and dark brown or gray, which may blend into each other. These lesions may appear as dark rings with light centers, similar to leaf lesions. The seed coat often cracks or flakes. As the discolored area on the seed increases in size, the seed germination rate decreases.

Fig. 61. Mostly circular spots, symptomatic of frogeye leaf spot (*Cercospora sojina*), on upper (left) and lower (right) leaf surfaces. (Courtesy J. B. Sinclair)

Causal Organism

Frogeye leaf spot is caused by *Cercospora sojina,* which is highly variable in colony color and rate of growth on culture media. Conidiophores arise in fascicles of two to 25 from thin stromata; they are light to dark brown, measure 4–6 × 52–120 µm, and have one to several septations with prominent bands (geniculations) and spore scars. Conidia are borne on the tips of conidiophores and pushed aside as the conidiophores continue their indeterminate growth. Generally, a single conidiophore produces one to three conidia.

Conidia are zero to 10 septate, hyaline when young, and elongate to fusiform, tapering toward the tip. The base of a conidium is usually rounded and bears circular scars that mark the attachment to the conidiophore. On infected leaves, conidia are 6–8 × 40–60 µm. The sizes and shapes of conidia and conidiophores vary with the substrate on which the fungus grows. Conidia of *C. sojina* are much shorter than those of *C. kikuchii* (see the previous section Cercospora Leaf Blight and Purple Seed Stain).

Conidia begin to germinate within 1 h in water, producing one to several germ tubes from the end cells and occasionally from the intervening cells. In culture, conidia frequently germinate in situ, forming short germ tubes that may in turn produce secondary conidia. Few conidia germinate after 6 months if kept dry and at room temperature, but they have been reported to withstand longer periods of drying.

Isolates sporulate best on V8 juice agar and lima bean agar. Seeds suspected of being infested with *C. sojina* can be identified by placing nonsurface-disinfested seeds on moist blotter paper or by placing disinfested seeds on agar. After 5–7 days,

Fig. 62. Sporulation of *Cercospora sojina,* the fungus that causes frogeye leaf spot. (Courtesy J. T. Yorinori)

Fig. 63. Reddish-brown lesions on pods, characteristic of frogeye leaf spot (*Cercospora sojina*). (Courtesy J. T. Yorinori)

the seeds can be examined under a dissecting microscope to observe whether fungal sporulation has occurred.

Many physiological races of *C. sojina* have been reported worldwide. The number of races reported in a given area varies because of the lack of standardized differential cultivars used in screening experiments. A review conducted in 2008 concluded that 93 isolates of *C. sojina* collected from Brazil, China, and the United States can be classified into 11 races based on their reactions on a set of 12 soybean differential cultivars. Cluster analyses of these isolates have identified some similarities among isolates from different countries, but diversity among isolates has also been found both within and among countries.

Disease Cycle and Epidemiology

C. sojina survives as mycelium in infected seeds and infested soybean residue. Germination of extensively discolored, infected seeds is usually poor, and infected seeds that do germinate may produce weak, stunted seedlings with lesions on the cotyledons. Sporulation on the cotyledon lesions and residue provides inoculum to infect young leaves. On inoculated plants, the first lesions are visible after 8–12 days, and if very moist conditions prevail, the first spores can be produced within the next 48 h. In warm, humid conditions, sporulation is profuse in the resulting lesions. Conidia are carried short distances by air currents and splashing rain, and under favorable conditions, secondary infections of leaves, stems, and pods may develop throughout the season. The link between seedling blight infection of the disease and the foliar infection process has yet to be determined.

Young leaves are infected more readily than older leaves, and the lesions that develop become progressively smaller and less uniform in size as the leaf ages. After the leaf has fully expanded, very few lesions develop, even with controlled inoculation under conditions ideal for invasion.

With adequate leaf moisture, new leaves become infected, and all the leaves on the plant may be affected. Alternatively, a layered pattern of heavily diseased leaves and leaves with few lesions can occur on the same plant if dry periods (which are unfavorable for infection) alternate with wet periods.

Initially, only seeds directly in contact with pod lesions become discolored, but discoloration may extend to other seeds as the pod matures. Seeds that are not directly infected may become contaminated with spores or mycelium during harvest.

Reactions of different cultivars vary from immunity to high susceptibility to a particular race of *C. sojina*. In the southern United States, the prevalent race or races at a given location may change dramatically from year to year. In northern China, however, the races at a given location tend to remain the same for several years; this consistency may result from the extremely limited long-distance seed movement in that region.

The single dominant genes *Rcs1, Rcs2,* and *Rcs3* have been recognized as conferring resistance to *C. sojina*. Resistance genes vary in the races to which they confer resistance. *Rcs3* conditions resistance to the known races found in the United States, and it also confers resistance to several races from Brazil and China that have been made available for testing.

Management

To prevent the development of frogeye leaf spot, adapted, resistant cultivars should be grown when possible. Commercial varieties may vary in the level of resistance and the source of genetic resistance. Commercially available seed may contain the *Rcs3* gene or combinations of *Rcs1* and *Rcs2*. High-quality seed that is free of the pathogen should be planted to avoid development of the seedling blight phase. If lower-quality seed is used, it should be treated with a fungicide that has activity against *C. sojina*.

Practices that encourage residue decomposition, such as tillage and crop rotation, will remove or reduce the source of inoculum available for subsequent soybean crops. If foliar fungicides are used, they should be applied at growth stages R2–R5. Fungicide timing, product choice, and application method all influence the efficacy of a fungicide for disease control.

Resistance of *C. sojina* to the quinone outside inhibitor (QoI) group of fungicides (Fungicide Resistance Action Committee [FRAC] group 11) has been confirmed in several states in the U.S. Midwest and South. The first report of chemical failure in this group was from Lauderdale County, Tennessee, in 2010. Repeated use of QoI fungicides (also known as "strobilurins") on susceptible cultivars in continuous soybean production has contributed to the development of resistance to this class of fungicides. To reduce selection pressure for resistance to QoI fungicides, growers should consider planting resistant cultivars and implement at least a 2-year rotation with other crops. If susceptible cultivars are selected and growers need to spray foliar fungicides to manage disease, they should select a product with an alternate mode of action from the QoI class.

Selected References

Arias, C. A. A., Yorinori, J. T., Toledo, J. F. F., and Kihl, R. A. S. 1996. Inheritance of resistance of soybean (*Glycine max* [L.] Merrill) to races 4 and 15 of frogeye leaf spot fungus (*Cercospora sojina* Hara). Braz. J. Genet. 19:295-304.

Kunwar, I. K., Singh, T., and Sinclair, J. B. 1985. Histopathology of mixed infections by *Colletotrichum truncatum* and *Phomopsis* spp. or *Cercospora sojina* in soybean seeds. Phytopathology 75:489-492.

Mengistu, A., Bond, J., Mian, R., Nelson, R., Shannon, G., and Wrather, A. 2011. Identification of soybean accessions resistant to *Cercospora sojina* by field screening, molecular markers, and phenotyping. Crop Sci. 51:1101-1109.

Mian, M. A. R., Missaoui, A. M., Walker, D. R., Phillips, D. V., and Boerma, H. R. 2008. Frogeye leaf spot of soybean: A review and proposed race designations for isolates of *Cercospora sojina* Hara. Crop Sci. 48:14-24.

Phillips, D. V., and Boerma, H. R. 1982. Two genes for resistance to race 5 of *Cercospora sojina* in soybeans. Phytopathology 72:764-766.

Singh, T., and Sinclair, J. B. 1985. Histopathology of *Cercospora sojina* in soybean seeds. Phytopathology 75:185-189.

(Prepared by K. A. Wise and M. E. Newman)

Fusarium Seed Rot and *Fusarium* Species Associated with Seed

Fusarium spp. may occur at high frequency in seed lots and are often associated with lower incidences of seed germination and seedling emergence. The first report of preharvest infection of soybean seeds by *Fusarium graminearum* was in the midwestern United States in 1986. A large proportion of seeds became infected in September of that year, just before harvest, after an unusually long period of warm, wet weather. Infected seeds were stained pink to red. The same year, *F. graminearum* and *F. sporotrichioides* were found in harvested seed in Canada. *F. graminearum* has also been found in seed in other soybean-producing countries, such as India.

Symptoms and Signs

Infected seeds may be moldy/chalky and/or light brown to pinkish-brown to reddish; germination from these seeds is delayed or reduced. Seedlings from infected plants do not emerge or are slow to emerge and unhealthy, and they do not compete with noninfected seedlings (Fig. 64).

Causal Organisms

Eleven *Fusarium* spp. have been isolated from soybean seeds: *F. acuminatum* (syn. *Gibberella acuminata*); *F. decemcellulare* (syn. *Albonectria rigidiuscula*); *F. equiseti* (syn. *G. intricans*); *F. graminearum* (syn. *G. zeae*); *F. incarnatum* (syn.

Fig. 64. Discoloration on seeds and seedlings, caused by *Fusarium graminearum.* Compare the diseased seedlings, which show seed rot and reduced vigor and growth, with the healthy seedling at the far left. (Reproduced, by permission, from Ellis et al., 2011)

F. semitectum); *F. lateritium* (syn. *G. baccata*); *F. oxysporum; F. solani* (syn. *Haemanectria haematococca*); *F. sporotrichioides; F. tricinctum* (syn. *G. tricincta*); and *F. verticillioides* (syn. *G. moniliformis*). All these species except *F. decemcellulare* and *F. lateritium* have been shown to be pathogenic on soybean, and in many cases, isolates within these species vary in pathogenicity when infecting soybean.

Disease Cycle and Epidemiology

In Brazil, strains of *F. graminearum* from soybean have been matched to lineages 1, 2, and 8; lineage 7 from the United States has also been shown to infect soybean. Infection of seed by *F. graminearum* is favored by freezing temperatures in the field.

F. semitectum has been reported as widespread in seeds in Mississippi. It has also been shown to infect pods.

Mycotoxin contamination is not usually an economic problem in soybean, but the mycotoxins zearalenone, deoxynivalenol (or vomitoxin), fusariotoxin, and HT-2 have been detected in seed infested with *F. graminearum* and *F. sporotrichioides.* Deoxynivalenol is found primarily in seeds stained red by *F. graminearum* and is concentrated in the seed coat. In Brazil, however, nivalenol is the major mycotoxin produced by strains from soybean. HT-2 toxin, produced primarily by *F. sporotrichioides,* can be detected in both stained and unstained seeds. *F. verticillioides* produces the mutagenic compound fusarin C on soybean seed under controlled conditions, and strains from soybean seeds produce fumonisins B1 and B2 under controlled conditions.

Management

Some degree of resistance to naturally occurring seed infection by *Fusarium* spp. has been observed in hard-seeded soybean lines. Seed treatment with specific fungicides, such as captan and fludioxonil, can reduce seed rot and seedling disease caused by seedborne *Fusarium* spp.

Selected References

Acuna, A., Lozano, M. C., Garcia, M. C., and Diaz, G. J. 2005. Prevalence of *Fusarium* species of the *Liseola* section on selected Colombian animal feedstuffs and their ability to produce fumonisins. Mycopathologia 160:63-66.

Avanzato, M. V. 2011. Characterization and interaction of *Pythium* and *Fusarium* spp. associated with soybean seed decay and root rot. Ph.D. thesis. University of Arkansas, Fayetteville.

Bacon, C. W., Marijanovic, D. R., Norred, W. P., and Hinton, D. M. 1989. Production of fusarin C on cereal and soybean by *Fusarium moniliforme.* Appl. Environ. Microbiol. 55:2745-2748.

Broders, K. D., Lipps, P. E., Paul, P. A., and Dorrance, A. E. 2007. Evaluation of *Fusarium graminearum* associated with corn and soybean seed and seedling disease in Ohio. Plant Dis. 91:1155-1160.

Clear, R. M., Nowicki, T. W., and Daun, J. K. 1989. Soybean seed discoloration by *Alternaria* spp. and *Fusarium* spp., effects on quality and production of fusariotoxins. Can. J. Plant Pathol. 11:308-312.

Ellis, M. L., Broders, K. D., Paul, P. A. and Dorrance, A. E. 2011. Infection of soybean seed by *Fusarium graminearum* and effect of seed treatments on disease under controlled conditions. Plant Dis. 95:401-407.

Jacobsen, B. J., Harlin, K. S., Swanson, S. P., Lambert, R. J., Beasley, V. R., Sinclair, J. B., and Wei, L. S. 1995. Occurrence of fungi and mycotoxins associated with field mold damaged soybeans in the Midwest. Plant Dis. 79:86-89.

Jordan, E. G., Manandhar, J. B., Thapliyal, P. N., and Sinclair, J. B. 1988. Soybean seed quality of 16 cultivars and four maturity groups in Illinois. Plant Dis. 72:64-67.

Martinelli, J. A., Bocchese, C. A. C., Xie, W., O'Donnell, K., and Kistler, H. C. 2004. Soybean pod blight and root rot caused by lineages of the *Fusarium graminearum* and the production of mycotoxins. Fitopatol. Bras. 29:492-498.

Miller, W. A., and Roy, K. W. 1982. Mycoflora of soybean leaves, pods, and seeds in Mississippi. Can. J. Bot. 60:2716-2723.

Osorio, J. A., and McGee, D. C. 1992. Effects of freeze damage on soybean seed mycoflora and germination. Plant Dis. 76:879-882.

Roy, K. W., and Rothnayake, S. 1997. Frequency of occurrence of *Fusarium pallidoroseum,* effects on seeds and seedlings, and associations with other fungi in soybean seeds and pods. Can. J. Plant Pathol. 19:188-192.

Wicklow, D. T., Bennett, G. A., and Shotwell, O. L. 1987. Secondary invasion of soybeans by *Fusarium graminearum* and resulting mycotoxin contamination. Plant Dis. 71:1146.

(Prepared by B. D. Nelson)

Leptosphaerulina Leaf Spot

Leptosphaerulina leaf spot was reported from India in 1972 and from the U.S. states of Maryland and Missouri in 1986 and 1987, respectively. Although the disease has since been mentioned, there are no focused reports about it or its causal fungus on soybean. Some reports have indicated that the disease may be geographically confined, given the lower temperature that is related to disease development (18°C optimal).

Symptoms include small, gray, circular, necrotic lesions, which appear on both young and old leaves. Later, the spots become irregularly shaped and dull white to gray in color.

The causal fungus, *Leptosphaerulina trifolii,* produces uniloculate ascostromata, which are ostiolate and dark brown, innate at first but later erumpent, and spherical (120–168 µm in diameter). Asci (36–39 × 64–80 µm) are black, obclavate to obovoid with truncate bases, and bitunicate. They arise in fascicles and are eight spored (with four to six spores in each locule). Ascospores (8–10 × 24–32 µm) are clavate to spindle shaped, hyaline, irregularly biseriate, and muriform; they usually have three transverse septa and one or two longitudinal septa.

L. trifolii affects a number of different species of leguminous plants.

Selected References

Anahosur, K. H., and Fazalnoor, K. 1972. Studies on follicular fungi of soybean crop. Indian Phytopathol. 25:504-508.

Graham, J. H., and Luttrell, E. S. 1961. Species of *Leptosphaerulina* on forage plants. Phytopathology 51:680-693.
Grybauskas, A. P. 1986. First report of soybean naturally infected with *Leptosphaerulina briosiana*. Plant Dis. 70:1159.
Li, X., and Yang, X. B. 2009. Similarity, pattern, and grouping of soybean fungal diseases in the United States: Implications for the risk of soybean rust. Plant Dis. 93:162-169.
Yang, X. B., and Feng, F. 2001. Ranges and diversity of soybean fungal diseases in North America. Phytopathology 91:769-775.

(Prepared by J. B. Sinclair; revised by G. L. Hartman)

Phomopsis Seed Decay

Phomopsis seed decay (PSD) can severely affect soybean seed quality and is a problem for many soybean growers. In the United States, the annual yield loss from PSD has been as high as 0.38 million metric tons, based on estimates from 1996 to 2007. Because of the hot and humid conditions that occurred in the U.S. South during the late season of 2009, PSD caused a total loss of more than 0.33 million metric tons across 16 states.

Symptoms and Signs

Infected soybean seeds display a range of symptoms, from being asymptomatic to severely shriveled, elongated, and cracked; they may also have a white, moldy appearance (Fig. 65). Affected seeds do not usually germinate or are slow to germinate. Seed infection results in pre- and postemergence damping-off, and under severe conditions, stands can be affected to the point of yield reduction.

Soybean pods can be infected at any time after being formed. The causal fungi colonize seed coats initially, followed by cotyledons and plumules. Mycelium invades the ovule and developing seeds through the funiculus and hilum. Within the seed, the fungi colonize all tissues of the seed coat and cotyledons and eventually the radicle and plumule.

Causal Organisms

Phomopsis seed decay can be caused by several different fungi in the *Diaporthe/Phomopsis* complex, but *D. longicolla* (syn. *P. longicolla*) is the primary cause. Colonies of *D. longicolla* on potato dextrose agar (PDA) are floccose, dense, and white and occasionally have greenish-yellow areas (Fig. 66). The undersides of colonies are pigment free, except for the large, black, spreading stromata. Conidiomata are pycnidial, black, stromatic, solitary or aggregated, and unilocular or multilocular; they have prominent necks (more than 200 µm long) and apical ostioles. Locules are uniostiolate or multiostiolate, globose, and up to 500 µm wide. Conidiophores are hyaline, usually branched, septate, and 1–4 × 3.5–24.0 µm. Conidiogenous cells are hyaline, filiform, and phialidic. Alpha-conidia are hyaline, ellipsoidal to fusiform, guttulate, and 1.5–3.5 × 5–9 µm; they are one celled and biguttulate. Beta-conidia (which rarely form) are hyaline, filiform, and hamate.

Studies of the aggressiveness of *D. longicolla* isolates from soybean and velvetleaf (*Abutilon theophrasti*) in eight U.S. states have revealed significant differences. In addition, a robust and reliable *Agrobacterium tumefaciens*-mediated transformation system has been developed as a foundation for future functional genomic research on *D. longicolla*.

Polymerase chain reaction- (PCR-) based molecular detection using specific primers is species specific and sensitive in detecting the fungi associated with seed decay. Even so, other methods are often used, including immunoassays and direct plating of seeds on agar media.

Disease Cycle and Epidemiology

Diseased seeds are an important factor in the long-range dissemination of *D. longicolla,* and infested crop debris and weeds commonly found in soybean fields are the major sources of primary inoculum. Soybean pods can become infected any time after they are formed, but significant seed infection does not occur before physiological maturity. Seed infection tends to be more severe when the harvest is delayed, when early maturing cultivars are grown, and when crops are grown in regions in which warm, humid conditions prevail at harvest time.

Seeds from pods at the bottom of the plant often have more Phomopsis seed decay than those at the top. More seed decay also occurs in plants that are deficient in potassium, infected with one or more viruses (e.g., *Soybean mosaic virus*), and heavily attacked by insects (see the section *Soybean mosaic virus* later in Part I). When harvest is delayed under wet conditions, seeds may be infected throughout the plant (Fig. 67). Alternating periods of wet and dry weather favor pod deterioration and splitting and facilitate infection of mature seeds. The fungus secretes enzymes, which degrade the seed coat proteins. Seed infection may be greater in densely populated fields and fields with excessive lodging of plants. Seed infection with *D. longicolla* decreases significantly after 1 or more years of storage; thus, seeds relatively free of the fungus can be obtained via long-term storage.

D. longicolla has been reported on other plant tissues, including roots and stems. However, the importance of these infections

Fig. 65. Symptoms of soybean seed decay, with some seeds supporting mycelial growth of *Diaporthe longicolla.* (Courtesy S. Li)

Fig. 66. Colonies of *Diaporthe longicolla,* the primary cause of Phomopsis seed decay, on acidified potato dextrose agar. (Courtesy S. Li)

Fig. 67. Pods and seeds infected by *Diaporthe longicolla,* the primary cause of Phomopsis seed decay, which occurred after plants were inoculated in the field. (Courtesy S. Li)

and the epidemiology related to these infections are not well understood.

Management

Few strategies are available for managing Phomopsis seed decay, and the strategies that have been applied have not completely protected soybean against the disease. The use of genetic resistance is the most practical and effective approach to controlling PSD. Different levels of resistance have been identified in diverse soybean germplasm, including plant introductions and cultivars. Resistance sources include Arksoy, Delmar, MO/PSD-0259, OX615, plant introduction (PI) 80837, PI 82264, PI 181550, PI 200501, PI 200510, PI 209908, PI 227687, PI 229358, PI 360841, and PI 417479. In 2011, eight additional sources of PSD-resistant germplasm (PI 424324B, PI 458130, PI 506647, PI 567270C, PI 567381B, PI 567521, PI 567635, and PI 594858A) were identified from a 4-year field screening trial in Mississippi. Breeders have been successful in developing breeding lines with low levels of seed infection derived from cultivars Arksoy, PI 80837, PI 200501, PI 360841, and PI 417479.

Additional management options include rotating soybean with maize or another nonlegume crop and plowing down infested residues in fields in which soil erosion is not a concern. Other recommendations are to plant high-quality seed that is relatively free of the pathogen and to apply a fungicide seed treatment if the seed infection level is greater than 15%. If the risk of seed infection is high, a foliar fungicide should be applied, as well.

If possible, growers should select planting times that allow maturation during a dry period. Harvesting soybean seeds promptly at maturity is recommended, as well.

Selected References

Anderson, T. R., Buzzell, R. I., and Buttery, B. R. 1995. Incidence of pod and seed infection in two soybean lines differing in resistance to Phomopsis seed decay. Can. J. Plant Sci. 75:543-545.

Balducchi, A. J., and McGee, D. C. 1987. Environmental factors influencing infection of soybean seeds by *Phomopsis* and *Diaporthe* species during seed maturation. Plant Dis. 71:209-212.

Chen, X., Pan, R., Xu, D., Ji, C., and Deng, M. 2013. First report of soybean stem blight caused by *Phomopsis longicolla* in Guangdong Province, Southern China. Plant Dis. 97:844.

Cross, C., Wrather, A., Fothergill, K., Tindall, K., Shannon, G., Li, S., Shumway, C., and Rupe, J. 2012. Effect of lactofen, azoxystrobin, and genotypes on charcoal rot, Phomopsis seed decay, and pod and stem blight in soybean. Plant Dis. 96:1154-1158.

Hobbs, T. W., Schmitthenner, A. F., and Kuter, G. A. 1985. A new *Phomopsis* species from soybean. Mycologia 77:535-544.

Jackson, E. W., Feng, C., Fenn, P., and Chen, P. 2009. Genetic mapping of resistance to Phomopsis seed decay in the soybean breeding line MO/PSD-0259 (PI562694) and plant introduction 80837. J. Hered. 100:777-783.

Li, S., Bradley, C. A., Hartman, G. L., and Pedersen, W. L. 2001. First report of *Phomopsis longicolla* from velvetleaf causing stem lesions on inoculated soybean and velvetleaf plants. Plant Dis. 85:1031.

Li, S., Hartman, L. G., and Boykin, D. L. 2010. Aggressiveness of *Phomopsis longicolla* and other *Phomopsis* spp. on soybean. Plant Dis. 94:1035-1040.

Li, S., Ridenour, J. B., Kim, H., Hirsch, R. L., Rupe, J. C., and Bluhm, B. H. 2013. *Agrobacterium tumefaciens*-mediated transformation of the soybean pathogen *Phomopsis longicolla*. J. Microbiol. Methods 92:244-245.

Li, S., Smith, J., and Nelson, R. 2011. Resistance to Phomopsis seed decay identified in maturity group V soybean plant introductions. Crop Sci. 51:2681-2688.

Minor, H. C., Brown, E. A., and Zimmerman, M. S. 1995. Developing soybean varieties with genetic resistance to *Phomopsis* spp. J. Am. Oil Chem. Soc. 72:1431-1434.

Rupe, J. C., and Ferriss, R. S. 1986. Effects of pod moisture on soybean seed infection by *Phomopsis* sp. Phytopathology 76:273-277.

Santos, J. M., Vrandecic, K., Cosic, J., Duvnjak, T., and Phillips, A. J. L. 2011. Resolving the *Diaporthe* species occurring on soybean. Persoonia 27:9-19.

Sinclair, J. B. 1993. Phomopsis seed decay of soybeans—A prototype for studying seed disease. Plant Dis. 77:329-334.

Thomison, P. R., Kenworthy, W. J., and McIntosh, M. S. 1990. Phomopsis seed decay in soybean isolines differing in stem termination, time of flowering, and maturity. Crop Sci. 30:183-186.

Wrather, J. A., Shannon, J. G., Stevens, W. E., Sleper, D. A., and Arelli, A. P. 2004. Soybean cultivar and fungicide effects on *Phomopsis* sp. seed infection. Plant Dis. 88:721-723.

Zhang, A. W., Hartman, G. L., Riccioni, L., Chen, W. D., Ma, R. Z., and Pedersen, W. L. 1997. Using PCR to distinguish *Diaporthe phaseolorum* and *Phomopsis longicolla* from other soybean fungal pathogens and to detect them in soybean tissues. Plant Dis. 81:1143-1149.

(Prepared by S. Li, P. Chen, and G. L. Hartman)

Phyllosticta Leaf Spot

Phyllosticta leaf spot (also called "Phyllosticta leaf blight") was reported in Italy in 1900 and in the United States in 1927. Subsequently, it has been observed in most soybean production areas throughout the world. In the United States, the disease occurs throughout the "soybean belt" (which includes Illinois, Indiana, Iowa, Minnesota, Missouri, Nebraska, and Ohio), and it is most common in areas in which cool, moist conditions favor disease development. The disease may cause severe defoliation early in the season but is generally considered of minor importance.

Symptoms and Signs

Plants of all ages are affected by Phyllosticta leaf spot, and lesions form primarily on leaves. Leaf spots are round to oval, irregularly shaped or V-shaped, and up to 2 cm in diameter (Fig. 68). Usually, only the first few trifoliolate leaves are affected. In young plants, infection starts at the leaf margin and progresses inward, forming a roughly V-shaped area. The spots are pale green at first and become dull gray to tan; they have narrow, dark-brown or purplish borders. Numerous small, black, speck-sized pycnidia form in older lesions.

The causal fungus may progress from the leaf blades into the petioles and then to the stipules and stem tissues at the leaf scar. Superficial lesions 1–4 mm wide and up to 12 mm long sometimes form on the petioles, stems, and pods (Fig. 69). These lesions are light gray, tan, or brownish with narrow, brown or purplish borders and may extend along the stem.

Lesions on pods are circular with reddish margins. The seeds beneath pod lesions may become infected.

Causal Organism

The causal fungus is *Pleosphaerulina sojicola* (syn. *Phyllosticta sojicola*). It produces amphiphyllous, globose pycnidia (which resemble those of *Ascochyta*); they are 75–200 μm in diameter and have ostioles 10–20 μm across. Conidia (1.9–3.5 × 4–10 μm) are one celled, narrowly elliptical with rounded ends, and hyaline; they have inconspicuous guttulae.

P. sojicola is seedborne in soybean.

Fig. 68. Lesions of various shapes and sizes of Phyllosticta leaf spot, caused by *Pleosphaerulina sojicola*. (Courtesy J. B. Sinclair)

Fig. 69. Superficial lesions on pods, characteristic of Phyllosticta leaf spot, caused by *Pleosphaerulina sojicola*. (Courtesy J. B. Sinclair)

Disease Cycle and Epidemiology

It is not uncommon to observe Phyllosticta leaf spot early in the season, soon after the unifoliolates fully expand. The disease may progress and be observed later in the season, as well, as the plant produces new leaves that are not infected.

Management

To prevent the development of Phyllosticta leaf spot, growers should plant high-quality, pathogen-free seeds. Other effective management strategies include rotating soybean with a nonhost crop and deep plowing crop residues in fields where soil erosion is not a concern.

Selected References

Crall, J. M. 1948. Defoliation of soybeans in southeast Missouri caused by *Phyllosticta glycineum*. Plant Dis. Rep. 32:184-186.

Irinyi, L., Kövics, G. J., and Sándow, E. 2009. Taxonomical re-evaluation of *Phoma*-like soybean pathogenic fungi. Mycol. Res. 113:249-260.

Jehle, R. A., Jenkins, A. E., Kreitlow, K. W., and Sherwin, H. S. 1952. An outbreak of Phyllosticta canker and leaf spot of soybeans in Maryland. Plant Dis. Rep. 36:155-158.

Kövics, G. J., De Gruyter, J., and van der Aa, H. A. 1999. *Phoma sojicola* comb. nov. and other hyaline-spored coelomycetes pathogenic on soybean. Mycol. Res. 103:1065-1070.

Olive, L. S., and Walker, E. A. 1944. A severe leafspot of soybean caused by *Phyllosticta sojaecola*. Plant Dis. Rep. 28:1122-1123.

Vernetti, F. J., Tomm, G. O., and Casela, C. R. 1985. Occurrence *of Mycosphaerella phaseolorum, Phyllosticta glycineum* and *Colletotrichum truncatum* in Mozambique. Fitopatol. Bras. 10:143-145.

Walters, H. J., and Martin, K. F. 1981. *Phyllosticta sojaecola* on pods of soybeans in Arkansas. Plant Dis. 65:161-162.

(Prepared by J. B. Sinclair; revised by G. L. Hartman)

Pod and Stem Blight

Pod and stem blight was first observed in the United States in 1920 and is now endemic throughout most soybean-growing areas of the world. When soybeans mature under warm, wet conditions and harvest is delayed, seed decay associated with pod and stem blight may lower seed viability.

Symptoms and Signs

Symptoms of pod and stem blight are most commonly observed on stems, but other tissues can be symptomatic, because the causal fungus infects all plant parts. When the fungus is found on seeds, the symptoms are not distinguishable from those of Phomopsis seed decay (see the previous section Phomopsis Seed Decay). When seeds are infected, lesions on the cotyledons of seedlings may form; they are bright red or brown and range in size from pinpoints to covering an entire cotyledon. The seed coat frequently remains attached after seedling emergence. Small, reddish-brown streaks up to 1.5 cm long occur on the hypocotyl at or below the soil line.

Seedling blight may occur from a seed infection. Buds may be deformed or completely decayed. Some reports have indicated that premature death of seedlings may occur under certain environmental conditions, causing substantial losses in yields.

The fungus that causes pod and stem blight initially infects the host early in the growing season without producing distinctive symptoms. It is not clear what affect, if any, a latent infection has on yield. The pathogen can be isolated from nonsymptomatic tissues by plating surface-disinfested tissue samples on culture media. Another technique for getting the fungus to sporulate on the tissue is to kill the tissue with desiccant herbicide and then incubate the sample pieces in a moist environment to allow the fungus to produce fruiting structures. The pathogen can first be detected in the field as pycnidia on the petioles of abscised leaves or on detached branches.

Under field conditions, no definite leaf or stem lesions are produced, yet signs of the pathogen predominate. Pycnidia can be found in dead tissues, such as *Phytophthora*-induced lesions, tissues damaged by hail injury, and plants killed by other pathogens. During wet seasons, latent infections result in pycnidia forming simultaneously over entire plants after they mature. Under dry conditions, however, pycnidia are confined to limited areas on the stems near the soil and generally cluster close to the nodes. Pycnidia are also scattered on dry, poorly developed pods. Not all infected pods produce pycnidia, but mature pods with pycnidia frequently contain infected seeds.

Dead stems may be covered with speck-sized pycnidia, usually arranged linearly (Fig. 70). The pycnidia may also be limited to small patches (generally near the nodes) and are sometimes confused with stem canker disease (see the section Stem Canker later in Part I).

Causal Organism

Colonies of *Diaporthe sojae* (syn. *Phomopsis phaseoli*) on acidified (pH 5) potato dextrose agar (PDA) (28°C) are floccose and ropy; they turn from white to tan to brown as the culture ages. The undersides of colonies are tan to dark brown and have black, pulvinate stromata. Conidiomata are pycnidial, black, stromatic, solitary or aggregated, and usually unilocular; they have no beaks or beaks less than 200 µm long, plus apical ostioles. Locules are uniostiolate to multiostiolate, lenticular, and up to 350 µm wide. Conidiophores are simple phialides; they are hyaline and up to 1.5–2.0 × 20 µm. Two types of conidia are produced (Fig. 71): Alpha-conidia are hyaline, usually fusiform, guttulate, and 1.5–3.5 × 5.5–10.5 µm, and beta-conidia (the more common type) are hyaline, filiform, and hamate.

Perithecia are produced on old agar cultures incubated in light or on overseasoned soybean stems. Mature perithecia are nearly spherical and slightly flat at the base (148–282 × 185–346 µm). Perithecia have long, tapered beaks (60–100 × 60–150 µm), which are usually solitary (unlike those of *D. caulivora* and *D. phaseolorum* var. *meridionalis,* which are clustered) (see the section Stem Canker). Asci (3.3–10.9 × 35–51 µm) are elongate and clavate and dissolve before ascospore liberation. In addition, asci are unitunicate and have distinct apical rings; they are released in a viscous fluid that oozes out ostiole tips.

Little is known about how ascospores are dispersed; the ascus walls disintegrate rapidly, leaving free ascospores in extrusion droplets. Ascospores are similar in shape to the alpha-conidia (fusiform–elliptical) but larger (2–6 × 9–13 µm) and bicellular. Ascospores are biguttulate in both cells; a single germ tube may be produced at the apex of each cell.

In addition to the pod and stem blight pathogen, fungi associated with dead stems include species of *Alternaria, Colletotrichum, Diaporthe, Phoma,* and *Septoria.*

Fig. 70. Pod and stem blight nodal lesion (left) and speck-sized pycnidia of *Diaporthe sojae* arranged linearly up and down the stem (right). (Courtesy J. B. Sinclair)

Disease Cycle and Epidemiology

D. sojae overseasons as dormant mycelium in soybean and other host debris and in infected seeds. Pycnidia are produced on overseasoned debris and on petioles of the current year's abscised leaves. Perithecia are produced in early summer on overseasoned stem debris.

The fungus colonizes plant tissue within 2 cm of the point of infection until the plant begins to senesce. At that time, the fungus spreads to tissue approximately 5 cm from the point of infection. Progressive spread in the plant is caused by infection from conidia that are dispersed by splashing water. Both alpha-conidia and ascospores can be splashed onto plants and initiate latent infections.

Only infections initiated in pods result in seed decay. Most seed infections occur during or after the yellow pod stage (R7). Prolonged wet periods during pod development and maturation and warm temperatures (above 20°C) favor the spread of the fungus from the pod to the seeds.

Management

The following plant introductions (PIs), with their corresponding maturity groups, are reported to be resistant to *D. sojae:* 80837 (IV), 204331 (VIII), 205089 (IV), 205907 (VIII), 205908 (VIII), 205912 (VIII), 259539 (VIII), 279088 (VIII), 341249 (IX), 360835 (II), 360841 (III), 385942 (IV), 417479 (IV), and 423903 (IV). Two complementary dominant genes control resistance in PI 417479.

Other management options include rotating with a nonhost crop (e.g., maize), burying crop residue (where applicable), and applying a foliar fungicide during the middle reproductive growth stages.

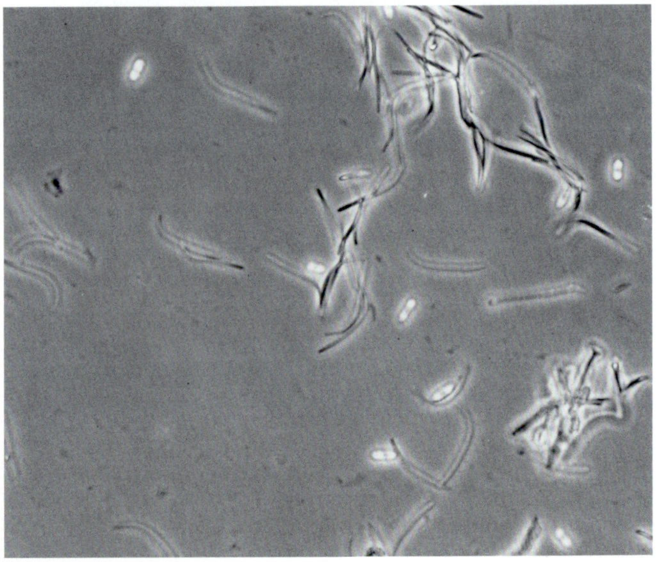

Fig. 71. Alpha-conidia (one-celled and biguttulate) and beta-conidia (elongate, filiform, and hamate) of *Diaporthe sojae,* which causes pod and stem blight. (Courtesy F. D. Tenne)

Selected References

Fernandez, F. A., and Hanlin, R. T. 1996. Morphological and RAPD analyses of *Diaporthe phaseolorum* from soybean. Mycologia 88:425-440.

Garzonio, D. M., and McGee, D. C. 1983. Comparison of seeds and crop residues as sources of inoculum for pod and stem blight of soybeans. Plant Dis. 67:1374-1376.

Gomes, R. R., Glienke, C., Videira, S. I. R., Lombard, L., Groenewald, J. Z., and Crous, P. W. 2013. *Diaporthe:* A genus of endophytic, saprobic and plant pathogenic fungi. Persoonia 31:1-41.

Hall, R., and Xue, A. G. 1995. Effectiveness of fungicidal seed treatments applied to smooth and shrivelled soybean seeds contaminated with *Diaporthe phaseolorum.* Phytoprotection 76:47-56.

Hilty, J. W., and Goodfellow, P. J. 1988. Evidence of toxin production by *Phomopsis phaseoli* (Desam.) Sacc. and its effect on germinating soybean seed. Seed Sci. Technol. 16:639-645.

Kulik, M. M. 1984. Symptomless infection, persistence, production of pycnidia in host and non-host plants by *Phomopsis batatae, Phomopsis phaseoli,* and *Phomopsis sojae,* and the taxonomic implications. Mycologia 76:274-291.

Kulik, M. M. 1988. Observations by scanning electron and bright-field microscopy on the mode of penetration of soybean seedlings by *Phomopsis phaseoli.* Plant Dis. 72:115-118.

Von Qualen, R. H., Abney, T. S., Huber, D. M., and Schrieber, M. M. 1989. Effect of rotation, tillage, and fumigation on premature dying of soybeans. Plant Dis. 73:740-744.

(Prepared by M. M. Kulik and J. B. Sinclair;
revised by G. L. Hartman)

Powdery Mildew

Powdery mildew was first reported on soybean in Germany in 1921 and has since been reported in Argentina, Australia, Brazil, Canada, China, India, Japan, Korea, Peru, Puerto Rico, South Africa, Venezuela, Vietnam, and the United States. The disease is common in greenhouse-grown plants and was reported in many fields in the U.S. Southeast and Midwest during the 1970s. In 1974, severe outbreaks were reported in the states of Delaware, Georgia, Illinois, Iowa, and Wisconsin. Yield losses of 10–35% have been recorded in susceptible cultivars.

Fig. 72. White, powderlike patches composed of mycelia and conidia of an *Erysiphe* sp., which causes powdery mildew. (Courtesy G. L. Hartman)

Symptoms and Signs

Powdery mildew is characterized by the formation of white, powderlike patches (composed of mycelia and conidia) on cotyledons, stems, pods, and leaf surfaces, which merge to cover all aboveground plant parts (Fig. 72). Different cultivars may develop different symptoms, such as chlorosis, asymptomatic green islands, rusty patches, defoliation, and combinations of these. Alternatively, the fungus may grow luxuriantly without producing visible symptoms. Some cultivars have susceptible reactions as seedlings but demonstrate resistance as mature plants. Chlorotic spots and veinal necrosis indicate a hypersensitive reaction.

Powdery mildew colonies may become darkened when parasitized by *Ampelomyces quisqualis.*

Causal Organisms

The obligate causal fungi are *Erysiphe diffusa* and *E. glycines*. Cleistothecia of *E. diffusa* (Fig. 73) may not be present on all soybean cultivars but develop abundantly in blue lupine (*Lupinus angustifolius*). Immature cleistothecia (before development of appendages) are light yellow to tan. At maturity, they are dark to rusty brown and have unbranched appendages. Mature cleistothecia are 55–126 µm in diameter, hemispherical, and black; each has four to 50 appendages, the lengths of which are 1.5–7.0 times the diameter of the cleistothecium (Fig. 74).

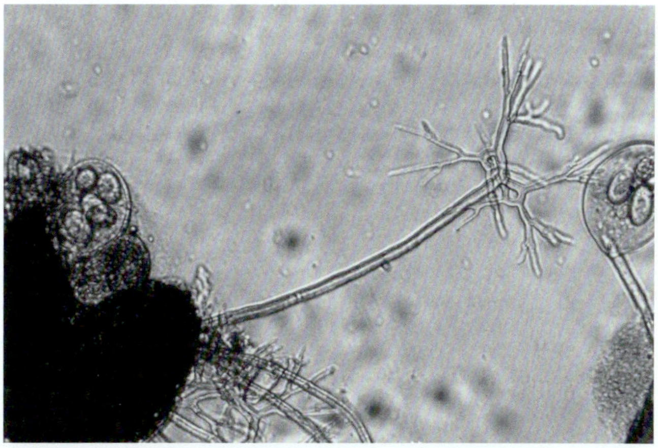

Fig. 73. Cleistothecium of *Erysiphe diffusa,* the cause of powdery mildew, with an escaping ascus containing ascospores and a dichotomously branched appendage. (Courtesy J. S. Mignucci)

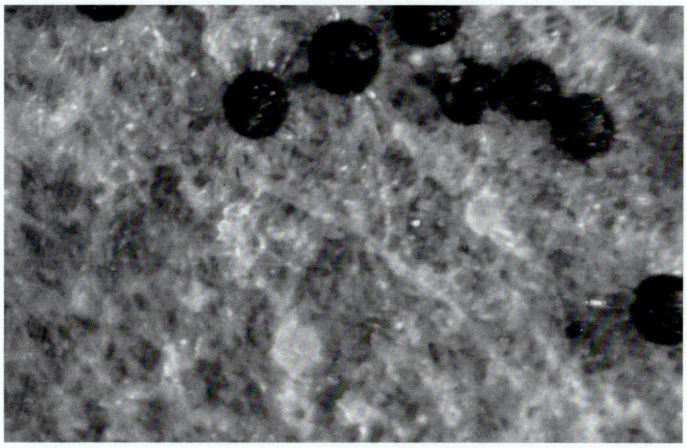

Fig. 74. Cluster of cleistothecia and mycelia of *Erysiphe diffusa,* the powdery mildew pathogen, on a soybean leaflet. (Courtesy J. S. Mignucci)

Each appendage is thick walled, hyaline to pale brown at the base, and dichotomously or subdichotomously branched three to five times at the apex. Multiple pyriform asci with short attachment stalks are released through broken cleistothecial walls. Each ascus contains up to six yellow, ovoid ascospores, measuring 9 × 18 µm. *E. glycines* commonly produces cleistothecia on soybean.

The fungi that cause powdery mildew produce mycelia and conidia on plant surfaces and intracellular haustoria in epidermal cells. Large numbers of conidia (17.1–21.1 × 27.7–54.1 µm) are produced in chains borne on short, simple conidiophores (Fig. 75). Conidia are slightly barrel shaped and have flattened ends. Dispersal of each fungus and rapid development and spread of the disease in the field are highly dependent on the asexual stage.

Disease Cycle and Epidemiology

Powdery mildew usually develops under cooler than normal conditions, when photosynthesis and transpiration are reduced. Under controlled conditions, temperatures between 18 and 24°C favor disease development, and the disease may be arrested at temperatures lower than 15°C and higher than 30°C. Leaf wetness of 8 h also enhances disease development, whereas leaf wetness of either 0 or 24 h drastically reduces infection. Susceptible greenhouse-grown plants often succumb to powdery mildew.

The causal fungi also infect bean, cowpea, pea, mung bean (*Vigna radiata*), and some wild *Glycine* spp., as well as some species in the Caprifoliaceae and the Solanaceae.

Management

In areas in which the risk for powdery mildew is high, resistant cultivars should be planted, if available. Resistance is governed by a dominant locus, *Rmd,* on chromosome 16, which has a number of alleles. Soybean resistance to powdery mildew is controlled by three alleles at the *Rmd* locus. Adult plant resistance is controlled by the gene *Rmd,* whereas *Rmd-c* and other single dominant genes found in plant introduction (PI) 243540 and PI 567301B provide both seedling and adult plant resistance. PI 243540 also confers resistance to *Aphis glycines,* the soybean aphid, with a single dominant gene, *Rag2*. Other sources of resistance include Burlison, Columbia, Sarosoy, PI 86150, and PI 90138. There appears to be a potential link to the *Rmd* locus with a gene providing resistance to the brown stem rot fungus, and additional sources of resistance have been reported from Brazil.

Application of foliar fungicides can control powdery mildew. It is not recommended, however, because the disease is usually not severe enough to warrant this treatment.

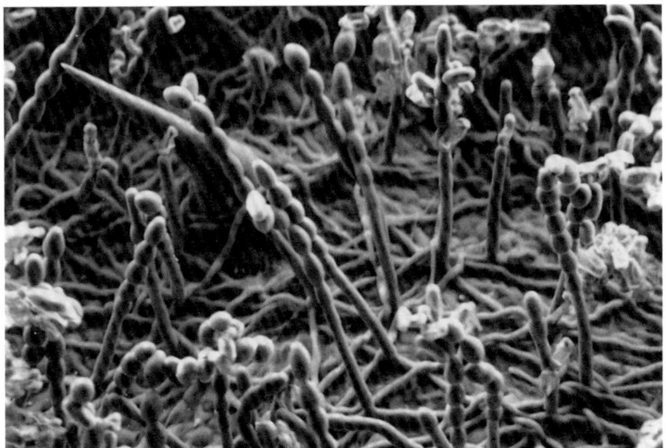

Fig. 75. Conidiophores and conidia of *Erysiphe diffusa,* which causes powdery mildew. (Courtesy N. M. Kokalis)

Selected References

Alves, M. D. C., Pozza, E. A., Costa, J. D. C. B., Ferreira, J. B., Araújo, D. V. D., Carvalho, L. G. D., Silva, F. M. D., and Sanches, L. 2011. Intensity of powdery mildew in soybean under changes of temperature and leaf wetness. Pages 183-192 in: Soybean Physiology and Biochemistry. H. El-Shemy, ed. InTech, Rijeka, Croatia.

Braun, U., and Cook, R. T. A. 2012. Taxonomic Manual of the *Erysiphales* (Powdery Mildews). CBS Biodivers. Ser. 11. CBS-KNAW Fungal Biodiversity Centre, Royal Netherlands Academy of Arts and Sciences, Utrecht.

Gonçalves, E. C. P., Mauro, A. O. D., and Centurion, M. A. P. C. 2002. Genetics of resistance to powdery mildew (*Microsphaera diffusa*) in Brazilian soybean populations. Genet. Mol. Biol. 25:339-342.

Jun, T.-H., Mian, M. R. A., Kang, S.-T., and Michel, A. P. 2012. Genetic mapping of the powdery mildew resistance gene in soybean PI 567301B. Theor. Appl. Genet. 125:1159-1168.

Kang, S.-T., and Mian, M. A. R. 2010. Genetic map of the powdery mildew resistance gene in soybean PI 243540. Genome 53:400-405.

Lohnes, D. G., and Bernard, R. L. 1992. Inheritance of resistance to powdery mildew in soybeans. Plant Dis. 76:964-965.

Lohnes, D. G., and Nickell, C. D. 1994. Effects of powdery mildew alleles *Rmd-c, Rmd,* and *rmd* on yield and other characteristics in soybean. Plant Dis. 78:299-301.

McTaggart, A. R., Ryley, M. J., and Shivas, R. G. 2012. First report of the powdery mildew *Erysiphe diffusa* on soybean in Australia. Australas. Plant Dis. Notes 7:127-129.

Mignucci, J. S., and Lim, S. M. 1980. Powdery mildew development on soybeans with adult-plant resistance. Phytopathology 70:919-921.

Phillips, D. V. 1984. Stability of *Microsphaera diffusa* and the effect of powdery mildew on yield of soybean. Plant Dis. 68:953-956.

Takamatsu, S., Shin, H.-D., Paksiri, U., Limkaisang, S., Taguchi, Y., Binh, N. T., and Sato, Y. 2002. Two *Erysiphe* species associated with recent outbreak of soybean powdery mildew: Results of molecular phylogenetic analysis based on nuclear rDNA sequences. Mycoscience 43:333-341.

(Prepared by G. L. Hartman)

Red Leaf Blotch

Red leaf blotch (also known as "Dactuliophora leaf spot," "Pyrenochaeta leaf blotch," and "Pyrenochaeta leaf spot") has been reported from central to southern Africa. In Zimbabwe and Zambia, it has caused estimated yield losses of 10–50%.

Symptoms and Signs

Symptoms of red leaf blotch include lesions on foliage, petioles, pods, and stems (Figs. 76 and 77). Lesions are often associated with primary leaf veins and may occur initially on unifoliolate leaves as dark-red to brown, circular to angular spots, 1–3 mm in diameter. When trifoliolate leaves are fully expanded, the lesions are dark red on the upper surfaces of leaves and reddish brown on the lower surfaces. Lesions enlarge and coalesce to form large necrotic blotches up to 2 cm in diameter. Within older lesions, sclerotia develop primarily on the lower surfaces of leaves and pycnidia develop primarily on the upper surfaces.

Heavily diseased plants defoliate and senesce prematurely.

Causal Organism

The fungus that causes red leaf blotch was renamed *Coniothyrium glycines* in 2012. Previous names have included *Phoma glycinicola, Dactuliochaeta glycines, Dactuliophora glycines, and Pyrenochaeta glycines.*

C. glycines resides in the soil but is easily isolated and grows on a wide range of culture media. It is known to infect glycine (*Neonotonia wightii*), a perennial legume that inhabits the woodlands and grasslands of southern Africa, and many other legumes experimentally. The fungus is known to occur in Cameroon, Ethiopia, Malawi, Nigeria, the Republic of the Congo, Rwanda, Uganda, Zambia, and Zimbabwe. It has not been reported in the United States.

Within lesions in the leaves of soybean and the perennial legume glycine, pycnidia are 87–298 μm in diameter and have aseptate to multiseptate setae clustered mostly around the ostioles. Conidiogenous cells are monophialidic and ampulliform and formed from the inner cells of the pycnidial wall (Fig. 78). Pycnidiospores are ellipsoidal, one celled, and 1–3 × 4–8 μm. Sclerotia are mostly spherical and dark brown to black; they have setae 5–36 μm long and range in size from 96 to 357 μm in diameter.

Spore germination is optimal at temperatures of 20–25°C and reduced at 5 and 35°C. Sclerotia germinate to form either pycnidia or secondary sclerotia (Fig. 79). Some sclerotia have been found to germinate after 18 months at 5°C and may be tolerant to high temperatures. The fungus produces sclerotia on substrates with organic nitrogen sources, such as oat and wheat seeds, wood, and detached leaves of some legume species.

Disease Cycle and Epidemiology

Sclerotia that reside in the upper layer of soil and that land on leaf surfaces may germinate to form pycnidia and/or mycelia. Pycnidia, in turn, form infectious conidia. Within older lesions, both pycnidia and sclerotia develop (Fig. 80). The soybean genotype affects the severity of disease and the number of sclerotia in the soil after harvest. Sclerotia from decaying leaf debris of infected soybeans and other legume hosts may reside in the upper soil matrix. Infection occurs when rain splashes soilborne sclerotia or conidia from pycnidia onto the leaf surfaces, where they germinate and infect the plant.

After the plant becomes infected, leaf lesions and blotches develop that vary in size but can be up to 2 cm in diameter.

Fig. 76. Lesions symptomatic of red leaf blotch (*Coniothyrium glycines*) on the underside of a leaf. (Reprinted from Hartman et al., 1987)

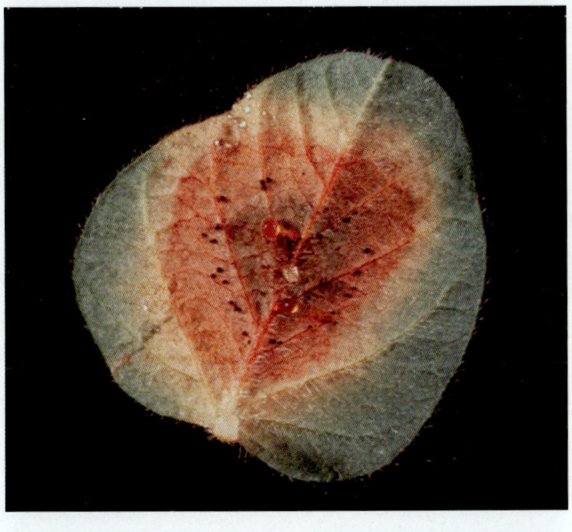

Fig. 77. Lesion of red leaf blotch (*Coniothyrium glycines*) on an inoculated leaf. Note the small, black sclerotia within the lesion. (Reprinted from Hartman et al., 1987)

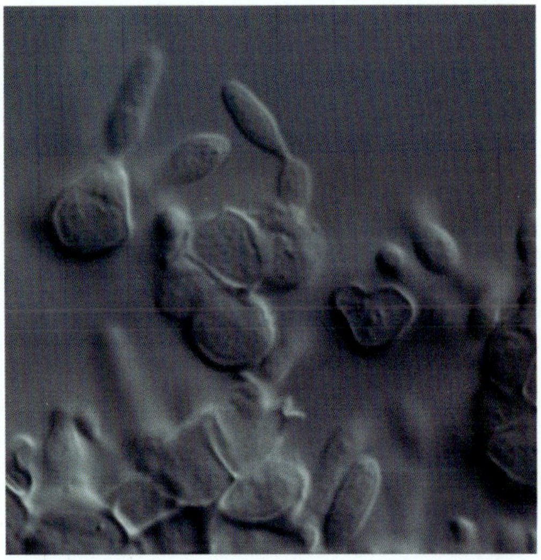

Fig. 78. Conidia and conidiogenous cells of *Coniothyrium glycines,* the pathogen that causes red leaf blotch. (Courtesy G. L. Hartman)

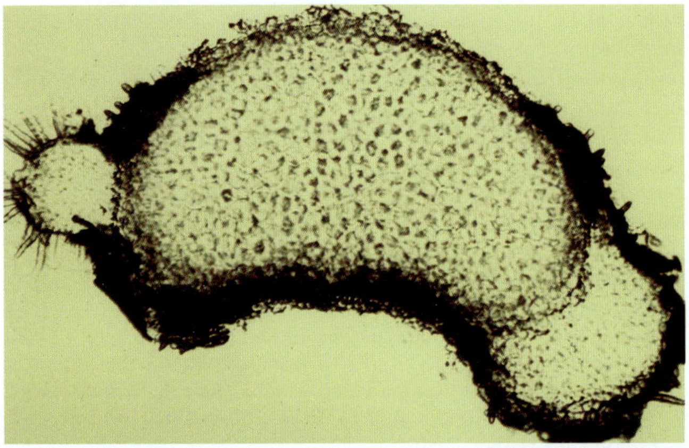

Fig. 79. Sclerotium of *Coniothyrium glycines,* which causes red leaf blotch, germinating to form a secondary sclerotium. (Courtesy G. L. Hartman)

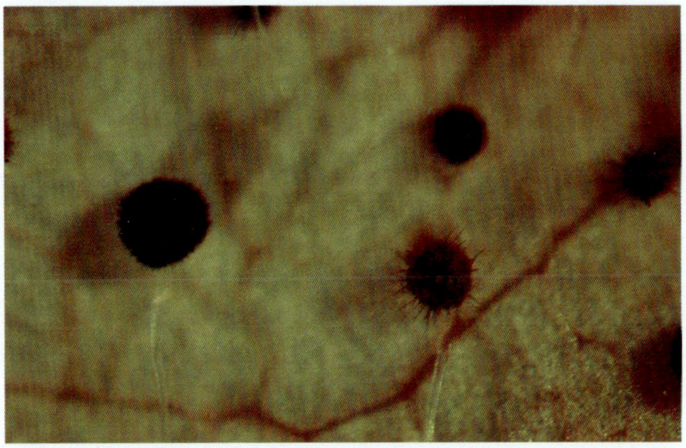

Fig. 80. Sclerotium (left) and pycnidia (right) of *Coniothyrium glycines,* the cause of red leaf blotch, on a leaf surface. (Courtesy G. L. Hartman)

Heavily diseased leaves senesce prematurely, and when foliage from diseased plants drops, newly formed sclerotia and pycnidia are released back into the soil, where they overseason and serve as initial inoculum for the next cycle. Sclerotia germinate to form pycnidia, secondary sclerotia, or infectious hyphae. Less is known about the overseasoning of pycnidia in leaf debris and soil.

The ability of *C. glycines* to survive and its potential host range are critical factors that may affect the outbreak of a disease epidemic, should the fungus enter and become established in a new area of the world. It is possible that the fungus might attack other important commercial crops in the United States, such as alfalfa, beans, and peanuts, along with other forage and timber legumes. Local spread of *C. glycines* occurs when rain showers, splashing water, and animal or human activities transport fungal propagules between plants and fields. Sclerotia residing in soil are the primary source of inoculum.

The economic impact of red leaf blotch on U.S. soybean production has not been studied. However, a risk assessment conducted by the U.S. Department of Agriculture, Animal and Plant Health Inspection Service, Plant Protection Quarantine has indicated that the regions at highest risk of disease establishment are the Mississippi River valley and states from Illinois east to the mid-Atlantic coast.

Management

Growers are advised to plant less susceptible cultivars (if available), to apply fungicides, and to mulch with grass and sawdust (which has been proven effective experimentally). Many foliar fungicides with different chemistries are registered for use on soybeans in the United States and may be highly efficacious, including diphenaconazol, which has been shown to be efficacious in Zambia.

Reports from the 1980s indicated that U.S.-grown commercial cultivars were susceptible to red leaf blotch in field tests conducted in Zambia and Zimbabwe. However, very little soybean germplasm has been evaluated for resistance.

Selected References

Akem, C. N., Akueshi, C. O., Uwala, O. A., and Dashiell, K. E. 1992. Outbreak of red leaf blotch, caused by *Dactuliochaeta glycines,* on soybeans in Nigeria. Plant Dis. 76:323.

Boerema, G. H., de Gruyter, J., Noordeloos, M. E., and Hamers, M. E. C., eds. 2004. Phoma Identification Manual: Differentiation of Specific and Infra-Specific Taxa in Culture. CABI, Wallingford, Oxfordshire, U.K.

Datnoff, L. E., Naik, D. M., and Sinclair, J. B. 1987. Effect of red leaf blotch on soybean yields in Zambia. Plant Dis. 71:132-135.

de Gruyter, J., Woudenberg, J. H. C., Aveskamp, M. M., Verkley, G. J. M., Groenewald, J. Z., and Crous, P. W. 2013. Redisposition of phoma-like anamorphs in *Pleosporales.* Stud. Mycol. 75:1-36.

Hartman, G. L., and Sinclair, J. B. 1988. *Dactuliochaeta,* a new genus for the fungus causing red leaf blotch of soybeans. Mycologia 80:332-343.

Hartman, G. L., and Sinclair, J. B. 1992. Cultural studies on *Dactuliochaeta glycines,* the causal agent of red leaf blotch of soybeans. Plant Dis. 76:847-852.

Hartman, G. L., and Sinclair, J. B. 1996. Red leaf blotch (*Dactuliochaeta glycines*) of soybeans (*Glycine max*) and its relationship to yield. Plant Pathol. 45:332-343.

Hartman, G. L., Datnoff, L. E., Levy, C., Sinclair, J. B., Cole, D. L., and Javaheri, R. 1987. Red leaf blotch of soybeans. Plant Dis. 71:113-118.

Hartman, G. L., Haudenshield, J. S., Smith, K. L., Tooley, P. W., Shelton, J., Bulluck, R., Engle, J., Magarey, R., Royer, M., Sutker, E., Cardwell, K., Herman, T., and Datnoff, L. 2009. Recovery Plan for Red Leaf Blotch of Soybean Caused by *Phoma glycinicola.* Gov. Publ. Rep. 242622. Online publication. U.S. Department of Agriculture, Agriculture Research Service, Washington, DC.

(Prepared by G. L. Hartman)

Rhizoctonia Aerial Blight and Web Blight

Rhizoctonia aerial blight and web blight (also known as "Rhizoctonia foliar blight" and "Rhizoctonia web blight") are found throughout tropical and subtropical regions. In the United States, the diseases occur in the Southeast as far west as southeast Texas and as far north as the Mississippi River Delta regions of Arkansas and Missouri. Observations suggest that up to 70% of soybean pod and foliage loss can be experienced in localized areas of fields, but controlled studies of yield loss are lacking.

Symptoms and Signs

Infected seedlings may have lesions on hypocotyls and shoot apices that are initially red but later turn brown and become dry. On leaves, lesions first appear water soaked and grayish-green (Fig. 81) and later become tan to brown; defoliation is common. Mycelium may adhere to detached leaves and re-infect leaves, pods, and stems. Reddish-brown to brown lesions can form on stems, pods, and petiole scars. Abortion or death of pods can occur if the disease is severe (Fig. 82), and plant senescence may be delayed.

Symptoms are accompanied by a weblike, spreading growth of fungal mycelium (Fig. 83), in which sclerotia form and attach to affected plant parts. The presence of sclerotia is usually the only sign of the fungus that remains during extended periods of low humidity (Fig. 84).

Causal Organism

Rhizoctonia aerial blight and web blight are typically caused by two intraspecific types of *Rhizoctonia solani* (syn. *Thanatephorus cucumeris*): AG 1-IA (aerial blight) and AG 1-IB (web blight). The two types are morphologically dissimilar. Larger (1–3 mm in diameter), spherical sclerotia are produced by AG 1-IA, whereas smaller (≤1 mm in diameter), irregularly shaped microsclerotia are produced by AG 1-IB.

R. solani AG 1-IA isolates typically cause extensive blighting of foliage, particularly within the soybean canopy. In contrast, the distinct leaf spots associated with AG 1-IB occur on more exposed areas of the canopy. Routine evaluations of disease severity are based on combined symptoms in some regions, because AG 1-IA and AG 1-IB isolates can occur in the same field and on the same plant. However, *R. solani* AG1-IA isolates are often predominant in isolates from diseased tissue.

Fig. 81. Grayish-green lesions on soybean leaves with aerial and web blight, caused by *Rhizoctonia solani* AG 1-IA. (Courtesy T. N. Spurlock)

Fig. 82. Soybean pod engrossed in mycelium of *Rhizoctonia solani* AG 1-IA, which causes Rhizoctonia aerial and web blight. The image demonstrates incomplete pod filling, dead tissue on the pod, and the formation of sclerotia that will serve as inoculum. (Courtesy T. N. Spurlock)

Fig. 83. Aerial mycelium of the aerial and web blight pathogen, *Rhizoctonia solani* AG 1-IA. (Courtesy T. N. Spurlock)

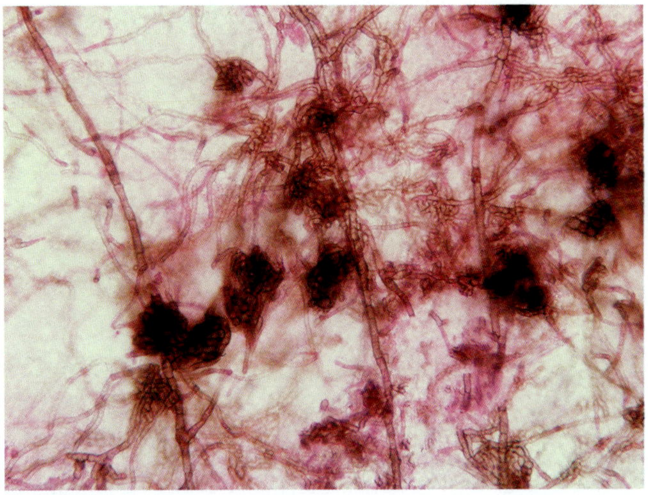

Fig. 84. Developing sclerotia of *Rhizoctonia solani.* (Courtesy J. B. Sinclair)

Neither AG 1-IA nor AG 1-IB has been reported to cause damping-off or root rot.

Disease Cycle and Epidemiology

Both AG 1-IA and AG 1-IB isolates of *R. solani* overseason as sclerotia or microsclerotia in soil and plant debris. Rhizoctonia aerial blight and web blight are described as having two stages. In stage 1, seedlings may be infected and inocula may be splashed onto stems and leaves of older plants. Stage 2 occurs following canopy closure, when mycelium spreads extensively on the surfaces of plants via contact between leaves, resulting in mats of webbed, diseased foliage. Often, symptoms cannot be seen from above the canopy. Localized disease foci develop rapidly, and detached, infected tissues serve as secondary inoculum.

Free moisture on plant surfaces is critical for disease development, which is favored by warm, wet, humid conditions. Evidence suggests that the number of days of rainfall is more critical to disease foci expansion than the amount of rainfall, because dry conditions will stop progress of the disease and the weblike mats of mycelium may not be visible.

Many crops serve as hosts of AG 1-IA and AG 1-IB isolates of *R. solani,* including bean, clover, corn, cowpea, fescue (*Festuca* spp.), peanut, and rice. Likewise, many weed species are hosts of these *R. solani* isolates. Seasonal rotation of rice and soybean ensures a continued abundance of inocula in organic matter from the previous crop, and isolates have been shown to infect both crops. Evidence suggests, however, that *R. solani* AG 1-IA isolates originating in soybean are more pathogenic on soybean than those originating in rice.

Sexual reproduction by *R. solani* AG 1-IA has been reported. The observance of the sexual stage appears to vary by region, and the role played by basidiospores in the disease cycle remains unclear. More research is necessary involving the geographical distribution of the perfect state of the fungus and the role of basidiospores in disease.

Management

Fungicides with known efficacy against Rhizoctonia aerial blight and web blight should be applied to leaves and pods during the early to middle reproductive stages. However, once either disease has progressed and extensive damage has occurred, applying fungicides is not likely an economical option.

Growers should plant the least susceptible cultivars (although most are highly susceptible to these diseases). Wide spacing of rows and maintaining a low plant population may reduce disease severity by increasing airflow within the canopy. Excessive irrigation should be avoided, especially in rotations with rice.

Selected References

Bernardes de Assis, J., Peyer, P., Rush, M. C., Zala, M., McDonald, B. A., and Ceresini, P. C. 2008. Divergence between sympatric rice- and soybean-infecting populations of *Rhizoctonia solani* anastomosis group-1 IA. Phytopathology 98:1326-1333.

Black, B. D., Griffin, J. L., Russin, J. S., and Snow, J. P. 1996. Weeds as hosts for *Rhizoctonia solani* AG-1, causal agent for Rhizoctonia foliar blight of soybean (*Glycine max*). Weed Technol. 10:865-869.

Fenille, R. C., Ciampi, M. B., Kuramae, E. E., and Souza, N. L. 2003. Identification of *Rhizoctonia solani* associated with soybean in Brazil by rDNA-ITS sequences. Fitopatol. Bras. 28:413-419.

Fenille, R. C., Ciampi, M. B., Souza, N. L., and Kuramae, E. E. 2002. Characterization of *Rhizoctonia solani* associated with soybean in Brazil. Eur. J. Plant Pathol. 108:783-792.

Jones, R. K., and Belmar, S. B. 1989. Characterization and pathogenicity of *Rhizoctonia* spp. isolated from rice, soybean, and other crops grown in rotation with rice in Texas. Plant Dis. 73:1004-1010.

Joye, G. F., Berggren, G. T., and Berner, D. K. 1990. Effects of row spacing and within-row plant population on Rhizoctonia aerial blight of soybean and soybean yield. Plant Dis. 74:158-160.

Muyolo, N. G., Lipps, P. E., and Schmitthenner, A. F. 1993. Anastomosis grouping and variation in virulence among isolates of *Rhizoctonia solani* associated with dry bean and soybean in Ohio and Zaire. Phytopathology 83:438-444.

Muyolo, N. G., Lipps, P. E., and Schmitthenner, A. F. 1993. Reactions of dry bean, lima bean, and soybean cultivars to Rhizoctonia root rot and hypocotyl rot and web blight. Plant Dis. 77:234-238.

Sneh, B., Jabaji-Hare, S., Neate, S., and Dijst, G., eds. 1996. *Rhizoctonia* Species: Taxonomy, Molecular Biology, Etiology, Pathology, and Disease Control. Kluwer Academic, London.

Yang, X. B., Berggren, G. T., and Snow, J. P. 1990. Effects of free moisture and soybean growth stage on focus expansion of Rhizoctonia aerial blight. Phytopathology 80:497-503.

Yang, X. B., Berggren, G. T., and Snow, J. P. 1990. Seedling infection of soybean by isolates of *Rhizoctonia solani* AG-1, causal agent of aerial blight and web blight of soybean. Plant Dis. 74:485-488.

Yang, X. B., Berggren, G. T., and Snow, J. P. 1990. Types of Rhizoctonia foliar blight on soybean in Louisiana. Plant Dis. 74:501-504.

Yang, X. B., Snow, J. P., and Berggren, G. T. 1989. Morphogenesis of microsclerotia and sasakii-type sclerotia in *Rhizoctonia solani,* anastomosis group 1, intraspecific groups IA and IB. Mycol. Res. 93:429-434.

Yang, X. B., Snow, J. P., and Berggren, G. T. 1990. Analysis of epidemics of Rhizoctonia aerial blight of soybean in Louisiana. Phytopathology 80:386-392.

(Prepared by J. C. Rupe and T. N. Spurlock)

Rust

Soybean rust is found in soybean-growing countries worldwide. In the Eastern Hemisphere, it occurs from Japan south to Australia and west to China, India, and central and southern Africa. In the Western Hemisphere, the pathogen has been reported in most countries in North and South America and in a number of countries in the Caribbean. The disease is often regional and seasonal within a given country. Significant yield losses of up to 90% have been reported in many countries, including Brazil and the United States. Heavily infected plants produce fewer pods, fewer seeds per pod, and seeds of reduced weight.

Symptoms and Signs

The most common symptom of soybean rust is the formation of gray-green, tan to dark-brown, or reddish-brown lesions with one to many erumpent, globose uredia, particularly on the undersides of leaflets. Lesions tend to be angular, restricted by leaf veins, and 2–5 mm in diameter; they may appear on petioles, pods, and stems. Telia form subepidermally among uredia and are dark brown to black at maturity; they are crustose, irregularly shaped to round, sparse to aggregated, and approximately 50–150 μm in diameter.

The color of the lesions depends on the interaction between the soybean genotype and the pathotype or race of the pathogen. A compatible interaction or a susceptible response results in the production of tan lesions with two or more uredinia, without extensive necrosis, and with abundant sporulation (Fig. 85). An incompatible interaction or an incomplete resistance response results in the production of dark reddish-brown lesions with zero, one, or two uredinia and extensive necrosis or in a complete resistance response with no lesions or sometimes small, grayish flecks. Lesions are frequently associated with leaf chlorosis, and high lesion density results in premature defoliation and early maturity (Fig. 86).

In the early stages of development (before the onset of sporulation), rust lesions may be confused with bacterial pustules (see the previous section Bacterial Pustule). Symptoms of the two diseases can be differentiated, however, by the presence of multiple uredinia in rust lesions and by the irregular cracks that usually appear in the host tissue of bacterial pustule lesions.

Causal Organisms

Soybean rust is caused by two species: *Phakopsora meibomiae* (syn. *Malupa vignae*) and *P. pachyrhizi* (syn. *M. sojae*). The anamorphic fruiting structures (uredinia) of the two species have cellular basal peridia terminating in paraphyses.

Teliospores of *P. meibomiae* are angularly subglobose, oblong to elliptical, and more or less regularly layered in rows or irregularly arranged; they are yellowish-brown to light chestnut brown and 6–12 × 12–26 μm, 1.5–2.0 μm thick, and thickened apically to 6.0 μm. Urediniospores are ellipsoid, 12–24 ×

Fig. 85. Abundant sporulation of *Phakopsora pachyrhizi,* which causes soybean rust. (Courtesy G. L. Hartman)

Fig. 86. Yellowing of the lower leaves of plants, characteristic of soybean rust (*Phakopsora pachyrhizi*). (Courtesy E. J. Sikora)

16–31 μm, densely echinulate, and colorless to pale yellowish-brown; they germinate between 16 and 28°C.

Teliospores of *P. pachyrhizi* are one celled, irregularly arranged, angularly subglobose, and oblong to ellipsoid; they are colorless to pale yellowish-brown, 6–12 × 15–26 μm, slightly thickened apically, and with walls uniformly 1 μm thick. The production of teliospores and their germination to produce basidiospores has been completed under laboratory conditions, but the infection of basidiospores has not been observed.

Urediniospores are colorless to pale yellowish-brown, nearly round to broadly elliptical, 15–24 × 18–34 μm, and 1.0 μm thick (Fig. 87); their walls are minutely and densely echinulate. Urediniospores germinate between 14 and 29°C within 3–6 h of incubation.

Neither fungus appears to have an alternative host. However, anastomosis of germ tubes and migration of nuclei may allow for some genetic exchange without sexual recombination.

Disease Cycle and Epidemiology

Soybean is susceptible to *P. meibomiae* and *P. pachyrhizi* at all growth stages. However, a rapid increase in disease incidence and severity usually coincides with canopy closure from bloom through pod fill, along with environmental conditions that favor disease development. The pathogens are not known to be seedborne in soybean.

Soybean rust epidemics are most severe during long periods of leaf wetness and mean daily temperatures lower than 28°C. Urediniospore germination and penetration require the presence of free water and occur between 8 and 28°C. Uredinia and urediniospores appear as early as 7–10 days after infection and become more abundant after 14 days. Urediniospores are the primary means of spread and can be carried long distances by wind to rust-free regions. However, urediniospores are easily killed by ultraviolet radiation; thus, the long-distance transport of viable urediniospores typically occurs only when spores are moved within a hurricane/tropical depression or in association with a multiday, midlatitude storm system.

Several studies from different countries have shown that field populations of *P. pachyrhizi* are composed of different races or pathotypes. Moreover, these races or pathotypes often have multiple virulence factors, as determined when purified isolates are inoculated on a set of soybean differentials.

The host ranges of both *P. meibomiae* and *P. pachyrhizi* are confined to the legume family, Fabaceae, and the subfamily Papilionoideae, which has an estimated 476 genera and more than 13,000 species. The hosts of *P. meibomiae* and *P. pachyrhizi* are similar and estimated to exceed 150 species, but the number could be much higher, as only a few of the 13,000 species within the subfamily have been evaluated. A 2012 report found *P. pachyrhizi* on a makamong tree (*Afzelia xylocarpa*) in Thailand, which is in the legume family but not the subfamily Papilionoideae. Hosts of *P. meibomiae* and *P. pachyrhizi* include soybean, certain cultivated crops (e.g., jicama [*Pachyrhizus erosus*]), all other *Glycine* spp. (see the Introduction), and many weedy plants. The fungus appears to have a long latent period in soybeans and perhaps in other hosts, as well; having this latent period may be a way for the fungus to avoid ultraviolet radiation.

Fig. 87. Urediniospores of *Phakopsora pachyrhizi*, which causes soybean rust. (Courtesy G. L. Hartman)

In the United States, kudzu is the most important weed host, because *P. pachyrhizi* can survive on this host when soybean is not grown continually. Thus, the role of kudzu has been shown to be critical in the overseasoning and early season build-up of the pathogen (Fig. 88).

P. pachyrhizi is considered the most important of the two rust pathogens, for several reasons: It has a wider geographic distribution, is known to overcome sources of resistance found in soybean, and causes generally more damage throughout its range of occurrence.

Management

A number of pathotype-specific resistance genes (*Rpp*) that control soybean rust have been identified and mapped based on flanking markers to five soybean chromosomes. Most of these genes for resistance to *P. pachyrhizi* are dominant and independently inherited. Based on these single genes for resistance, many soybean germplasm lines have been registered for resistance to *P. pachyrhizi*. With the spectrum of virulence genes in *P. pachyrhizi*, there are no sources of resistance in soybean that appear to confer resistance to all populations of the fungus. In Brazil, growers are advised to plant resistant cultivars and to make foliar applications of fungicides.

Sources of resistance also occur among wild *Glycine* spp. from Australasia (see the Introduction). Additional studies on these perennials have identified potential genes by using both traditional and differentially expressed genes in leaves of *G. tomentella*. Efforts to transfer resistance from perennial *Glycine* spp. to soybean have produced experimental lines with resistance to rust, but the genes from these sources have yet to be used commercially. Other types of resistance, such as rate-reducing resistance and/or tolerance, have been demonstrated, but breeding for these traits has proven more difficult.

Soybean rust can be managed with well-timed fungicide applications made prior to infection with *P. pachyrhizi* or shortly thereafter (Fig. 89). Fungicide spray programs are more economical when coordinated efforts across regions are initiated to monitor the disease and to determine the need for and/or timing of fungicide applications. The triazole and strobilurin

Fig. 88. Leaves of a kudzu plant infected with *Phakopsora pachyrhizi*. (Courtesy E. J. Sikora)

Fig. 89. Soybean plots to which fungicide has (left) and has not (right) been applied. Note the rapidly senescing plants in the untreated plot, caused by infection with *Phakopsora pachyrhizi*. (Courtesy E. J. Sikora)

classes of fungicides and their mixtures have been used successfully in conjunction with monitoring programs in a number of countries (including the United States) to prevent the occurrence of major epidemics. Additional studies have shown that application timing of the fungicide is critical and often dependent on the phase of the epidemic, the prevailing weather conditions, and the type of chemical applied. In general, applications at growth stage R3 and again at growth stage R5 are effective in minimizing yield losses. However, in Brazil, isolates are known to lack sensitivity to triazoles; this could have a broader impact in other regions in which foliar fungicides are used to control soybean rust.

Other management techniques include cultural practices that may reduce the impact of the disease. In Brazil, to reduce the impact of soybean rust, planting-date restrictions have been implemented that include a period of 60–90 days in the offseason when soybean planting is restricted (see the Introduction, the section Soybean Diseases in Brazil).

Selected References

Burdon, J. J. 1988. Major gene resistance to *Phakopsora pachyrhizi* in *Glycine canescens,* a wild relative of soybean. Theor. Appl. Genet. 75:923-928.

Chang, H.-X., Miller, L. A., and Hartman, G. L. 2014. Melanin-independent accumulation of turgor pressure in appressoria of *Phakopsora pachyrhizi*. Phytopathology 104:977-984.

Diers, B. W., Kim, K. S., Frederick, R. D., Hartman, G. L., Unfried, J. R., Schultz, S. J., and Cary, T. R. 2013. Registration of eight soybean germplasm lines resistant to soybean rust. J. Plant Reg. 8:96-101.

Hartman, G. L., Hill, C. B., Twizeyimana, M., Miles, M. R., and Bandyopadhyay, R. 2011. Interaction of soybean and *Phakopsora pachyrhizi*, the cause of soybean rust. CAB Rev. 6:1-13.

Hartman, G. L., Miles, M. R., and Frederick, R. D. 2005. Breeding for resistance to soybean rust. Plant Dis. 89:664-666.

Hartman, G. L., Wang, T. C., and Hymowitz, T. 1992. Sources of resistance to soybean rust in perennial *Glycine* species. Plant Dis. 76:396-399.

Hartman, G. L., Wang, T. C., and Tschanz, A. T. 1991. Soybean rust development and the quantitative relationship between rust severity and soybean yield. Plant Dis. 75:596-600.

Isard, S. A., Dufault, N. S., Miles, M. R., Hartman, G. L., Russo, J. M., De Wolf, E. D., and Morel, W. 2006. The effect of solar irradiance on the mortality of *Phakopsora pachyrhizi* urediniospores. Plant Dis. 90:941-945.

Miles, M. R., Bonde, M. R., Nester, S. E., Berner, D. K., Frederick, R. D., and Hartman, G. L. 2011. Characterizing resistance to *Phakopsora pachyrhizi* in soybean. Plant Dis. 95:577-581.

Miles, M. R., Levy, C., Morel, W., Mueller, T., Steinlage, T., van Rij, N., Frederick, R. D., and Hartman, G. L. 2007. International fungicide efficacy trials for the management of soybean rust. Plant Dis. 91:1450-1458.

Mueller, T. A., Miles, M. R., Morel, W., Marois, J. J., Wright, D. L., Kemerait, R. C., Levy, C., and Hartman, G. L. 2009. Effect of fungicide and timing of application on soybean rust severity and yield. Plant Dis. 93:243-248.

Ono, Y., Buritica, P., and Hennen, J. F. 1992. Delimitation of *Phakopsora, Physopella,* and *Cerotelium* and their species on Leguminosae. Mycol. Res. 96:825-850.

Paul, C., Bowen, C. R., Bandyopadhyay, R., Tefera, H., Adeleke, R., Sikora, E., Pegues, M. D., and Hartman, G. L. 2010. Registration of three soybean germplasm lines resistant to *Phakopsora pachyrhizi* (soybean rust). J. Plant Reg. 4:244-248.

Paul, C., Hartman, G. L., Marois, J. J., Wright, D. L., and Walker, D. R. 2013. First report of *Phakopsora pachyrhizi* adapting to soybean genotypes with *Rpp1* or *Rpp6* rust resistance genes in field plots in the United States. Plant Dis. 97:1379.

Schoen, D. J., Burdon, J. J., and Brown, A. H. D. 1992. Resistance of *Glycine tomentella* to soybean leaf rust *Phakopsora pachyrhizi* in relation to ploidy level and geographic distribution. Theor. Appl. Genet. 83:827-832.

Sikora, E. J. 2014. Kudzu, the vine harboring the soybean rust pathogen in the United States: Invasive weed supports soybean rust pathogen through winter months in southeastern United States. Outlooks Pest Manage. 25:175-179.

Sikora, E. J., Allen, T. W., Wise, K. A., Bergstrom, G., Bradley, C. A., Bond, J., Brown-Rytlewski, D., Chilvers, M., Damicone, J., DeWolf, E., Dorrance, A., Dufault, N., Esker, P., Faske, T. R., Giesler, L., Goldberg, N., Golod, J., Gomez, I. R. G., Grau, C., Grybauskas, A., Franc, G., Hammerschmidt, R., Hartman, G. L., Henn, A., Hershman, D., Hollier, C., Isakeit, T., Isard, S., Jacobson, B., Jardine, D., Kemerait, B., Koenning, S., Langham, M., Malvick, D., Markell, S., Marois, J. J., Monfort, S., Mueller, D., Mueller, J., Mulroony, R., Newman, M., Osborne, L., Padgett, G. B., Ruden, B. E., Rupe, J., Schneider, R., Schwartz, H., Shaner, G., Singh, S., Stromberg, E., Sweets, L., Tenuta, A., Vaiciunas, S., Yang, X. B., Young-Kelly, H., and Zidek, J. 2014. A coordinated effort to manage soybean rust in North America: A success story in soybean disease monitoring. Plant Dis. 98:864-875.

Sikora, E. J., Delaney, D. P., Delaney, M. A., Lawrence, K. S., and Pegues, M. 2009. Evaluation of sequential fungicide spray programs for control of soybean rust. Plant Health Progress. Plant Management Network. doi:10.1094/PHP-2009-03XX-01-RS.

Slaminko, T. L., Miles, M. R., Frederick, R. D., Bonde, M. R., and Hartman, G. L. 2008. New legume hosts of *Phakopsora pachyrhizi* based on greenhouse evaluations. Plant Dis. 92:767-771.

Soria-Guerra, R., Rosales-Mendoza, S., Chang, S., Haudenshield, J. S., Zheng, D., Suryadevara, S. R., Hartman, G. L., Ghabrial, S. A., and Korban, S. S. 2010. Identifying differentially expressed genes in leaves of *Glycine tomentella* in the presence of the fungal pathogen *Phakopsora pachyrhizi*. Planta 232:1181-1189.

Suneerat, S., Basabutra, T., Beasley, D. R., Tan, Y. P., and Shivas, R. G. 2012. A new host genus and species (*Afzelia xylocarpa*) for *Phakopsora pachyrhizi* found in Thailand. Australas. Plant Dis. Notes 7:125-126.

Twizeyimana, M., and Hartman, G. L. 2010. Culturing *Phakopsora pachyrhizi* on detached leaves and urediniospore survival at different temperatures and relative humidities. Plant Dis. 94:1453-1460.

Twizeyimana, M., and Hartman, G. L. 2012. Pathogenic variation of *Phakopsora pachyrhizi* isolates on soybean in the United States from 2006 to 2009. Plant Dis. 96:75-81.

Twizeyimana, M., Ojiambo, P. S., Hartman, G. L., and Bandyopadhyay, R. 2011. Dynamics of soybean rust epidemics in sequential plantings of soybean cultivars in Nigeria. Plant Dis. 95:43-50.

Twizeyimana, M., Ojiambo, P. S., Haudenshield, J. S., Caetano-Anollés, G., Pedley, K. F., Bandyopadhyay, R., and Hartman, G. L. 2011. Genetic structure and diversity of *Phakopsora pachyrhizi* isolates from soybean. Plant Pathol. 60:719-729.

Vittal, R., Haudenshield, J. S., and Hartman, G. L. 2012. A multiplexed immunofluorescence method identifies *Phakopsora pachyrhizi* urediniospores and determines their viability. Phytopathology 102:1143-1152.

Vittal, R., Yang, H., and Hartman, G. L. 2012. Anastomosis of germ tubes and nuclear migration of nuclei in germ tube networks of the soybean rust pathogen, *Phakopsora pachyrhizi*. Eur. J. Plant Pathol. 132:163-167.

(Prepared by G. L. Hartman, E. J. Sikora, and J. C. Rupe)

Scab

Scab was discovered in Japan in 1947 and later named *Kokuto-byo*. It has been reported in China, in Indonesia (where the disease has been observed yearly since 2006 on vegetable soybean), and in Korea.

Symptoms and Signs

Lesions are frequently visible on both leaf surfaces and are commonly centered on veins. Lesions are circular to irregular in shape, up to 4 mm in diameter, and usually more or less raised above the leaf surface. They are initially buff colored and often fade to a drab pale gray.

Stem lesions vary in size from minute, elliptical–elongate spots to large areas up to 2 cm long, which develop as a result of the coalescence of smaller lesions. Stem lesions are buff and sometimes have reddish-brown margins.

Young lesions on pods are red to brown (Fig. 90). At maturity, these lesions turn dark olive to black; they have pale centers and reddish-brown margins. Seeds fail to develop in severely infected pods.

Causal Organism

The causal fungus, *Sphaceloma glycines,* infects soybean leaves, stems, and pods. Acervuli are solitary or confluent, subcuticular and later erumpent, and pseudoparenchymatous at the stromalike bases. Conidiophores are hyaline at first and become dusky or dark with age. They are unbranched, blunted or pointed at the tips, continuous or one septate, and closely compacted; they measure up to 4 × 26 µm and bear conidia on the tips. Conidia are ovoid or oblong–elliptical, hyaline, biguttulate, and 2–5 × 4–13 µm; they become dusky over time.

Disease Cycle and Epidemiology

Direct infection occurs through the epidermises of leaf blades, petioles, and stems by means of an infection peg formed beneath an appressorium. After entering the leaves, hyphae grow both among and within cells in the palisade and spongy mesophyll tissues. Chloroplasts and nuclei collapse, and cytoplasm coagulates and turns brown.

In both young and old leaves of resistant cultivars, infection is confined to the epidermal cells. In mature leaves, abnormal cell division in the palisade tissue results in a cork layer three to five sclerenchymatous cells thick; it acts as a barrier against further extension of the developing mycelium.

Fig. 90. Scab lesions on stem, petioles, and pods, caused by *Sphaceloma glycines.* (Courtesy K. Nashi)

Management

Some sources of resistance to the pathogen have been reported, but resistant commercial cultivars have not been developed.

Selected References

Jenkins, A. E. 1951. Sphaceloma scab, a new disease of soybeans discovered by plant pathologists in Japan. Plant Dis. Rep. 35:110-111.

Kurata, H., and Kuribayashi, K. 1954. Soybean scab caused by *Sphaceloma glycines* sp. nov. Ann. Phytopathol. Soc. Jpn. 18:119-121.

Tasugi, H., and Mogi, S. 1958. Resistance of soybean leaves to the scab, caused by *Sphaceloma glycines*. Ann. Phytopathol. Soc. Jpn. 23:159-164.

(Prepared by J. B. Sinclair;
revised by S. Hardaningsih and G. L. Hartman)

Sclerotinia Stem Rot

Sclerotinia stem rot (often referred to as "white mold") of soybean was first reported in Hungary in 1924, and the disease has since been reported in Argentina, Brazil, Canada, India, Nepal, South Africa, and the United States. Each year, losses occur in northern Argentina, central and southern Brazil, and the upper Midwest of the United States. Crop yields are inversely correlated with the incidence of Sclerotinia stem rot; an estimated of 0.25 metric ton per ha is lost for each 10% increment of diseased plants.

Symptoms and Signs

Plants become infected during the flowering stages. The first symptoms appear at growth stage R2 or later. Initially, the upper leaves wilt, and as the pathogen infects more of the stem, these leaves die. Leaves become grayish-green as necrosis begins; they eventually turn brown but frequently remain attached to the stems. Foliar symptoms of Sclerotinia stem rot can be mistaken for late-season symptoms of brown stem rot, Fusarium wilt, Phytophthora root rot, and stem canker (see sections on these diseases elsewhere in Part I).

Foliar symptoms result from stem lesions that develop 7–14 days earlier. Infection originates at stem nodes and usually develops 10–50 cm above the soil line. Blossoms are the primary sites of infection, although the infection of tissues such as leaves, petioles, and internodes can occur when a plant comes in contact with adjacent diseased plants. Lesions completely encircle the stem and disrupt the transport of water, mineral nutrients, and photosynthates to developing pods. Pod development and fill above stem lesions is reduced. Stem infections are rarely observed at the soil line, although infections can occur at the lower stem nodes (Fig. 91). Initially, water-soaked lesions develop at the nodes, change from tan to white, and spread acropetally and basipetally from the nodes. Side branches and pods may also be infected. Cottony mycelial growth on all diseased plant parts is a characteristic sign of the disease (Fig. 92).

By crop maturity, stem tissues have turned white and tissues have a shredded appearance if disturbed. A reddish discoloration is scattered within the diseased stem tissues and at the borders of lesions. At harvest, diseased plants are characterized by poor pod development, white or bleached stems, and an abundance of sclerotia. Diseased pods are outwardly white in appearance, and mycelium and sclerotia can be readily observed inside pods. Infected seed ranges in appearance from being white and moldy to symptomless; the level of seed infection is generally very low, however.

Sclerotia are returned to the soil at harvest or as contaminants associated with seed at planting. Large, black, round to irregularly shaped sclerotia of varying sizes form on stems, which are partially covered with dense, white mycelium. Sclerotia also form in the stem pith and are conspicuous when the stem is opened (Fig. 93). Seeds may become infected within diseased pods (Fig. 94). If infected early, seeds become flattened and shriveled and are sometimes replaced by black sclerotia.

One hazard associated with Sclerotinia stem rot is the contamination of seed with sclerotia during harvest (Fig. 95). A small center of infection within a field may contaminate a large volume of seeds. When seed is shipped and checked at a foreign port of entry, the detection of even a few sclerotia in a shipment intended for human consumption is grounds for rejection. Furthermore, sclerotia can germinate if free water is present, resulting in seed-decay problems in storage.

Causal Organism

The causal fungus, *Sclerotinia sclerotiorum,* produces hard, black, irregularly shaped sclerotia (2–22 mm in diameter). These sclerotia are much larger than those of *Macrophomina phaseolina,* which causes charcoal rot, and are easily distinguished from the smaller, round, reddish to tan-brown sclerotia of *Sclerotium rolfsii.* (See discussions of these diseases in the section Fungal Diseases of Lower Stems and Roots, later in Part I.) Soybean is also susceptible to *S. minor.*

A single sclerotium may produce one to many cup-shaped apothecia, borne on slender stipes (Fig. 96). Apothecia are 0.5–2.0 mm or more in diameter, funnel shaped to discoid, and light tan to brown. A hymenium is composed of closely packed, eight-spored asci interspersed with slender, simple, hyaline paraphyses. Asci are narrow, 4–22 × 81–252 μm, and cylindrical to clavate.

S. sclerotiorum grows on potato dextrose agar (PDA) at 1–31°C (optimum 24°C). The fungus is highly variable in culture, producing white to chocolate-brown aerial mycelium and black sclerotia of diverse sizes and shapes. Spherical microconidia are formed in culture but rarely germinate in water or on nutrient agar. Apothecia and ascospores develop in alternating light and darkness at 12–15°C, provided the sclerotia are incubated in a water-saturated medium. Apothecia are produced between 18 and 25°C when moisture levels are at half or full field capacity.

S. sclerotiorum has a wide host range that includes numerous dicotyledonous plants. Many broad-leaf weeds are hosts and contribute to inoculum replenishment in fields planted to a nonhost.

Disease Cycle and Epidemiology

Sclerotia of *S. sclerotiorum* survive in the soil for long periods. They are highly resistant to dry heat (up to 70°C) and

Fig. 91. Mycelia and sclerotia of *Sclerotinia sclerotiorum,* which causes Sclerotinia stem rot, on the lower part of a soybean stem. (Courtesy G. L. Hartman)

Fig. 92. Cottony mycelium of *Sclerotinia sclerotiorum,* which causes Sclerotinia stem rot, on soybean stems, petioles, and pods. (Courtesy D. D. Hoffman)

Fig. 93. Sclerotia of *Sclerotinia sclerotiorum* inside a soybean stem. (Courtesy L. M. Kull)

Fig. 94. Seed discoloration caused by *Sclerotinia sclerotiorum.* (Courtesy G. L. Hartman)

to prolonged periods of freezing and thawing. Sclerotia within 5.0 cm of the soil surface germinate in the field by producing apothecia. Prolonged periods of low soil temperatures (5–15°C) and soil matric potentials (less than –5 bars for 10–14 days) provide the conditions most favorable for apothecial development. Apothecia usually appear after the crop canopy has at least partially closed to shade the soil surface.

Ascospores are forcibly ejected from asci and transported by wind to nearby stems, branches, flowers, and pods—sometimes moving as far as 50 m from the source. Blossoms that adhere to emerging pods are colonized and provide nutrients for the infection of stems. Infection of leaves, petioles, and internodes can occur through contact with adjacent diseased plants and through wounds.

Sclerotinia stem rot is a disease of environments with high yield potential. It is favored by moderate air temperatures and frequent rain events from just prior to flowering to the pod development stage of growth. This disease is responsive to cultural practices that promote moderate air temperatures in the crop canopy, cool and moist soil conditions, and a shaded canopy floor. Development of Sclerotinia stem rot is a function of the timing of canopy closure, crop flowering, development of apothecia, and climatic conditions that favor infection and colonization of stems by the pathogen.

The optimal environmental conditions for infection of soybeans by *S. sclerotiorum* occur when crop flowering coincides with extended periods of below-average air temperatures and moist weather. Specifically, optimal conditions include canopy temperatures less than 28°C and plant surface wetness for 12–16 h on a daily basis or continuous surface wetness for 42–72 h. These conditions may occur in the north-central United States when soybean is flowering. In contrast, warmer and drier conditions tend to prevail in southern areas of the country; this may account for the lack of Sclerotinia stem rot in the South, despite the fact that the causal fungus is widely distributed in that region. The disease has also been observed during periods of low rainfall when the crop has been planted with row spacing of less than 38 cm.

The occurrence of Sclerotinia stem rot is favored by inclusion of another host crop in rotation with soybean. Broad-leaf weeds may serve as hosts and thereby as sources of inoculum. Planting seeds contaminated with sclerotia is the most common means of introducing the pathogen into a previously uninfested field.

Since infection occurs at flowering, it follows that significant delays in planting date might influence infection and disease development. Soybean that is planted early generally grows taller and has a denser canopy than soybean planted later. Thus, planting soybean early may encourage development of the disease because more favorable conditions exist. Early maturing cultivars tend to escape infection because of their usually short stature and early flowering. In contrast, late-maturing cultivars are believed to be more prone to disease because of their lush vegetative growth and later flowering.

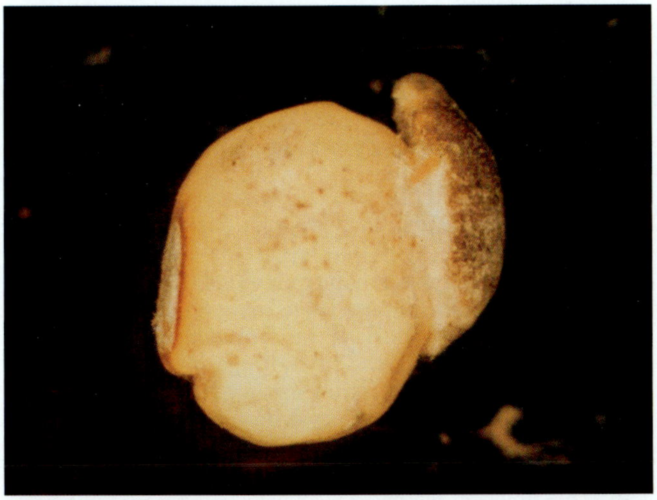

Fig. 95. Sclerotia of *Sclerotinia sclerotiorum* developing on a seed. (Courtesy J. B. Sinclair)

Fig. 96. Apothecia of *Sclerotinia sclerotiorum,* which causes Sclerotinia stem rot. (Courtesy D. S. Mueller)

Management

Soybean cultivars differ in their reaction to *S. sclerotiorum,* and careful cultivar selection can reduce the risks associated with specific management systems. Thus, the reaction of a particular soybean cultivar to *S. sclerotiorum* should determine the types of cultural practices used in a field with a history of Sclerotinia stem rot. Production practices designed to increase the yield potential frequently promote fast and complete canopy closure—for example, early planting dates, narrow row-width systems, soil tillage, high plant populations, and high soil fertility. Even though each of these practices can increase yield, it has an associated risk of increasing the incidence and severity of Sclerotinia stem rot.

Growers can modify several agronomic practices to reduce the risk of yield loss. Tilling should be avoided, because the number of viable sclerotia declines if the soil is not tilled. The number of viable sclerotia also declines in the absence of a host crop. Growers should also avoid planting soybean directly after common bean, sunflower, and other susceptible crops. In a field with a history of Sclerotinia stem rot, growers should avoid planting highly susceptible cultivars if row spacing is less than 76 cm. In addition, excessive irrigation should be avoided until flowering has ceased, and good weed control practices should be followed.

Growers should plant seeds free of sclerotia and treat seeds with fungicides if seedborne infection is suspected. Several fungicides are registered for control of Sclerotinia stem rot, but studies have shown that when disease incidence is greater than 25%, consistent control is not obtainable.

Selected References

Boland, G. J., and Hall, R. 1987. Evaluating soybean cultivars for resistance to *Sclerotinia sclerotiorum* under field conditions. Plant Dis. 71:934-936.

Boland, G. J., and Hall, R. 1988. Epidemiology of Sclerotinia stem rot of soybean in Ontario. Phytopathology 78:1241-1245.

Calla, B., Voung, T., Radwin, O., Hartman, G. L., and Clough, S. J. 2009. Gene expression profiling soybean stem tissue early response to *Sclerotinia sclerotiorum* and in silico mapping in relation to resistance markers. Plant Genome 2:149-166.

Diers, B. W., Kopisch-Obuch, F. J., Hoffman, D. D., Hartman, G. L., Pedersen, W. L., Grau, C. R., and Wang, D. 2006. Registration of

AxN-1-55 soybean germplasm with partial resistance to Sclerotinia stem rot. Crop Sci. 46:1403-1404.
Grau, C. R. 1988. Sclerotinia stem rot of soybean. Pages 56-66 in: Soybean Diseases of the North Central Region. T. D. Wyllie and D. H. Scott, eds. American Phytopathological Society, St. Paul, MN.
Hartman, G. L., Kull, L., and Huang, Y. H. 1998. Occurrence of *Sclerotinia sclerotiorum* in soybean fields in east-central Illinois and enumeration of inocula in soybean seed lots. Plant Dis. 82:560-564.
Hoffman, D. D., Diers, B. W., Hartman, G. L., Nickell, C. D., Nelson, R. L., Pedersen, W. L., Cober, E. R., Graef, G. L., Steadman, J. R., Grau, C. R., Nelson, B. D., del Rio, L. E., Helms, T., Anderson, T. T., Poysa, V., Rajcan, I., and Stienstra, W. C. 2002. Selected soybean plant introductions with partial resistance to *Sclerotinia sclerotiorum*. Plant Dis. 86:971-980.
Hoffman, D. D., Hartman, G. L., Mueller, D. S., Leitz, R. A., Nickell, C. D., and Pedersen, W. L. 1998. Yield and seed quality of soybean cultivars infected with *Sclerotinia sclerotiorum*. Plant Dis. 82:826-829.
Kim, H. S., Hartman, G. L., Manandhar, J. B., Graef, G. L., Steadman, J. R., and Diers, B. W. 2000. Reaction of soybean cultivars to Sclerotinia stem rot in field, greenhouse, and laboratory evaluations. Crop Sci. 40:665-669.
Kull, L. S., Vuong, T. D., Powers, K. S., Eskridge, K. M., Steadman, J. R., and Hartman, G. L. 2003. Evaluation of resistance screening methods for Sclerotinia stem rot of soybean and dry bean. Plant Dis. 87:1471-1476.
Mueller, D. S., Dorrance, A. E., Derksen, R. C., Ozkan, E., Kurle, J. E., Grau, C. R., Gaska, J. M., Hartman, G. L., Bradley, C. A., and Pedersen, W. L. 2002. Efficacy of fungicides on *Sclerotinia sclerotiorum* and their potential for control of Sclerotinia stem rot on soybean. Plant Dis. 86:26-31.
Mueller, D. S., Hartman, G. L., and Pedersen, W. L. 1999. Development of sclerotia and apothecia of *Sclerotinia sclerotiorum* from infected soybean seed and its control by fungicide seed treatment. Plant Dis. 83:1113-1115.
Mueller, D. S., Hartman, G. L., and Pedersen, W. L. 2002. Effect of crop rotation and tillage on *Sclerotinia sclerotiorum* on soybean. Can. J. Plant Path. 24:450-456.
Nelson, B. D., Helms, T. C., and Kural, I. 1991. Effect of temperature and pathogen isolate on laboratory screening of soybean resistance to *Sclerotinia sclerotiorum*. Can. J. Plant Sci. 71:347-352.
Vuong, T. D., Diers, B. W., and Hartman, G. L. 2008. Identification of QTL for resistance to Sclerotinia stem rot (*Sclerotinia sclerotiorum*) in plant introduction 194639. Crop Sci. 48:2209-2214.
Vuong, T. D., Hoffman, D. D., Diers, B. W., Miller, J. F., Steadman, J. R., and Hartman, G. L. 2004. Evaluation of soybean, dry bean, and sunflower for resistance to *Sclerotinia sclerotiorum*. Crop Sci. 44:777-783.

(Prepared by C. R. Grau and G. L. Hartman)

Stemphylium Leaf Blight

Stemphylium leaf blight was reported from India in 1972 and from Turkey in 1984. Few new reports have been made since then, except for two that identified the causal pathogen in Taiwan and in Argentina.

Symptoms include small, circular, necrotic spots with dark-brown margins and gray centers, which appear on leaves of all ages. The spots enlarge and often coalesce, particularly along the leaf margins. Fungal growth (conidiophores and conidia) is often observed on the surfaces of lesions that appear dark. Later, the leaves dry up and severe defoliation may occur.

The pathogen, *Stemphylium botryosum* (syn. *Pleospora tarda*), produces subhyaline to brown, septate, endophytic hyphae 4–6 μm wide. Conidiophores are olivaceous-brown to dark brown and septate; they emerge from the epidermis in clusters and proliferate, each bearing a single conidium at the tip. Conidia are 10–12 × 24–28 μm, dark brown, ovoid to ellipsoid, and muriform; each usually has three transverse septa and one to two longitudinal septa. *S. botryosum* has many hosts.

Stemphylium leaf blight on soybean is also caused by *Pleospora herbarum*.

Selected References

Anahosur, K. H., and Fazalnoor, K. 1972. Studies on follicular fungi of soybean crop. Indian Phytopathol. 25:504-508.
Chen, L. S., Chu, C., Liu, C. D., Chen, R. S., and Tsay, J. G. 2006. PCR-based detection and differentiation of anthracnose pathogens, *Colletotrichum gloeosporioides* and *C. truncatum*, from vegetable soybean in Taiwan. J. Phytopathol. 154:654-662.
Kalekar, A. R., and Patil, D. L. 1972. *Pleospora herbarum* (Fr.) Rab., on soybean—A new record for India. Res. J. Mahatma Phule Agric. Univ. 3:146-147.
Larrán, S., and Mónaco, C. 2010. Status and progress of research in endophytes from agricultural crops in Argentina. Pages 149-161 in: Management of Fungal Plant Pathogens. A. Arya and A. E. Perelló, eds. CABI, Wallingford, Oxfordshire, U.K.
Zayed, M. A., Darrag, I. E., Aly, A. Z., and El-Gantiry, S. M. 1984. Reaction of soybean varieties to *Stemphylium vesicarium* (Wallr.) Simmons and its relation to the chemical constituents of the plant. Agric. Res. Rev. Zagazig Univ. 60:105-116.

(Prepared by J. B. Sinclair; revised by G. L. Hartman)

Target Spot

Since 1945, when target spot was first reported in the United States, it has been found in most soybean-growing states. Yield losses of 18–32% have been recorded for very susceptible lines of soybean grown in Mississippi during years when rainfall was above normal during August and September. The disease has also been reported in most other countries that grow soybean. There are some indications that the incidence of target spot has increased in recent years in certain areas of Brazil and the United States.

Symptoms and Signs

Target spot affects leaves, stems, pods, seeds, hypocotyls, and roots. Leaf lesions are round to irregular in shape, reddish-brown in color, and from specks to mature spots (10–15 mm or more) in diameter. Lesions are frequently surrounded by dull-green or yellowish-green halos. Larger spots are often distinctly zonate (hence, the common name "target spot") (Figs. 97 and 98). Narrow, elongated spots develop along the veins on the upper surfaces of leaves of susceptible cultivars. Severely affected leaves drop prematurely.

Infected areas on the petioles and stems are dark brown and range in shape and size from specks to elongated, spindle-shaped

Fig. 97. Zonate lesion surrounded by a yellow halo, characteristic of target spot (*Corynespora cassiicola*). (Courtesy C. V. Godoy)

lesions. Pod spots are generally circular and approximately 1 mm in diameter; they have slightly depressed, purple-black centers and brown margins. During periods of extended rainfall and/or high humidity, spots may coalesce and cover the entire pod. In some cases, the causal fungus penetrates the pod wall and produces small, blackish-brown lesions on the seeds.

Dark reddish-brown, circular to oval lesions form on the hypocotyl, taproot, and large lateral roots, sometimes girdling the smaller lateral roots. Lesions turn dark violet-brown when the fungus sporulates; they elongate and coalesce as the plant ages, frequently girdling the taproot and the stem. The entire root system may be discolored. Dark masses of conidiophores and conidia of the causal fungus form on root surfaces, as well as other infected plant parts.

Causal Organism

Target spot is caused by *Corynespora cassiicola,* and two strains of the fungus have been identified. Isolates of one strain (which infects the hypocotyl, roots, and stem) differ pathologically and morphologically from isolates of the other strain (which infects the leaves, pods, and seeds), which is probably a different species. The causal fungus is widespread and infects more than 350 host plants. Isolates from one host frequently infect other unrelated hosts, although there is some indication of host specialization among isolates, as well.

Erect, branched, brown conidiophores arise singly or in clusters from swollen basal cells; conidiophores are one to 20 septate and 4–11 × 44–135 µm (most commonly, three to five septate and 8 × 125–200 µm). Conidia are borne singly or in chains of two to six and are typically smooth and brownish; they vary greatly in shape, being cylindrical or broadest at the base and tapering slightly toward the tip. Conidia are three to 20 pseudoseptate, straight or slightly curved, and 7–22 × 39–520 µm (on average, 8 × 135 µm); each has a central hilum at the basal end. Conidia germinate readily over a wide range of temperatures on potato dextrose agar (PDA) and Czapek–Dox agar, producing germ tubes at one or both ends. Chlamydospores, which are formed in older cultures, are hyaline, terminal or intercalary, oval, and 14–20 × 16–30 µm.

C. cassiicola grows and sporulates well on PDA, Czapek–Dox agar, and V8 juice agar at pH 6.5 under continuous light. The optimal temperature range for the growth of most isolates is 18–21°C; some isolates grow best at 28°C. The mycelium in culture is white and flocculent or appressed at first; later, it turns dark gray and forms an olivaceous black mat. This dark color eventually extends throughout the medium.

Fig. 98. Leaf with multiple lesions, characteristic of target spot (*Corynespora cassiicola*). (Courtesy C. V. Godoy)

Disease Cycle and Epidemiology

C. cassiicola overseasons in diseased stems, roots, and seeds and can survive in fallow soil for more than 2 years. The fungus can colonize a wide range of plant residues in soil, and it can also colonize cysts of *Heterodera glycines* (see later in Part I, the section Soybean Cyst Nematode). Conidia are dispersed via air.

Leaf infection takes place only when the relative humidity is 80% or higher and/or when free moisture is present on leaves. Dry conditions inhibit the fungus in both leaves and roots.

Stems and roots are first infected during the early seedling stage. Underground symptoms can be seen as early as 3 days after emergence. Soil temperatures of 15–18°C are optimal for infection and disease development. Seedlings that grow at 15°C develop large lesions on the primary roots, and growth of secondary roots is retarded. At 20°C, these symptoms are less severe, and seedlings produce nearly normal root systems. No symptoms develop at temperatures greater than 20°C.

Management

Most adapted, high-yielding soybean cultivars grown in the southeastern United States are tolerant to target spot. In Brazil, fungicides containing fluxapyroxad and prothioconazole have been found most efficient in controlling the disease and are recommended for susceptible cultivars. Resistance to methyl 2-benzimidazole carbamate fungicides has been reported on isolates of soybean *C. cassiicola* in Brazil.

Selected References

Carris, L. M., Glawe, D. A., and Gray, L. E. 1986. Isolation of the soybean pathogens *Corynespora cassiicola* and *Phialophora gregata* from cysts of *Heterodera glycines* in Illinois. Mycologia 78:503-506.

Dixon, L. J., Schlub, R. L., Pernezny, K., and Datnoff, L. E. 2009. Host specialization and phylogenetic diversity of *Corynespora cassiicola*. Phytopathology 99:1015-1027

Seaman, W. L., Shoemaker, R. A., and Peterson, E. A. 1965. Pathogenicity of *Corynespora cassiicola* on soybean. Can. J. Bot. 43:1461-1469.

Spencer, J. A., and Walters, H. J. 1969. Variations in certain isolates of *Corynespora cassiicola*. Phytopathology 59:58-60.

Xavier, S. A., Canteri, M. G., Barros, D. C. M., and Godoy, C. V. 2013. Sensitivity of *Corynespora cassiicola* from soybean to carbendazim and prothioconazole. Trop. Plant Pathol. 38:431-435.

(Prepared by C. V. Godoy)

Yeast Spot of Pods and Seeds

Yeast spot was first observed in soybean in the United States in 1928. It was reported as a serious disease affecting soybean seeds in North Carolina and Oklahoma in 1943 and in Kansas and Missouri in 1964. Since then, it has been reported in soybean-growing areas throughout the world. Badly damaged seeds are downgraded in commercial trade. The yeast has also been reported on more than 40 other hosts, including species of *Citrus* and *Coffea*.

Symptoms and Signs

Yeast spot is primarily a disease of soybean seeds, although pods may show small areas of discoloration where they have contacted infected seeds. When infection occurs during pod formation, seeds fail to develop and pods drop prematurely. Diseased seeds are discolored and have reduced oil content. Diseased tissue of young seeds is pale yellow to grayish-brown and slightly sunken (Fig. 99); the sunken spots are light or cream colored on more mature seeds. Affected areas of cotyledons are off white and cheesy in texture. The seed coat remains intact except for minute, discolored punctures, which may be seen upon close examination. Seeds that are infected early are small, shriveled, and lightweight. Severely infected seeds do not mature and are greatly shrunken and wrinkled.

Causal Organisms

Eremothecium ashbyi and *E. coryli* both cause yeast spot of pods and seeds. The vegetative growth of each yeast consists of individual cells or septate strands that resemble mycelium and are 2.3–3.5 × 90–140 µm. Buds arise at the cross-walls of strands. Individual cells are usually elliptical (6–10 × 8–14 µm) and spherical (12–20 µm in diameter), but occasionally, they are slender and tapering and shaped like tennis rackets.

Asci and ascospores are formed on seeds and in nutrient broth. Asci are cylindrical (10–12 × 60–85 µm) and have broadly rounded ends; each ascus contains eight ascospores in two groups of four. Ascospores (2.5–3.5 × 40–60 µm) are slender and two celled; they are slightly constricted at the septa and have sharp tips. The base of the ascospore extends into a slender, nonmotile whip that is approximately 1.25 times as long as the spore.

Ascospores germinate by forming a germ tube or by budding. Usually, the basal cell swells at the septum, forming a sphere (approximately 6 µm in diameter) from which a germ tube protrudes. The germ tube develops into a septate, branched, mycelium-like strand. On occasion, round to elliptical cells bud from the spherical basal cell.

Fig. 99. Soybean seeds with whitish or cheesy spots, symptoms of yeast spot (*Eremothecium* spp.), and an asymptomatic seed (upper left). (Courtesy J. T. Yorinori)

The yeasts grow well at 30–40°C in nutrient broth supplemented with 1.0% glucose, 0.5% yeast extract, and 1,000 µm per ml of streptomycin sulfate. Maximum growth is reached in 24 h at 40°C; at 18°C, the yeasts grow poorly, and asci and ascospores develop slowly. Buffered media of pH 6.0 are optimal for growth. Each yeast lowers the pH of nonbuffered media to a level at which growth is inhibited.

Disease Cycle and Epidemiology

E. coryli is usually transmitted by the green stink bug, *Chinavia hilaris* (see Part II, Arthropod Pests, the section Stink Bugs). Many other species of stink bug are also capable of transmitting the yeast to soybean—among them, *Euschistus servus euschistoides, E. tristigmus, E. variolarius, Riptortus clavatus,* and *Thyanta custator.*

For seeds to become infected, a yeast-infested stink bug must pierce the pod wall and feed on the developing embryo, forcing yeast cells into it. Only yeast cells near the insect's mouthparts are transmitted. Nymphs cannot transmit the yeast after molting. Nymphs and newly emerged adults must feed on a source of yeast after each molt to become transmitters. The yeast survives in the adult throughout the insect's life and thus overwinters in this life stage.

Management

To manage yeast spot, growers should plant healthy, pathogen-free seeds. They should also use insecticides to control stink bugs when the population exceeds the threshold for the local area.

Selected References

Jordan, E. G., Manandhar, J. B., Thapliyal, P. N., and Sinclair, J. B. 1988. Soybean seed quality of 16 cultivars and four maturity groups in Illinois. Plant Dis. 72:64-67.

Kimura, S., Tokumaru, S., and Kuge, K. 2008. *Eremothecium ashbyi* causes soybean yeast-spot and is associated with stink bug, *Riptortus clavatus*. J. Gen. Plant Pathol. 74:275-280.

Kurtzman, C. P., and de Hoog, G. S. 2004. *Eremothecium* Borzi emend Kurtzman (1995). Pages 405-412 in: The Yeasts—A Taxonomic Study. 5th ed. Vol. 2. C. P. Kurtzman, J. W. Fell, and T. Boekhout, eds. Elsevier, Amsterdam.

Russin, J. S., Orr, D. B., Layton, M. B., and Boethel, D. J. 1988. Incidence of microorganisms in soybean seeds damaged by stink bug feeding. Phytopathology 78:306-310.

Sinclair, J. B. 1995. Reevaluation of grading standards and discounts for fungus-damaged soybean seeds. J. Amer. Oil Chem. Soc. 72:1415-1419.

Yeh, C. C., and Sinclair, J. B. 1980. Effect of *Chaetomium cupreum* on seed germination and antagonism to other seedborne fungi of soybean. Plant Dis. 64:468-470.

(Prepared by M. M. Kulik and J. B. Sinclair; revised by G. L. Hartman)

Diseases of Lower Stems and Roots

Brown Stem Rot

Brown stem rot is widespread in the upper-midwestern region of the United States and occurs to a lesser extent in the southeastern region. It has also been reported in Argentina, Brazil, Canada, Egypt, Mexico, Japan, and Yugoslavia. The symptoms and the extent of yield reduction depend on the environmental conditions, the cultivar used, and the pathogen type. Reductions in the number and size of seeds produced by infected plants may cause yield losses of up to 30%. Losses are greatest when wet and cool conditions prevail during the pod-fill stage and are followed by hot, dry conditions.

Two major subtypes of the brown stem rot pathogen have been identified; although they cause different patterns of symptoms, both can cause reductions in yield. In addition to soybean, the causal fungus has been reported to infect azuki (*Vigna angularis*) and mung bean (*V. radiata*).

Symptoms and Signs

Soybean plants with brown stem rot have brown discoloration in the pith and vascular tissues and may or may not develop foliar symptoms. A light- to dark-brown discoloration of the pith and vascular tissues, which extends up from the base of the plant or the roots, is visible in a longitudinally split stem (Figs. 100 and 101). Vascular browning develops first in the lower stem and is most prominent at the nodes. As the plant matures, the browning of the pith develops throughout the stem. Pith and vascular browning in the stem are not diagnostic, however; other pathogens can cause similar symptoms.

Foliar symptoms on diseased plants are variable and may not develop in some environments and geographic locations. Leaf tissues between the veins turn yellow and brown, and a narrow, green border may outline the veins (Fig. 102). These leaf symptoms of brown stem rot can mimic those associated with sudden death syndrome (SDS) (see later in Part I, the section Sudden Death Syndrome), but plants with SDS do not develop brown piths. Plants infected with the brown stem rot pathogen show no symptoms before early pod set. Leaf symptoms become apparent at or after the R4 (full-pod) growth stage (Fig. 103).

Under some conditions and for some cultivars, the leaves on infected plants turn brown and dry but remain attached to the petioles, and the characteristic interveinal chlorosis does not develop. These symptoms may develop 5–20 days before host maturation begins. The affected plants, when viewed from the edge of the field, may appear in patches to be frost damaged and clearly contrast the yellow-green color of normally maturing soybean plants.

Soybean growth stage R5 is the optimum period for assessing symptoms that predict yield loss.

Causal Organism

Brown stem rot is caused by the fungus *Cadophora gregata* (syn. *Phialophora gregata*). Pathogen identification is often assisted with molecular assays, as well as agar media containing soybean stem or green bean extract The fungus grows slowly on these media, averaging approximately 2 mm per day at 20–25°C. Hyphae are 1.2–4.7 mm in diameter, hyaline, septate, and branched.

Radial growth rate is a good initial test to distinguish *C. gregata* from other common soybean stem-infecting fungi that the pathogen resembles. For instance, isolates of *Acremonium* and *Plectosporium* spp. can be confused with *C. gregata,* but they grow much more quickly and sporulate sooner on agar containing either soybean stem or green bean extract. *C. gregata* does not sporulate on potato dextrose agar (PDA), and this characteristic can be used to distinguish it from many other fungi

Fig. 100. Split soybean stem showing brown, decomposing pith of a plant (lower) with brown stem rot (*Cadophora gregata*), compared with white pith in a healthy, noninfected stem of a plant (upper) of the same cultivar and age. (Courtesy D. K. Malvick)

Fig. 101. Split soybean stem showing inconsistent brown pith in a plant with developing brown stem rot (*Cadophora gregata*). (Courtesy D. K. Malvick)

Fig. 102. Soybean leaves from a plant infected with brown stem rot (*Cadophora gregata*), exhibiting foliar symptoms of interveinal chlorosis and necrosis. (Courtesy D. K. Malvick)

Fig. 103. Leaf yellowing and early senescing, symptoms of brown stem rot (*Cadophora gregata*), in a susceptible soybean cultivar bordered on both sides by a resistant cultivar. This photograph also shows how damage from brown stem rot can be confused with the effects of premature senescence. (Courtesy D. K. Malvick)

commonly isolated from soybean stems. *C. gregata* can also be clearly and quickly differentiated from other stem-infecting fungi by using specific standard and quantitative polymerase chain reaction (PCR) assays.

Conidia of *C. gregata* generally form after 10–15 days on agar media containing either soybean stem or green bean extract when isolates are incubated at 20°C. One-celled conidia are ovoid to elliptical and hyaline, and they measure 1.7–3.4 × 3.4–7.6 mm when produced on these media. Sporulation is optimal from 15 to 26°C and does not occur above 28°C. Conidia are produced in succession at the tips of phialids and aggregate in slime droplets, forming irregular, noncatenulate heads (Fig. 104). The conidiogenous cells have distinct collarettes and are phialidic (4–15 mm), hyaline, aseptate or septate, and straight or flask shaped. They are borne singly, lateral or terminal, and frequently in clusters, arising from recent hyphae, hyphal coils, and aerial strands.

Isolates of *C. gregata* vary in cultural characteristics and the severity and types of symptoms they cause. Two subtypes of *C. gregata* are known to occur across the U.S. Midwest and in Canada and can be distinguished by the types of symptoms expressed by infected plants and by pathogen genotypes. Type A (also termed "genotype A") causes internal stem browning, as well as leaf chlorosis and necrosis. Type B (also termed "genotype B" and "genotype M") typically produces only internal stem browning. Both types of *C. gregata* cause premature defoliation and may infect a single plant or multiple plants in a field. Because type B (and sometimes type A) causes slight or no foliar symptoms, the incidence and severity of brown stem rot in a field can be determined only by splitting stems to detect internal symptoms. Types A and B can readily be differentiated using PCR assays based on sequence differences in the intergenic spacer regions of their rDNA or using DNA fingerprinting.

Types A and B also interact differently with susceptible and resistant soybean cultivars. The two types may infect different parts of the soybean vascular system. Type B is often detected in resistant soybean cultivars at a higher frequency than type A, whereas susceptible cultivars are more often infected similarly by both genotypes. Different sources of resistance to brown stem rot also appear to influence infection by types A and B, and the resistance of the soybean cultivars planted can influence the levels of aggressiveness and soil populations of *C. gregata*. Monocultures of cultivars that are resistant or susceptible to brown stem rot have been found to enhance the aggressiveness and population density of type B isolates in a field.

Disease Cycle and Epidemiology

C. gregata survives in soybean residue and in soil. Mycelium enters the plant through the main and lateral roots, grows and spreads upward via the xylem, and then grows laterally in the stem. The spread of conidia through the xylem stream has been postulated as a means of rapid spread and damage within the plant. Mycelium within xylem vessels varies in quantity from a few strands to masses of hyphae occupying the entire lumen. *C. gregata* types A and B appear to colonize the xylem vessels in the vascular systems of resistant and susceptible cultivars differently. The fungus also colonizes the leaf vascular system and can be isolated from leaf tissue with greater frequency after flowering; it can be detected with PCR assays isolated from stem tissue before stem or foliar symptoms develop. *C. gregata* produces gregatins, which are phytotoxins that may influence symptom development.

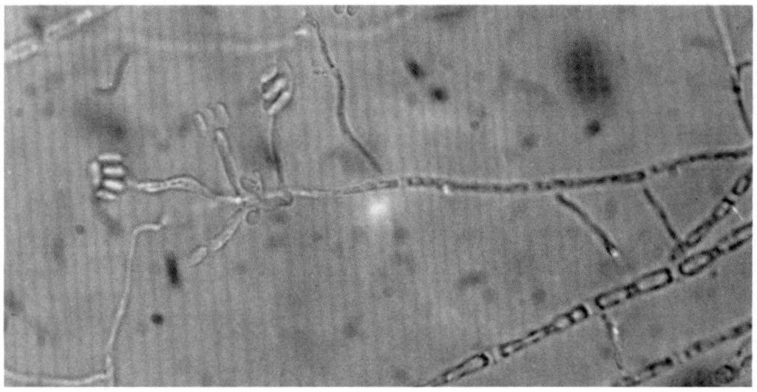

Fig. 104. Hyphae and conidia of *Cadophora gregata* on soybean stem decoction agar. (Courtesy L. E. Gray)

Environmental conditions and crop growth stage strongly influence the development of brown stem rot. Disease development is greatest at temperatures of 15–27°C. Internal pith and vascular browning can be reduced as the air temperature approaches 27°C, and few or no stem symptoms develop at 32°C. Foliar symptom development is greatest when conditions are cool and wet during the reproductive growth stages. Leaf symptoms do not develop if the air temperature exceeds 35°C during growth stages R3 and R4. Leaf symptoms are more severe on plants that receive heavy precipitation or irrigation that maintains the soil moisture near field capacity after flowering. Low soil pH (<6.0) may also enhance symptom expression. Foliar symptoms of brown stem rot are more prominent after flowering.

The survival of *C. gregata* is closely associated with rate of soybean residue decomposition. The severity of brown stem rot and yield reduction has been related to the quantity of propagules of the fungus in stem residue. The population in stem residue can be determined in the spring before the crop is planted by using a quantitative PCR assay or a semiselective medium to isolate the fungus. However, *C. gregata* does not sporulate on the semiselective medium, so suspect colonies must be transferred to plates of green bean or soybean stem agar for identification.

Brown stem rot can interact with other pathogens—most notably, the soybean cyst nematode, *Heterodera glycines*. For some cultivars with resistance to brown stem rot, infection by *H. glycines* may reduce resistance and increase the development of symptoms. In addition, both brown stem rot and infection by *H. glycines* occur in many of the same fields across the northern United States. These two diseases can simultaneously infect the same plant and cause symptoms of both diseases, but the result of this interaction on plant growth and yield is unknown.

Management

Brown stem rot can be managed by rotating crops and planting resistant cultivars. Three genes for resistance to brown stem rot in soybean have been identified—*Rbs1, Rbs2,* and *Rbs3*—and they have been mapped to chromosome 16. Resistance has been reported in many soybean plant introductions (PIs), including PI 88788, which is widely used for management of *H. glycines* and has been incorporated into many commercial cultivars. However, if used for brown stem rot, the PI 88788 source of resistance is effective only with specific modifier genes. Other sources of resistance to *H. glycines* may not provide resistance to brown stem rot. Cultivars with resistance to brown stem rot may produce more xylem vessels in response to infection by *C. gregata* and restrict the pathogen from the vascular system, compared with susceptible cultivars. Resistance to symptom development may not correlate with resistance to pathogen colonization in soybean stems. Some isolates of *C. gregata* can cause severe symptoms on cultivars with the *Rbs1* gene. The long-term use of cultivars with specific resistance genes can lower the frequency of *C. gregata* type A and increase the frequency of type B in field soils.

Growing a nonhost crop for a minimum of 3 years usually reduces the incidence and severity of brown stem rot. In many areas, soybean is rotated annually with a nonhost crop (e.g., corn), but this crop rotation sequence often does not provide long-term reduction in the incidence and severity of brown stem rot.

The pathogen survives longer and at higher population densities under no-tillage conditions than in buried residue. Therefore, burying infested crop residue can be help in fields where soil erosion is not a concern. Managing *H. glycines* may also contribute to management of brown stem rot.

Fungicides are not registered and have not been shown to be effective for the management of brown stem rot.

Selected References

Adee, E. A., Oplinger, E. A., and Grau, C. R. 1994. Severity of brown stem rot and soybean yield as influenced by tillage, rotation sequence and cultivar. J. Prod. Agric. 7:341-347.

Adee, E. A., Oplinger, E. S., and Grau, C. R. 1995. Inoculum density of *Phialophora gregata* related to severity of brown stem rot and yield of soybean in microplot studies. Plant Dis. 79:68-73.

Chen, W., Grau, C. R., Adee, E. A., and Meng, X. 2000. A molecular marker identifying subspecific populations of the soybean brown stem rot pathogen, *Phialophora gregata*. Phytopathology 90:875-883.

Harrington, T. C., and McNew, D. L. 2003. Phylogenetic analysis places the *Phialophora*-like anamorph genus *Cadophora* in the *Helotiales*. Mycotaxon 87:141-151.

Harrington, T. C., Steimel, J., Workneh, F., and Yang, X. B. 2003. Characterization and distribution of two races of *Phialophora gregata* in the north-central United States. Phytopathology 93:901-912.

Hughes, T. J., and Grau, C. R. 2010. Influence of soybean monoculture on *Phialophora gregata* f. sp. *sojae* IGS-genotype B isolate aggressiveness. Plant Dis. 94:69-74.

Hughes, T. J., Atallah, Z. K., and Grau, C. R. 2009. Real-time PCR assays for the quantification of *Phialophora gregata* f. sp. *sojae* IGS genotypes A and B. Phytopathology 99:1008-1014.

Hughes, T. J., Chen, W., and Grau, C. R. 2002. Pathogenic characterization of genotypes A and B of *Phialophora gregata* f. sp *sojae*. Plant Dis. 86:729-735.

Hughes, T. J., Koval, N. C, Esker, P. D., and Grau, C. R. 2009. Influence of monocropping brown stem rot-resistant and -susceptible soybean accessions on soil and stem populations of *Phialophora gregata* f. sp *sojae*. Plant Dis. 93:1050-1058.

Impullitti, A. E., and Malvick, D. K. 2014. Anatomical response and infection of soybean during latent and pathogenic infection by type A and B of *Phialophora gregata*. PLoS One 9(5):e98311.

Impullitti, A. E., Malvick, D. K., and Grau, C. R. 2009. Characterizing reaction of soybean to *Phialophora gregata* using pathogen population density and DNA quantity in stems. Plant Dis. 93:734-740.

Malvick, D., and Grunden, E. 2008. Association between genotypes of the brown stem rot pathogen *Phialophora gregata* and resistant and susceptible soybean cultivars in the north-central United States and Ontario. Can. J. Plant Pathol. 30:581-587.

Malvick, D., and Impullitti, I. 2007. Detection and quantification of *Phialophora gregata* in plant and soil samples with a quantitative, real-time PCR assay. Plant Dis. 91:736-742.

Mengistu, A., Tachibana, H., and Grau, C. R. 1991. Selective medium for isolation and enumeration of *Phialophora gregata* from soybean straw and soil. Plant Dis. 75:196-199.

Patzoldt, M. E., Carlson, S. R., and Diers, B. W. 2005. Characterization of resistance to brown stem rot of soybean in five accessions from central China. Crop Sci. 45:1092-1095.

Perez, P. T., Diers, B. W., Lundeen, P., Tabor, G. M., and Cianzio, S. R. 2010. Genetic analysis of new sources of soybean resistance to brown stem rot. Crop Sci. 50:2431-2439.

Sills, G. R., Gritton, E. T., and Grau, C. R. 1991. Differential reactions of soybean genotypes to isolates of *Phialophora gregata*. Plant Dis. 75:687-690.

Tabor, G. M., Tylka, G. L., and Bronson, C. R. 2006. Soybean stem colonization by genotypes A and B of *Cadophora gregata* increases with increasing population densities of *Heterodera glycines*. Plant Dis. 90:1297-1301.

Tabor, G. M., Tylka, G. L., Behm, J. E., and Bronson, C. R. 2003. *Heterodera glycines* infection increases incidence and severity of brown stem rot in both resistant and susceptible soybean. Plant Dis. 87:655-661.

Taylor, S. L., Peterson, R. E., and Gray, L. E. 1985. Isolation of gregatin A from *Phialophora gregata* by preparative high-pressure liquid chromatography. Appl. Environ. Microbiol. 50:1328-1329.

Waller, R. S., Nickell, C. D., and Gray, L. E. 1992. Environmental effects on the development of brown stem rot in soybean. Plant Dis. 76:454-457.

Workneh, F., Tylka, G. L., Yang, X. B., Faghihi, J., and Ferris, J. M. 1999. Regional assessment of soybean brown stem rot, *Phytophthora sojae,* and *Heterodera glycines* using area-frame sampling: Prevalence and effects of tillage. Phytopathology 89:204-211.

(Prepared by D. K. Malvick, C. R. Grau, and L. E. Gray)

Charcoal Rot

Charcoal rot is prevalent throughout the world, with reports from tropical to temperate regions. It is also referred to as "dry-weather wilt" and "summer wilt," because it often occurs when plants are under heat and drought stresses. These stresses can also occur in irrigated soybean crops if water is withheld during critical stages of plant growth after the beginning of flowering. Charcoal rot symptoms are evident in soybean plants of all ages, and the pathogen attacks plants throughout the season, often causing progressive debilitation of the host. The symptoms associated with this disease, which affect the roots and the lower stems, are often confounded with the effects of hot and dry soil conditions.

In the tropics, where the pathogen causes a blight of emerging seedlings, plant losses of up to 77% have been reported. When severe, the disease can reduce both yield and seed quality. Based on estimated yield losses in the United States, charcoal rot is often ranked as one of the most important diseases affecting soybean.

Symptoms and Signs

Infected seeds may be asymptomatic or symptomatic, and microsclerotia often develop in the fissures and cracks in seed coats (Fig. 105). Fungal hyphae penetrate inter- and intracellularly in seed coat tissues and form microsclerotia in the fissures and cracks in the seeds. The fungus remains dormant until the seeds germinate. Infected seeds may not germinate, or they may germinate and be symptomatic or asymptomatic.

An infected seedling may show a reddish-brown discoloration at the emerging portion of the hypocotyl. If infection occurs through the roots, discoloration is evident at the soil line and above, as the hypocotyl turns dark brown to black; the infected seedling may die, particularly under hot, dry conditions. If wet, cool conditions persist, an infected seedling can survive into the reproductive stages. Symptoms may reappear later with hot, dry weather. The pathogen produces an exotoxin (phaseolinone), which is responsible for germination inhibition and seedling wilt.

Plants infected after the seedling stage generally show no aboveground symptoms until after flowering, and symptoms

Fig. 105. Shriveled seed from a plant with charcoal rot (*Macrophomina phaseolina*). Note the development of microsclerotia in the fissures and cracks. (Courtesy J. B. Sinclair)

commonly appear during periods of hot, dry weather. Initially, an infected plant produces smaller than normal leaflets and displays a subtle loss of vigor. Later, the leaves turn yellow and wilt but remain attached (Fig. 106). After flowering, a light-gray or silvery discoloration of the epidermal and subepidermal tissues develops in the taproot and the lower part of the stem.

Microsclerotia on aboveground parts of the plant are first visible in the stem nodes as profuse, small, black, randomly distributed specks. The fungus causes a dark discoloration in the vascular tissues of the taproot and progresses up into the stem, discoloring the vascular and pith tissues. The woody portion of the root and the base of the stem may show black streaks when split (Fig. 107). When the lower stem and root samples from an infected plant are ground and plated on a selective medium, distinct colony-forming units (CFUs) are produced after 3 days of incubation (Fig. 108). The best diagnostic symptom is found when the epidermis is peeled away from the stem, exposing numerous small, black bodies that give infected tissue a grayish-black color, resembling a sprinkling of finely powdered charcoal (hence, the name "charcoal rot"). Microsclerotia (usually 50–75 μm in diameter) are frequently produced in the xylem and pith of the stem and may block water flow (Fig. 109).

Causal Organism

The charcoal rot pathogen, *Macrophomina phaseolina,* is the only species in the fungal genus *Macrophomina.* This pathogen has a wide host range, infecting approximately 500 plant species in more than 100 families throughout the world, including economically important crops such as common bean, maize, peanut, and sesame (*Sesamum indicum*), among others. The diversity of host species and the geographic range over which this fungus is found make it quite heterogeneous. *M. phaseolina* isolates have been subgrouped based on morphology, pathogenesis, and chlorate utilization in culture medium.

Cultural and morphological characteristics may differ among and within isolates and with the age of the culture. The fungus grows well on any standard solid cultural medium (e.g., potato dextrose agar [PDA]), and it produces abundant microsclerotia, depending on the nutritional level of the substrate. Isolates of *M. phaseolina* that do not produce microsclerotia are rare. Colonies of *M. phaseolina* are characteristically gray to black, depending on the nutrient level of the medium. Aerial mycelium with completely or partially appressed growth may or may not be produced. Some isolates form concentric growth rings. Hyphal branches generally arise at right angles to the parent hyphae, but branching at acute angles is common, as well. Isolates of the fungus also vary in microsclerotial size and the ability to produce pycnidia. Microsclerotial morphology is a key taxonomic characteristic in identification of *M. phaseolina.* Microsclerotia are jet black in color and round to oblong or irregular in shape; both size and shape vary with the isolate and the medium. Microsclerotia are uniformly reticulate and formed from aggregates of hyphal cells joined by a melaninlike material. Individual microsclerotia are composed of 50–200 or more individual cells united by a septal pore in each cell.

The pycnidial stage is common on garden bean and jute bean, but it is less common on soybean. Some strains of the

Fig. 106. Dead leaves and petioles remain attached to the stems of plants killed by charcoal rot (*Macrophomina phaseolina*). (Courtesy A. Mengistu)

Fig. 107. Soybean stem split in half, revealing microsclerotia of *Macrophomina phaseolina,* the charcoal rot pathogen. (Courtesy A. Mengistu)

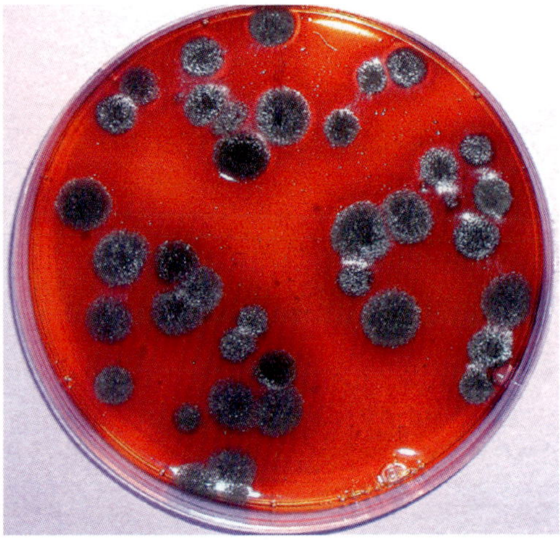

Fig. 108. Colonies of *Macrophomina phaseolina,* the fungus that causes charcoal rot, produced on selective media. (Courtesy A. Mengistu)

Fig. 109. Microsclerotia of *Macrophomina phaseolina*, the fungus that causes charcoal rot. (Courtesy A. Mengistu)

pathogen produce pycnidia on culture media, and a variety of techniques have been reported for induction of pycnidia, differing primarily in the substratum and light regime. Pycnidia, which are initially immersed in host tissues, are erumpent at maturity. They are more or less globose; membranous or subcarbonaceous; initially dark to grayish but become black with age; and generally 100–250 µm in diameter. The small, truncate ostiole may be inconspicuous or have a definite opening. Pycnidiospores are hyaline, single celled, and elliptic or oval; they have a length-to-width ratio of 3:1 and measure 6–12 × 14–33 µm.

The optimal temperature for growth in culture ranges from 28 to 35°C. The fungus can be readily isolated from host tissues or soil on a semiselective medium of PDA amended with chloroneb and streptomycin sulfate or with rifampicin and a nonionic surfactant.

Disease Cycle and Epidemiology

M. phaseolina survives in soil and soybean debris as microsclerotia, which serves as the primary source of inoculum. Seed infection occurs, as well, with the pathogen residing in the seed coat and causing seed rot or later seedling infection. The fungus can survive up to 3 years as mycelium in asymptomatic seeds and as microsclerotia in symptomatic seeds.

Infection of soybean from inoculum in the soil or seed can occur at emergence and at the cotyledonary stage, producing an 80–100% incidence of infected plants 2–3 weeks after planting. The pathogen usually remains latent in the host until environmental stresses (e.g., low soil moisture and high ambient temperature) coincide with the plant's reproductive growth stages. The levels of colonization in the root and lower stem tissues increase with stress and as the soybean plant matures. Seed yields often negatively correlate with the density of *M. phaseolina* propagules in host tissues. Colonized soybean tissues contribute microsclerotia to the soil after harvest. In dry soils, microsclerotia survive in soil or embedded in host residue for 2 or more years. In wet soils, microsclerotia cannot survive more than 7–8 weeks and mycelia cannot survive more than 7 days.

M. phaseolina survives even when soil nutrient levels are low and temperatures are greater than 30°C. Growth in soil is initiated by nutrients stored in microsclerotia and continues when soil nutrient levels are insufficient for fungal competitors. The population density of microsclerotia in the soil increases when soybean or another host species is grown continuously in the same field; the disease becomes more severe in successive soybean crops. Microsclerotia must germinate either on the surface of or in close proximity to the roots for infection to occur. They produce numerous germ tubes, and penetration generally occurs from appressoria formed over anticlinal walls of epidermal cells or through natural openings. The fungal hyphae grow first intercellularly and then intracellularly through the xylem and form microsclerotia, which plug the vessels. Microsclerotia form in green and juvenile tissues, usually as a result of moribundity and the release of nutrients. *M. phaseolina*, in part, causes disease via the mechanical plugging of xylem vessels by microsclerotia, toxin production, enzymatic action, and mechanical pressure during the penetration of the middle lamellae.

Management

All commercial soybean cultivars are considered susceptible to *M. phaseolina,* although a few lines show partial resistance. When possible, growers should not plant cultivars that have performed poorly in the past because of charcoal rot.

Because charcoal rot is triggered by stress, soybean management practices should be followed that minimize stress, such as adjusting seeding rates to minimize the impact of drought, maintaining good weed control, and irrigating (if available) during the later reproductive growth stages. Planting high-quality seed that is free from charcoal rot and other pathogens will produce vigorous seedlings, which tend to escape or minimize infection by the charcoal rot fungus. Planting at high seeding rates should be avoided, because high plant density often predisposes plants to drought stress (due to competition) and makes them more vulnerable to disease. Drought stress can be reduced using no-till and conservation tillage (versus conventional tillage), thereby reducing charcoal rot severity. Altering planting dates and cultivar maturities to avoid periods of drought stress can be beneficial, as well.

No chemicals are currently available for the control of charcoal rot. However, if seeds are infected, a seed treatment may be helpful.

Selected References

Bhattucharya, D., Dhar, T. K., and Siddiqui, K. A. I. 1994. Inhibition of seed germination by *Macrophomina phaseolina* is related to phaseolinone production. J. Appl. Bacteriol. 77:129-133.

Collins, D. J., Wyllie, T. D., and Anderson, S. H. 1991. Biological activity of *Macrophomina phaseolina* in soil. Soil Biol. Biochem. 23:495-496.

Dhingra, O. D., and Sinclair, J. B. 1978. Biology and pathology of *Macrophomina phaseolina*. Universidade Federal de Viçosa, Brazil.

Ma, J., Hill, C. B., and Hartman, G. L. 2010. Production of *Macrophomina phaseolina* conidia by multiple soybean isolates in culture. Plant Dis. 94:1088-1092.

Mayék-Pérez, N., López-Castañeda, C., González-Chavira, M., Garcia-Espinosa, R., Acosta-Gallegos, J., Martínez de la Vega, O., and Simpson, J. 2001. Variability of Mexican isolates of *Macrophomina phaseolina* based on pathogenesis and AFLP genotype. Physiol. Mol. Plant Path. 59:257-264.

Mengistu, A., Ray, J. D., Smith, J. R., and Paris, R. L. 2007. Charcoal rot disease assessment of soybean genotypes using a colony forming unit index. Crop Sci. 47:2453-2461.

Mihail, J. D., and Taylor, S. J. 1995. Interpreting variability among isolates of *Macrophomina phaseolina* in pathogenicity, pycnidium production, and chlorate utilization. Can. J. Bot. 73:1596-1603.

Smith, G. S., and Carvil, O. N. 1997. Field screening of commercial and experimental soybean cultivars for their reaction to *Macrophomina phaseolina*. Plant Dis. 81:363-368.

Twizeyimana, M., Hill, C. B., Pawlowski, M., Paul, C., and Hartman, G. L. 2012. A cut-stem inoculation technique to evaluate soybean for resistance to *Macrophomina phaseolina*. Plant Dis. 96:1210-1215.

(Prepared by A. Mengistu, A. Wrather, and J. C. Rupe)

Fusarium Blight or Wilt, Root Rot, and Pod and Collar Rot

Fusarium-induced diseases of soybean occur in most of the world's soybean-growing areas and have been attributed to different species, *formae speciales,* and physiological races of *Fusarium*. For example, sudden death syndrome (SDS) is caused by at least four species of *Fusarium* (see later in Part I, the section Sudden Death Syndrome). Fusarium blight or wilt of soybean was first reported in 1917, Fusarium root rot in 1961, and Fusarium pod and collar rot in 1972.

Fusarium diseases cause seed rots, damping-off, and seedling blights. On adult plants, these diseases cause root rot, reductions in plant height and weight, and reductions in root weight and volume, and they also affect root morphology, infect pods and seeds, and reduce seed yield. Economic losses of up to 59% have resulted from Fusarium blight or wilt, of up to 64% from root rot, and of up to 50% from reduced pod formation.

Since the 1970s, an increasing amount of evidence has led researchers and growers to conclude that Fusarium root rot is a significant disease in North America—especially in the primary soybean-growing regions—and that multiple species are involved. At least 22 *Fusarium* spp. have been isolated from soybean plants, but the role of many species in disease development has not been fully investigated. The etiology of Fusarium root rots is poorly understood. Most of the research on pathogenicity has shown that *Fusarium* spp. can cause disease on seedlings and a root rot in adult plants, but there can be significant differences among isolates of a particular species in their ability to cause disease.

F. acuminatum, *F. graminearum*, *F. oxysporum*, and *F. solani* appear to be the most common species isolated from soybean roots in North America, and researchers presume that these species occur in most soybean production regions throughout the world. One study from Canada identified nine *Fusarium* spp. from soybean; *F. oxysporum* was the most prevalent.

Symptoms and Signs

Fusarium Blight or Wilt. *F. oxysporum* causes blight or wilt. Symptoms appear approximately midseason during periods of hot conditions (temperatures higher than 28°C), particularly on plants growing in sandy soils. The disease has not been reported commonly in seedlings. The key symptom is blackening or browning of the vascular system in the roots and stems, which is evident when stems are split. Leaves of affected plants may become chlorotic and then wither and eventually drop. Flaccid leaves and wilted stem tips are the most common symptoms on young plants. Pods of infected plants are often poorly developed.

F. oxysporum is also reported to cause substantial damping-off and can cause root rot, but the strains that produce these diseases may be different from those that cause wilt. Fusarium wilt is not commonly reported from the north-central soybean production area of the United States.

Fusarium Root Rot and Seedling Disease. Root rot is caused by a number of *Fusarium* spp. One of those species is *F. oxysporum,* which usually causes root rot on seedlings and young plants under cool conditions (14°C). Older plants are generally less susceptible to root rot than younger ones. When the disease is severe, seedling emergence is slow and poor, and affected seedlings are stunted and weak. Symptoms are generally confined to the roots and lower stems (Fig. 110). Cotyledons of diseased seedlings are chlorotic and later become necrotic. The lower part of the taproot and lateral root system may be destroyed. The pathogen is usually confined to the cortex, but vascular elements are invaded in advanced stages of root rot. When soil moisture is low, infected seedlings or plants may wilt, and in some instances, an entire field of plants may be wilted. Adventitious roots commonly develop from the older portions of the taproot, producing a shallow, fibrous root system. Infected plants beyond the seedling stage seldom die, but their seeds are usually small and shriveled. Although the pathogen is normally confined to the roots, nearly mature pods may be invaded under prolonged humid, wet conditions. Pod infection may result in seed transmission of the pathogen.

A second *Fusarium* sp., *F. solani,* causes preemergence damping-off and root rot with limited postemergence damping-off. Lesions on the taproots and lateral roots of seedlings are light to dark brown or reddish-brown to black and often sunken, as well. Lesions also occur on the hypocotyls and cotyledons, but the primary damage occurs on the roots. On older plants, lesions on the roots and the bases of stems are dark brown to black and girdled. There is also extensive decay of the plant's cortex (Fig. 111), and there can be extensive decay of the lateral roots. Early infection and extensive root rot can result in the plant's death before flowering. Foliar symptoms can be associated with root rot on adult plants. The primary symptom is chlorosis that begins at the leaf edges and progresses inward. Veins may remain green from 1 to several weeks. Leaves may eventually become completely chlorotic and then necrotic; when they fall, the petioles remain attached to the stems. Foliar symptoms do not begin with the chlorotic mottling that is characteristic of SDS, but in advanced stages, symptoms may be confused with those of SDS. Plant stands, heights, weights, root volumes, and other root characteristics and seed yields are all reduced by root rot. Disease incidence and severity may be greater when poor-quality seeds are planted.

F. graminearum causes a root rot and seedling disease that is relatively new, with pathogenicity first verified in 2004. Symptoms can occur on both seedlings and adult plants in the reproductive stages. Light-brown, discolored, elongate lesions on shoots and roots become necrotic and turn brown to black, causing plants to wilt and die. Irregularly shaped, necrotic lesions may appear on cotyledons and leaves, and interveinal chlorosis of leaves may develop. Pod infection can result in the appearance of circular, chocolate-brown lesions that spread,

Fig. 110. Distinct brown lesions on the lower stems of seedlings, characteristic of Fusarium root rot (*Fusarium* spp.). (Courtesy J. B. Sinclair)

Fig. 111. Blackening or browning of the vascular system of a split root and stem (left), characteristic of Fusarium root rot (*Fusarium* spp.), and a healthy root and stem (right). (Courtesy B. D. Nelson)

causing young pods to yellow, wither, and die within 1 week. Seeds may not develop in infected pods, and seeds that do develop may be infected. The pathogen can also cause preemergence damping-off. In Brazil, strains of *F. graminearum* from soybean were identified as being from lineages 1, 2, and 8, and lineage 7 from the United States has also been shown to infect soybean. The disease is also reported in other countries, including Canada and India.

Researchers have identified a number of other *Fusarium* spp. that cause damping-off and/or root rot and thereby reduce plant growth and yield: *F. acuminatum* (syn. *Gibberella acuminata*); *F. avenaceum* (syn. *G. avenacea*); *F. commune; F. equiseti* (syn. *G. intricans*); *F. proliferatum* (syn. *G. intermedia*); *F. redolens; F. sporotrichioides;* and *F. tricinctum* (syn. *G. tricincta*). In the U.S. state of Iowa, both *F. proliferatum* and *F. sporotrichioides* were found to cause severe root rot under greenhouse conditions, and *F. equiseti* was found to increase root rot in the field. *F. heterosporum* (syn. *G. cyanea*) and *F. verticillioides* (syns. *F. moniliforme* and *G. moniliformis*) have also been reported as pathogenic on soybean. Field research with many of the *Fusarium* spp. has been limited; thus, the effects of many species on plant growth in the field are not clearly understood. For most *Fusarium* spp. that are pathogenic on soybean, little information is available on the etiology of disease. The following additional species have been isolated from soybean, but there is no solid evidence of their pathogenicity: *F. culmorum; F. merismoides* (syn. *Fusicolia merismoides*); *F. poae* (syn. *Sporotrichum poae*); *F. sambucinum* (syn. *G. pulicaris*); *F. subglutinans* (syn. *G. subglutinans*); and *F. udum* (syn. *G. indica*).

Pod and Collar Rot (see also the previous section Fusarium Seed Rot and *Fusarium* Species Associated with Seed). *F. incarnatum* (syns. *F. pallidoroseum* and *F. semitectum*) causes pod and collar rot. It produces reductions in emergence and causes depressed, water-soaked, cream-colored, serrated lesions to form on the cotyledons and hypocotyls of emerging and slightly older seedlings. As seedlings mature, these areas turn dark brown to black and eventually coalesce to form large lesions. Invaded hypocotyls are thin and soft, and shoot weight and root volume are both reduced. In advanced stages of pod and collar rot, fungal growth is evident on infected tissues.

Pods may be infected and dry prematurely; the drying begins at the pod tip and progresses toward the base. Pods eventually turn dark brown or black, and severely infected pods produce no seeds. The seed coats and cotyledons of diseased seeds turn dark brown to black; this is particularly apparent in seeds that contact infected pod tissue. Plants with decayed pods often show no other symptoms.

Moldy and/or chalky seeds can also be associated with a high level of seed infection. In the U.S. state of Mississippi, *F. incarnatum* was reported as widespread and prevalent in seeds and pods. In Brazil and Puerto Rico, the pathogen was shown to be seedborne and to render seeds nonviable. In Iowa, *F. incarnatum* was reported to cause a root rot and to reduce various parameters of root growth.

Causal Organisms

Fusarium spp. can be difficult to identify. Useful references for preliminary morphological identification include *The* Fusarium *Laboratory Manual* (Leslie and Summerell, 2006) and *The Genus* Fusarium—*A Pictorial Atlas* (Gerlach and Nirenberg, 1982). DNA sequence comparisons are needed, however, to thoroughly identify species.

F. oxysporum causes wilt and root rot of soybean. Two *formae speciales*—*F. oxysporum* f. sp. *tracheiphilum* and *F. oxysporum* f. sp. *vasinfectum*—also cause blight or wilt. A number of physiological races of *F. oxysporum* f. sp. *tracheiphilum* (primarily a cowpea pathogen) have been reported, but only race 1 is known to cause blight or wilt in soybean. Race 2 of *F. oxysporum* f. sp. *vasinfectum* (which causes vascular wilt of cotton) can also infect and cause blight or wilt in soybean.

F. oxysporum is highly variable in culture, producing either aerial mycelium or a pionnotal form, which appears wet and slimy. In general, mycelium on potato dextrose agar (PDA) is floccose and septate. Colonies are white to pale violet. Isolates commonly produce a pale- to dark-violet or dark-magenta pigment in the agar. Conidiophores are branched or unbranched monophialides and usually short and doliiform. Sporodochia are usually abundant and pale orange. Chlamydospores are intercalary or terminal and form singly or in pairs. Microconidia are single celled and oval to kidney shaped, and they are produced in false heads. Macroconidia are hyaline, usually three septate, and straight to slightly curved; a single macroconidium has a foot-shaped to pointed basal cell and a tapered and curved apical cell, sometimes with a slight hook.

The different *formae speciales* of *Fusarium* are indistinguishable morphologically and culturally. *F. oxysporum* is difficult to distinguish from *F. solani* and *F. subglutinans.*

F. solani (syns. *F. haematococcum* and *Nectria haematococca*) is variable in culture, has rapid growth, and is usually white to cream colored on PDA. Sporodochia are common and may appear cream colored or less commonly blue or green on carnation leaf agar (CLA). The undersurface is generally colorless, but some strains of *F. solani* produce a dark-violet to maroon pigment. Conidiophores are branched or unbranched monophialides. Chlamydospores are intercalary or terminal, form singly and in pairs, and are usually abundant. Microconidia are generally single celled and oval to kidney shaped; they may be sparse to abundant. Macroconidia are abundant, three to seven septate, relatively wide, and generally straight (although some are slightly curved). The basal cell is rounded or foot shaped or notched, whereas the apical cell is blunt and rounded.

F. graminearum (syn. *Gibberella zeae*) on PDA grows rapidly and can vary in color from white to orange to yellow. Sporodochia are reddish to orange, and red pigments form in the agar. Microconidia are absent. Macroconidia are five or six septate and moderately curved to straight; they have tapered apical cells and well-developed, foot-shaped basal cells. Chlamydospores form most commonly in macroconidia but may occur in mycelia; they can be single or form in clusters or chains. The fungus is homothallic, and the perithecia (*G. zeae*) form readily on CLA, on carrot agar, and on corn kernels. Subdivisions of *F. graminearum* have been proposed based on molecular characteristics.

F. incarnatum (syns. *F. semitectum* and *F. pallidoroseum*) causes primarily pod and collar rot of soybean. The fungus is variable in culture but usually appears as an off-white mycelium that becomes beige or brown with age on PDA. Sporodochia, if present, are orange. The agar undersurface varies from peach to tan to brown. Conidiophores are unbranched or branched monophialides and polyphialides. Chlamydospores (which are uncommon) are intercalary or terminal. Microconidia (also uncommon) are usually one septate and pyriform to obovate. Mesoconidia are abundant in aerial mycelia and distinctive on CLA. Macroconidia are usually three to five septate, are relatively slender, and have curved dorsal and straighter ventral surfaces. The apical cell is curved and tapers to a point, whereas the basal cell is foot shaped.

F. proliferatum (syn. *G. intermedia*) causes a root rot and seedling disease. In culture, the aerial mycelium is generally abundant and white, but it may become purple-violet with age. Sporodochia, if present, may be tan to pale orange. In macroconidia, the apical cells are curved and the basal cells are poorly developed; they are usually three to five septate. Microconidia are usually abundant, single celled, and club shaped with flattened bases; they may form chains. Conidiogenous cells are monophialides and polyphialides. Chlamydospores are absent.

F. sporotrichioides causes a root rot and a seedling disease. In culture, mycelia are dense and white or pale red; red pigments are found in older cultures. Orange sporodochia may be present. Macroconidia are abundant, have curved and tapering apical cells, and have indistinctly foot-shaped or notched basal cells; macroconidia are generally three to five septate. Pyriform microconidia have no septa, whereas ellipsoid to fusoid forms have one septum. Fusoid mesoconidia may also form and have up to five septa. Conidiogenous cells are monophialides and polyphialides. Chlamydospores are abundant, globose, and smooth walled; they are initially hyaline and become light brown with age.

F. equiseti (syn. *G. intricans*) causes a root rot. In culture, it has abundant white mycelium that becomes brown with age. Sporodochia may form but be covered by mycelium. The spore mass may be pale orange to light brown, and a dark-brown pigment may appear on the agar surface. Macroconidia from sporodochia are five to seven septate, are long and slender, and have tapered and elongate, whiplike apical cells. The basal cell has a prominent foot shape that may be elongated. Microconidia are not observed, but chlamydospores with brown pigment and verrucose walls (when mature) are abundant.

Disease Cycle and Epidemiology

F. oxysporum inhabits the soil by colonizing various plant residues, and it overseasons as chlamydospores or mycelia. Seedborne infection by *F. oxysporum* has been reported and occasionally reduces the rate of seed germination. *F. oxysporum* f. sp. *tracheiphilum* penetrates soybean plants through wounds. In the early stages of infection, it is confined to xylem vessels near the pith. Later, most of the xylem vessels become filled with mycelia, and the parenchymal tissues are invaded, as well. Penetration occurs directly through epidermal cells and indirectly through lenticels, from wounds caused by secondary root development, and via the stomates on the hypocotyl. Initial intercellular penetration is followed by intercellular growth of hyphae in the cortex. The stele generally is not affected; however, in advanced stages, the xylem may be invaded.

Disease development is influenced greatly by the environment. *F. oxysporum* is most destructive when the soil is saturated and temperatures are 14–23°C; infection is greatest at temperatures in lower end of the range. Drought can aggravate disease development.

F. oxysporum has been isolated from 16 dicotyledonous weeds representing 10 plant families that are commonly found in soybean fields. Symptoms of Fusarium blight were not observed on the weeds, but all the isolates were pathogenic on soybean. *F. oxysporum* interacts with *Rhizoctonia solani* in causing a root rot of soybean (see later in Part I, the section Rhizoctonia Damping-Off and Root Rot).

The biology of soybean root rot caused by *F. solani, F. graminearum,* and other *Fusarium* spp. known to infect soybean roots is poorly understood. Primary inoculum originates in the soil. *F. solani* can colonize soybean immediately following planting, and mycelia or conidia directly penetrate the roots and invade the cortex. *F. solani* can infect soybean roots under both wet and dry soil conditions. Root rot is favored by conditions that reduce root growth and plant vigor, such as increased bulk density of the soil. Isolates of *F. solani* can be aggressive and cause extensive cortical root rot. Infection by *Macrophomina phaseolina* and *F. solani* together can increase the damage caused by root rot. *F. solani* and other *Fusarium* spp. are often members of a complex of fungi that invade soybean roots. Isolates of *F. solani* from soybean are pathogenic on roots of common bean.

Several nematodes—namely, *Heterodera glycines, Meloidogyne* spp., and *Belonolaimus longicaudatus*—predispose seedlings and young soybean plants to infection by *F. oxysporum* (see later in Part I, Diseases Caused by Nematodes), as do dinitroaniline herbicides. Seedlings growing in soil infested with *H. glycines* and *F. oxysporum* frequently develop symptoms of Fusarium wilt. Soybean plants are less predisposed to infection by *Meloidogyne* spp. than by *H. glycines,* because infection by *Meloidogyne* spp. is limited mainly to the root-tip region, whereas juveniles of *H. glycines* migrate intercellularly, resulting in slight cellular destruction. In contrast, larvae of *H. glycines* invade more mature root tissue, migrate intracellularly, and cause extensive wounding, thus making soybean plants more susceptible to attack by *Fusarium* spp. In Iowa, *H. glycines* was shown to increase the severity of root rot caused by *F. oxysporum, F. graminearum, F. semitectum, F. solani,* and *F. sporotrichioides.*

Management

If cultivars resistant to *Fusarium* spp., *H. glycines,* and *Meloidogyne* spp. are available, growers are encouraged to plant them. Finding a cultivar that is resistant to all *Fusarium* spp. is not likely, but some cultivars seem to have partial resistance to more than one species.

Growers are also advised to plant high-quality seeds in warm, well-drained soil and to use cultivation practices that prevent or reduce soil compaction and promote favorable soil moisture. Maintaining good soil fertility is important, as well. In fields with a history of Fusarium root rot, soil should be ridged around the bases of plants to promote the development of adventitious roots from stem bases. Crop rotation to nonhost plants may reduce pathogen inoculum levels, especially when the previous crop was infected by a *Fusarium* sp. pathogenic to soybean.

Seed treatments with fungicides can reduce damage from seedling diseases caused by *Fusarium* spp. Seed treatment with *Bacillus subtilis* has shown promise in reducing damage from root rot.

Selected References

Aoki, T., O'Donnell, K., and Geiser, D. 2014. Systematics of key phytopathogenic *Fusarium* species: Current status and future challenges. J. Gen. Plant Pathol. 80:1-11.

Bienapfl, J. C., Malvick, D. K., and Percich, J. A. 2010. First report of *Fusarium redolens* causing root rot of soybean in Minnesota. Plant Dis. 94:1069.

Bradley, C. A. 2008. Effect of fungicide treatments on stand establishment, seedling disease, and yield of soybean in North Dakota. Plant Dis. 92:120-125.

Carson, M. L., Arnold, W. E., and Todt, P. E. 1991. Predisposition of soybean seedlings to Fusarium root rot with trifluralin. Plant Dis. 75:342-347.

Datnoff, L. E., and Sinclair, J. B. 1988. The interaction of *Fusarium oxysporum* and *Rhizoctonia solani* in causing root rot of soybeans. Phytopathology 78:771-777.

Díaz-Arias, M. M., Leandro, L. F., and Munkvold, G. P. 2013. Aggressiveness of *Fusarium* species and impact of root infection on growth and yield of soybean. Phytopathology 103:822-832.

Díaz-Arias, M. M., Munkvold, G. P., Ellis, M. L., and Leandro, L. F. S. 2013. Distribution and frequency of *Fusarium* species associated with soybean roots in Iowa. Plant Dis. 97:1557-1562.

Díaz-Arias, M. M., Tylka, G. L., Leandro, L., and Munkvold, G. P. 2011. Interactions between Fusarium root rot pathogens and *Heterodera glycines.* Phytopathology 101:S42.

Ellis, M. L., Broders, K. D., Paul, P. A., and Dorrance, A. E. 2011. Infection of soybean seed by *Fusarium graminearum* and effect of seed treatments on disease under controlled conditions. Plant Dis. 95:401-407.

Gerlach, W., and Nirenberg, H. I. 1982. The Genus *Fusarium*—A Pictorial Atlas. G. Hagedorn, M. Burhenne, and H. I. Nirenberg, eds. Biologischen Bundesanstalt für Land-und Forstwirtschaft, Berlin, Germany.

Jensen, C. E., Kurle, J. E., and Percich, J. A. 2004. Integrated management of edaphic and biotic factors limiting yield of irrigated soybean and dry bean in Minnesota. Field Crop Res. 86:211-224.

Killebrew, J. F., Roy, K. W., and Abney, T. S. 1993. Fusaria and other fungi on soybean seedlings and roots of older plants and interrelationships among fungi, symptoms and soil characteristics. Can. J. Plant Pathol. 15:139-146.

Killebrew, J. F., Roy, K. W., Lawrence, G. W., McLean, K. S., and Hodges, H. H. 1988. Greenhouse and field evaluation of *Fusarium solani* pathogenicity to soybean seedlings. Plant Dis. 72:1067-1070.

Leslie, J. F., and Summerell, B. A. 2006. The *Fusarium* Laboratory Manual. Blackwell, Oxford, U.K.

Leslie, J. F., Pearson, C. A. S., Nelson, P. E., and Toussoun, T. A. 1990. *Fusarium* spp. from corn, sorghum, and soybean fields in the central and eastern United States. Phytopathology 80:343-350.

Martinelli, J. A., Bocchese, C. A. C., Xie, W., O'Donnell, K., and Kistler, H. C. 2004. Soybean pod blight and root rot caused by lineages of the *Fusarium graminearum* and the production of mycotoxins. Fitopatol. Bras. 29:492-498.

Meyer, P. W., Clancey, M., Brose, I. E., and Kurle, J. E. 2010. First report of Fusarium root rot in soybean caused by *Fusarium tricinctum* in Minnesota. Phytopathology 100:S188.

Nelson, B. D., Hansen, J. M., Windels, C. E., and Helms, T. C. 1997. Reaction of soybean cultivars to isolates of *Fusarium solani* from the Red River Valley. Plant Dis. 81:664-668.

Roy, K. W., and Ratnayake, S. 1997. Frequency of occurrence of *Fusarium pallidoroseum,* effects on seeds and seedlings, and associations with other fungi in soybean seeds and pods. Can. J. Plant Pathol. 19:188-192.

Schlub, R. L., Lockwood, J. L., and Komada, H. 1981. Colonization of soybean seeds and plant tissue by *Fusarium* species in soil. Phytopathology 71:693-697.

Sumner, D. R., and Minton, N. A. 1987. Interaction of Fusarium wilt and nematodes in Cobb soybean. Plant Dis. 71:20-23.

Zhang, J., Xue, A., Cober, E., Morrison, M., Zhang, H., Zhang, S., and Gregorich, E. 2013. Prevalence, pathogenicity and cultivar resistance of *Fusarium* and *Rhizoctonia* species causing soybean root rot. Can. J. Plant Sci. 93:221-236.

Zhang, J. X., Xue, A. G., Zhang, H. J., Nagasawa, A. E., and Tambong, J. T. 2010. Response of soybean cultivars to root rot caused by *Fusarium* species. Can. J. Plant Sci. 90:767-776.

(Prepared by B. D. Nelson)

Phytophthora Root and Stem Rot

Phytophthora root and stem rot occurs predominately in fields in which soils drain slowly and readily become saturated for days, creating the ideal environmental conditions for infection. The more prevalent of the two *Phytophthora* spp. found on soybeans is *P. sojae*. In the United States, yield losses attributed to *P. sojae* were as great as 100% (121,406 ha) in one state in the late 1960s; more recent losses have been set at 1–2 million metric tons a year. Significant yield losses attributed to *P. sojae* have also been reported from Argentina, Australia, Brazil, China, and Japan.

A second *Phytophthora* spp. that has been documented in a number of surveys from diseased seedlings was described in 2009 as *P. sansomeana*. This species had previously been isolated sporadically from seedlings with symptoms of damping-off and from surveys using soybean as a bait host. In one survey from the U.S. state of Illinois, 70% of the isolates recovered from two fields with a long history of seedling blight were *P. sansomeana*. In 2010, this species was reported to infect soybean in China. Moreover, it has a broader host range than *P. sojae,* including corn, Douglas fir, and several types of weeds.

Symptoms and Signs

For both *Phytophthora* spp., reduced stands and damping-off are the primary symptoms in the early part of the season on highly susceptible soybean (Fig. 112). Empty spaces in fields, as well as skips and remnants of dying seedlings, can be observed scattered throughout fields when environmental conditions favor infection (Fig. 113). Flooding injury is often mistaken for Phytophthora root and stem rot, but in flooding situations, the epidermis is killed and the root stele is still present. With Phytophthora root and stem rot, all the roots are brown and soft or missing.

Mid- to late-season symptoms include stunting, root rot, and stem rot on highly susceptible cultivars. Stem rot, in which a chocolate-brown canker develops up the stem, is the most diagnostic feature (Fig. 114). Internally, both the cortex and the

Fig. 112. Postemergence damping-off, caused by *Phytophthora sojae*. (Courtesy A. E. Dorrance)

Fig. 113. Cultivars resistant (left) and susceptible (right) to Phytophthora root and stem rot (*Phytophthora sojae*). Note the empty spaces and remnants of dead plants. (Courtesy A. E. Dorrance)

Fig. 114. Linear stem lesion and girdling of the stem, characteristic of Phytophthora stem rot (*Phytophthora sojae*). (Courtesy A. E. Dorrance)

vascular tissue have a brown discoloration. This has been observed at all growth stages and develops on highly susceptible cultivars 1–2 weeks after saturating rain. Affected plants wilt and the leaves turn yellow as the cankers develop on the main stems. The leaves droop and remain attached after the plants die. Plants that die after the R1 growth stage may be observed at harvest. In some instances, the canker develops up one side of the stem on cultivars with high levels of resistance. Only *P. sojae* has been recovered from stem cankers in the field.

In older plants with high levels of partial resistance, the only symptoms of Phytophthora root and stem rot are discoloration of the secondary roots and the taproot. These plants are not killed by the disease but may be stunted in comparison to plants with effective resistance genes.

Causal Organisms

P. sojae and *P. sansomeana* are similar in many ways to other *Phytophthora* spp. in the *P. megasperma* complex. All members of the complex produce simple, indeterminate sporangiophores. Typically, terminal sporangia are obpyriform (32–53 × 42–65 µm) and nonpapillate (Fig. 115). Sporangia germinate indirectly by extruding fully formed zoospores into thin, delicate, membranous, evanescent vesicles, which quickly expand and rupture. Zoospores are sometimes trapped within sporangia and germinate in place. The resulting germ tubes penetrate the sporangia's walls. Empty sporangia commonly proliferate internally, forming new sporangia terminally or directly. Hyphal swellings are also commonly formed.

The optimum temperature for direct germination is 25°C; for indirect germination, it is 14°C, and for zoospore germination, it is 20°C. Zoospores are ovoid, bluntly pointed at one or both ends, and flat on the sides. A single zoospore has two flagella: a short one directed anteriorly and another four to five times as long directed posteriorly. At the end of the motile period, which may last up to several days, zoospore movement becomes sluggish and jerky and encystment occurs. Encystment may be triggered prematurely by agitation or by bivalent or trivalent cations. Cysts often germinate immediately and directly, producing germ tubes, which generally swell to form appressoria when they contact solid surfaces. After a short time, normal growth of mycelia is resumed from the appressoria. Cysts sometimes germinate by producing secondary zoospores, leaving the cyst membranes behind. On rare occasions, a miniature sporangium is formed at the tip of the germ tube. Cysts germinate more vigorously in nutrient solutions than in distilled water.

P. sojae and *P. sansomeana* are both homothallic. For *P. sojae,* antheridia and oogonia develop in abundance on agars containing corn meal, lima bean, V8 juice, and one-quarter-strength potato dextrose (PDA). Antheridia are diclinous and generally paragynous, although amphigynous antheridia can be found. Oogonia (29–58 µm) are thin walled and spherical or subspherical. An oospore develops after an antheridium fertilizes an oogonium. For *P. sansomeana,* antheridia and oogonia develop in abundance on the same media but also on full-strength PDA. Using this method is one of the ways that *P. sansomeana* can be distinguished from *P. sojae.* The antheridia are predominately paragynous, and the oogonia tend to be somewhat larger (37–45 µm).

A dormant oospore has thick, smooth inner and outer walls; a fine-grained cytoplasm; a spherical, refractive body in the center; a well-developed reserve granule; and a pair of pellucid bodies at the outer edge of the cytoplasm (Fig. 116). Germination may occur after the pellucid bodies merge and fuse. At germination, the pellucid bodies are no longer detectable, the smooth inner wall erodes, the central refractive body is absorbed, and the spore has the appearance of a sporangium. When the oospore germinates, the inner wall is absorbed and the germ tube produces either a sporangium or a mycelium.

Oospores may germinate approximately 30 days after formation. Germination occurs in distilled water and is enhanced by low levels of nutrients and root exudates. Light increases the percentage of germination but inhibits subsequent formation of sporangia. The inhibition of sporangial formation is reversed by root exudates. Oospore germination is initiated within 2 days of separation from mycelium but is nonsynchronous and may continue for 30 days or more. The optimum temperature for formation and germination of oospores is 24°C.

P. sojae sporangia develop in dilute extracts of lima bean or other material; they can also be induced to form on solid agar (e.g., V8 juice) if washed repeatedly with water. Sporangia will form abundantly in young cultures grown on lima bean extract that are washed repeatedly with Chen–Zentmyer salt solution. The medium and the age of the culture affect oospore size and sporangial shape and size. *P. sansomeana* sporangia will develop in water.

Mycelium is coencytic when young and becomes septate with age. It branches mostly at right angles, and there is a slight constriction at the base of each branch. Hyphae are 3–9 µm wide and frequently slightly curled. The optimum temperature for the growth of most isolates is 25–28°C.

Isolates of *P. sojae* vary widely in cultural characteristics, morphology, and virulence to specific resistance (*R*) genes. There are 55 described races of *P. sojae.* Instead of being assigned to races, isolates are often described based on their virulence formulas. A virulence formula is based on a series of 15 *Rps* genes found in soybean differentials. Virulence on all 15

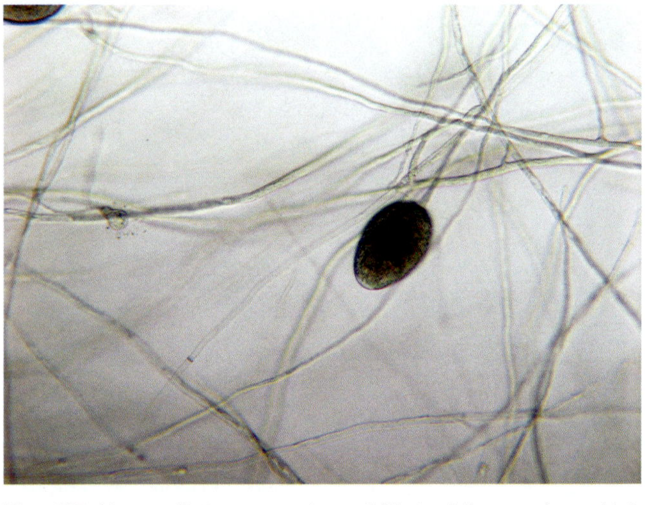

Fig. 115. Nonpapillate sporangium of *Phytophthora sojae,* which causes Phytophthora root and stem rot. (Courtesy A. E. Dorrance)

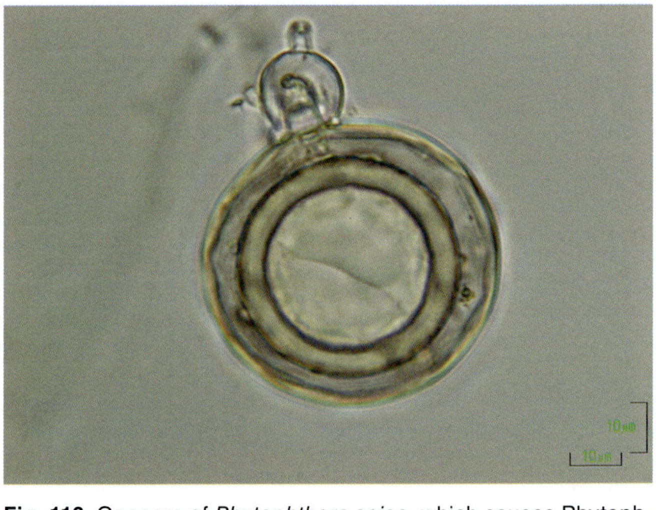

Fig. 116. Oospore of *Phytophthora sojae,* which causes Phytophthora root and stem rot. (Courtesy A. E. Dorrance)

differentials or on a subset of those differentials (*rps, Rps*1a, *Rps*1b, *Rps*1c, *Rps*1k, *Rps*3a, *Rps*6) can be used to define the virulence formula of any given isolate.

Three common selective media are used to isolate both *P. sojae* and *P. sansomeana* from plants: P10VP, PARPH, and PBNIC. The first two have a cornmeal agar base and include pimaricin (10 mg/L) and quintazone (100 mg/L) for selective inhibition of non-oomycetes. The first utilizes vancomycin (200 mg/L) and the second ampicillin (250 mg/L) and rifampicin (10 mg/L) for bacterial control. The PBNIC medium utilizes benomyl (10 mg/L), quintazone (40 mg/L), and iprodione (20 mg/L) for control of non-oomycetes and neomycin sulfate (100 mg/L) and chloramphenicol (10 mg/L) for bacterial control. Hymexazol (20 mg/L) can be used in all three media for partial control of *Pythium* spp. *P. sojae* is usually not directly isolated from soil. A semiquantitative baiting method has been described in which soybean leaf disks are floated on water flooded over soil for 2 h and then plated on selective media.

The genome of *P. sojae* has been sequenced. Simple sequence repeat markers have been used in population studies, which have included the mapping, identification, and cloning of several of the avirulence loci.

Disease Cycle and Epidemiology

P. sojae persists in soil as oospores in crop residues or free in soil after residues decompose. Many oospores are formed in roots and stems of susceptible and moderately resistant soybean cultivars (Fig. 117). Oospores can survive for many years without a host. The factors that control dormancy and germination in soil have not been determined.

P. sojae cannot be recovered from soil directly. Soil that is incubated for 1–2 weeks at 25°C can then be baited using soybean seedlings or leaf disks. Little oospore germination occurs until soil temperatures exceed 15°C, and germination is delayed at that temperature. Based on these and other observations, oospores are purported to germinate whenever the temperature and soil-water conditions are suitable for forming sporangia. Sporangia form after repeated saturation events, at which time zoospores are released. Sporangia form on the roots of infected plants, providing secondary inoculum, but their role in spread is thought to be minimal. Zoospores can swim short distances (1 cm or less) in saturated soil but can also be moved via flooding water. Phytophthora root rot is essentially a monocyclic disease.

Zoospores are attracted to seeds and roots by genistein and other isoflavonoid exudates. They encyst on the root surfaces and immediately germinate and penetrate the roots. Hyphae grow intercellularly in root tissues. Globular and fingerlike haustoria penetrate the host cells in susceptible and moderately resistant cultivars. Colonization of the cortex is similar in susceptible and *R*-gene-mediated resistance for approximately 14 h. Colonization then ceases in *R*-gene-mediated resistance, as *P. sojae* fails to reach the vascular tissue. Oogonia will form in cultivars that are susceptible, are moderately resistant, or have *R*-gene-mediated resistance. However, many more oospores will form in cultivars that are highly susceptible.

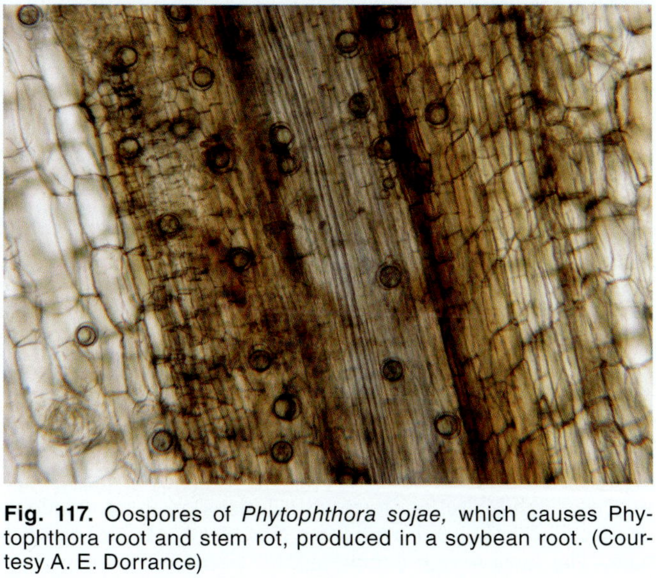

Fig. 117. Oospores of *Phytophthora sojae,* which causes Phytophthora root and stem rot, produced in a soybean root. (Courtesy A. E. Dorrance)

Phytophthora root and stem rot occur most commonly in heavy, tightly compacted, fine (clay) soil and soil that has an impervious shallow hardpan and is subject to saturation and flooding. Flooding rains within 1 week after planting create the conditions most favorable for disease development.

Management

Host resistance and fungicide seed treatments are the two primary strategies for management of *P. sojae*. Key management strategies for *P. sansomeana* have yet to be defined.

Resistance to *P. sojae* is expressed in at least three types. The first is *R*-gene mediated, which is expressed as immunity when effective. There are currently more than 14 single *Rps* genes (*Rps*1a, *Rps*1b, *Rps*1c, *Rps*1d, *Rps*1k, *Rps*2, *Rps*3a, *Rps*3b, *Rps*3c, *Rps*4, *Rps*5, *Rps*6, *Rps*7, and *Rps*8) at more than seven loci, but only a few of them (*Rps*1a, *Rps*1c, *Rps*1k, *Rps*3a, and *Rps*6) have been deployed in commercial cultivars. In some instances, virulence exists in the *P. sojae* population before the widespread use of an *R* gene. In addition, some genes are no longer effective in certain regions where cultivars in maturity groups I, II, and III are grown (e.g., *Rps*1a in Ohio, northern Indiana, southern Michigan) because of the development of a virulent *P. sojae* population able to overcome the resistance.

A second type of race-specific resistance is referred to as "root resistance." The hypocotyls have a susceptible or intermediate response (50% are killed or develop elongated lesions), but the roots are highly resistant. Soybean cultivars or lines with *Rps*2 often express this type of resistance.

A third type of resistance, which is nonrace specific and quantitatively inherited, is referred to as "partial resistance," "field resistance," and "tolerance." This type of resistance is not expressed in the seed but is effective once the first trifoliolate forms, based on greenhouse assays and in the field. Conrad is one of the best-studied cultivars with partial resistance, and commercial soybean cultivars have been identified as having partial resistance. Several different mechanisms have been associated with partial resistance, including increased amounts of preformed suberin and multiple defense pathways that are up-regulated following infection.

For areas in which *P. sojae* has been a consistent problem, cultivars that have both *Rps*-mediated resistance and high levels of partial resistance should be used in combination with a seed treatment. Several surveys have been conducted throughout both the U.S. Midwest and some regions in the South to assess virulence patterns within *P. sojae* populations. These regional surveys provide data that assists seed companies in knowing which gene stacks will be critical for certain regions; however, in some cases, none of the genes is effective against all the *P. sojae* populations (e.g., in Ohio, northern Indiana, southern Michigan). For an individual field, if a cultivar develops stem rot, that is clear evidence that the *Rps* gene is not effective and that the partial resistance level is not high enough to prevent disease. It is important to scout fields 1–2 weeks after saturating rain to assess the effectiveness of the resistance in soybean cultivars.

Metalaxyl/Mefenoxam is an acylalanine fungicide specific for oomycetes and is often applied to cultivars with partial resistance as a seed treatment. Used as an in-furrow application, this product is water soluble and taken up by the roots. The in-furrow application lasts longer than the seed treatment, and higher rates of seed treatment are necessary when conditions are favorable and inoculum levels are high.

In some regions, additional controls are needed to limit yield losses even further. Improved drainage is important to reduce the incidence and severity of flooding through tillage and narrow spacing of drain tiles. For irrigated soybean, it is critical to monitor soil saturation levels and timing irrigation for later growth stages. No-till production practices result in higher inoculum levels in the top 4–5 cm of soil and maintain soil moisture for longer periods. Crop rotation should be used to avoid the build-up of inoculum, and the application of potash fertilizer and animal manure immediately before planting should be avoided. Very early planting, when soil temperatures are 15°C, may allow soybean to avoid significant damage but increases the likelihood of infection by *Pythium* spp. (see the following section, Pythium Seed Rot, Damping-Off, and Root Rot).

Selected References

Canaday, C. H., and Schmitthenner, A. F. 2010. Effects of chloride and ammonium salts on the incidence of Phytophthora root and stem rot of soybean. Plant Dis. 94:758-765.

Dong, S., Qutob, D., Tedman-Jones, J., Kuffu, K., Wang, Y., Tyler, B. M., and Gijzen M. 2009. The *Phytophthora sojae* avirulence locus Avr3c encodes a multi-copy RXLR effector with sequence polymorphisms among pathogen strains. PLoS One 4:e5556.

Dorrance, A. E., and Grünwald, N. 2009. *Phytophthora sojae:* Diversity among and within populations. Pages 197-212 in: Oomycete Genetics and Genomics: Diversity, Interactions, and Research Tools. K. Lamour and S. Kamoun, eds. Wiley–Blackwell, Hoboken, NJ.

Dorrance, A. E., and McClure, S. A. 2001. Beneficial effects of fungicide seed treatments for soybean cultivars with partial resistance to *Phytophthora sojae*. Plant Dis. 85:1063-1068.

Dorrance, A. E., Berry, S. A., Anderson, T. R., and Meharg, C. 2008. Isolation, storage, pathotype characterization, and evaluation of resistance for *Phytophthora sojae* in soybean. Plant Health Progress. Plant Management Network. doi:10.1094/PHP-2008-0118-01-DG.

Dorrance, A. E., Jia, H., and Abney, T. S. 2004. Evaluation of soybean differentials for their interaction with *Phytophthora sojae*. Plant Health Progress. Plant Management Network. doi:10.1094/PHP-2004-0309-01-RS.

Dorrance, A. E., Robertson, A. E., Cianzo, S., Giesler, L. J., Grau, C. R., Draper, M. A., Tenuta, A. U., and Anderson, T. R. 2009. Integrated management strategies for *Phytophthora sojae* combining host resistance and seed treatments. Plant Dis. 93:875-882.

Ferro, C. R., Hill, C. B., Miles, M. R., and Hartman, G. L. 2006. Evaluation of soybean cultivars with *Rps1k* gene for partial resistance or field tolerance to *Phytophthora sojae*. Crop Sci. 46:2427-2436.

Gijzen, M., and Qutob, D. 2009. *Phytophthora sojae* and soybean. Pages 303-329 in: Oomycete Genetics and Genomics: Diversity, Interactions, and Research Tools. K. Lamour and S. Kamoun, eds. Wiley–Blackwell, Hoboken, NJ.

Hansen, E. M., Wilcox, W. F., Reeser, P. W., and Sutton, W. 2009. *Phytophthora rosacearum* and *P. sansomeana,* new species segregated from the *Phytophthora megasperma* "complex." Mycologia 101:129-135.

Jackson, T. A., Kirkpatrick, T. L., and Rupe, J. C. 2004. Races of *Phytophthora sojae* in Arkansas soybean fields and their effects on commonly grown soybean cultivars. Plant Dis. 88:345-351.

Malvick, D. K., and Grunden, E. 2004. Traits of soybean-infecting *Phytophthora* populations from Illinois agricultural fields. Plant Dis. 88:1139-1145.

Mideros, S., Nita, M., and Dorrance, A. E. 2007. Characterization of components of partial resistance, *Rps2,* and root resistance to *Phytophthora sojae* in soybean. Phytopathology 97:655-662.

Nelson, B. D., Mallik, I., McEwen, D., and Christianson, T. 2008. Pathotypes, distribution, and metalaxyl sensitivity of *Phytophthora sojae* from North Dakota. Plant Dis. 92:1062-1066.

Qutob, D., Tedman-Jones, J., Dong, S., Kuflu, K., Pham, H., Wang, Y., Dou, D., Kale, S. D., Arredondo, F. D., Tyler, B. M., and Gijzen, M. 2009. Copy number variation and transcriptional polymorphisms of *Phytophthora sojae* RXLR effector genes *Avr1a* and *Avr3a*. PLoS One 4:e5066.

Schmitthenner, A. F. 1985. Problems and progress in control of Phytophthora root rot of soybean. Plant Dis. 69:362-368.

Tang, Q. H., Gao, F., Li, G. Y., Wang, H., Zheng, X. B., and Wang, Y. C. 2010. First report of root rot caused by *Phytophthora sansomeana* on soybean in China. Plant Dis. 94:378.

Thomas, R., Fang, X., Ranathunge, K., Anderson, T. R., Peterson, C. A., and Bernards, M. A., 2007. Soybean root suberin: Anatomical distribution, chemical composition, and relationship to partial resistance to *Phytophthora sojae*. Plant Physiol. 144:299-311.

Tyler, B. M., Tripathy, S., Zhang, X., Dehal, P., Jiang, R. H. Y., Aerts, A., Arredondo, F. D., Baxter, L., Bensasson, D., Beynon, J. L., Chapman, J., Damasceno, C. M. B., Dorrance, A. E., Dou, D., Dickerman, A., Dubchak, I. L., Garbelotto, M., Gijzen, M., Gorgon, S. G., Govers, F., Grunwald, N. J., Huang, W., Ivors, K. L., Jones, R. W., Kamoun, S., Krampis, K., Lamour, K. H., Lee, M., McDonald, W. H., Medina, M., Miejer, H. G. G., Nordber, E. K., Maclean, D. J., Ospina-Giraldo, M. D., Morris, P. F., Phuntumart, V., Putnam, N. H., Rash, S., Rose, J. K. C., Sakihama, Y., Salamov, A. A., Savidor, A., Schuering, C. F., Smith, B. M., Sobral, B. W. S., Terry, A., Torto-Alalibo, T., Win, J., Xu, Z., Zhang, H., Grigoriev, I. V., Rokhsar, D. S., and Boore, J. L. 2006. Phytophthora genome sequences uncover evolutionary origins and mechanisms of pathogenesis. Science 313:1261-1266.

Wang, H., Waller, L., Tripathy, S., St. Martin, S. K., Zhou, L., Krampis, K., Tucker, D. M., Mao, Y., Hoeschele, I., Saghai Maroof, M. A., Tyler, B. M., and Dorrance, A. E. 2010. Analysis of genes underlying soybean quantitative trait loci conferring partial resistance to *Phytophthora sojae*. Plant Gen. 3:23-40.

(Prepared by A. F. Schmitthenner;
revised by A. E. Dorrance)

Pythium Seed Rot, Damping-Off, and Root Rot

Pythium spp. are a key part of the seedling disease complex on soybean and a wide range of other crops. These pathogens cause stand-establishment problems wherever soybeans are grown. As a part of the disease complex, *Pythium* spp. reduce emergence of the crop and kill emerged seedlings, resulting in lower populations and reduced vigor of surviving plants. Compensation by neighboring plants minimizes, in part, the potential yield loss from reduced stands, but loss can occur nonetheless.

The genus *Pythium,* like the genus *Phytophthora,* is in the kingdom Chromalveolata, and *Pythium* spp. are a commonly found component of the soil microflora. Because *Pythium* spp. are commonly found in soils, environmental conditions at planting and for the first several days after planting are important in determining losses from this group of organisms.

Symptoms and Signs

Pythium spp. are oomycetes and cause a range of symptoms on soybean. Soybean seed may be infected within hours after planting, preventing germination and causing a seed rot. *Pythium* spp. may also be associated with death of the developing seedling—a symptom known as "damping-off." Preemergence damping-off (i.e., death of seedlings prior to emergence) (Fig. 118) and seed rot result in the absence of plants in the row. *Pythium* spp. may also cause postemergence damping-off (i.e., death of seedlings after emergence). Symptoms on affected seedlings may include yellow to tan to brown lesions or general discoloration and rotting of the hypocotyls (Fig. 119) and roots (Fig. 120). Affected tissue is often water soaked. Other seedlings affected by *Pythium* spp. may exhibit few symptoms other than reduced root development, stunted roots, and reduced root branching. *Pythium* spp. often infect roots near the tips, and thus, much of the affected tissue may slough off when the roots are removed from the soil. Seedlings affected by *Pythium* spp. may also exhibit yellow to reddish lesions on the cotyledons or so-called bald-head symptoms (i.e., death of the terminal bud). The most severe symptoms are associated with germination and the early stages of seedling development, because *Pythium* spp. infect juvenile plant tissue. However, root pruning and necrosis caused by *Pythium* spp. have been reported to cause stunting and wilting of soybean plants later in the season.

Signs of a *Pythium* pathogen may be observed in the infected tissue as oospores (Fig. 121) or after floating the tissue in water as the production of sporangia. Isolation using water agar or one of a number of selective media may help identify the pathogen, as well. The presence of a *Pythium* sp. may or may not indicate its role in seedling disease in any given situation, since a number of pathogens (including *Fusarium* spp., *Phytophthora sojae,* and *Rhizoctonia solani*) contribute to the seedling disease complex and frequently occur simultaneously within the same field and on the same seedlings. Diagnosis of the problem may require knowledge of the environmental conditions prior to observing symptoms in the field and the prevalence of *Pythium* spp. in infected tissue compared to healthy seedlings.

Causal Organisms

More than 30 *Pythium* spp. and/or species in closely related genera have been isolated and found pathogenic on soybean from various areas of the world. As more techniques have been developed to assist in identification and more sites have been examined, the number of *Pythium* spp. associated with seedling disease on soybean has grown. Some of the most common and virulent species include *P. aphanidermatum, P. dissotocum, P. inflatum, P. myriotylum,* and *P. torulosum*. In addition, *Globisporangium irregulare* (syn. *P. irregulare*), *G. sylvaticum* (syn. *P. sylvaticum*), and *G. ultimum* (syn. *P. ultimum*) are commonly recovered from diseased soybean seedlings.

Several *Pythium* spp. that have been isolated from soybean are not pathogenic. Moreover, species and isolates within a species may differ greatly in their virulence on soybean.

Disease Cycle and Epidemiology

Pythium spp. share the same basic life cycle (Fig. 122), although the importance of specific stages of the life cycle may differ across species. Aid in identifying species is provided by the presence and appearance of sexual structures (oogonium, antheridium, and oospore) and asexual structures (sporangium and zoospore), along with features such as sporangial shape and size, oogonial size and ornamentation, antheridial number, and mode of attachment. These structures may be observed by culturing isolates on corn meal agar or other nutrient media or by growing isolates on boiled grass leaves or other plant material immersed in sterile water. The development of several molecular techniques has also assisted in identification of *Pythium* spp.

The sexual spore of the genus is a thick-walled oospore (Fig. 122, part H); it results from the mating of an antheridium and an oogonium (Fig. 122, part G). Most species of *Pythium* are homothallic, which means an individual isolate can form oospores. However, opposite mating types may be needed for oospore formation in heterothallic species, such as *P. sylvaticum*. The sporangium (Fig. 122, part B) generally produces a short tube that is connected to a membrane-bound vesicle (Fig. 122, part C), where numerous biflagellate zoospores are formed and released (Fig. 122, part D). In some species, sporangia may not form zoospores but instead serve as survival structures and

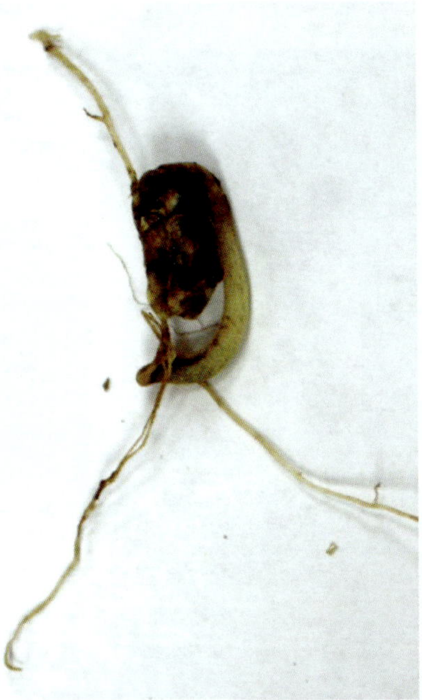

Fig. 118. Preemergence damping-off, characteristic of Pythium damping-off and root rot (*Pythium* spp.). (Courtesy M. V. Avanzato)

Fig. 119. Lesions on hypocotyls, characteristic of Pythium damping-off and root rot (*Pythium* spp.). (Courtesy M. V. Avanzato)

Fig. 120. Severe root discoloration and root rot, characteristic of Pythium damping-off and root rot (*Pythium* spp.). (Courtesy M. V. Avanzato)

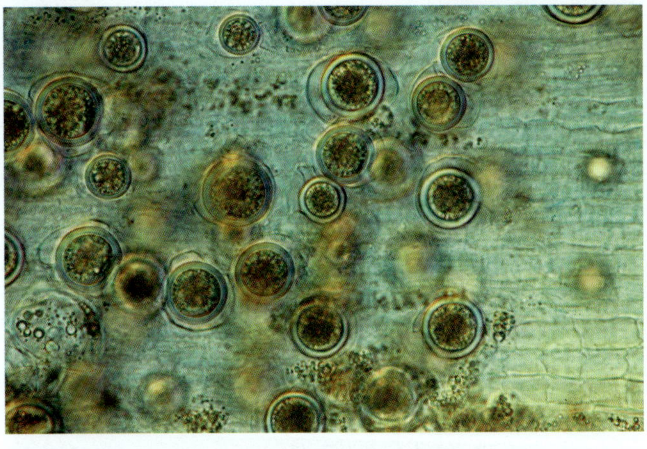

Fig. 121. Oospores of a *Pythium* sp. in a colonized soybean root—a sign of Pythium damping-off and root rot. (Courtesy M. V. Avanzato)

germinate directly, forming hyphal swellings. This process occurs in *G. irregulare, G. sylvaticum,* and *G. ultimum.*

Oospores or hyphal swellings (depending on the species) are the primary survival structures found in the soil. Saprophytic growth on organic matter and parasitic activity on other crops, weeds, and volunteer plants may also be important factors in the persistence of *Pythium* spp. As a group, *Pythium* spp. may grow saprophytically, acting as "pioneer" colonizers of crop residue and organic matter in the soil. In addition, many *Pythium* spp. can parasitize a very large range of plant species, causing disease symptoms or reproducing in the absence of any symptoms. In the U.S. Midwest, where rotating corn and soybean is common, a number of *Pythium* spp. have been isolated and found pathogenic on both crops, ensuring their survival.

Infection of soybean may occur soon after planting, when dormant propagules of the pathogen germinate in response to seed and root exudates from the host under environmental conditions that favor the pathogen. Stressful conditions for plant germination and growth increase the amounts of plant exudates and favor disease development. The pathogen can reach the host by hyphal growth or by producing sporangia and then zoospores that swim to the seed, root tip, or developing hypocotyl. The pathogen infects the plant tissue directly and grows quickly, both intracellularly and intercellularly, colonizing the tissue and causing a wet rot. Reproduction as sporangia or the production of oospores may occur quickly in colonized tissue. Secondary infections are limited, however.

Environmental conditions (primarily, soil temperature and the availability of water) are key factors in the severity of seedling disease. Temporary flooding and saturation of soils produce increases in stand losses, root discoloration, and the presence of *Pythium* spp. on seedlings. Oxygen levels are reduced in saturated soils, increasing the amounts of plant exudates and favoring the diffusion of plant exudates and *Pythium* spp. A wide range of temperatures are associated with activity among the *Pythium* spp. pathogenic to soybean; thus, *Pythium* spp. can cause disease over a diversity of planting conditions. The greatest losses are often associated with cool, wet soils, however. Cultural practices such as no tillage and conservation tillage may enhance the damage caused on soybean as a result of cooler, wetter soils in the spring (compared with conventional tillage).

In addition, the nature of the soil environment may be associated with the regions in which particular *Pythium* spp. are important. The importance and prevalence of these pathogens may also be determined by soil chemical and physical properties, including pH, calcium and magnesium levels, and field capacity.

Management

Management practices that are effective for other diseases on soybean (e.g., cultivar selection and crop rotation) have little impact on controlling *Pythium* diseases. Even though resistance to *Pythium* spp. has been reported on several occasions, no commercial soybean cultivars are advertised as having resistance to this disease. Crop rotation has limited effectiveness because of the many species of *Pythium* that can cause disease, the wide host ranges of many of these pathogens (including many weed species), and the long-term resistant survival structures produced by members of this genus.

The first step in managing seedling diseases caused by *Pythium* spp. is to plant under conditions favorable for soybean germination and emergence. Growers should avoid planting when the soil will be cool and wet for a few days, because such conditions favor development of the seedling diseases caused by *Pythium* spp., as well as abiotic problems (e.g., soil anoxia). Cool, wet conditions may occur more commonly in fields with fine-textured soils and poorly drained areas and those in which conservation tillage is practiced. Soil drainage should be improved, when possible. Planting on beds may prevent saturated soil from surrounding the seed. Cultural practices such as planting high-quality seed and avoiding deep planting may also improve stands.

Seed treatments with fungicides have been effective in improving stands by reducing damage from *Pythium* spp. Metalaxyl/Mefenoxam is the most widely used seed treatment for control of the seedling diseases caused by *Pythium* spp. Sensitivity to this chemical may differ among species; thus, maximum protection may not be achieved by using it alone. Adding a broad-spectrum fungicide seed treatment to the use of metalaxyl/

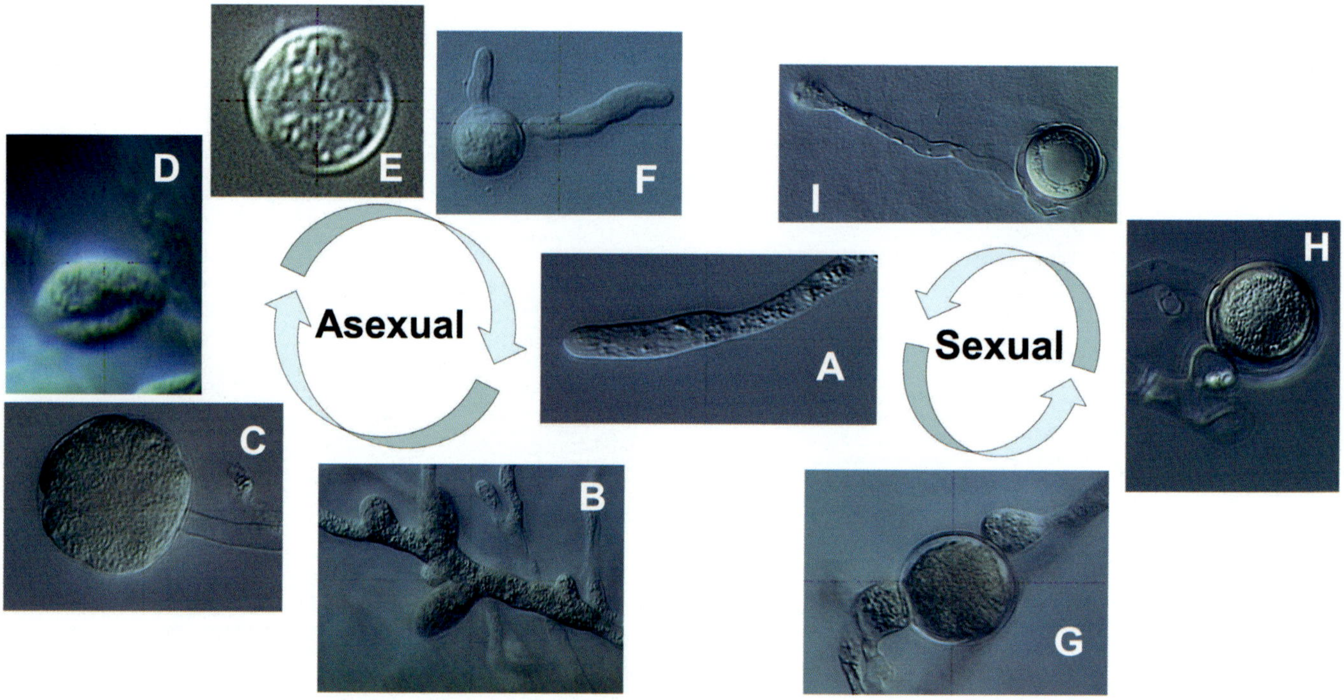

Fig. 122. Life cycle of *Pythium aphanidermatum,* one of the pathogens that cause Pythium damping-off and root rot. **A,** Hypha. **B,** Sporangium containing cytoplasm. **C,** Vesicle with tube exiting sporangium. **D,** Zoospore. **E,** Encysted zoospore. **F,** Germinating zoospore. **G,** Oogonium and antheridia. **H,** Oogonium containing oospore. **I,** Germinating oospore. (Courtesy J. H. Taylor)

mefenoxam may improve control of *Pythium* spp. Studies have been conducted over multiple years and at multiple locations to assess the value of seed treatments in controlling diseases caused by *Pythium* spp. These studies have shown the importance of *Pythium* pathogens in the seedling disease complex and demonstrated the consistent economic return that comes from the small investment required for seed treatments. Increases in seed costs and technology fees have made the purchase of seed a larger percentage of the overall operating cost, thereby making the protection of this investment more important.

Selected References

Broders, K. D., Lipps, P. E., Paul, P. A., and Dorrance, A. E. 2007. Characterization of *Pythium* spp. associated with corn and soybean seed and seedling disease in Ohio. Plant Dis. 91:727-735.

Broders, K. D., Wallhead, M. W., Austin, G. D., Lipps, P. E., Paul, P. A., Mullen, R. W., and Dorrance, A. E. 2009. Association of soil chemical and physical properties with *Pythium* species diversity, community composition, and disease incidence. Phytopathology 99:957-967.

Brown, G. E., and Kennedy, B. W. 1965. Pythium pre-emergence damping-off of soybean in Minnesota. Plant Dis. Rep. 49:646-647.

Brown, G. E., and Kennedy, B. W. 1966. Effect of oxygen concentration on Pythium seed rot of soybean. Phytopathology 56:407-411.

Ellis, M. L., McHale, L. K., Paul, P. A., St. Martin, S. K., and Dorrance, A. E. 2013. Soybean germplasm resistant to *Pythium irregular* and molecular mapping of resistance quantitative trait loci derived from the soybean accession PI 424354. Crop. Sci. 53:1008-1021.

Guy, S. O., and Oplinger, E. S. 1989. Soybean cultivar performance as influenced by tillage system and seed treatment. J. Prod. Agric. 2:57-62.

Jiang, Y. N., Haudenshield, J. S., and Hartman, G. L. 2012. Characterization of *Pythium* spp. from soil samples in Illinois. Can. J. Plant Pathol. 34:448-454.

Kirkpatrick, M. T., Rothrock, C. S., Rupe, J. C., and Gbur, E. E . 2006. The effect of *Pythium ultimum* and soil flooding on two soybean cultivars. Plant Dis. 90:597-602.

Kirkpatrick, M. T., Rupe, J. C., and Rothrock, C. S. 2006. Soybean response to flooded soil conditions and the association with soilborne plant pathogenic genera. Plant Dis. 90:592-596.

Poag, P. S., Popp, M., Rupe, J., Dixon, B., Rothrock, C., and Boger, C. 2005. Economic evaluation of soybean fungicide seed treatments. Agron. J. 97:1647-1657.

Rosso, M. L., Rupe, J. C., Chen, P., and Mozzoni, L. A. 2008. Inheritance and genetic mapping of resistance to *Pythium* damping-off caused by *Pythium aphanidermatum* in 'Archer' soybean. Crop Sci. 48:2215-2222.

Thomson, T. B., Athow, K. L., and Laviolette, F. A. 1971. The effect of temperature on the pathogenicity of *Pythium aphanidermatum, P. debaryanum,* and *P. ultimum* on soybean. Phytopathology 61:933-935.

Zhang, B. Q., and Yang, X. B. 2000. Pathogenicity of *Pythium* populations from corn–soybean rotation fields. Plant Dis. 84:94-99.

(Prepared by C. S. Rothrock, M. V. Avanzato, and J. C. Rupe)

Red Crown Rot

Red crown rot (also referred to as "Cylindrocladium root rot," "black root rot," and "Calonectria root rot") occurs in warm-temperate and tropical regions worldwide. In the United States, a 25–30% yield loss was recorded from a uniformly diseased field in Mississippi, and a theoretical yield loss of 50% was predicted for 100% disease incidence on a susceptible cultivar in Louisiana. The actual disease incidence rarely exceeds 30%, however.

Symptoms and Signs

Foliar and stem symptoms of red crown rot commonly appear during the middle to late reproductive growth stages. Infected plants are usually distributed randomly in small groups scattered throughout the field. The initial foliar symptom is the appearance of small (1–2 cm diameter), interveinal, chlorotic blotches in the upper canopy. When conditions are favorable for disease development, these areas enlarge and become necrotic (Fig. 123). Desiccation and defoliation (including loss of petioles) can follow within 2 weeks.

Stem symptoms of red crown rot are first evident as a reddish discoloration on the stem, beginning at the soil line and extending upward a few centimeters. As the disease develops, red-orange perithecia develop on the stem surfaces near the soil (Fig. 124). These structures are most abundant after periods of high moisture late in the growing season—usually, when the plant is near maturity.

Extensive root colonization can occur without the development of aboveground signs or symptoms. The roots are deteriorated and black during the advanced stages of disease (Fig. 125). Vascular tissue in the lower few centimeters of the stem may have a gray discoloration, but browning of the pith does not occur, as it does with brown stem rot.

The foliar symptoms of red crown rot resemble those of brown stem rot, stem canker, and sudden death syndrome (SDS) (see elsewhere in Part I, the sections Brown Stem Rot; Stem Canker; and Sudden Death Syndrome). Red crown rot can be distinguished from stem canker by the lack of stem cankers and the lesser amounts of interveinal necrosis of foliage and defoliation. SDS has foliar symptoms similar to those of red crown rot but lacks the external stem discoloration described in the previous paragraph.

In addition to soybean, the fungus that causes red crown rot attacks species of *Acacia, Arachis, Carica, Crotalaria, Eucalyptus, Leea, Medicago, Nerium,* and *Vaccinium.* It also causes a serious disease of peanut.

Fig. 123. Interveinal leaf chlorosis and necrosis, characteristic of red crown rot (*Calonectria ilicicola*). (Courtesy G. B. Padgett)

Fig. 124. Soybean stem encrusted with perithecia of *Calonectria ilicicola*, the fungus that causes red crown rot. (Courtesy G. B. Padgett)

Causal Organisms

The perithecia of *Calonectria ilicicola* (syn. *Cylindrocladium parasiticum*) are red-brown and superficial and usually occur in groups; they measure 280–400 × 300–550 µm. Asci are clavate and 12–19 × 90–140 µm. Ascospores are fusoid to falcate, mostly one septate, 4.5–6.5 × 30–65 µm, and slightly or not constricted at the septum (Fig. 126). Macroconidia are cylindrical, straight, hyaline, 4–7 × 45–90 µm, and mostly three septate. Microconidia are cylindrical, straight, hyaline, one septate, and 4–7 × 16–47 µm. Microsclerotia are brown and vary in diameter from 22 to 133 µm (Fig. 127). Only perithecia, ascospores, and microsclerotia have been observed on infected plant samples.

Fig. 125. Healthy roots (right) and deteriorated roots (left), associated with red crown rot (*Calonectria ilicicola*). (Courtesy G. B. Padgett)

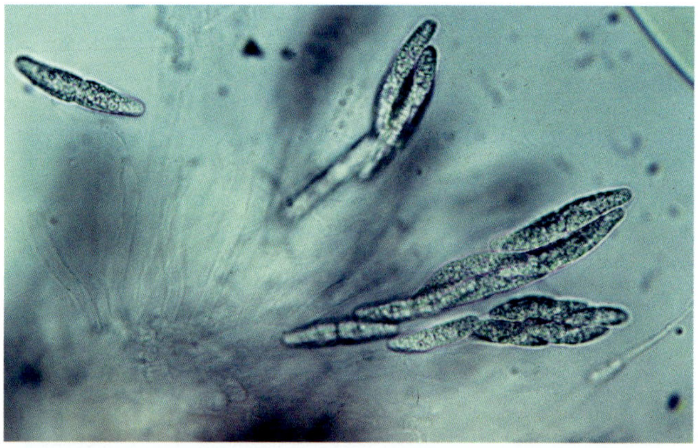

Fig. 126. Asci and ascospores of *Calonectria ilicicola*. (Courtesy J. S. Russin)

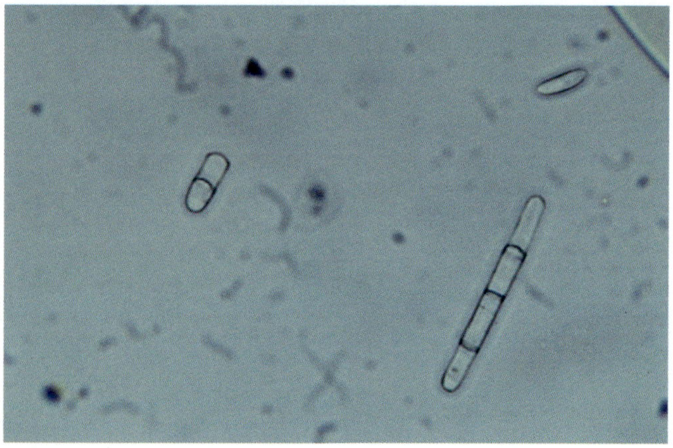

Fig. 127. Macroconidia and microconidia of *Calonectria ilicicola*. (Courtesy J. S. Russin)

Soybean is also susceptible to *Calonectria morganii* (syn. *Cylindrocladium scoparium*). This fungus has been reported in several countries around the world, but it has not been as problematic as *C. ilicicola* in the United States.

Disease Cycle and Epidemiology

The colonization of soybean roots may occur soon after planting. Early root colonization has been shown to be critical for the development of red crown rot, which suggests that the initial pathogen population in the soil may be a significant factor in disease development. Maximum root infection occurs when the soil temperature is 25–30°C. The frequency of root infection decreases at temperatures greater than 30°C and is negligible at 40°C. Evidence suggests that delayed planting results in low initial root colonization and that colonization remains low throughout the season. This delay in planting may be associated with soil temperatures that are sufficient to inhibit root colonization.

Disease severity is often greater in poorly drained, heavy soils, and disease distribution is usually clumped within a given field. Microsclerotia can survive several years in soil and plant debris. The role of ascospores and conidia in the disease cycle is not known.

Management

Growers should use less susceptible cultivars, if available. Rotations with nonhost plants for 2 or more years may reduce inoculum levels, but rotations with peanut should be avoided. Delayed planting and the use of glyphosate as a preplant herbicide have both been reported to reduce the incidence of red crown rot. Managing nematodes that affect soybean may also reduce the risks associated with this disease.

Selected References

Berner, D. K., Berggren, G. T., and Snow, J. P. 1991. Effects of glyphosate on *Calonectria crotalariae* and red crown rot of soybean. Plant Dis. 75:809-813.

Berner, D. K., Berggren, G. T., Snow, J. P., and White, E. P. 1988. Distribution and management of red crown rot of soybean. Appl. Agric. Res. 3:160-166.

Crous, P. W. 2002. Taxonomy and Pathology of *Cylindrocladium* (*Calonectria*) and Allied Genera. American Phytopathological Society, St. Paul, MN.

Dianese, J. C., Ribeiro, W. R. C., and Urben, A. F. 1986. Root rot of soybean caused by *Cylindrocladium clavatum* in central Brazil. Plant Dis. 70:977-980.

Kuruppu, P. U., Schneider, R. W., and Russin, J. S. 2004. Effects of soil temperature on microsclerotia of *Calonectria ilicicola* and soybean root colonization by this fungus. Plant Dis. 88:620-624.

Kuruppu, P. U., Schneider, R. W., and Russin, J. S. 2004. Factors affecting soybean root colonization by *Calonectria ilicicola* and development of red crown rot following delayed planting. Plant Dis. 88:613-619.

(Prepared by G. B. Padgett, P. U. Kuruppu, and J. S. Russin)

Rhizoctonia Damping-Off and Root Rot

Rhizoctonia root rot is one of the most common soilborne diseases of soybean. It is worldwide in distribution, and in North America, it is a major component of the soybean root rot complex. Regional epidemics often occur sporadically over the years. The pathogen causes pre- and postemergence damping-off and seedling blight, resulting in stand reduction; in severe cases, fields must be replanted. In older plants, the pathogen causes root rot and stunting. Damage varies from minimal to yield losses of up to 50% in Brazil and 48% in the United States.

Symptoms and Signs

Preemergence symptoms are typical of the seed rot that is not visible in the field and thus often goes unrecognized. The most commonly observed symptom is postemergence damping-off at the seedling stage before the first trifoliolate leaf develops (growth stage V1). Infected plants have reddish-brown lesions on the hypocotyls at the soil line and become most evident after seedlings have been removed from the soil (Fig. 128). Infected seedlings may or may not die, depending on the virulence of the isolates and the existing environmental conditions.

As the season progresses, root symptoms can be found in plants at stages of advanced vegetative to reproductive growth. These plants may have been infected as seedlings, and the lesions will continue to enlarge and girdle the stems. Infected plants are stunted and yellowing and have poor root systems; the lateral roots often decay, leaving only the taproot.

The wilt symptoms associated with root rot occur throughout the early reproductive stages, when the disease coincides with warm, dry conditions, as well as herbicide injury; damage from *Heterodera glycines,* the soybean cyst nematode (see later in Part I, the section Soybean Cyst Nematode); and other stresses. Diseased plants first appear in areas such as hillsides, where water and nutrients are limited, and may be scattered in lightly infested fields or form circular, discolored areas in heavily infested fields. Patches of symptomatic plants are sometimes elongated, following the direction of tillage.

Causal Organism

Rhizoctonia solani (syn. *Thanatephorus cucumeris*) differs from other *Rhizoctonia* spp. in hyphal morphology. The fungus is characterized by dolipore septa and hyphal branches at right angles, which are constricted at their points of origin (Fig. 129). Hyphae are brown and have multinucleate cells. Some isolates produce basidiospores under field conditions and in the laboratory. Although these spores do not serve as inoculum in the disease cycle of damping-off and root rot, they are important in the development of web blight epidemics. Other characteristics of *R. solani* that may vary among isolates are mycelial growth rate, monilioid cells, and pseudosclerotia that are brown to black, of varying size, and composed of compacted monilioid cells that do not have rinds and are the primary survival structure of the fungus.

R. solani is readily isolated on water agar from freshly infected tissues. The success of isolation decreases sharply when plant tissues have been stored for a long time. The fungus can also be isolated from soil by sieving and plating debris on selective media. This method is used as a semiquantitative means of determining inoculum density of the fungus in soil. In potato dextrose agar (PDA) culture, young colonies are whitish but gradually turn brown and display an increase in aerial mycelia. Colonies produce dark secretions in places in which sclerotia form.

Anastomosis reactions between hyphal cells and the number of nuclei in a single cell of vegetative hyphae are used to group *R. solani* isolates. Fourteen anastomosis groups (AGs) in *R. solani* have been identified and differ in ecology and host range. Some genetic groups (especially AG 1 and AG 2) are further categorized into subgroups that feature distinct degrees of host specificity. AG 1-IB is one of the six subgroups of AG 1, and isolates in this group are pathogenic on soybean and other hosts. In addition, *R. solani* isolates from soybeans have been reported in other AG groups, such as AG 4, in which isolates often cause lesions on the hypocotyls. The draft genome sequence of *R. solani* AG 1-IB has been completed, and more is known about its gene structure, function, and content. Simple sequence repeats (SSRs) have become increasingly important in the analyses of population genetics.

Disease Cycle and Epidemiology

R. solani is a facultative saprophyte and survives in the soil in the absence of host plants as sclerotia or in colonized plant debris as mycelium. The food base in the soil also affects parasitism and inoculum potential. *R. solani* AG 4 can survive in a wide range of temperature and moisture conditions; an intermediate water potential ranging from –2 to –15 bars is optimal. The pathogen population declines sharply when a field is flooded for a long period of time or is irrigated. The pathogen survival rate is reduced by high soil temperature, solarization of soil, and fallow practices for 1 year. Forms of survival include pseudosclerotia, hyphae in saprophytically colonized organic matter, and hyphae in host debris following the parasitic phase. *R. solani* is often aggregated in fields, which is indicated by patchy disease distribution by midseason.

Rhizoctonia damping-off and root rot is monocyclic. The fungus grows well under aerated conditions and is distributed mainly within the top 10 cm of soil. Infection occurs soon after

Fig. 128. Reddish-brown, sunken lesions girdling soybean hypocotyls, characteristic of Rhizoctonia root rot (*Rhizoctonia solani*). (Courtesy J. B. Sinclair)

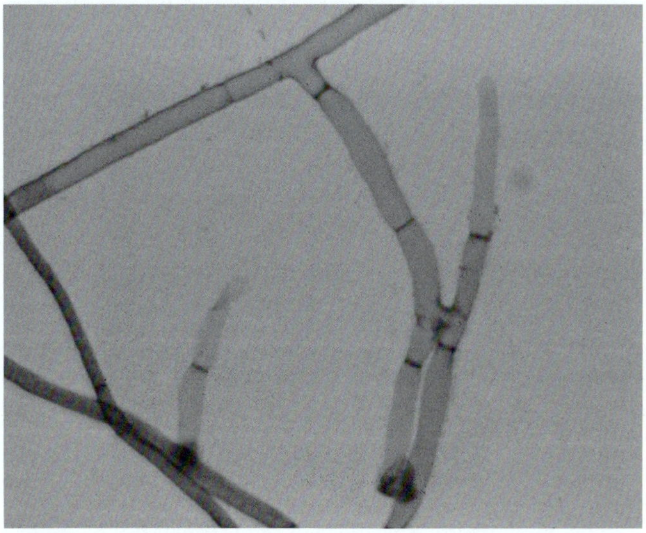

Fig. 129. Hyphae of *Rhizoctonia solani,* with right-angle branching characteristic of Rhizoctonia root rot. (Courtesy X.-B. Yang)

seeds have been planted. After emergence, apparent points of invasion are in the hypocotyl area at the soil line, which is characteristic of isolates in AG 4. If seedlings survive the damping-off phase, infection may continue to develop into root rot when the lesions expand down into the root system.

The temperature requirements for infection are different for different AGs. The optimum soil temperatures for the isolates that cause root rot are 26–32°C for AG 1 and 24–26°C for AG 4. Soil moisture at 30–60% water-holding capacity is optimal for the growth of root rot isolates. The growth rate is reduced if the water-holding capacity is greater than 70% and eliminated at 90% water-holding capacity because of reduced soil aeration.

Disease severity is also associated with soil type. Sandy and light soils favor seedling blight, as do pH levels greater than 6.6. Soil moisture of approximately 70% water-holding capacity also favors infection of seedlings. The damage caused by *R. solani* can be exacerbated by herbicide stress.

Injury caused by soil insects, such as the seedcorn maggot (*Delia platura*), enhances the severity of seedling blight. Interaction of this disease with *H. glycines* is frequently observed in the north-central region of the United States. *Fusarium oxysporum,* another root rot pathogen, interacts synergistically with *R. solani.*

R. solani has a very broad host range. Some isolates of AG 2-2 are highly virulent to both soybean and maize. The type of rotation crop or sequence of crop rotation affects the inoculum potential, because the carbon-to-nitrogen ratio of crop residue affects the long-term survival of the fungus. The effects of reduced tillage on the disease are not clear, although increased herbicide stress favors disease development.

Management

Soybean cultivars vary in their response to *R. solani.* However, cultivars with absolute resistance are not available, and most commercial cultivars are not advertised as being resistant to *R. solani.* A more practical suggestion is to treat seeds with an appropriate fungicide or fungicide package that may contain antagonists to the fungus.

Growers should use good cultural practices to promote seedling health at early stages. In row-planted soybean, mid-season cultivation is recommended to promote the new growth of roots, which can reduce root rot damage. Efforts should be made to reduce other stresses, such as herbicide injury, poor soil quality, insect damage, and feeding of *H. glycines.*

Selected References

Arakawa, M., and Inagaki, K. 2014. Molecular markers for genotyping anastomosis groups and understanding the population biology of *Rhizoctonia* species. J. Gen. Plant Pathol. doi:10.1007/s10327-014-0536-0.

Baird, R. E., Carling, E. D., and Mullinix, B. G. 1996. Characterization and comparison of isolates of *Rhizoctonia solani* AG-7 from Arkansas, Indiana, and Japan, and select AG-4 isolates. Plant Dis. 80:1421-1424.

Bauske, E. M., and Kirby, H. W. 1992. Effects of dinitroaniline herbicides, carboxin-pentachloronitrobenzene seed treatment, and Rhizoctonia disease on soybean. Plant Dis. 76:236-239.

Bradley, C. A., Hartman, G. L., Mueller, D. S., Hoffman, D. D., Nickell, C. D., and Pedersen, W. L. 2005. Genetic analysis of partial resistance to *Rhizoctonia solani* in the soybean cultivar 'Savoy.' Can. J. Plant Pathol. 27:137-142.

Bradley, C. A., Hartman, G. L., Nelson, R. L., Mueller, D. S., and Pedersen, W. L. 2001. Response of ancestral soybean lines and commercial cultivars to Rhizoctonia root and hypocotyl rot. Plant Dis. 85:1091-1095.

Bradley, C. A., Hartman, G. L., Wax, L. M., and Pedersen, W. L. 2002. Influence of herbicides on Rhizoctonia root and hypocotyl rot of soybean. Crop Prot. 21:679-687.

Carling, D. E., Baird, R. E., Gitaitis, R. D., Brainard, K. A., and Kuninaga, S. 2002. Characterization of AG-13, a newly reported anastomosis group of *Rhizoctonia solani.* Phytopathology 92:893-899.

Ciampi, M. B., Meyer, M. C., Costa, M. J. N., Zala, M., McDonald, B. A., and Ceresini, P. C. 2008. Genetic structure of populations of *Rhizoctonia solani* anastomosis group-1 IA from soybean in Brazil. Phytopathology 98:932-941.

Datnoff, L. E., and Sinclair, J. B. 1988. The interaction of *Fusarium oxysporum* and *Rhizoctonia solani* in causing root rot of soybeans. Phytopathology 78:771-777.

Dorrance, A. E., Kleinhenz, M. D., McClure, S. A., and Tuttle, N. T. 2003. Temperature, moisture, and seed treatment effects on *Rhizoctonia solani* root rot of soybean. Plant Dis. 87:533-583.

Fenille, R. C., Souza, N. L. D., and Kuramae, E. E. 2002. Characterization of *Rhizoctonia solani* associated with soybean in Brazil. Eur. J. Plant Pathol. 108:783-792.

Harikrishnan, R., and Yang, X. B. 2002. Effects of herbicides on root rot and damping-off caused by *Rhizoctonia solani* in glyphosate-tolerant soybean. Plant Dis. 86:1369-1373.

Harikrishnan, R., and Yang, X. B. 2004. Recovery of anastomosis groups of *Rhizoctonia solani* from different latitudinal positions and influence of temperatures on their growth and survival. Plant Dis. 88:817-823.

Liu, Z., and Sinclair, J. B. 1992. Genetic diversity of *Rhizoctonia solani* anastomosis group 2. Phytopathology 82:778-787.

Muyolo, N. G., Lipps, P. E., and Schmitthenner, A. F. 1993. Anastomosis grouping and variation in virulence among isolates of *Rhizoctonia solani* associated with dry bean and soybean in Ohio and Zaire. Phytopathology 83:438-444.

Nelson, B., Helms, T., Christianson, T., and Kural, I. 1996. Characterization and pathogenicity of *Rhizoctonia solani* from soybean. Plant Dis. 80:74-80.

Sneh, B., Jabaji-Hare, S., Neate, S., and Dijst, G. 1996. *Rhizoctonia* Species: Taxonomy, Molecular Biology, Ecology, Pathology, and Disease Control. Kluwer Academic, London.

Wibberg, D., Jelonek, L., Rupp, O., Hennig, M., Eikmeyer, F., Goesmann, A., Hartmann, A., Borriss, R., Grosch, R., Pühler, A., and Schlüter, A. 2013. Establishment and interpretation of the genome sequence of the phytopathogenic fungus *Rhizoctonia solani* AG1-IB isolate 7/3/14. J. Biotech. 167:142-155.

Wibberg, D., Jelonek, L., Rupp, O., Kröber, M., Goesmann, A., Grosch, R., Pühler, A., and Schlüter, A. 2014. Transcriptome analysis of the phytopathogenic fungus *Rhizoctonia solani* AG1-IB 7/3/14 applying high-throughput sequencing of expressed sequence tags (ESTs). Fungal Biol. 118:800-813.

Yang, G., and Li, C. 2012. General description of *Rhizoctonia* species complex. Chapter 2 in: Plant Pathology. C. J. Cumagum, ed. InTech China, Shanghai.

Yang, X. B., Snow, J. P., and Berggren, G. T. 1989. Morphogenesis of microsclerotia and *sasakii*-type sclerotia in *Rhizoctonia solani,* anastomosis group one. Mycol. Res. 93:429-434.

Yu, G. Y., Sinclair, J. B., Hartman, G. L., and Bertagnolli, B. L. 2002. Production of iturin A by *Bacillus amyloliquefaciens* suppressing *Rhizoctonia solani.* Soil Biol. Biochem. 34:955-963.

Zhang, J. X., Xue, A. G., Cober, E. R., Morrison, M. J., Zhang, H. J., Zhang, S. Z., and Gregorich, E. 2013. Prevalence, pathogenicity and cultivar resistance of *Fusarium* and *Rhizoctonia* species causing soybean root rot. Can. J. Plant Pathol. 93:221-236.

Zhao, G., Ablett, G. R., Anderson, T. R., Rajcan, I., and Schaafsma, A. W. 2005. Anastomosis groups of *Rhizoctonia solani* associated with soybean root and hypocotyl rot in Ontario and resistance of accession PI 442031 to different anastomosis groups. Can. J. Plant Pathol. 27:108-117.

Zhao, G., Ablett, G. R., Anderson, T. R., Rajcan, I., and Schaafsma, A. W. 2005. Inheritance and genetic mapping of resistance to Rhizoctonia root and hypocotyl rot in soybean. Crop Sci. 45:1441-1447.

(Prepared by X. B. Yang; revised by G. L. Hartman)

Sclerotium Blight

Sclerotium blight (also known as "southern blight," "southern stem rot," and "white mold") is a minor disease of soybean. It has been reported on 500 species in 100 plant families and is usually found in warm regions worldwide. In Brazil, a disease incidence of 12% is common, and yield reductions of 59%

have been reported in Nigeria. The causal pathogen was first described on tomato in 1893 in the U.S. state of Florida.

There are differences in the level of infection among cultivars. Under certain conditions, significant yield losses can occur in monoculture and in short rotations of soybean with other crops susceptible to the pathogen.

Symptoms and Signs

Soybean is susceptible to the fungus that causes Sclerotium blight from seedling emergence through pod fill. Generally, the disease develops on isolated plants scattered throughout a field. Under favorable conditions, spread to adjacent plants within the same row can occur rapidly (Fig. 130). Infection usually occurs at or just below the soil surface. Light-brown lesions, which quickly darken, enlarge until the hypocotyl or stem is girdled. A sudden yellowing or wilting of the plant is usually the first symptom. Leaves of infected plants turn brown and dry and often cling to the dead stem. A leaf spot phase is characterized by circular, tan to brown, zonate lesions with dark-brown margins. Twin-stem abnormality is associated with seedling infection.

The most characteristic sign of Sclerotium blight is the white, fanlike mat of fungal mycelia that forms on stem bases, leaf debris, and the soil surface around infected plants (Fig. 131). The mycelial mat may extend several centimeters up the stem above the soil line. Hot, humid conditions favor development of the mycelial mat, which commonly disappears during drought. Numerous tan to brown, spherical sclerotia (about the size of mustard seed) form on infested plant material and on the soil surface.

Fig. 130. Dying soybean plants, associated with Sclerotium blight (*Athelia rolfsii*). (Courtesy J. C. Rupe)

Fig. 131. Sclerotia and mycelia of *Sclerotium rolfsii,* the fungus that causes Sclerotium blight, on soybean stems. (Courtesy D. E. Hershman)

Causal Organism

The causal fungus, *Athelia rolfsii* (syn. *Sclerotium rolfsii*), produces cordlike, septate, hyaline hyphae that are 5–9 µm in diameter and have regular clamp connections. Hyphal branching typically occurs at right angles to the points of origin. Asexual spores are not produced. The sexual stage rarely develops on crop debris and culture media. Clavate basidia, bearing hyaline, pyriform basidiospores and measuring 1.0–1.7 × 6–12 µm, are produced on exposed hymenia. Spherical sclerotia, 1–2 mm in diameter, develop 6–12 days after formation of the mycelial mat. The outer covering, or rind, of the sclerotia is initially white and later turns dark brown; the center remains white, however.

Disease Cycle and Epidemiology

Sclerotium blight generally occurs under hot, wet conditions. Maximum disease development occurs at temperatures close to the 25–35°C that is optimum for mycelial growth and sclerotial germination. High moisture levels both in the soil and under a dense crop canopy favor disease development and damage. Drought conditions often precede a severe disease outbreak. Sclerotium blight occurs most commonly and is most damaging in well-aerated sandy or sandy loam soils.

A. rolfsii survives from one season to the next by producing sclerotia. Their survival is influenced by factors such as temperature and moisture, as well their proximity to a susceptible host and their depth in the soil. Sclerotia on or near the soil surface survive longer than those buried in the soil. Because sclerotia are left near the soil surface, the development of Sclerotium blight is often favored by conservation tillage and no-till production, especially when debris is heavy. Sclerotia are readily dispersed with soil moved by water, wind, and tillage systems and equipment. Seed transmission of the pathogen has been reported but is much less significant than transmission by soilborne sclerotia.

Germination of sclerotia in water-saturated soil is stimulated by volatile organic exudates from crop residue. Sclerotia are deficient in energy; thus, the fungus gains energy by colonizing undecomposed organic matter before attacking a susceptible host. The fungus kills host tissue in advance of its invading mycelia by producing enzymes and oxalic acid, which degrade pectin and cellulose.

Management

Soybeans with resistance to Sclerotium blight are not available, but there are differences in susceptibility among genotypes, such that some are more susceptible than others. Alternating soybean or another susceptible crop with a nonhost crop (such as maize, grain sorghum, wheat, or pasture grasses) or clean fallow for 2 years may prevent inoculum from building up to damaging levels. In fields that are heavily infested with *S. rolfsii,* a 3- to 4-year rotation with a nonhost crop may be needed. Burying crop debris and sclerotia to a depth of 15–25 cm with a moldboard plow may reduce inoculum carryover and delay disease development. In no-till fields, keeping debris away from plants' stems should help reduce the incidence of Sclerotium blight. Movement of soil around the stems during cultivation should be avoided.

Disease incidence may be reduced by the application of calcium and nitrogen fertilizers.

Selected References

Akem, C. N., and Dashiell, K. E. 1991. First report of southern blight caused by *Sclerotium rolfsii* on soybeans in Nigeria. Plant Dis. 75:537.

Beute, M. K., and Rodriguez-Kabana, R. 1981. Effects of soil moisture, temperature, and field environment on survival of *Sclerotium rolfsii* in Alabama and North Carolina. Phytopathology 71:1293-1296.

Punja, Z. K. 1985. The biology, ecology, and control of *Sclerotium rolfsii*. Annu. Rev. Phytopathol. 23:97-127

Punja, Z. K., Carter, J. D., Campbell, G. M., and Rossell, E. L. 1986. Effects of calcium and nitrogen fertilizers, fungicides, and tillage practices on incidence of *Sclerotium rolfsii* on processing carrots. Plant Dis. 70:819-824.

(Prepared by J. C. Rupe and A. M. R. Almeida)

Stachybotrys Root Rot

The fungus that causes Stachybotrys root rot is a well-known human pathogen that produces mycotoxins, including a host of immunosuppressant compounds, and it is one of the causes of what is known as "sick-building syndrome." The pathogen has been found in soil and on substances rich in cellulose, such as cereal grains, cotton, hay, paper, plant debris, straw, and wood pulp. Although it is rarely known as a plant pathogen, it has been isolated from the stems of sea holly (*Acanthus ilicifolius*), barley seeds, and tobacco leaves, and it causes leaf spot disease of sunn hemp (*Crotalaria juncea*). The pathogen was reported for the first time in soybean roots in 2000.

Symptoms and Signs

The fungus that causes Stachybotrys root rot produces brown to black root lesions in soybean (Fig. 132). Lesion development appears somewhat restricted, and the fungus does not colonize the entire root system. Root lesions vary in length from 7 to 25 mm. Both the taproot and lateral roots may be infected.

Foliar disease symptoms have not been observed.

Causal Organism

Stachybotrys root rot is caused by *Stachybotrys chartarum*—a cellulytic, saprophytic, hyphomycete fungus. *S. chartarum* can be cultured on cornmeal agar, oatmeal agar (OMA), and potato dextrose agar (PDA). Colonies on PDA appear dark brown to black and have dark masses of conidia. In some cases, the colony margins may be white to gray and have compact hyphae without sporulation. Colonies on cefoperazone MacConkey agar (CMA) are appressed and uncolored at first but turn dull/dark and form concentric rings in response to light. Colonies on OMA are similar to those on PDA but uncolored on the reverses of the plates. Growth of *S. chartarum* at 25°C on CMA, OMA, and PDA after 10 days was 33 (±4), 47 (±2), and 33 (±3) mm in diameter, respectively. Growth on PDA at 25°C was greater than growth at 20 and 30°C.

An isolate from soybean produced obovate or ellipsoidal phialides in whorls that arose from solitary or groups of determinate, macronematous, dark-olivaceous conidiophores. Conidiophores were simple or irregularly branched, one through four (but mostly two or three) septate, often rough walled, hyaline near the base, and dark olivaceous on the upper part (Figs. 133 and 134). Phialides were terminal, determinate, discrete, unicellular, and 4–6 × 9–14 µm; they were initially hyaline but later turned dark olivaceous. Conidia were unicellular, round or ellipsoidal, and 4–7 × 5–13 µm; they arose singly and successively, were acrogenous, and aggregated in slime to form globose conidial heads. Conidia were initially hyaline and had smooth walls, but they turned dark brown and had rough or ridged walls when mature. Warts occasionally formed on the surfaces of conidia. Mycelia were septate and branched and measured 2–5 µm wide; they were hyaline when young and black when mature.

Biochemical analysis indicated that the soybean isolate produced satratoxins G and H along with roridin L-2, as well as the spirocyclic lactones and lactams in rice culture. Polymerase

Fig. 132. Root lesions and reduced root mass (left), compared with healthy roots (right). Symptoms are characteristic of Stachybotrys root rot, caused by the soybean isolate of *Stachybotrys chartarum,* SC-SB1. (Courtesy S. Li)

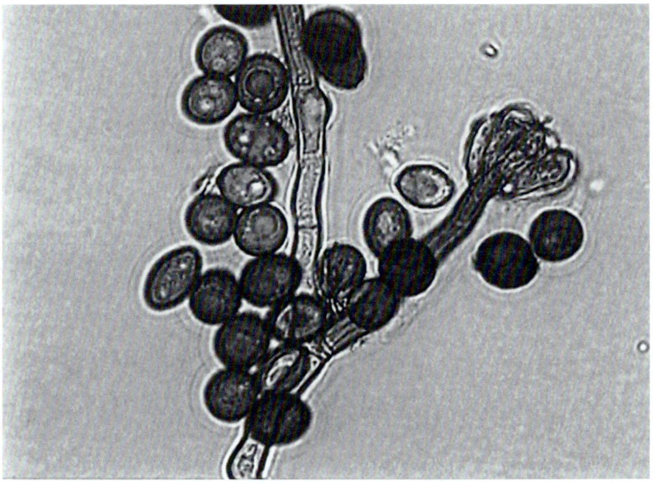

Fig. 133. Light micrograph of conidia, conidiophore, conidial head, and phialides of a soybean isolate of *Stachybotrys chartarum*, SC-SB1, which causes Stachybotrys root rot. (Courtesy S. Li)

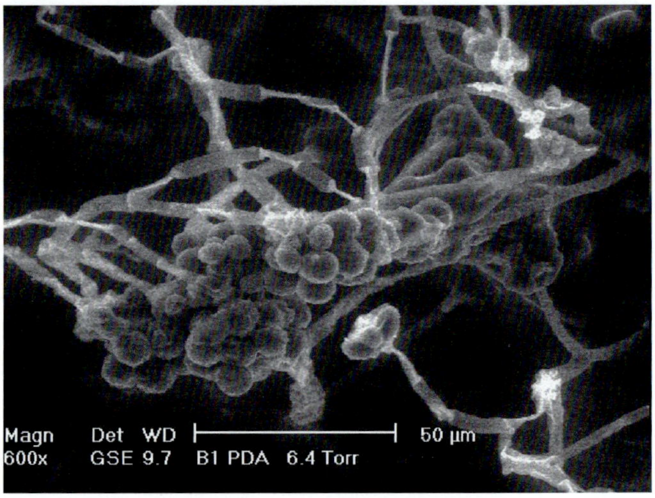

Fig. 134. Environmental scanning electron micrograph of conidia, conidiophore, conidial head, and phialides of a soybean isolate of *Stachybotrys chartarum,* SC-SB1. (Courtesy S. Li)

chain reaction (PCR) using a *S. chartarum*-specific primer StacR3 and IT51 amplified a 198-base pair (-bp) DNA fragment from the total genomic DNA. The DNA sequence of the internal transcriber spacer (ITS) region from the soybean isolate was 100% identical to *S. chartarum* strain ATCC 9182, one nucleotide mismatch with *S. chartarum* strain UAMH 7900, and different from all published sequences of 12 other species of *Stachybotrys* and two species of *Memnoniella* in GenBank (with genetic divergence ranging from 5.26 to 9.98%).

Disease Cycle and Epidemiology

Stachybotrys root rot is a relatively new disease found on soybean; thus, the disease cycle and epidemiology are not clear. Infested soil and debris may be the sources of primary inoculum.

With the increasing concern about mycotoxins in human and animal health, specific and effective detection and identification of *S. chartarum* are important. The soybean isolate of *S. chartarum* was found to produce macrocyclic trichothecene mycotoxins, but more research is needed to determine if the macrocyclic trichothecene mycotoxins play any role in the infection of soybean roots.

Management

No specific management techniques have been studied or implemented for Stachybotrys root rot. General production practices that promote plant growth—and more specifically, those practices that promote root growth and reduce soil waterlogging—are believed to reduce the impact of this disease.

Selected References

Fung, F., Clark, R., and Williams, S. 1998. *Stachybotrys,* a mycotoxin-producing fungus of increasing toxicologic importance. J. Toxicol. Clin. Toxicol. 36:79-86.

Haugland, R. A., and Heckman, J. L. 1998. Identification of putative sequence specific PCR primers for detection of the toxigenic fungal species *Stachybotrys chartarum*. Mol. Cell. Probes 12:387-396.

Jarvis, B. B., Hinkley, S. F., and Nielsen, K. F. 2000. *Stachybotrys:* An unusual mold associated with water-damaged buildings. Mycotoxin Res. 16A:105-108.

Li, S., and Hartman, G. L. 2000. First report of *Stachybotrys chartarum* causing soybean root rot. Plant Dis. 84:100.

Li, S., Hartman, G., Jarvis, B. B., and Tak, H. 2002. A *Stachybotrys chartarum* isolate from soybean. Mycopathologia 154:41-49.

Nejat, S. A., and Ershad, D. 1994. An investigation on mycoflora of barley seeds in Iran. Iran. J. Plant Pathol. 30:23-28.

Nikulin, M., Pasanen, A.-L., Berg, S., and Hintikka, E.-L. 1994. *Stachybotrys atra* growth and toxin production in some building materials and fodder under different relative humidities. Appl. Environ. Microbiol. 60:3421-3424.

Pathak, P. D., and Chauhan, R. K. S. 1976. A leaf spot disease of *Crotalaria juncea* L., caused by *Stachybotrys atra* Corda. Curr. Sci. 45:567.

Sadaba, R. B., Vrijmoed, L. L. P., Jones, E. B. G., and Hodgkiss, I. J. 1995. Observations on vertical distribution of fungi associated with standing senescent *Acanthus ilicifolius* stems at Mai Po Mangrove, Hong Kong. Hydrobiologia 295:119-126.

(Prepared by S. Li and G. L. Hartman)

Stem Canker

Stem canker has been reported in Europe, South America, and most soybean-growing areas of Canada and the United States. There are two distinct stem canker diseases, which are referred to as "northern stem canker" and "southern stem canker" based initially on their occurrence in the northern and southern regions of the United States, respectively. Both diseases have similar symptoms on soybean plants and are capable of producing a 100% yield loss under favorable environmental conditions.

During the 1950s, when susceptible cultivars such as Blackhawk and Hawkeye were grown, northern stem canker was widespread, causing yield losses of up to 50% in the north-central United States. Since the removal of extremely susceptible cultivars from production, northern stem canker has become less significant, although sporadic outbreaks still occur in some years in the states in this region (i.e., Illinois, Iowa, Minnesota, North Dakota, South Dakota, and Wisconsin).

Southern stem canker was first identified in 1973 and caused widespread losses during the early 1980s until resistant cultivars were developed. When environmental conditions are favorable, the disease still causes severe yield losses throughout the U.S. South on susceptible cultivars. There have also been reports since the late 1990s of the southern stem canker pathogen from Illinois and Wisconsin. Moreover, both the northern and southern types of stem canker have been found in Santa Fe Province in Argentina, indicating that the causal fungi of the two diseases overlap geographically.

Symptoms and Signs

For northern stem canker, a single canker usually forms per plant, occurring on one node in the lower stem. The position of the canker on the stem may vary within a field, but typically, the canker on the majority of stems occurs at a particular node or restricted group of nodes in a given year (Fig. 135). Expansion of the stem canker can occur rapidly and be related to plant maturity, beginning predictably in the last 2 weeks of August and continuing into the first 2 weeks of September, when soybeans plants are in the middle to late reproductive growth stages. Once the stems have been girdled, plants may be rapidly killed and easily spotted, as the wilted, darkened, shriveled leaves remain attached to the stems. Cankers on stems appear as dark, blackened zones. Small, circular stromata develop within the canker and give rise to perithecia on stem residue during the following spring (Fig. 136).

Under some conditions, the foliar symptoms of northern stem canker may precede the emergence and girdling activity of cankers on stems. If this occurs, leaf blades and petioles may become chlorotic, and tissue at the bases of the petioles may become necrotic and collapse, allowing leaves to droop. Leaves may remain attached after wilting and turn dark. In some cases, the leaf blades fall off, leaving behind attached petioles (referred to as "spurs").

The symptoms of southern stem canker occur later in the season, during the reproductive growth stages, from infections that develop during the vegetative growth stages (V1–V7). The pathogen remains latent for an extended period. The first external symptoms are observed on the stem near a leaf petiole on a lower node during the early reproductive stage. Initially, cankers are small, slightly sunken, and reddish brown with

Fig. 135. Symptoms of stem canker, including browning and internal and external browning of stems, caused by *Diaporthe phaseolorum.* (Courtesy J. B. Sinclair)

grayish-brown centers. As the season progresses, cankers expand longitudinally, become more sunken, and have reddish-brown margins and grayish-brown centers. Underneath the canker, the pith and vascular tissue are discolored. Cankers become longer but rarely girdle the stem, unlike those produced by *Phytophthora sojae,* which are dark brown and usually girdle the stem from below the soil line upward (see the previous section Phytophthora Root and Stem Rot).

Leaf symptoms of southern stem canker may include only tip dieback, but more characteristically, interveinal chlorosis and necrosis (without wilting) develop, as well (Fig. 137). The leaf symptoms are similar to those of several other diseases, such as brown stem rot, red crown rot, and sudden death syndrome (SDS) (see elsewhere in Part I, the sections Brown Stem Rot; Red Crown Rot; and Sudden Death Syndrome). Foliar symptoms and plant death are caused, in part, by a low-molecular-weight phytotoxin, which may be related to another toxin found in a species of *Phomopsis* that causes wilt in pine trees. Occasionally, necrosis of the terminal meristem develops with the characteristic shepherd's crook curling. Dead leaves are retained at harvest. Yield loss results from reductions in seed number and size because of premature plant death (Fig. 138).

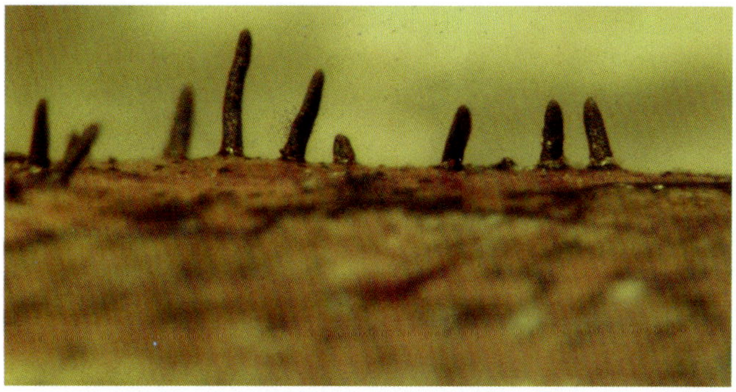

Fig. 136. Perithecial beaks of *Diaporthe phaseolorum* var. *caulivora*, the fungus that causes northern stem canker. (Courtesy J. B. Sinclair)

Fig. 137. Large, one-sided stem lesion and foliar interveinal chlorosis and necrosis, caused by *Diaporthe phaseolorum* var. *meridionalis,* the fungus that causes southern stem canker. (Courtesy D. Ploper)

With both types of stem canker, stromata and perithecia are often observed on mature stems. Perithecia generally do not form on plants infected with the pathogen that causes northern stem canker during the growing season or following harvest; instead, small, circular stromata form that overwinter on the residue, giving rise to perithecia on residue the following spring. Pycnidia and perithecia of the southern stem canker pathogen may form in cankers during the season in very susceptible cultivars, but pycnidial formation is not consistent. Disease severity varies from year to year according to planting date, weather conditions, and cultivar selection. Cool, moist conditions during the early vegetative stages greatly enhance the chances for infection.

Causal Organisms

Diaporthe phaseolorum var. *caulivora* causes northern stem canker, and *D. phaseolorum* var. *meridionalis* causes southern stem canker. Perithecia are commonly observed on overseasoned stems as perithecial beaks that emerge from stem surfaces. Perithecia of *D. phaseolorum* var. *caulivora* are black and globose, form either singly or in caespitose groups of two to 12, and are embedded between epidermal and cortical tissues. Perithecia of *D. phaseolorum* var. *meridionalis* are almost always borne singly. Perithecia of both fungi are 165–340 × 282–412 μm and have protruding beaks, which are highly variable in length and width.

D. phaseolorum var. *caulivora* produces white colonies with interspersed, cottony tufts of mycelium on potato dextrose agar (PDA). Perithecia form in irregular to strongly valsoid clusters in small, poorly developed stromata. The eight-spored asci (5.6–5.8 × 25.7–28.7 μm) are elongate clavate and have thin, evanescent walls and distinct refractive rings at the apexes. Ascospores (2.3–2.5 × 8.1–8-4 μm) are hyaline, ellipsoidal to bursiform, two celled, slightly constricted at the septa, and often biguttulate in each cell. Alpha- and beta-conidia are rarely produced under natural conditions, although they can sometimes be produced under laboratory conditions (e.g., on autoclaved soybean stems in water agar). Alpha-conidia (2.1 × 5.2 μm) are hyaline, nonseptate, fusiform, ellipsoidal or oblong ellipsoidal, and usually guttulate with an oil droplet in each end. Beta-conidia (1.7 × 9.3 μm) are hyaline, nonseptate, filiform, and hamate or curved.

D. phaseolorum var. *meridionalis* produces a lanose colony on PDA that is white-buff colored when young and becomes light to dark brown with age. Some isolates produce large stromata in which perithecia form, whereas other isolates do not form stromata or perithecia. Perithecial formation can be induced by incorporating sterilized wooden toothpicks or sterilized pieces of senescent soybean stem in the agar. Asci and

Fig. 138. Premature dieback, associated with southern stem canker (*Diaporthe phaseolorum* var. *meridionalis*). (Courtesy D. E. Hershman)

ascospores are similar to those of *D. phaseolorum* var. *caulivora* but larger. Asci are 6.7–7.0 × 35.8–37.0 µm, and ascospores are 3.1–3.4 × 9.5–9.8 µm. Only alpha-conidia are produced (sometimes inconsistently) on PDA or on water agar with soybean stem pieces. However, many isolates produce abundant pycnidia on small pieces of sterilized soybean leaflets on the surface of lima bean agar.

The stem canker pathogens are part of a species complex that includes the pod and stem blight and seed decay pathogens, *D. sojae* and *D. longicolla*. Nucleic acid-based diagnostic techniques have been developed for detection and discrimination of different species and varieties among these species.

Disease Cycle and Epidemiology

The severity of stem canker depends on when plants become infected. In general, severity is more profound when plants are infected early in the season during the vegetative growth stages and is promoted by extended periods of rainy, wet conditions. Although infection of plants often occurs during early vegetative growth, external symptoms appear much later during reproductive growth. Both stem canker fungi overwinter in debris, and conservation tillage practices (especially when rotation is short or not used) may increase the occurrence of disease. Sporulation usually occurs after harvest and requires adequate moisture and temperatures above 20°C, which normally occur during the spring. Although the fungi have been reported to sporulate on fallen petioles, the production of secondary inoculum during the season probably does not affect disease development and yield loss.

Seed infection has been proposed as a means of spreading the stem canker fungi. Levels of infection by *D. phaseolorum* var. *caulivora* are 10–20%, whereas the level by *D. phaseolorum* var. *meridionalis* rarely exceeds 1%. Seed infection may be a means of introducing the pathogen to the field, but inocula for epidemics come mostly from infested debris.

Most of the epidemiological work on stem canker has been conducted on southern stem canker. The development of infection from this pathogen requires a leaf-wetness period of 24–96 h following rain and temperatures of 22–30°C. In the field, disease development is associated with the occurrence of early infection (growth stages V1–V7) and the level of cultivar resistance. Cultivars with moderate resistance must be infected very early to develop extensive disease; cultivars with little resistance can be infected much later and still develop severe disease. Some early season cultivars mature before developing the disease. Spread of the disease from a point source is fairly limited, occurring within 1.0–1.5 m of the inoculum source, and the maximum spread is 4–5 m. The distribution pattern of the disease reflects the wind direction during rain events.

The epidemiology of northern stem canker is assumed to be fairly similar to that of southern stem canker in many respects, although questions have been raised about the origin of the initial inoculum. For example, an outbreak of 50–90% disease incidence was reported even though the fields lacked a recent history of soybean cultivation. (These fields had been planted primarily in a continuous corn, alfalfa, or grass/alfalfa rotation for up to 5 years.) In at least one case, a high-incidence northern stem canker outbreak was observed in a soybean field in South Dakota that had been in the Conservation Reserve Program for the previous 11 years and had been maintained as a mixture of intermediate wheat grass and alfalfa.

In studies of host range, the isolates that cause southern stem canker in the United States were found to produce symptoms after inoculation on hairy indigo (*Indigofera hirsuta*) and hemp sesbania (*Sesbania cannabina*). However, black nightshade, entire-leaf morning glory, small-flowered morning glory, tall morning glory, spiny amaranth (*Amaranthus spinosus*), prickly sida (*Sida spinosa*), red-weed (*Rumex acetosella*), sicklepod (*Arabis canadensis*), and wild poinsettia were infected but asymptomatic.

Management

Southern stem canker is managed primarily through the use of resistant cultivars. Four resistance genes have been described and designated *Rdc1, Rdc2, Rdc3,* and *Rdc4.* (They may also be referred to as *Rdm1, Rdm2, Rdm3,* and *Rdm4.*) These genes have different loci and give high levels of resistance. Races of *D. phaseolorum* var. *meridionalis* have been reported in Argentina but not in North America. Levels of resistance to southern stem canker range from moderately to highly resistant. Even though moderate resistance may be effective under some conditions, severe losses have occurred with moderately resistant cultivars when the environmental conditions have favored the development of disease. Genes that provide resistance to southern stem canker are not necessarily effective in providing resistance to northern stem canker.

Delayed planting usually reduces the incidence of southern stem canker. Conservation tillage practices in problem fields should be avoided, where permissible. The application of foliar fungicides has been inconsistent and is not effective once foliar symptoms have begun to develop.

Given the unpredictable nature of northern stem canker and the renewed occurrence of outbreaks (beginning in the late 1980s), there are no clear management guidelines for growers. Additionally, the widespread and endemic nature of the pathogen as a seedborne, endophytic, and saprophytic organism suggests that ecological and environmental conditions, as well as host susceptibility, are more important factors than the availability of inoculum in the epidemiology of northern stem canker.

Selected References

Backman, P. A., Weaver, D. B., and Morgan-Jones, G. 1985. Soybean stem canker: An emerging disease problem. Plant Dis. 69:641-647.

Black, B. D., Padgett, G. B., Russin, J. S., Griffin, J. L., Snow, J. P., and Berggren, G. T., Jr. 1996. Potential weed hosts for *Diaporthe phaseolorum* var. *caulivora,* causal agent of stem canker. Plant Dis. 80:763-765.

Bowers, G. R., Ngeleka, K., and Smith, O. D. 1993. Inheritance of stem canker resistance in soybean cultivars Crocket and Dowling. Crop Sci. 33:67-70.

Chase, T. E. 2011. Occurrence of Northern stem canker in first soybean plantings following conservation reserve in South Dakota. Phytopathology 101:S31.

Fernandez, F. A., and Hanlin, R. T. 1996. Morphological and RAPD analyses of *Diaporthe phaseolorum* from soybean. Mycologia 88:425-440.

Gravert, C. E., Li, S., and Hartman, G. L. 2001. Occurrence of *Diaporthe phaseolorum* var. *meridionalis* on soybean in Illinois. Plant Dis. 85:1211.

Keeling, B. L. 1985. Soybean cultivar reactions to soybean stem canker caused by *Diaporthe phaseolorum* var. *caulivora* and pathogenic variation among isolates. Plant Dis. 69:132-133.

Keeling, B. L. 1988. Influence of temperature on growth and pathogenicity of geographic isolates of *Diaporthe phaseolorum* var. *caulivora.* Plant Dis. 72:220-222.

Lalitha, B., Snow, J. P., and Berggren, G. T. 1989. Phytotoxin production by *Diaporthe phaseolorum* var. *caulivora,* the causal organism of stem canker of soybean. Phytopathology 79:499-504.

Li, S., Kurtzweil, N. C., Grau, C. R., and Hartman, G. L. 2004. Occurrence of soybean stem canker (*Diaporthe phaseolorum* var. *meridionalis*) in Wisconsin. Plant Dis. 88:576.

Lu, X., Robertson, A. E., Byamukama, E. Z., and Nutter, F. W., Jr. 2010. Evaluating the importance of stem canker of soybean in Iowa. Plant Dis. 94:167-173.

McGee, D. C., and Biddle, J. A. 1987. Seedborne *Diaporthe phaseolorum* var. *caulivora* in Iowa and its relationship to soybean stem canker in the southern United States. Plant Dis. 71:620-622.

Pioli, R. N., Morandi, E. N., Martínez, M. C., Lucca, F., Tozzini, A., Bisaro, V., and Hopp, H. E. 2003. Morphologic, molecular, and pathogenic characterization of *Diaporthe phaseolorum* variability in the core soybean-producing area of Argentina. Phytopathology 93:136-146.

Smith, E. F., and Backman, P. A. 1988. Soybean stem canker: An overview. Pages 47-55 in: Soybean Diseases of the North Central Region. T. D. Wyllie and D. H. Scott, eds. American Phytopathological Society, St. Paul, MN.

Zhang, A. W., Hartman, G. L., Curio-Penny, B., Pedersen, W. L., and Becker, K. B. 1999. Molecular detection of *Diaporthe phaseolorum* and *Phomopsis longicolla* from soybean seeds. Phytopathology 89:796-804.

Zhang, A. W., Hartman, G. L., Riccioni, L., Chen, W. D., Ma, R. Z., and Pedersen, W. L. 1997. Using PCR to distinguish *Diaporthe phaseolorum* and *Phomopsis longicolla* from other soybean fungal pathogens and to detect them in soybean tissues. Plant Dis. 81:1143-1149.

Zhang, A. W., Riccioni, L., Pedersen, W. L., Kollipara, K. P., and Hartman, G. L. 1998. Molecular identification and phylogenetic grouping of *Diaporthe phaseolorum* and *Phomopsis longicolla* isolates from soybean. Phytopathology 88:1306-1314.

(Prepared by J. C. Rupe)

Sudden Death Syndrome

Sudden death syndrome (SDS) was first observed in the U.S. state of Arkansas in 1971, and it was not reported in other states until 1984, when it was reported in Kentucky, Mississippi, Missouri, and Tennessee. In 1985, SDS was reported in Illinois and Indiana, and it has subsequently become widespread throughout the north-central United States. The disease also occurs in Argentina and Brazil and was reported from Malaysia in 2014.

Yield losses from SDS, which range from slight to nearly 100%, are dependent on the growth stage of the host at the time of disease onset and severity. Infected plants often prematurely defoliate and die, resulting in significant reductions in yield.

Symptoms and Signs

Foliar symptoms of sudden death syndrome first appear on leaflets as scattered, interveinal, chlorotic spots (Fig. 139). They may become necrotic or enlarge and coalesce into chlorotic streaks that become necrotic, leaving only the midvein and major lateral veins green (Fig. 140). Severely affected leaves detach from the petioles, which often remain attached to the stem—sometimes, up to harvest (Fig. 141). Severe foliar symptoms give affected areas in the field a tan to brown cast when viewed from a distance (Fig. 142). In some cases, the early symptoms of SDS resemble those of viral diseases: namely, mosaic and mottling symptoms on newly formed leaves. The leaves may also be slightly misshaped, cupped, and rugose. The effect of SDS on yield depends on when the initial foliar symptoms develop, how severe they are, and how quickly they progress. When severe foliar symptoms occur in the early reproductive stages, the effects may include flower and pod abortion. The early appearance of symptoms may also be associated with seed-size reduction, resulting in the greatest loss in yield. Most commonly, SDS occurs in the early to middle reproductive growth stages and becomes most noticeable as plants reach the full-pod stage.

The root mass of SDS-affected plants is reduced. In addition, the roots appear discolored, often preceding foliar symptoms (Fig. 143). Severely infected plants can easily be pulled from the ground. Longitudinal sections of taproots and lower stems of diseased plants have gray- to brown-colored xylem vessels; the discoloration extends up the lower stems, but the pith remains white. The fungus can be isolated from the roots and uncommonly from the lower stem near the soil line.

The foliar symptoms of SDS are similar to those of brown stem rot, red crown rot, and stem canker (see elsewhere in Part I, the sections Brown Stem Rot; Red Crown Rot; and Stem Canker). Brown stem rot is distinguished from SDS by its distinct discoloration of the pith. Red crown rot produces red perithecia at the soil line, which are absent in plants with SDS, and plants affected by stem canker have cankers on the lower stems, which are not found on plants with SDS. Plants that are infected with multiple pathogens at the same time are more difficult to diagnose; it may be common for a plant to be infected simultaneously by the pathogens that cause brown stem rot and SDS.

Fig. 140. Interveinal chlorosis and necrosis, severe symptoms of sudden death syndrome (*Fusarium virguliforme*). (Courtesy T. K. Herman)

Fig. 139. Mottled and chlorotic leaves, characteristic of sudden death syndrome (*Fusarium virguliforme*). (Courtesy J. C. Rupe)

Fig. 141. Premature defoliation and pod abortion, characteristic of sudden death syndrome (*Fusarium virguliforme*). (Courtesy J. C. Rupe)

Causal Organisms

In the United States, sudden death syndrome is caused by the soilborne fungus *Fusarium virguliforme* (previously known as *F. solani* f. sp. *glycines*). However, *F. brasiliense, F. crassistipitatum, F. tucumaniae,* and *F. virguliforme* have been reported to cause SDS in South America. *F. cuneirostrum* was initially thought to be a causal agent of SDS, but in 2012, it was characterized as a bean root rot pathogen that does not cause SDS. Quantitative DNA assays are available to detect and quantify *F. virguliforme.* Several publications have compared the species that cause SDS, but the description of the SDS pathogen is based on *F. virguliforme,* as the literature regarding it is more voluminous than that about other species and it is the only species reported to occur in the United States.

In culture, *F. virguliforme* is appressed to the agar, producing abundant macroconidia in a slimy, pionnotal, blue to blue-green or blue-purple mass. Mycelium is a grayish white and may be visible on the outer edges of the culture. Macroconidia are three to five septate, measure 3.8–5.0 × 34–66 µm, and are produced on short, simple to branched conidiophores on monophialides (Fig. 144). Microconidia are not abundant and may be produced only rarely. Chlamydospores are globose to subglobose; they form terminally or intercalary in hyphae and may form in or adjacent to macroconidia (Figs. 145 and 146). Molecular studies of isolates from various U.S. states have indicated that the fungus is genetically homogeneous and distinct from other non-SDS-causing *Fusarium* isolates.

The optimum temperature for growth of *F. virguliforme* on agar media is 25°C. Distinctive foliar symptoms have been

Fig. 142. Patchy areas of plants displaying premature senescence, characteristic of sudden death syndrome (*Fusarium virguliforme*). (Courtesy T. K. Herman)

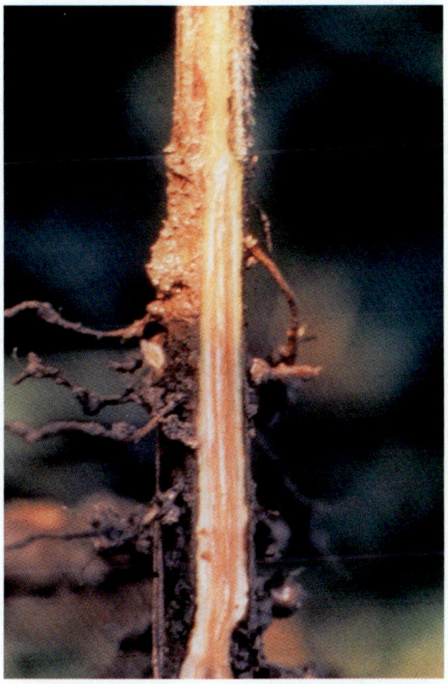

Fig. 143. Root rot and vascular discoloration, symptoms of sudden death syndrome (*Fusarium virguliforme*). (Courtesy J. C. Rupe)

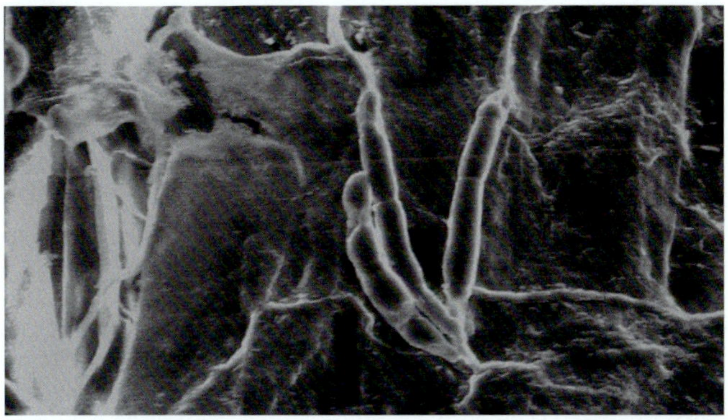

Fig. 144. Macroconidia of *Fusarium virguliforme,* the fungus that causes sudden death syndrome. (Courtesy S. Li)

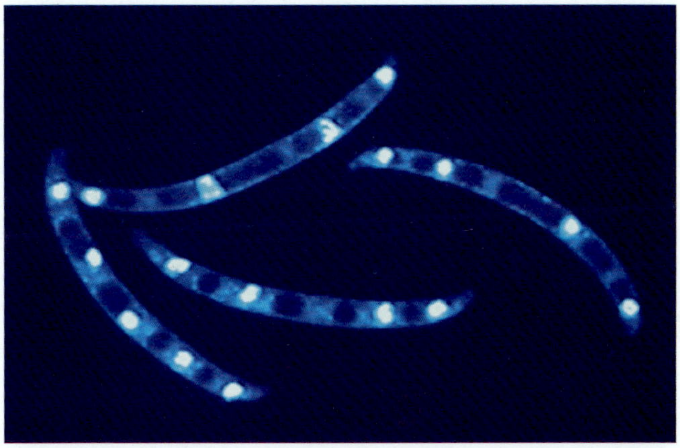

Fig. 145. Chlamydospores of *Fusarium virguliforme,* which causes sudden death syndrome. (Courtesy S. Li)

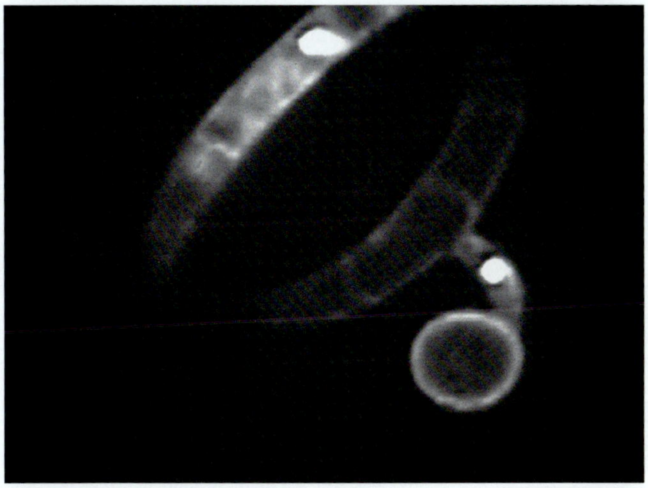

Fig. 146. Conidium of *Fusarium virguliforme,* which causes sudden death syndrome, germinating to form a chlamydospore. (Courtesy S. Li)

shown to be caused by one or more toxins produced by the fungus and translocated to the foliage. Methods that examine stem cuttings and detached leaflets have successfully been used to research the toxins and toxic culture filtrates. The fungus can be reisolated from soil in infested fields, from infected soybean roots, and from cysts and eggs of *Heterodera glycines,* the soybean cyst nematode, using Nash and Snyder's medium or a semiselective medium.

Disease Cycle and Epidemiology

In the field, sudden death syndrome is favored more by wet conditions than by dry conditions during the growing season, although the timeliness of the wet conditions may be important. Hot and especially dry conditions may arrest development of the disease. SDS can be severe in irrigated fields and in saturated soils, such as those in low spots and areas prone to waterlogging and compaction. Transition zones between areas in the field with severe SDS versus little or no SDS can be quite abrupt and may be related to changes in soil conditions.

Fields in which SDS has been problematic in the past are more likely to develop the disease in the future. Long rotations and cropping sequences with cotton, maize, rice, and sorghum appear to have little impact on SDS severity when soybean is planted again; it is possible, however, that rotations with other crops and using some cover crops may reduce the occurrence of SDS. Research findings have indicated that the presence of corn residue in the field may actually maintain the inoculum level of the fungus.

Moderate to high populations of *H. glycines* may be associated with SDS and may increase the severity of the disease, especially in cultivars that are susceptible to the type of *H. glycines* present. Other factors associated with an increased incidence of SDS include high soil fertility and soil compaction.

Management

Most commercial companies provide soybean cultivars with some resistance to sudden death syndrome. Many sources of resistance have been reported—including public cultivars such as Forrest and Ripley (maturity groups V and IV, respectively) and plant introductions (PIs) such as PI 520733, 567374, 567315, 567441C, 567650B, and 567664—that may potentially be used in breeding programs. Additional sources of resistance and better phenotyping techniques have also been addressed in research completed to improve the level of resistance found in soybean cultivars. Before considering other management strategies, growers that have problems with SDS are advised first to make an effort to plant less susceptible cultivars or cultivars advertised as having SDS resistance or tolerance.

Delayed planting and/or planting earlier-maturing cultivars may reduce both the incidence and the severity of SDS. However, in some cases, late-planted, double-cropped soybean planted after wheat may develop SDS at a greater severity level than soybean planted for the full season (but this is atypical). Other recommendations for delaying the onset and/or reducing the severity of SDS include using cultural practices that improve internal and surface drainage in fields, reducing the soybean cyst nematode population, and removing layers of compacted soil.

Selected References

Aoki, T., O'Donnell, K., and Scandiani, M. M. 2005. Sudden death syndrome of soybean in South America is caused by four species of *Fusarium: Fusarium brasiliense* sp. nov., *F. cuneirostrum* sp. nov., *F. tucumaniae,* and *F. virguliforme.* Mycoscience 46:162-183.

Aoki, T., Scandiani, M. M., and O'Donnell, K. 2012. Phenotypic, molecular phylogenetic, and pathogenetic characterization of *Fusarium crassistipitatum* sp. nov., a novel soybean sudden death syndrome pathogen from Argentina and Brazil. Mycoscience 53:167-186.

Brar, H., Swaminathan, S., and Bhattacharyya, M. K. 2011. The *Fusarium virguliforme* toxin FvTox1 causes foliar sudden death syndrome-like symptoms in soybean. Mol. Plant-Microbe Interact. 24:1179-1188.

Chehri, K., Salleh, B., and Zakaria, L. 2014. *Fusarium virguliforme,* a soybean sudden death syndrome fungus in Malaysian soil. Australas. Plant Dis. Notes. doi:10.1007/s13314-014-0128-z.

Díaz-Arias, M. M., Leandro, L. F., and Munkvold, G. P. 2013. Aggressiveness of *Fusarium* species and impact of root infection on growth and yield of soybeans. Phytopathology 103:822-832.

Farias Neto, A. F., Hartman, G. L., Pedersen, W. L., Li, S., Bollero, G. A., and Diers, B. W. 2006. Irrigation and inoculation treatments that increase the severity of soybean sudden death syndrome in the field. Crop Sci. 46:2547-2554.

Gao, X., Jackson, T. A., Lambert, K. N., Li, S., Hartman, G. L., and Niblack, T. L. 2004. Detection and quantification of *Fusarium solani* f. sp. *glycines* in soybean roots with real-time quantitative polymerase chain reaction. Plant Dis. 88:1372-1380.

Gongora-Canul, C. C., and Leandro, L. F. S. 2011. Plant age affects root infection and development of foliar symptoms of soybean sudden death syndrome. Plant Dis. 95:242-247.

Hartman, G. L., Chang, H.-X., and Leandro, F. F. 2015. Research advances and management of sudden death syndrome. Crop Prot. doi:10.1016/j.cropro.2015.01.017.

Hartman, G. L., Huang, Y. H., Nelson, R. L., and Noel, G. R. 1997. Germplasm evaluation of *Glycine max* for resistance to *Fusarium solani,* the causal organism of sudden death syndrome. Plant Dis. 81:515-518.

Hartman, G. L., Noel, G. R., and Gray, L. E. 1995. Occurrence of soybean sudden death syndrome in east-central Illinois and associated yield losses. Plant Dis. 79:314-318.

Jin, H., Hartman, G. L., Nickell, C. D., and Widholm, J. M. 1996. Characterization and purification of a phytotoxin produced by *Fusarium solani,* the causal agent of soybean sudden death syndrome. Phytopathology 86:277-282.

Li, S., and Hartman, G. L. 2003. Molecular detection of *Fusarium solani* f. sp. *glycines* in soybean roots and soil. Plant Pathol. 52:74-83.

Li, S., Gray, L. E., and Hartman, G. L. 1998. Chlamydospore formation, production, and nuclear status in *Fusarium solani* f. sp. *glycines* soybean sudden death syndrome-causing isolates. Mycologia 90:414-421.

Roy, K. W., Hershman, D. E., Rupe, J. C., and Abney, T. S. 1997. Sudden death syndrome of soybean. Plant Dis. 81:1100-1111.

Rupe, J. C., Gbur, E. E., and Marx, D. M. 1991. Cultivar responses to sudden death syndrome of soybean. Plant Dis. 75:47-50.

Scherm, H., and Yang, X. B. 1996. Development of sudden death syndrome of soybean in relation to soil temperature and soil water matric potential. Phytopathology 86:642-649.

Tang, E., Hill, C. B., and Hartman, G. L. 2010. Carbon utilization profiles of *Fusarium virguliforme* isolates. Can. J. Microbiol. 56:979-986.

(Prepared by G. L. Hartman, L. F. Leandro, and J. C. Rupe)

Thielaviopsis Root Rot

Thielaviopsis root rot (also called "black root rot") was first reported in Canada in 1925 and then later reported in Australia, China, Germany, Russia, and the United States. The disease is widespread in the U.S. state of Michigan and in the southwestern part of Ontario, Canada, and it may occur more widely than reported because of its wide host range. The importance of this disease in reducing soybean yields is unknown, but the pathogen has a wide host range and is an important disease in tobacco.

Symptoms and Signs

The cortices of hypocotyls and taproots turn dark brown to black, sometimes resulting in severe root rot (Fig. 147). The causal fungus frequently severs the taproot 2.5–5.0 cm below the soil surface. An infected soybean plant responds by forming shallow, adventitious roots, which support the plant when soil moisture is ample but cannot sustain it under drought conditions.

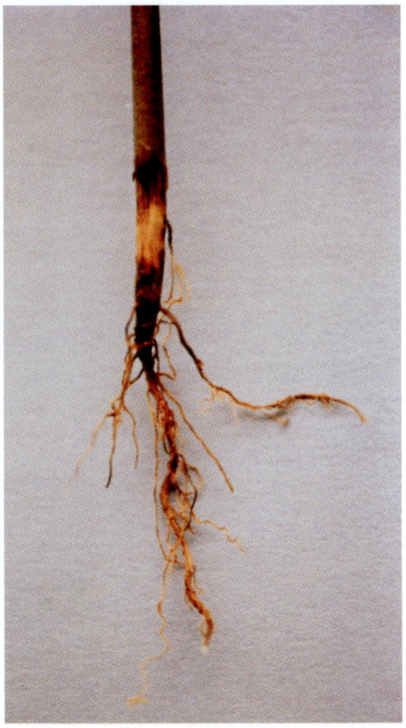

Fig. 147. Taproot rot, caused by *Thielaviopsis basicola,* the fungus that causes Thielaviopsis root rot. (Courtesy J. C. Rupe)

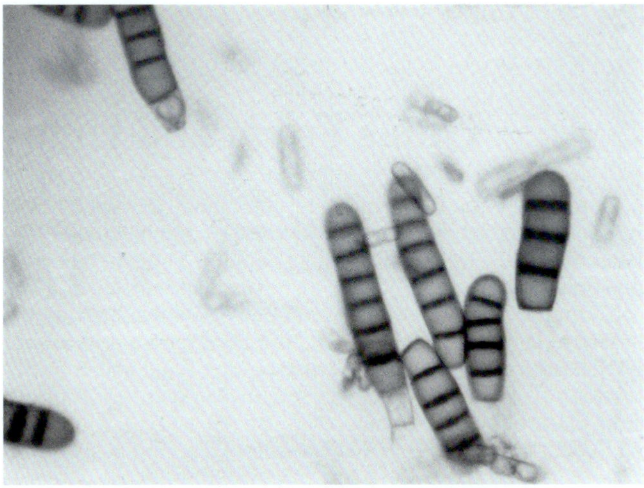

Fig. 148. Chlamydospore chains of *Thielaviopsis basicola*, which causes Thielaviopsis root rot. (Courtesy J. L. Lockwood)

Causal Organism

The causal fungus, *Thielaviopsis basicola,* is primarily a soil inhabitant. Mycelia are almost hyaline. Conidia form within short, hyaline conidiophores that are swollen at the base and taper toward the apex. Numerous cylindrical conidia are borne within a single conidiophore tip and are pushed out slowly in chains. The fungus produces characteristic chlamydospores within infected tissues (Fig. 148). Chlamydospores form on short, clavate, lateral hyphae in a series of short, brown, thick-walled cells, which soon separate. Chlamydospores survive for long periods in the soil, whereas conidia are short lived.

T. basicola is not readily isolated from soil or diseased plant tissue. It can be selectively isolated on fresh carrot disks and on specific selective agar media. Molecular techniques have been used to detect as few as three conidia per reaction.

The fungus has a wide host range and also causes black root rot of tobacco. Isolates from soybean have been found to be more pathogenic on cotton than on soybean. Rotation with soybean may contribute to carryover of the fungus in cotton farming systems.

Disease Cycle and Epidemiology

Thielaviopsis root rot is most severe early in the season, when soils are cool (16–20°C). Diseased seedlings may be killed by the disease, and older plants that survive may wilt under drought stress. Seedborne infection has been reported, as well. Thielaviopsis root rot often occurs together with Phytophthora root and stem rot (see the section Phytophthora Root and Stem Rot).

T. basicola survives in the soil for long periods of time because of the thick-walled chlamydospores. The pathogen adheres to root surfaces and can be transferred to new areas on seeds or by insects.

Management

Tests have indicated that commercial cultivars vary considerably in susceptibility. The herbicide chloramben should not be used in soils in which *T. basicola* is present, because greenhouse tests and limited field tests have shown that it increases the severity of root rot. Tillage and delayed planting may help reduce the incidence of this disease.

Selected References

Anderson, T. R. 1984. Thielaviopsis root rot in Ontario and susceptibility of commercial cultivars to root inoculation. Can. J. Plant Pathol. 6:71-74.

Hood, M. E., and Shew, H. D. 1997. Reassessment of the role of saprophytic activity in the ecology of *Thielaviopsis basicola*. Phytopathology 87:1214-1219.

Huang, J., and Kang, Z. 2010. Detection of *Thielaviopsis basicola* in soil with real-time quantitative PCR assays. Microbiol. Res. 165:411-417.

Lee, M., and Lockwood, J. L. 1977. Enhanced severity of *Thielaviopsis basicola* root rot induced in soybean by the herbicide chloramben. Phytopathology 67:1360-1367.

Mondal, A. H., Nehl, D. B., and Allen, S. J. 2004. First report of *Thielaviopsis basicola* on soybean in Australia. Australas. Plant Pathol. 33:451-452.

Paulin-Mahady, A. E., Harrington, T. C., and McNew, D. L. 2002. Phylogenetic and taxonomic evaluation of *Chalara, Chalaropsis,* and *Thielaviopsis* anamorphs associated with *Ceratocystis*. Mycologia 94:62-72.

Punja, Z. K., and Sun, L.-J. 1999. Morphological and molecular characterization of *Chalara elegans* (*Thielaviopsis basicola*), cause of black root rot on diverse plant species. Can. J. Plant Pathol. 77:1801-1812.

Specht, L. P., and Griffin, G. J. 1985. A selective medium for enumerating low populations of *Thielaviopsis basicola* in tobacco field soils. Can. J. Plant Pathol. 7:438-441.

(Prepared by J. B. Sinclair; revised by G. L. Hartman)

Other Fungi and Oomycetes Reported in Soybean

Numerous other fungi and oomycetes have been reported to infect, to be isolated from, or to be closely associated with soybean. The U.S. Department of Agriculture Fungal Databases puts that number at 1,970 records (as of February 2015). The following sections describe several pathogens that are not discussed at length in this book but have been reported in several publications. Table 6 includes a list of other fungi and oomycetes whose names may be encountered when making isolations from diseased soybean plants. (See also Appendix I, which lists additional names of fungi and oomycetes reported in soybean.)

Selected References

Cho, W. D., and Shin, H. D., eds. 2004. List of Plant Diseases in Korea. 4th ed. Korean Society of Plant Pathology. Suwon, South Korea.

Farr, D. F., and Rossman, A. Y. Fungal Databases, Systematic Mycology and Microbiology Laboratory. U.S. Department of Agriculture, Agricultural Research Service, Washington, DC. Available online at http://nt.ars-grin.gov/fungaldatabases/

Gai, Y., Pan, R., Xu, D., Ji, C., Deng, M., and Chen, W. 2011. First report of soybean Neocosmospora stem rot caused by *Neocosmospora nasinfecta* var. *vasinfecta* in China. Plant Dis. 95:1031.

Gray, F. A., Rodrigues-Kabana, R., and Adams, J. R. 1980. Neocosmospora stem rot of soybeans in Alabama. Plant Dis. 64:321-322.

Gray, L. E. 1978. *Mycoleptodiscus terrestris* root rot of soybeans. Plant Dis. Rep. 62:72-73.

Harrington, T. C., Steimel, J., Workneh, F., and Yang, X. B. 2000. Molecular identification of fungi associated with vascular discoloration of soybean in the north central United States. Plant Dis. 84:83-89.

Irinyi, L., Kovics, G. J., and Sandor, E. 2009. Taxonomical re-evaluation of *Phoma*-like soybean pathogenic fungi. Mycol. Res. 113:249-260.

Mashkina, E. S., Nickolaeva, M. I., and Uspenskaya, G. D. 1982. *Marssonina sojaecola* E. Mashk., M. Kikol, and G. Ipsen, sp. nov., a new species found in the central Chernozem region. Mikol. Fitopatol. 16:110-111.

Mendes, M. A. S., da Silva, V. L., et al. 1998. Fungos em Plants no Brasil. (In Portuguese.) Embrapa-SPI/Embrapa-Cenargen, Brasilia, Brazil.

Mueller, J. D., and Sinclair, J. B. 1986. Occurrence and role of *Gliocladium roseum* in field-grown soybeans in Illinois. Trans. Br. Mycol. Soc. 86:297-301.

Ries, E. M., dos Santos, H. P., Pereira, L. R., and Denardin, J. L. 1985. *Helminthosporium sativum* propagules detected in soybean (*Glycine max*) crop litter and their relationship to years of rotation and winter crops. Fitopatol. Bras. 10:455-460.

TABLE 6. Other Fungi and Oomycetes, Reported by Nation

Species	Nation(s)
Ascochyta sojaecola	Poland; former Yugoslavia
Bipolaris sorokiniana	Brazil
Cladosporium tenuissimum	United States
Exserohilum rostratum	India
Gaeumannomyces graminis	United States
Gliocladium roseum	United States
Marssonina sojaecola	Russia
Mycosphaerella phaseolorum	Mozambique
Olpidium viciae	Russia
Phoma exigua	Poland; Rwanda
Phyllosticta glycineum	Mozambique
Phymatotrichopsis omnivora	United States
Plectosphaerella cucumerina	United States
Pseudorobillarda sojae	United States
Setosphaeria glycinea	India
Verticillium nigrescens	United States

Roy, K. W., Abney, T. S., Huber, D. M., and Keeler, R. 1982. Isolation of *Gaeumannomyces graminis* var. *graminis* from soybeans in the Midwest. Plant Dis. 66:822-825.

Smith, R. R., Grau, C. R., and Gray, L. E. 1998. First report of *Mycoleptodiscus terrestris* infecting forage legumes and soybeans in Wisconsin. Plant Dis. 82:126.

Sneh, R., and Stack, J. 1990. Selective medium for isolation of *Mycoleptodiscus terrestris* from soil sediments of aquatic environments. Appl. Environ. Microbiol. 56:3273-3277.

Srivastava, L. S., Vishwa, D., Shambi, H. S., and Chaudhary, D. N. 1987. A new leaf spot disease of soybean in India. Indian J. Mycol. Plant Pathol. 12:240-242.

Uecker, F. A., and Kulik, M. M. 1986. *Pseudorobillarda sojae,* a new pycnidial coelomycete from soybean stems. Mycologia 78:449-453.

Vernetti, F. J., Tomm, G. O., and Casela, C. R. 1985. Occurrence of *Mycosphaerella phaseolorum, Phyllosticta glycineum,* and *Colletotrichum truncatum* in Mozambique. Fitopatol. Bras. 10:191-193.

Vesper, S. J., Turner, J. T., Jr., and Phillips, D. V. 1983. Incidence of *Verticillium nigrescens* in soybeans. Phytopathology 73:1338-1341.

Xin, Z. S., Xiong, C. L., and Zhang, Y. H. 1984. A preliminary study of blister disease of broad-bean (*Olpidium viciae* Kusano) and its control. Acta Phytopathol. Sin. 14:165-173.

Yang, X. B., and Feng, F. 2001. Ranges and diversity of soybean fungal diseases in North America. Phytopathology 91:769-775.

Zhu, X. F., Pan, Y., Chen, L. J., Duan, Y. X., and Wang, Y. Y. 2012. First report of leaf spot of soybean caused by *Aristastoma guttulosum* in China. Plant Dis. 96:1694.

Zhuang, W.-Y., ed. 2005. Fungi of Northwestern China. Mycotaxon, Ithaca, NY.

(Prepared by G. L. Hartman)

Gliocladium roseum

Gliocladium roseum, a common soilborne fungus, is frequently found associated with soybean roots and vascular exudates. It has a wide host range. *Gliocladium* spp. are identical in growth habit to *Verticillium* spp. Verticillate conidiophores are present in immature colonies and branch repeatedly as the colony ages to form penicillate structures in which phialides are all present on the same level. *G. roseum* produces conidia in mucilaginous drops, which may fuse to form larger balls of conidia enveloped in slime. The presence of verticillate conidiophores in immature colonies can lead to the misidentification of a *Gliocladium* sp. as a *Verticillium* sp.

(For additional information, see the Selected References provided with the introduction to the section Other Fungi and Oomycetes Reported in Soybean.)

(Prepared by G. L. Hartman)

Mycoleptodiscus terrestris

Mycoleptodiscus terrestris causes a crown and root rot of soybean in India, northern Africa, and certain areas of the United States, and it has a wide host range. Distinct symptoms are difficult to characterize, because the pathogen is usually associated with other root rot fungi in the field. *M. terrestris* causes a postemergence damping-off of soybean seedlings, which is characterized by a reddish-brown to black cortical decay of the crown and root tissues. The pathogen also causes a dark decay of the lateral root system and the taproots of older plants, similar to that produced by *Rhizoctonia solani* (see the section Rhizoctonia Damping-Off and Root Rot).

(For additional information, see the Selected References provided with the introduction to the section Other Fungi and Oomycetes Reported in Soybean.)

(Prepared by G. L. Hartman)

Neocosmospora vasinfecta

Neocosmospora vasinfecta (syn. *Acremonium* sp.) causes a dark discoloration of soybean stems and pith, along with occasional chlorosis and defoliation of the lower leaves. This pathogen has been reported in Nigeria and the United States, and it has a wide host range. The symptoms it produces may be confused with those caused by the brown stem rot pathogen, *Cadophora gregata* (syn. *Phialophora gregata*) (see the section Brown Stem Rot). Bright-field microscopy has revealed mycelia in the pith and xylem vessels of stem sections of young plants infected with *C. gregata*, whereas plants infected with *N. vasinfecta* have mycelia in the pith tissues only.

(For additional information, see the Selected References provided with the introduction to the section Other Fungi and Oomycetes Reported in Soybean.)

(Prepared by G. L. Hartman)

Diseases Caused by Nematodes

Nematodes are unsegmented roundworms that inhabit almost all ecological niches. The phylum Nematoda includes parasites of plants and animals. Nematodes are the most numerous multicellular animals on Earth, accounting for 80% of all individual animals. Vast numbers of various soil-inhabiting forms can be found in every field, regardless of the crop. A soil sample of 250 cm^3 may contain 20,000 or more nematodes. An estimated 7.5 billion nematodes live within the top 15–20 cm of 1 ha of typical soil.

Some 25,000 species of nematodes have been described, and of them, approximately 10,000 are parasitic on animals or plants. Most plant-parasitic nematodes are microscopic. However, some species—such as the adult female and cyst stage of *Heterodera glycines* (soybean cyst nematode)—are readily observed on plant roots. Most juveniles and adults of plant-parasitic species are vermiform, but females of some nematodes—including cyst-forming *Heterodera* and *Globodera* spp., *Meloidogyne* spp. (root-knot nematodes), *Nacobbus* spp. (false root-knot nematodes), and *Rotylenchulus* spp. (reniform nematodes)—become swollen and saclike.

Plant-parasitic nematodes are isolated using a variety of methods: sieving, sugar centrifugation, elutriation, and the Baermann funnel. The specific extraction method is often optimized for the target nematode. Routine identification of plant-parasitic nematodes is typically accomplished by microscopic examination of morphological features. A standard dissecting microscope with transmitted light and 50× magnification is sufficient for genus-level identification. Species-level identification requires the use of a compound microscope or, more commonly, the application of molecular diagnostics.

The life cycles of most plant-parasitic nematodes are relatively simple (Fig. 149). Following embryogenesis, the nematode develops through four juvenile stages (J1–J4) and then progresses into the adult reproductive stage. Most plant-parasitic nematodes hatch as the J2 stage from an egg. The mobility of nematodes in soil is limited; they rarely migrate more than 75 cm per year. However, they move passively via water, windblown soil, infected plant parts, and infected insect vectors, as well as on tools, vehicles, and machinery contaminated with infested soil. Under optimal conditions, most plant-parasitic species complete their life cycles in 3–4 weeks within a temperature range of 15–32°C. The reproductive potential varies greatly among species and is influenced by the environment and host suitability. Depending on the species, a different life stage (i.e., egg, J2, J4, or adult) may serve as the survival stage.

All nematodes that attack plants are obligate parasites and must feed on living tissue to complete their life cycles; therefore, Koch's postulates cannot be strictly fulfilled. Ectoparasitic nematodes browse on epidermal cells and may sometimes partially penetrate the plant's root. Endoparasitic nematodes complete most of their life cycles inside the root. Some nematodes, such as *Pratylenchus* spp. (lesion nematodes), may be migratory inside roots but also feed as ectoparasites on root hairs and epidermal cells. Members of genera such as *Meloidogyne* and *Heterodera* establish feeding sites and are classified as "sedentary endoparasites."

Nematodes feed on plant cells by puncturing the cell wall with a stylet (which resembles a minute hypodermic needle), injecting secretory products into the cell via the stylet, and then ingesting the partially digested contents. Some secretory products alter the root's physiology, as in the case of the giant cells produced by *Meloidogyne* spp. and the syncytia produced by *Heterodera* spp. Nematodes that feed on or in the roots reduce the root system's efficiency in absorbing and transporting water and nutrients to the shoot. Disruption of normal root functions suppresses the plant's growth and is often accompanied by stunting, chlorosis, and other symptoms of nutrient deficiency.

The degree of crop damage caused by nematodes is related to the cropping history of the field, the production practices used, the species and number of nematodes in the soil at the time of planting, and the environmental conditions following the emergence of seedlings. Usually, young plants are the most susceptible to nematode injury, and plants become more tolerant with age, as a larger root system develops and compensates for feeding by nematodes. Symptoms of nematode damage often mimic those induced by low or unbalanced fertility, soil hardpans, poor drainage, drought, soil-inhabiting insects, root rot fungi, and herbicides. Nematodes can cause significant crop loss in the absence of obvious symptoms. Thus, diagnosis of nematode disease should never be based solely on aboveground plant symptoms. Diagnosis requires proper sample collection and handling of soil and roots, followed by nematode identification.

Economically significant crop injury caused by nematodes is influenced by the species and quantity of nematodes in the soil and roots relative to the age and physiological condition of the plants. The presence of additional biotic and abiotic stresses may also influence the effects of nematodes. Damage often reflects the additive effects of feeding by several species on a root system, as well as their interactions with other root-infecting pathogens. Thresholds for nematode damage should therefore be interpreted with caution.

More than 100 species of nematodes have been associated with the roots of soybean plants, but only a few genera and species are of economic importance. Generally, the management of nematode parasites in soybean fields relies on crop rotation with nonhosts, planting of resistant cultivars, and seed treatment.

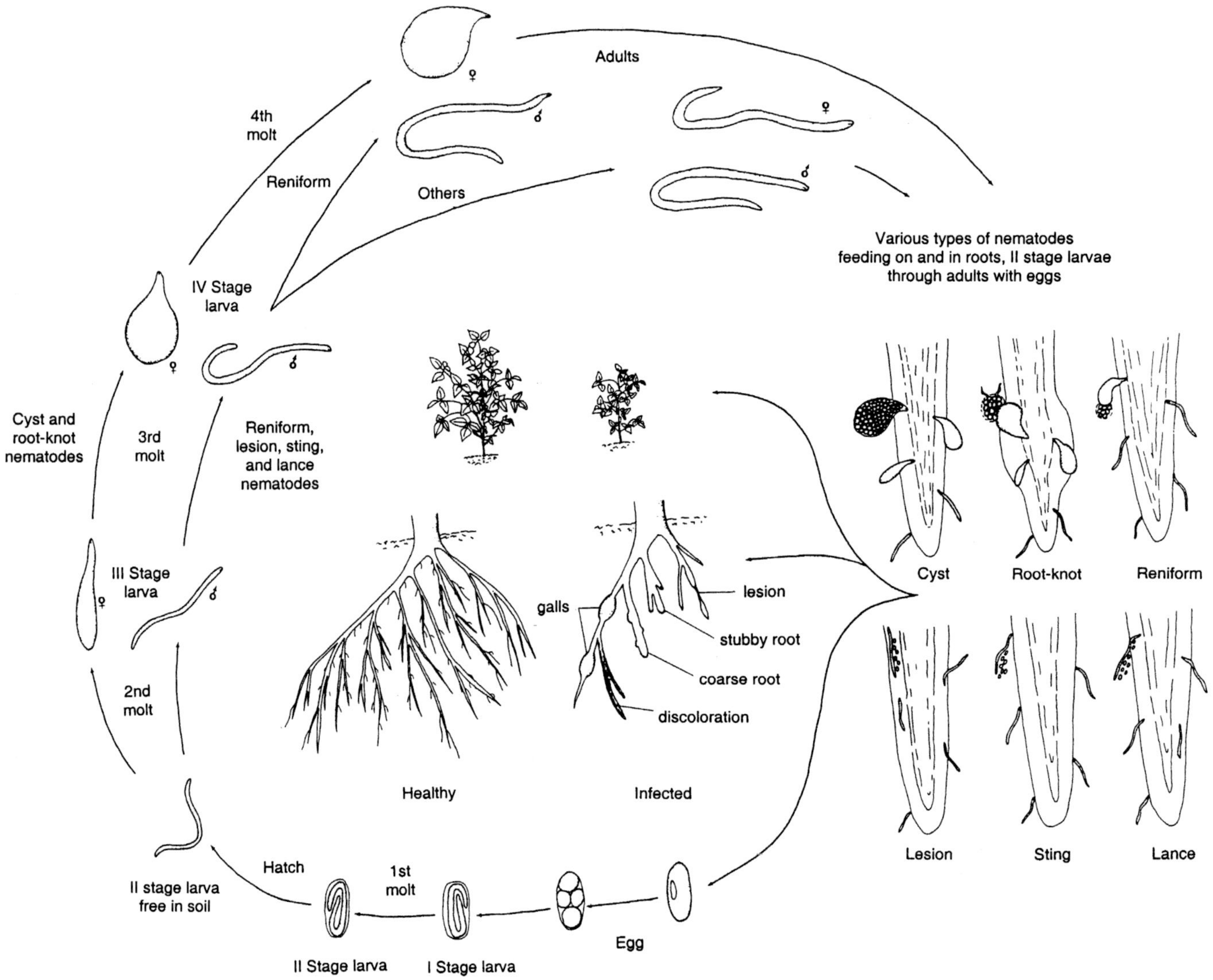

Fig. 149. Generalized life drawings of six genera of plant-parasitic nematodes that attack soybeans and the symptoms they cause. (Developed by R. B. Malek; drawing by D. J. Royce)

Selected References

Baldwin, J. G., Nadler, S. A., and Wall, D. H. 2000. Nematodes: Pervading the Earth and linking all life. Pages 176-191 in: Nature and Human Society: The Quest for a Sustainable World. P. H. Raven, ed. National Academy Press, Washington, DC.

Nematode Descriptions and Identification. n.d. Plant and Insect Parasitic Nematodes. Nematology Lab, University of Nebraska–Lincoln. Available online at http://nematode.unl.edu/nemaid.htm

Niblack, T. L., Tylka, G. L., and Riggs, R. D. 2004. Nematode pathogens of soybean. Pages 821-851 in: Soybeans: Improvement, Production, and Uses. 3rd ed. Agronomy Series No. 16. H. R. Boerma and J. E. Specht, eds. American Society of Agronomy, Madison, WI.

Perry, R. N., and Moens, M. 2006. Plant Nematology. CABI, Wallingford, Oxfordshire, U.K.

Platt, H. M. 1994. Foreword in: The Phylogenetic Systematics of Free-Living Nematodes. Ray Society, London.

Powers, T. 2004. Nematode molecular diagnostics: From bands to barcodes. Annu. Rev. Phytopathol. 42:367-383.

Starr, J. L., Cook, R., and Bridge, J. 2002. Plant Resistance to Parasitic Nematodes. CABI, Wallingford, Oxfordshire, U.K.

(Prepared by G. R. Noel; revised by N. E. Schroeder)

Lance Nematodes

Lance nematodes, which belong to the species *Hoplolaimus,* are an important group of migratory ecto-endoparasites, which feed on the roots of many economically important plants worldwide. Reports of lance nematodes causing damage to soybean are limited to the southeastern United States.

Symptoms and Signs

Feeding by *H. columbus* results in chlorotic, stunted plants. Damage to susceptible cultivars the following growing season may also occur if the fall population of *H. columbus* exceeds 60–100 nematodes per 100 cm^3 of soil, especially when soil moisture is limited. Infected roots may exhibit excessive forking and necrosis to the extent that the taproot is extremely abbreviated and difficult to identify. *H. columbus* favors coarse-textured soils and is most commonly found in those with greater than 85% sand content.

Causal Organisms

Four species of *Hoplolaimus* are associated with soybean: *H. columbus, H. galeatus, H. magnistylus,* and *H. stephanus.*

Infection by and damage from *H. columbus* were first noted during the early 1960s in central South Carolina and are now found in cotton, maize, and soybean in Alabama, Georgia, Louisiana, North and South Carolina, and Texas. Soybean roots are also infected by *H. galeatus* and *H. stephanus* in the southeastern United States and by *H. magnistylus* in Arkansas, Mississippi, and Tennessee, but these latter species do not seem to cause severe damage.

Disease Cycle and Epidemiology

H. columbus occurs in great numbers in soybean roots (up to 500 per gram of fresh roots). *H. columbus* often feeds in the plant's cortex, where all life stages of the nematode can be found. Older feeding sites may show evidence of vascular damage, especially in the phloem and xylem parenchyma. Most of the damage from endoparasitic feeding is associated with the migration of nematodes through the root tissue and the resulting lack of integrity of the cortex (Fig. 150). Ectoparasitic feeding is common, as well (Fig. 151). Individual nematodes may be located completely in or out of root tissues, where rootlets emerge or other nematodes have penetrated.

The life cycle of *H. columbus* takes approximately 28 days under normal conditions and occurs more rapidly at 30°C than at either 25 or 20°C. Eggs of *H. columbus* are unusual in that they have a terminal filament.

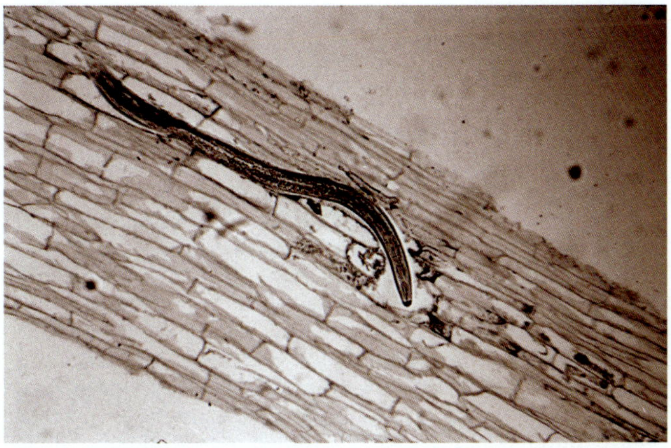

Fig. 150. Endoparasitic feeding by a lance nematode (*Hoplolaimus* sp.). (Courtesy J. D. Mueller)

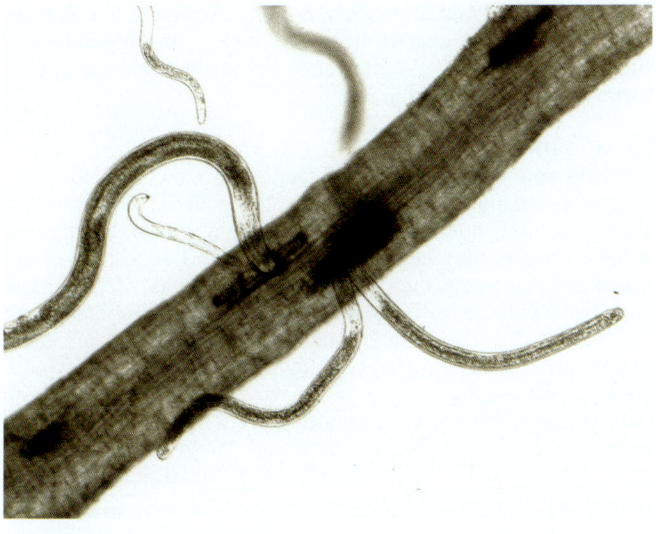

Fig. 151. Ectoparasitic feeding by lance nematodes (*Hoplolaimus* sp.). (Courtesy J. D. Mueller)

Correlation of *H. columbus* to yield losses is highest for juveniles at 3 and 30 weeks, for adults at 12 and 20 weeks, and for adults plus juveniles at 3 and 20 weeks after planting. The number of nematodes per gram of roots (dry weight or fresh weight) at 8 weeks after planting strongly correlates with yield.

Management

Tolerant cultivars should be planted (e.g., Centennial, Dillon, Hagood, Santee, and Stonewall), and intolerant cultivars should be avoided (e.g., Braxton, Bryan, and Perrin). Some cultivars that are tolerant to *Heterodera glycines* (soybean cyst nematode) are also tolerant to *H. columbus,* but tolerance to *H. columbus* varies by location and year.

Use of a nematicide such as aldicarb in furrow at planting, even at low to moderate rates, can enhance crop yield. (Aldicarb is not labeled for soybean in all states, however.) The best response occurs when a nematicide is applied to a tolerant cultivar.

To alleviate some of the impact of nematode feeding, in-row subsoiling or residual tillage from a previous small-grain crop is advised in areas with hardpans. Feeding by *H. columbus* may increase the incidence and severity of soilborne fungi, such as *Calonectria pyrochroa, Rhizoctonia solani,* and *Sclerotium rolfsii.*

Selected References

Arslan, M. 1998. Effect of Columbia lance nematode (*Hoplolaimus columbus* Sher) on growth and development of tolerant and intolerant soybean cultivars. Ph.D. thesis. Clemson University, Clemson, SC.

Boerma, H. R., and Hussey, R. S. 1984. Tolerance to *Heterodera glycines* in soybean. J. Nematol. 16:289-296.

Davis, R. F., and Noe, J. P. 2000. Extracting *Hoplolaimus columbus* from soil and roots: Implications for treatment comparisons. J. Cotton Sci. 4:105-111.

Ma, X., Agudelo, P., Mueller, J. D., and Knap, H. T. 2011 Molecular characterization and phylogenetic analysis of *Hoplolaimus stephanus*. J. Nematol. 43:25-34.

Mueller, J. D., Schmidt, D. P., Weiser, G. C., Shipe, E. R., and Musen, H. L. 1988. Performance of soybean cultivars in *Hoplolaimus Columbus*-infested fields. Ann. Appl. Nematol. 2:65-69.

Perez, E. E., Mueller, J. D., and Lewis, S. A. 1996. Effect of planting date on population densities of *Hoplolaimus columbus* and yield of soybean. J. Nematol. 28:569-574.

Schmitt, D. P., and Imbriani, J. L. 1987. Management of *Hoplolaimus columbus* with tolerant soybean and nematicides. Ann. Appl. Nematol. 1:56-63.

Supramana, S., Lewis, S. A., Mueller, J. D., Fortnum, B. A., and Ballard, R. E. 2001. Biology of Columbia lance nematode (*Hoplolaimus columbus* Sher) on soybean excised root culture. Nematropica 31:281-293.

Supramana, S., Lewis, S. A., Mueller, J. D., Fortnum, B. A., and Ballard, R. E. 2001. Evaluation of *Hoplolaimus columbus* Sher reproduction on selected soybean cultivars using greenhouse and excised root cultural techniques. Nematropica 31:265-269.

(Prepared by J. D. Mueller and P. Agudelo)

Lesion Nematodes

Lesion nematodes, which are members of the genus *Pratylenchus,* are found throughout the world and attack a wide range of plants. Several species of *Pratylenchus* feed on the roots of soybean and may cause considerable losses in crop yields.

Symptoms and Signs

Pratylenchus spp. are considered migratory endoparasites and attack the root cortex of the soybean plant. However, they may also feed on root hairs and epidermal cells. Their feeding causes dark lesions and an overall browning of the roots, which may decrease root growth by 25% (Fig. 152). The epidermis and the cortex of a severely infected root may slough away from the stele.

The degree of damage caused by lesion nematodes is generally related to the population level. Crop loss can occur without obvious symptoms of chlorosis and stunting. However, when nutrients and soil moisture are deficient, yield reduction is greater and accompanied by symptoms typical of nematode damage.

Causal Organisms

Lesion nematodes reported on soybean include *P. agilis, P. alleni, P. brachyurus, P. coffeae, P. crenatus, P. hexincisus, P. neglectus, P. penetrans, P. pratensis, P. scribneri, P. sefaenis, P. teres, P. thornei, P. vulnus,* and *P. zeae*. Members of the genus *Pratylenchus* are among the smallest nematodes that feed on plants; adults range in length from 400 to 800 µm.

Some species, such as *P. brachyurus* and *P. zeae,* are found primarily in tropical and subtropical regions. Others, such as *P. penetrans* and *P. scribneri,* occur commonly in heavy soils and are most abundant in temperate regions.

Disease Cycle and Epidemiology

Lesion nematodes complete their life cycle in 4–8 weeks. All juvenile and adult stages are worm shaped and motile, and all stages (except the egg and J1) infect soybean roots and migrate intracellularly, mainly throughout the cortex. The nematodes move in and out of the roots, and as they migrate, zones of necrosis or lesions develop.

Lesion nematodes overwinter in infested plant parts or in soil at any life stage. They migrate only 1–2 m, and their spread within the field usually occurs via cultural practices or water moving through an infested area.

Management

Soybean cultivars vary in susceptibility to species of *Pratylenchus,* but cultivars with resistance to other plant-parasitic nematodes are not necessarily resistant to *Pratylenchus* spp. For instance, cultivars with Fayette in their pedigrees (i.e., plant introduction [PI] 88788)—the source of resistance used to manage *Heterodera glycines*—are susceptible to *P. scribneri.*

Many *Pratylenchus* spp. have wide host ranges, including various monocot and dicot species; thus, crop rotation may by ineffective for controlling these nematodes. Several seed treatments are currently marketed for protection against lesion nematodes in soybean, including the contact nematicide abamectin and a strain of the natural soil bacterium *Bacillus firmus.*

Fig. 152. Root lesions caused by *Pratylenchus scribneri,* a lesion nematode. (Courtesy N. Acosta)

Selected References

Acosta, N. 1982. Vertical distribution of *Pratylenchus alleni* and *P. scribneri* in soybean roots. J. Agric. Univ. P. R. 66:60-64.

Acosta, N., and Malek, R. B. 1981. Symptomatology and histopathology of soybean roots infected by *Pratylenchus scribneri* and *P. alleni*. J. Nematol. 13:6-12.

Agu, C. M. 2006. Susceptibility of soybean cultivars to root nematodes in southwestern Ethiopia. Trop. Sci. 46:143-146.

Ferraz, L. C. C. B. 1996. Reaction of soybean cultivars to *Pratylenchus brachyurus*. Nematol. Bras. 20:22-31.

Fourie, H., McDonald, A. H., and Loots, G. C. 2001. Plant-parasitic nematodes in field crops in South Africa 6. Soybean. Nematology 3:447-454.

Frederick, J. J., and Tarjan, A. C. 1986. A compendium of the genus *Pratylenchus* Filipjev, 1936 (Nemata: Pratylenchidae). Rev. Nematol. 12:243-256.

Koenning, S. R., and Schmitt, D. P. 1986. Integrated pest management of *Pratylenchus brachyurus* on soybean. Nematropica 16:237-244.

Lawn, D. A., and Noel, G. R. 1985. Field interrelationships among *Heterodera glycines, Pratylenchus scribneri,* and three other nematode species associated with soybean. J. Nematol. 18:99-106.

Norton, D. C., Donald, P., Kimpinski, J., Myers, R. F., Noel, G. R., Noffsinger, E. M., Robbins, R. T., Schmitt, D. P., Sosa-Moss, C., and Vrain, T. C. 1984. Distribution of Plant-Parasitic Nematodes in North America. Society of Nematologists, Marceline, MO.

Rebois, R. V., and Huettel, R. V. 1986. Population dynamics, root penetration and feeding behavior of *Pratylenchus agilis* in monoxenic root cultures of corn, tomato, and soybean. J. Nematol. 18:392-397.

(Prepared by G. R. Noel; revised by N. E. Schroeder)

Reniform Nematode

Infection of the roots of soybean plants by *Rotylenchulus reniformis*, a reniform nematode, was first reported from western Africa in 1956 and then from the U.S. state of South Carolina in 1967. In the United States, the greatest incidence of *R. reniformis* occurs in the Gulf Coast states of Alabama, Louisiana, and Mississippi; in addition, this nematode is found in Arkansas, California (although extremely limited), Georgia, Missouri, North Carolina, South Carolina, Tennessee, and Virginia. Distribution of this reniform nematode has increased rapidly throughout the southern United States since the 1990s. Its spread into the agricultural lands of Arizona, California, and New Mexico has been limited by the imposition of strict quarantines.

R. reniformis also has a worldwide distribution and is found in many tropical and subtropical regions. Although it is thought to be incapable of overwintering in regions at temperate latitudes, it is potentially important to soybeans produced in both tropical and subtropical areas.

Symptoms and Signs

Symptoms of infection by *R. reniformis* include chlorosis, stunting, and unthrifty growth of plants, along with production of a large number of empty pods. The root systems of affected plants may be severely stunted and necrotic—often, much more severely than is the case with either *Meloidogyne* spp. (root-knot nematodes) or *Heterodera glycines* (soybean cyst nematode). The most common sign of *R. reniformis* is the appearance of soil-covered egg masses on the roots, which are sticky to the touch.

Soybean yields decrease significantly with increasing population levels of *R. reniformis* in the soil. Yield losses as high as 30–60% have been reported within individual fields, depending on the cultivar planted.

Causal Organism

Although nine species have been identified in the genus *Rotylenchulus,* only *R. reniformis* is important on soybean. It is a

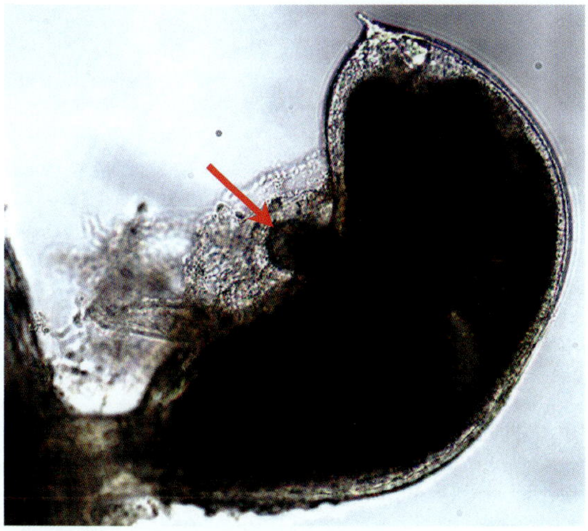

Fig. 153. Female reniform nematode (*Rotylenchulus reniformis*), a semiendoparasite. The arrow indicates a single egg protruding from the vulva. (Courtesy E. C. McGawley)

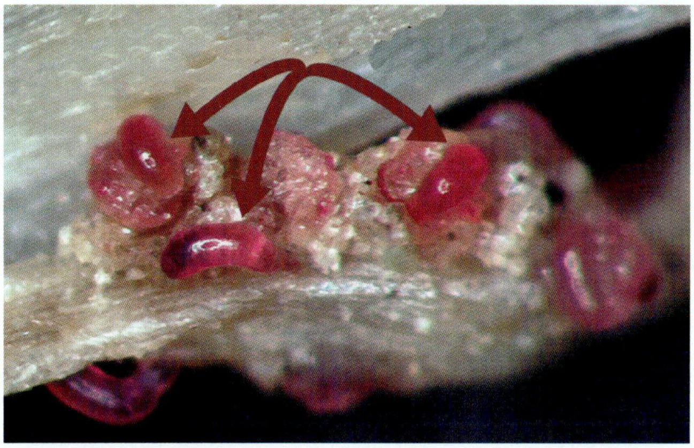

Fig. 154. Stained female reniform nematodes (*Rotylenchulus reniformis*) on the root of a soybean plant. The arrows indicate individual females. (Courtesy E. C. McGawley)

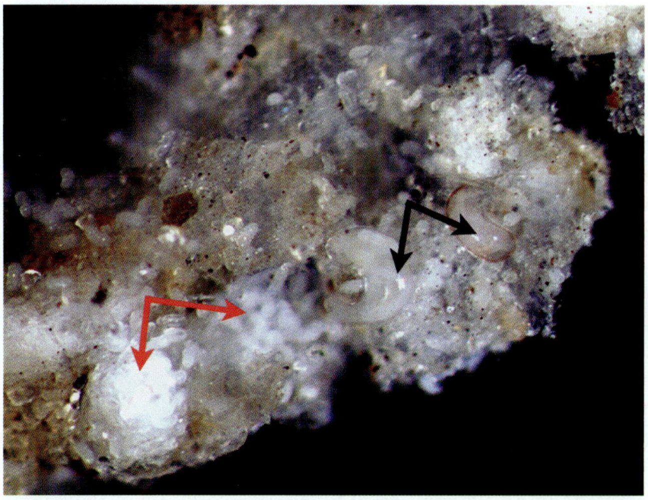

Fig. 155. Female reniform nematodes (*Rotylenchulus reniformis*) and eggs on the root of a soybean plant. The red arrows indicate egg masses, and the black arrows indicate nonstained females. (Courtesy E. C. McGawley)

semiendoparasite and partially embeds itself in the roots, with the head near the plant's vascular system (Fig. 153). Because *R. reniformis* is small and soil particles adhere to the portion of its body outside the root, it can very easily be overlooked.

Disease Cycle and Epidemiology

Eggs of *R. reniformis* are elongate (33–44 × 72–100 µm), and unlike the eggs of most other nematodes, they hatch in water within 24 h, yielding a male-to-female sex ratio of approximately 1:1. The juvenile is 345–560 µm and has a well-developed stylet 16–22 µm long. The juvenile molts three times in the soil without feeding and then becomes an adult. Each molt produces a slightly smaller and shorter stage of the life cycle. The fourth-stage juvenile (i.e., preadult) is the infective stage.

Only females parasitize the plant roots; males remain in the rhizosphere or in the gelatinous matrices of the females. The female penetrates the epidermis and cortical parenchyma of the roots and reaches the stelar region within 2 days; there, she establishes a feeding site called a "syncytium." The syncytium develops through partial cell wall lysis and separation and coalescence of the cytoplasmic components. The posterior of the female remains outside the root and swells to the characteristic kidney shape (Fig. 154). Males remain vermiform. Mature females are 410–655 µm long and lay 70–80 eggs outside the roots, many of which are infertile (Fig. 155).

The life cycle of *R. reniformis* is completed in 17–23 days at a soil temperature of 29°C—the optimum for root invasion and nematode development. Infection is most likely to occur when the soil moisture level is just below field capacity; extremely wet and extremely dry conditions reduce the likelihood of root invasion.

Two races of *R. reniformis,* designated "A" and "B," have been described. Great differences in fecundity, aggressiveness, and morphometrics have been identified among various populations of reniform nematode on both cotton and soybean cultivars.

Management

Resistant soybean cultivars (e.g., Forrest) and several resistant breeding lines have been reported. Resistance to *R. reniformis* in soybean is quantitative and controlled by two pairs of genes (*Rrn*) with unequal effects. There is a linkage in resistance between *R. reniformis* and *H. glycines.* Cultivars that derive their resistance to *H. glycines* from cultivar Peking or plant introduction (PI) 437654 or 90763 are also resistant to *R. reniformis.* However, cultivars that derive their resistance to *H. glycines* from PI 88788 are not resistant to *R. reniformis.* Very few cultivars appear to be resistant to *R. reniformis.*

A number of agronomic crops (e.g., maize, oats, rice, grain sorghum, sugarcane, and wheat) are not hosts of *R. reniformis* and thus are usually effective in a rotation with soybean. Production of these crops for 2 or more years usually reduces the reniform nematode population sufficiently to allow production of a susceptible soybean crop for 1 year. Soybean cultivars that are resistant to *R. reniformis* have been used successfully to reduce the nematode population to allow successful rotation with a susceptible crop, such as cotton.

Selected References

Agudelo, P., Robbins, R. T., Stewart, J. M., and Szalanski, A. L. 2005. Intraspecific variability of *Rotylenchulus reniformis* from cotton-growing regions in the United States. J. Nematol. 37:105-114.

Lewis, S. A., Drye, C. E., Saunders, J. A., Shipe, E. R., and Halbrendt, J. M. 1993. Plant-parasitic nematodes on soybean in South Carolina. J. Nematol. 25 (suppl.):890-894.

McGawley, E. C., and Overstreet, C. 1995. Reproductive and pathological variation in populations of the reniform nematode *Rotylenchulus reniformis.* (Abstr.) J. Nematol. 27:508.

Robbins, R. T., Shipe, E. R., Arelli, P., Chen, P., Shannon, G., Koenning, S., Rakes, L., Jackson, L. E., Gbur, E. E., and Dombek, D. G. 2009.

Reniform nematode reproduction on soybean cultivars and breeding lines in 2008. Pages 104-114 in: 2009 Beltwide Cotton Conferences. S. Boyd, M. Huffman, D. Richter, and B. Robinson, eds. National Cotton Council, Cordova, TN.

Robbins, R. T., Shipe, E. R., Rakes, L., Jackson, L. E., Gbur, E. E., and Dombek, D. G. 2002. Host suitability of soybean cultivars and breeding lines to reniform nematode in tests conducted in 2001. J. Nematol. 34:378-383.

Robinson, A. F. 2002. Reniform nematodes: *Rotylenchulus* species. Pages 153-174 in: Plant Resistance to Parasitic Nematodes. J. L. Starr, R. Cook, and J. Bridge, eds. CABI, Wallingford, Oxfordshire, U.K.

Westphal, A., and Scott, A. W., Jr. 2005. Implementation of soybean in cotton cropping sequences for management of reniform nematode in South Texas. Crop Sci. 45:233-239.

(Prepared by E. C. McGawley and C. Overstreet)

Root-Knot Nematodes

Root-knot disease of soybean, caused by several species of *Meloidogyne,* has been reported from many soybean-growing areas of the world, including both North and South America. Three *Meloidogyne* spp. account for most of the damage reported in areas with long growing seasons and warm climates.

A large portion of the hectarage in the U.S. Southeast (especially the Mississippi River Delta region), in which soybean in maturity groups VI–VIII have traditionally been grown, now grow maturity group V or earlier to avoid the effects of drought. Cultivars in maturity groups IV and earlier are generally extremely susceptible to most *Meloidogyne* spp., whereas many later-maturing cultivars in groups V–VIII have high levels of resistance to one or more species of root-knot nematode.

Since the 1990s in many areas of the South, growing different soybean cultivars has resulted in higher losses associated with *Meloidogyne* spp. Yield losses vary from inconsequential to complete, depending on the species and population density of the nematode, along with cultivar susceptibility and the intensity of drought stress to which the host is subjected. Because of the effects of drought stress, the greatest yield losses are likely to occur in fields in which soybean is produced in sandy, light-textured soils. In addition, a high incidence of root-knot nematode infection decreases the formation of nodules by *Bradyrhizobium* spp. and increases host susceptibility to vascular fungal pathogens.

Symptoms and Signs

The primary symptom for identifying *Meloidogyne* spp. is the presence of galls on infected roots (Fig. 156). Galls are produced by the increased size and number of cells in infected and adjacent tissues in response to feeding by *Meloidogyne* spp. The galls vary in number and size depending on the intensity of the infection and the nematode species involved. When galls are numerous, they tend to coalesce, and the plant's entire root system may become greatly swollen. The vascular elements of infected tissues are disrupted, and the flow of water and nutrients through the plant can be inhibited. As a result, secondary symptoms may develop on infected plants, including chlorosis and stunting and a tendency to wilt in the heat of the day. Patches of infected plants usually occur irregularly in the field and tend to spread in the direction of cultivation and surface water flow.

The population density of *Meloidogyne* spp. is typically related to the soil type and texture within a field, and populations are generally greatest in sandy areas of fields. Infested sites frequently display dense weed growth, caused by the reduced crop canopy.

Causal Organisms

Root-knot disease on soybean is caused by four species of *Meloidogyne: M. arenaria, M. hapla, M. incognita,* and *M. javanica.*

M. arenaria and *M. incognita* are the most ubiquitous parasites; management of these species is necessary in Brazil and in much of the southern United States. At least two races of *M. arenaria* and at least four races of *M. incognita* are identified by host specificity. Populations of *M. incognita* vary in their ability to reproduce on soybean, but reproductive ability may not be correlated with race. Race 2 of *M. arenaria* tends to be a more aggressive pathogen of soybean than race 1 (the peanut race)—a characteristic that has considerable importance in the management of these nematodes.

M. hapla is perhaps the least damaging of the root-knot nematodes found on soybean; it forms small galls and causes increased lateral branching of the root system. *M. hapla* occurs most commonly in the northern regions of the United States.

M. javanica, which is possibly the most damaging of the root-knot species, is the most common species found in Brazil and other soybean-producing countries in the world's subtropical and tropical regions. In the United States, *M. javanica* occurs in the most southerly areas of soybean production. In particular, it is found on soybean grown in areas associated with tobacco production, although this association does not restrict its distribution.

M. enterolobii (syn. *M. mayaguensis*) has been found extensively in Florida and also reported in several countries in Africa, Asia, Central America, and South America. It can reproduce on soybean, although populations vary in their ability to parasitize root-knot-resistant cultivars.

Disease Cycle and Epidemiology

The vermiform second-stage juveniles of *Meloidogyne* spp. invade the roots of soybean plants, mostly at or near the tips (Fig. 157). Infection can occur throughout the growing season. Migration within a root does not occur, except in massive infections by many juveniles. Juveniles normally settle at a feeding site close to the point of entry, and galling is initiated within 2–3 days. Following a series of three molts, the infecting juveniles become adults. Only a few of the adults are males, which

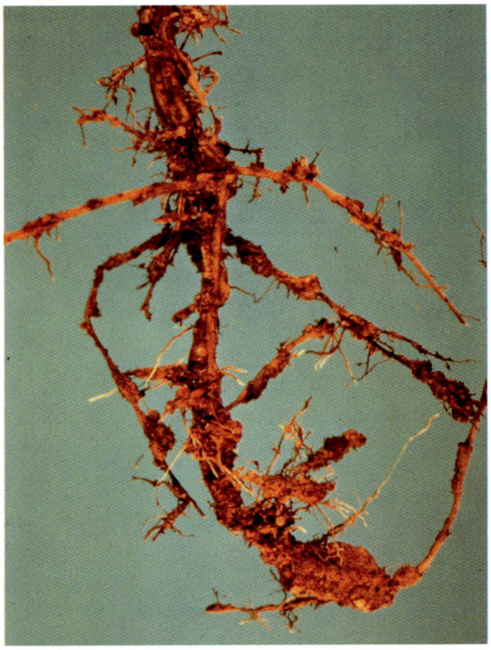

Fig. 156. Galls produced on soybean roots by a root-knot nematode (*Meloidogyne* sp.). (Courtesy J. B. Sinclair)

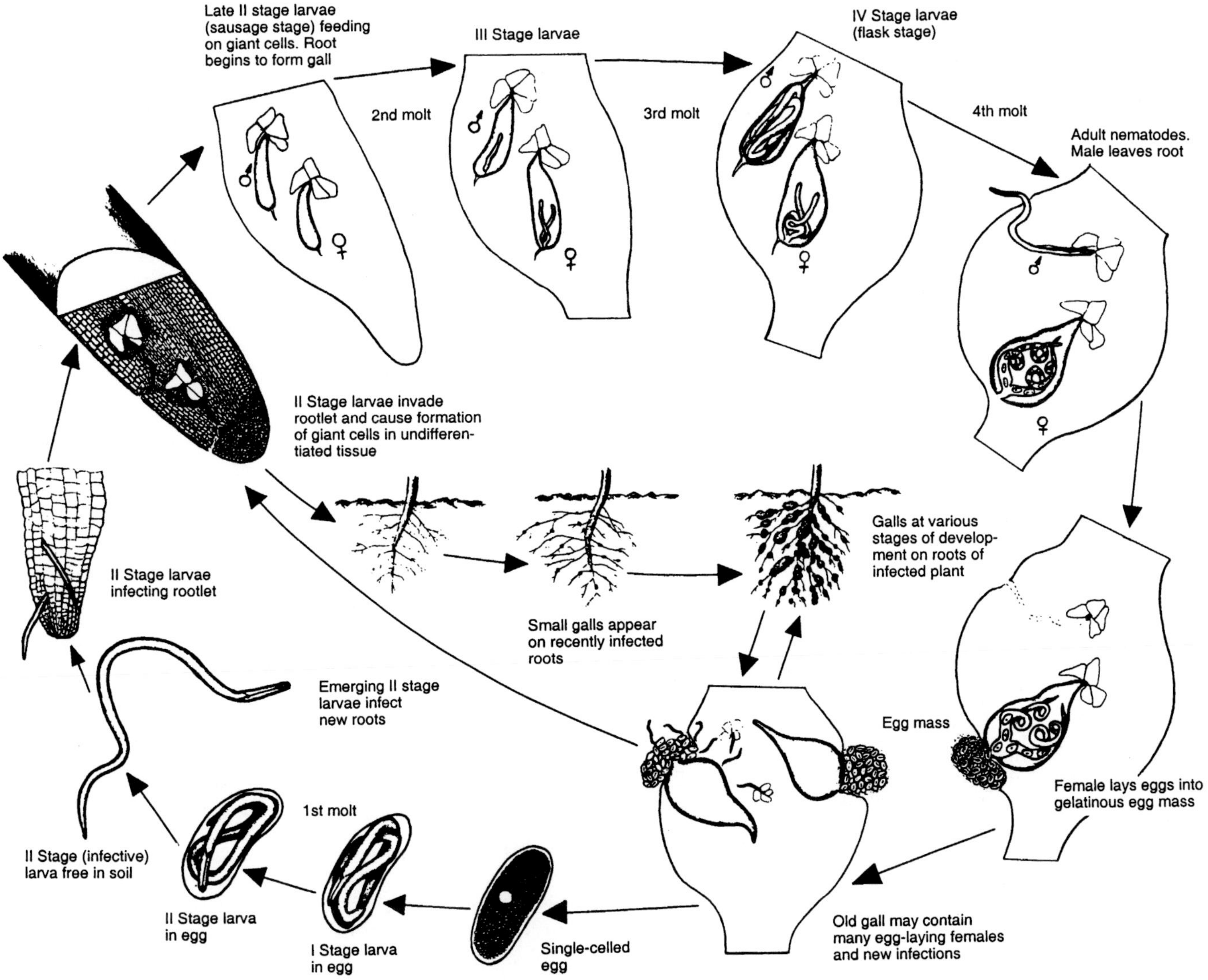

Fig. 157. Life cycle of the root-knot nematode (*Meloidogyne* spp.). (Adapted from Agrios, 1988)

are elongate (1.0–1.5 mm) and migratory. The majority are females, which become swollen and pear shaped as they develop.

In latitudes at which soil temperatures at the time of soybean maturity are sufficient for egg hatch (>15°C), *Meloidogyne* spp. survive the winter in soil as infective juveniles. When temperatures at soybean maturity are lower, *Meloidogyne* spp. are likely to survive as eggs. The number of nematodes in the soil reaches its maximum at soybean maturity. The population declines slowly through the winter and then precipitously as soil temperatures rise in the spring. This rapid decrease in the number of infective nematodes can occur several weeks before planting. It may be caused by the increased activity of the nematodes and depletion of their food reserves and/or by an increase in the predatory and parasitic activities of the soil arthropod and microbial communities, which are detrimental to the nematodes' survival. By planting time, the number of nematodes in the soil is at its lowest—usually less than 10% of the population at soybean maturity.

Management

To help manage *Meloidogyne* spp., growers should plant resistant and tolerant cultivars. The genotypes used in breeding for resistance to *M. arenaria* are Forrest, plant introductions (PI) 200538, PI 230977, Jackson, and G93-9106; to *M. javanica,* PI 80466, PI 595099, PI 230977, Gordon, G93-9223, and Coodetec 201; and to *M. incognita,* Avery, D3-3349, D86-3429, Forrest, G93-9009, and PIs 417444 and 96354. The cultivars Forrest, Kirby, and Palmetto are also sources of resistance. Several cultivars in maturity groups V–VIII are tolerant to some *Meloidogyne* spp. and should be grown when soils are suspected of harboring these nematodes. Resistance to the nematode is horizontal, and some yield loss occurs in these cultivars when the soil population of *Meloidogyne* spp. is high.

Damaging levels of root-knot nematodes usually occur following several years of continuous planting of soybean or another host in an infested field. Consequently, both preventive and corrective management measures are needed. Planting and cultivating equipment must be thoroughly washed to be free of contaminated soil, and fields should be maintained free of weeds (many of which are hosts of *Meloidogyne* spp.). Gramineaceous crops are less susceptible hosts, and when grown in rotation with soybean, they reduce the residual population of juveniles. When the soil infestation level of juveniles reaches a damaging density, summer planting with a graminaceous crop is beneficial, irrespective of the *Meloidogyne* sp. present.

The application of nematicides is rarely cost effective for managing *Meloidogyne* spp.

Selected References

Carpenter, A. S., and Lewis, S. A. 1991. Aggressiveness and reproduction of four *Meloidogyne arenaria* populations on soybean. J. Nematol. 23:232-238.

Castagnone-Sereno, P. 2012. *Meloidogyne enterolobii* (= M. *mayaguensis*): Profile of an emerging, highly pathogenic, root-knot nematode species. Nematology 14:133-138.

Cetintas, R., Brito, J. A., and Dickson, D. W. 2008. Virulence of four Florida isolates of *Meloidogyne mayaguensis* to selected soybean genotypes. Nematropica 38:127-135.

Herman, M., Hussey, R. S., and Boerma, H. R. 1990. Response of resistant soybean plant introductions to *Meloidogyne incognita* in field microplots. J. Nematol. 22:237-241.

Herman, M., Hussey, R. S., and Boerma, H. R. 1991. Penetration and development of *Meloidogyne incognita* on roots of resistant soybean genotypes. J. Nematol. 23:155-161.

Humphreys-Pereira, D. A., and Elling, A. A. 2015. Mitochondrial genome plasticity among species of the nematode genus *Meloidogyne* (Nematoda: Tylenchina). Gene. doi:org/10.1016/j.gene.2015.01.065.

Hussey, R. S., and Janssen, G. J. W. 2002. Root-knot nematodes: *Meloidogyne* species. Pages 43-70 in: Plant Resistance to Parasitic Nematodes. J. L. Starr, R. Cook, and J. Bridge, eds. CABI, Wallingford, Oxfordshire, U.K.

Kinloch, R. 1992. Management of root-knot nematodes in soybean. Pages 147-154 in: Pest Management in Soybean. L. E. Copping, M. B. Green, and R. T. Rees, eds. Elsevier Applied Science, London.

Luzzi, B. M., Boerma, H. R., and Hussey, R. S. 1994. Inheritance of resistance to the southern root-knot nematode in soybean. Crop Sci. 34:1240-1243.

Luzzi, B. M., Boerma, H. R., and Hussey, R. S. 1995. Inheritance of resistance to the peanut root-knot nematode in soybean. Crop Sci. 35:50-53.

Minton, H. A., and Parker, M. B. 1987. Root-knot nematode management and yield of soybean as affected by winter cover crops, tillage systems, and nematicides. J. Nematol. 19:38-43.

Mitkowski, N. A., and Abawi, G. S. 2011. Root-Knot Nematode. Plant Health Instructor. Plant Management Network. doi:10.1094/PHI-I-2003-0917-01.

Rodriguez-Kabana, R., Robertson, D. G., Backman, P. A., and Ivey, H. 1988. Soybean peanut rotations for the management of *Meloidogyne arenaria*. Ann. Appl. Nematol. 2:81-85.

Silva, J. F. V., Ferraz, L. L. C. B., and Arias, C. A. 2001. Inheritance of resistance to *Meloidogyne javanica* in soybean. Nematropica 31:211-219.

Walters, S. A., and Barker, K. R. 1994. Current distribution of five major *Meloidogyne* species in the United States. Plant Dis. 78:772-774.

Weaver, D. B., Rodriguez-Kabana, R., Robertson, D. G., Akridge, R. L., and Cardere, E. L. 1988. Effect of crop rotation on soybean in a field infested with *Meloidogyne arenaria* and *Heterodera glycines*. Ann. Appl. Nematol. 2:106-109.

Wrather, J. A., and Koenning, S. R. 2009. Effects of diseases on soybean yields in the United States 1996 to 2007. Plant Health Progress. Plant Management Network. doi:10.1094/PHP-2009-0401-01-RS.

Young, L. D. 1998. Breeding for nematode resistance and tolerance. Pages 187-207 in: Plant Nematode Interactions. K. R. Barker, G. A. Pederson, and G. L. Windham, eds. American Society of Agronomy, Crop Science Society of America, Soil Science Society of America, Madison, WI.

(Prepared by S. R. Koenning)

Soybean Cyst Nematode

In the United States, *Heterodera glycines,* the soybean cyst nematode, causes greater losses than any other single pathogen of soybean. *H. glycines* is also known to occur in all major soybean production areas in the world, and in many of them, it ranks among the most economically destructive pathogens.

H. glycines probably coevolved with leguminous and other host plants and was well established in Southeast Asia by the time agricultural soybean production began. A soybean disease called "moon night" and "yellow dwarf" was observed in Japan and in China in the 1880s. The symptoms of the disease—namely, circular patches of stunted, yellow soybean plants—suggest that it was caused by the soybean cyst nematode, but proof of the nematode as a soybean pathogen was not published for several decades.

Exactly where *H. glycines* occurred in nature before soybean production became popular in the Western Hemisphere during the twentieth century is still a matter for debate; it may have been introduced once or several times. Nonetheless, this nematode has behaved as an invasive pest, in terms of its spread and apparent dearth of natural enemies, since it was first identified in the United States in 1954. Its occurrence in the United States is widespread (Fig. 158).

Symptoms and Signs

Aboveground symptoms of infection by *H. glycines* may include slight to severe stunting and chlorosis that occurs in circular to oblate patches (Fig. 159), but symptoms are not typically apparent even when a 15–30% crop loss occurs. Although symptoms can develop suddenly, their appearance likely suggests that *H. glycines* has affected yield for several cropping seasons. Many times, the sudden appearance of field symptoms of *H. glycines* indicates the presence of another root-affecting problem or pathogen and results from a disease interaction.

On the root system, symptoms of *H. glycines* infection range from slight discoloration to severe necrosis. In a heavily infected plant, both nodulation by *Bradyrhizobium japonicum* and nitrogen fixation in the nodules that do form are reduced; some populations of *H. glycines* cause such damage at relatively low infection levels. Because the symptoms are nondescript, confirmation of the presence of *H. glycines* must be based on the presence of white or yellow females attached to the plant's roots (Fig. 160).

The lack of or nonspecific nature of aboveground symptoms interferes with diagnosis of infection by *H. glycines.* A variety of soilborne pathogens and abiotic stresses (e.g., nutrient deficiencies, drought, herbicide injury) can produce similar symptoms.

Causal Organism

The young female is the only life stage of *H. glycines* that is easily visible without magnification. A young female is white or yellow and turns brown with age. After death, the female's body becomes a dark-brown, lemon-shaped cyst, 350–590 × 560–850 µm (Fig. 161). The cyst wall has a pattern of short, irregular, zigzag lines. The morphological characteristics of the cyst vulval cone—such as the size and shape of the brown bullae (i.e., internal projections within the cone)—are important identifying features. Several other species of cyst nematodes are known to occur in the same geographic range as *H. gly-*

TABLE 7. Indicator Lines for HG Type Classification of Genetically Diverse Populations of *Heterodera glycines*[a]

Number	Indicator Line
0	—[b]
1	PI[c] 548402 (Peking)
2	PI 88788
3	PI 90763
4	PI 437654
5	PI 209332
6	PI 89772
7	PI 548316 (Cloud)

[a] Each HG type is designated by the number of the indicator line on which a soybean cyst nematode population can develop cyst numbers equal to or greater than 10% of the number of cysts produced on the standard susceptible check cultivar Lee 74. HG type 0 cannot reproduce on any source of resistance; HG type 1 is able to develop on indicator line 1 (PI 548402); HG type 2 is able to develop on indicator line 2 (PI 88788); and so forth. An HG type designated by more than one number (such as the common HG type 2.5.7) is able to develop on indicator lines 2, 5, and 7 (PIs 88788, 209332, and 548316).

[b] No reproduction on any indicator line.

[c] PI = plant introduction.

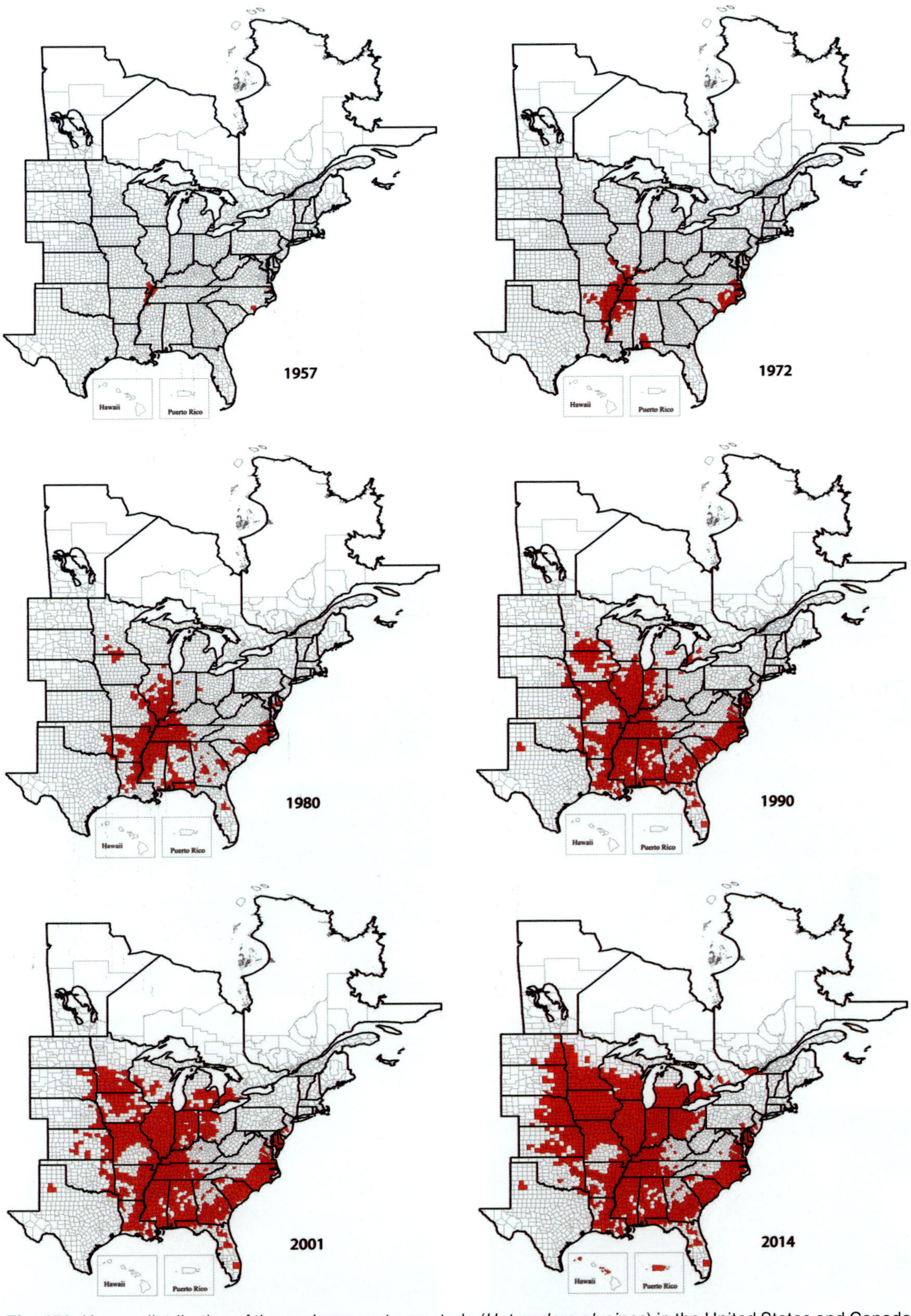

Fig. 158. Known distribution of the soybean cyst nematode (*Heterodera glycines*) in the United States and Canada in selected years from 1957 to 2014. Known infested counties are indicated in red. (Reprinted, with permission, from Tylka and Marett, 2014. Maps © 2014 C. C. Marett and G. L. Tylka, Iowa State University)

cines; therefore, species identification may be crucial to diagnosis if cyst numbers are low.

Populations of *H. glycines* differ in their ability to parasitize resistant or tolerant soybean cultivars. These populations can be distinguished along a set of indicator lines that represent the seven sources of resistance used in U.S. breeding programs. These designations are virulence profiles known as "HG types" (Table 7).

Disease Cycle and Epidemiology

The life cycle of *H. glycines* is similar to the life cycles of other plant-parasitic nematodes. The eggs contain early stage

Fig. 159. Roughly circular patches of soybean plants exhibiting stunting and chlorosis, associated with a high level of infection by the soybean cyst nematode (*Heterodera glycines*). (Courtesy T. L. Niblack)

Fig. 160. Female soybean cyst nematodes (*Heterodera glycines*) attached to the root system of a soybean plant. (Courtesy G. R. Noel)

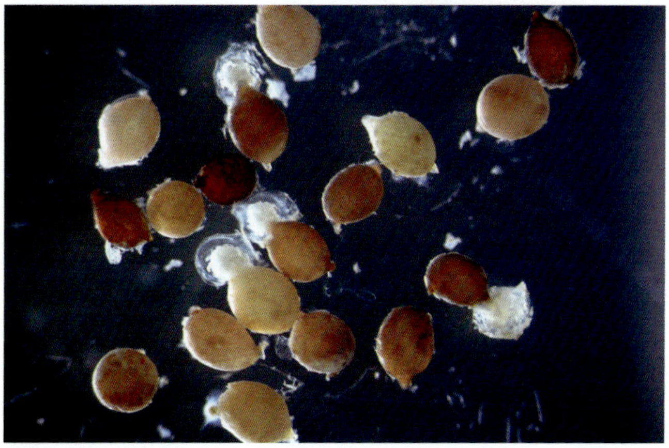

Fig. 161. Soybean cyst nematode (*Heterodera glycines*) females (white to yellow) and cysts (tan to dark brown), some with egg-containing, gelatinous matrices attached (translucent to opaque masses). (Courtesy L. Wen)

embryos, which develop quickly into first-stage juveniles; these juveniles molt once within the egg into second-stage juveniles (J2), which are 291–585 µm in length. A J2 either remains dormant or emerges (i.e., hatches) from the egg (Fig. 162). Several factors affect hatching, including temperature, dormancy, time, and the presence of host-produced hatching factors.

J2s penetrate the plant's roots behind the root tip and usually migrate to the vascular tissue (only occasionally remaining in the cortex) (Fig. 163). They induce the formation of a feeding site called a "syncytium," which consists of large, distinctive, metabolically active cells. When feeding begins, the nematodes enlarge, become sedentary, and undergo three more molts before maturing into adults (Fig. 164). Nematode development can occur at temperatures ranging from 18 to 32°C, but 24–28°C is optimal. Females become so large that they rupture the cortex and epidermis while remaining attached to the feeding site. Males, which develop faster than females, regain their vermiform shape (1.0–1.5 mm long) and exit the roots at maturity. Both sexes are required for reproduction. Under optimum conditions, the life cycle can take as little as 22 days.

The female produces an external gelatinous matrix (Fig. 165), into which 50–200 eggs may be deposited. The remaining 50–400 eggs produced by the female are retained within her body and can remain viable for 12 or more years. The eggs'

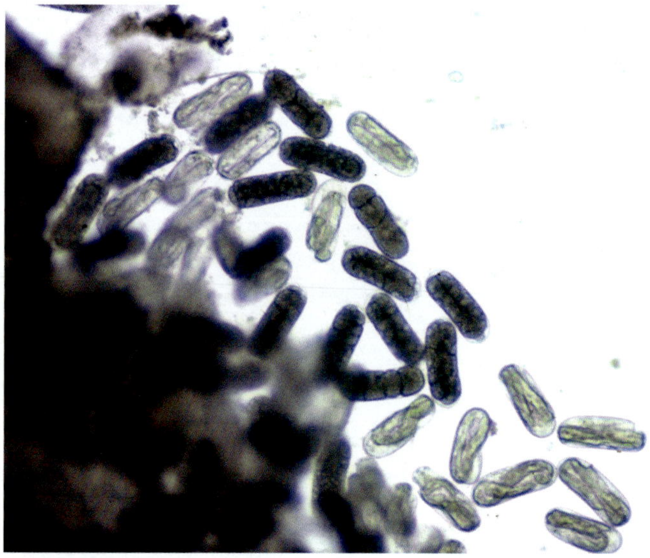

Fig. 162. Soybean cyst nematode (*Heterodera glycines*) eggs at various developmental stages. The eggs represent the inoculum potential of infested soil. (Courtesy T. L. Niblack)

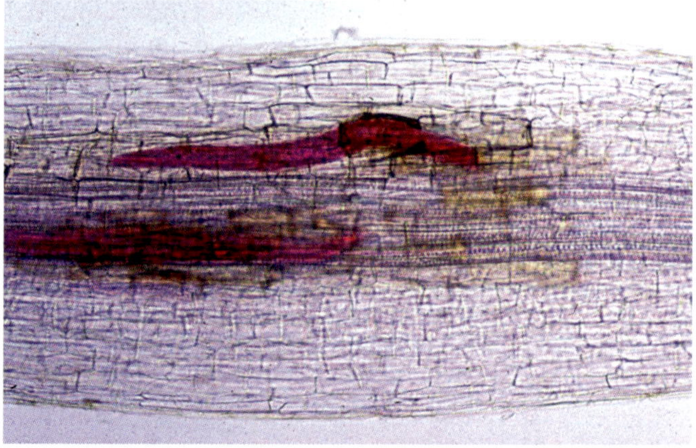

Fig. 163. Soybean cyst nematode (*Heterodera glycines*) infective juvenile initiating a syncytium adjacent to the vascular tissue in a soybean root. (To make this stage visible, the root tissue was cleared with sodium hypochlorite [i.e., bleach], and the nematode was stained with acid fuchsin.) (Courtesy T. L. Niblack)

survival may depend, in part, on protection from the cyst wall and on a genetically controlled dormancy.

Propagules of *H. glycines* can be disseminated by birds, water, wind, machinery, and soil peds (i.e., clumped particles of soil) in improperly cleaned seed. Virtually anything that can move soil can disseminate this nematode.

In general, soybean yields are inversely and linearly related to the number of *H. glycines* eggs in the soil at planting. The number of eggs represents the inoculum potential of the soil, in that each egg represents one potentially infective unit. In some cases in which the yield loss and the nematode population are both high, a point might be reached beyond which an increase in population is not accompanied by an increase in yield loss. Host tolerance of low population levels of *H. glycines* is dependent on soil and physical factors, especially those related to moisture and nutrition.

A number of root-infecting pathogens are known to interact with *H. glycines,* which may result in greater disease either additively or synergistically. For at least one disease—brown stem rot of soybean—infection by *H. glycines* can result in loss of the expression of resistance to the causal fungus (see earlier in Part I, the section Brown Stem Rot). The interaction between *H. glycines* and *Fusarium virguliforme* has varied across different studies from synergism to the lack of any synergism, but in general, both pathogens reduce the growth of soybean. The expression of foliar symptoms caused by the fungus does not necessarily increase with *H. glycines* infection, however (see earlier in Part I, the section Sudden Death Syndrome).

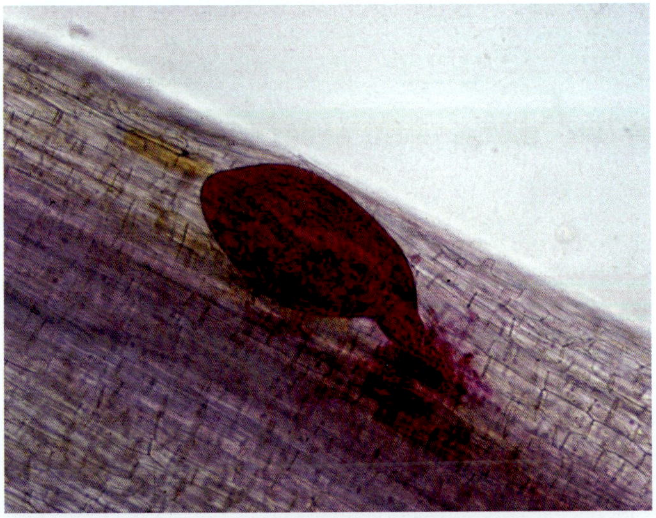

Fig. 164. Soybean cyst nematode (*Heterodera glycines*) at the late juvenile stage, after the nematode has become immobile. (To make this stage visible, the root tissue was cleared with sodium hypochlorite [i.e., bleach], and the nematode was stained with acid fuchsin.) (Courtesy T. L. Niblack)

Fig. 165. Adult female soybean cyst nematode (*Heterodera glycines*) producing a gelatinous matrix. (Courtesy R. D. Riggs)

Management

Management of *H. glycines* begins with confirming the presence and population density of the nematode in the field. Because young females may be visible on plants' roots for only a short segment of the growing season, the use of soil samples is recommended for confirmation. Egg counts (i.e., numbers of eggs per unit of soil) can be used to predict the likelihood and extent of yield loss from *H. glycines.*

Growers should have three goals for management of *H. glycines*: (1) to improve soybean yield; (2) to reduce *H. glycines* infestation levels; and (3) to preserve the future yield potential of *H. glycines*-resistant cultivars. The most effective strategy for achieving these goals and thus for long-term management involves a crop rotation system that utilizes nonhost crops and *H. glycines*-resistant soybean cultivars.

The most commonly used source of genes for resistance to the nematode in soybean is plant introduction (PI) 88788; in fact, however, soybean cultivars derived from PI 88788 vary widely in levels of resistance. Other sources of resistance that have been used to develop soybean cultivars include PIs 548402 (also known as Peking), 90763, 437654, and 209332; in addition, at least 118 more sources have been identified. Cultivars with resistance from PIs other than 88788 are not commercially available in most soybean-producing countries. Consistent use of the same source of resistance results in the selection of a *H. glycines* population that is adapted to and able to damage that source of resistance and any cultivar developed from it. Periodic HG-type testing can aid soybean growers in long-term nematode management using resistant cultivars with different resistance levels or sources of resistance.

Cultural practices that reduce plant stress—including practicing effective weed control and maintaining adequate fertility and soil moisture—should be used to reduce yield losses caused by *H. glycines.* The use of synthetic chemical nematicides to reduce the nematode population density is rare, because the chemicals are expensive, cause environmental damage, and may have labeling issues in various regions of the world. Biological and chemical seed treatments for nematode management may be effective.

Selected References

Barker, K. R., and Koenning, S. R. 1998. Developing sustainable systems for nematode management. Annu. Rev. Phytopathol. 36:165-205.

Gao, X., Jackson, T. A., Hartman, G. L., and Niblack, T. L. 2006. Interactions between the soybean cyst nematode and *Fusarium solani* f. sp. *glycines* based on greenhouse factorial experiments. Phytopathology 96:1409-1415.

Niblack, T. L. 2005. Soybean cyst nematode management reconsidered. Plant Dis. 89:1020-1026.

Niblack, T. L., Arelli, P. R., Noel, G. R., Opperman, C. H., Orf, J., Schmitt, D. P., Shannon, J. G., and Tylka, G. L. 2002. A revised classification scheme for genetically diverse populations of *Heterodera glycines.* J. Nematol. 34:279-288.

Niblack, T. L., Colgrove, A. L., Colgrove, K., and Bond, J. P. 2008. Shift in virulence of soybean cyst nematode is associated with use of resistance from PI 88788. Plant Health Progress. Plant Management Network. doi:10.1094/PHP-2008-0118-01-RS.

Niblack, T. L., Lambert, K. N., and Tylka, G. L. 2006. A model plant pathogen from the kingdom Animalia: *Heterodera glycines,* the soybean cyst nematode. Annu. Rev. Phytopathol. 44:283-303.

Noel, G. R. 1992. History, distribution, and economics. Pages 1-13 in: Biology and Management of the Soybean Cyst Nematode. R. D. Riggs and J. A. Wrather, eds. American Phytopathological Society, St. Paul, MN.

Schmitt, D. P., Wrather, J. A., and Riggs, R. D., eds. 2004. Biology and Management of the Soybean Cyst Nematode. 2nd ed. Schmitt & Associates, Marceline, MO.

Tylka, G. L., and Marett, C. C. 2014. Distribution of the soybean cyst nematode, *Heterodera glycines,* in the United States and Canada: 1954 to 2014. Plant Health Progress. Plant Management Network. doi:10.1094/php-BR-0006.

Wrather, J. A., and Koenning, S. R. 2006. Estimates of disease effects on soybean yields in the United States 2003 to 2005. J. Nematol. 38:173-180.

Wrather, J. A., Anderson, T. R., Arsyad, D. M., Tan, Y., Ploper, L. D., Porta-Puglia, A., Ram, H. H., and Yorinori, J. T. 2001. Soybean disease loss estimates for the top ten soybean-producing countries in 1998. Can. J. Plant Pathol. 23:115-121.

(Prepared by T. L. Niblack and R. D. Riggs)

Sting Nematodes and Other Nematodes

Belonolaimus spp., known as "sting nematodes," are associated with soils consisting of 90% or more sand. Nine species are recognized, and of them, *B. longicaudatus* is the most ubiquitous and has the widest host range. In the United States, *B. longicaudatus* is found primarily in the states along the Atlantic and Gulf Coasts, from Connecticut to Texas, but it has also been found in the interior states of Arkansas, Kansas, and Oklahoma. This nematode has also been reported from the Pacific Coast of Costa Rica. As a pathogen of soybean, *B. longicaudatus* has been reported from the United States to cause crop losses of up to 50% in specific fields.

Belonolaimus spp. are migratory ectoparasites. The initial symptom they cause is the appearance of minute, dark, sunken lesions along the root axes or at the root tips. These lesions may enlarge to girdle the roots or extend along the longitudinal axis. The plant's terminal growth often ceases, the meristematic tissues are destroyed, and the roots usually break off at the sites of girdling, resulting in a stubby-root appearance. Root proliferation is stimulated above the points of attack, and a many-branched root system develops—a feature characteristic of plants parasitized by a sting nematode. Top growth becomes chlorotic and severely stunted.

In addition to *Heterodera glycines, Pratylenchus* spp., *Rotylenchulus reniformis,* and *Meloidogyne* spp., more than 20 other genera of plant-parasitic nematodes have been reported in soybean fields (see Appendix I). *Xiphinema americanum,* a dagger nematode, is one of the few species known to transmit viruses (see later in Part I, the sections Soybean severe stunt virus and *Tobacco ringspot virus*). Whether all the reported species of nematode can reproduce on soybean is uncertain. In some cases, a given species may be reported to cause serious yield losses on other crops, but little is known about its relationship to soybean. In the field, symptoms of damage caused by these nematodes are difficult to differentiate from those caused by other pathogens.

Selected References

McSorley, R., and Dickson, D. W. 1990. Vertical distribution of plant-parasitic nematodes in sandy soil under soybean. J. Nematol. 22:90-96.

Smart, G. C., and Nyugen, K. B. 1991. Sting and awl nematodes *Belonolaimus* spp. and *Dolichodorus* spp. Pages 627-667 in: Manual of Agricultural Nematology. W. R. Nickel, ed. Marcel Dekker, New York.

(Prepared by G. R. Noel)

Diseases Caused by Viruses

More than 1,000 plant viruses have been described, and they cause a wide range of diseases in plants. Viruses do not have a cellular form and live only within certain living cells as obligate parasites. They are macromolecular particles composed of genetic information in the form of nucleic acid—either RNA (ribonucleic acid) or DNA (deoxyribonucleic acid) surrounded by a protective protein or lipoprotein coat. Virus particles are visible only with an electron microscope, and each virus has a characteristic size and shape. The most common forms of virus particles are isometric (roughly spherical, 20–70 nm in diameter) and elongated rods (rigid or flexuous, up to 2 μm long, 10–20 nm in diameter).

Viruses exist mainly in the cells of their hosts. They are transmitted to other hosts through wounds created by arthropods and in some cases by fungal and nematode vectors. Viruses also are transmitted by mechanical inoculation, pollination, and human activities related to cultural practices in growing plants.

The diseases caused by viruses are manifested by a variety of symptoms. Some infections are latent and cause no obvious symptoms, whereas others cause symptoms on all plant parts or on specific plant parts. For instance, leaves may display mosaic patterns, become rugose, turn yellow or red, and become necrotic. Stems may become discolored and turn necrotic. Plants may be stunted and seeds misshapen and discolored. If two or more viruses infect the same plant simultaneously, the effect can be additive, synergistic, or cross-protective (i.e., one strain of a virus protects plants from infection by a second strain of the same virus).

Identifying virus and viruslike diseases in the field using symptoms alone can be unreliable. The same virus strain often produces different symptoms based on the host or even the specific cultivar of the same host. In addition, the symptoms caused by a virus may be altered by environmental conditions, age of the plant host, and infection with another pathogen. Symptoms on soybean may be quite similar even though induced by different viruses. Furthermore, plants in the field may become infected by a new virus strain or by other viruses. Infection with multiple viruses often induces symptoms different from those produced by any one virus.

Given the variety of possible factors that can affect symptomatology, symptoms alone are seldom diagnostic. To confirm the diagnosis of a virus, it is necessary to perform immunological assays (e.g., enzyme-linked immunosorbent assay [ELISA]) and nucleic acid detection (e.g., polymerase chain reaction [PCR]) using antibodies or oligonucleotide primers specific for particular viruses. In addition, new viruses and virus diseases continue to be identified.

Viruses are classified within a hierarchical scheme, similar to other biological entities. Viruses with common characteristics are grouped into families, genera, and species. Family and genus names are Latin terms and based on the nature, organization, and expression of the nucleic acid genome. Assignment of a virus to taxa is based on nucleotide sequence, as well as particle structure, physical stability, serological relationships, type of vector (if any), and host specificity. Species names of viruses are English terms. A plant virus is generally named for the original host from which it was isolated or by the major crop that it affects, followed by a descriptor of a distinctive symptom it causes. Not all viruses are assigned to a genus or to a family because of insufficient characterization.

A list of viruses reported to infect soybean naturally (or by inoculation) is presented in Table 8. The table is organized according to the current classifications of the International Committee on Taxonomy of Viruses (ICTV). Names of viruses recognized

TABLE 8. Viruses Reported to Infect Soybean Naturally in the Field

Family	Genus	Species	Family	Genus	Species
Single-Stranded, Positive-Sense RNA Genomes			***Single-Stranded, Ambisense RNA Genomes***		
Betaflexiviridae	*Carlavirus*	*Cowpea mild mottle virus*	*Bunyaviridae*	*Tospovirus*	Soybean vein necrosis virus[b]
Bromoviridae	*Alfamovirus*	*Alfalfa mosaic virus*			*Tomato spotted wilt virus*
	Bromovirus	*Cowpea chlorotic mottle virus*			
	Cucumovirus	*Cucumber mosaic virus*	***Double-Stranded DNA Genomes***		
		Peanut stunt virus	*Caulimoviridae*	*Caulimovirus*	Soybean Putnam virus[b]
	Ilarvirus	*Tobacco streak virus*		*Soymovirus*	*Soybean chlorotic mottle virus*
Luteoviridae	*Luteovirus*	*Indonesian soybean dwarf virus*[a]	*Nanoviridae*	*Nanovirus*	*Milk vetch dwarf virus*
		Soybean dwarf virus			
Potyviridae	*Potyvirus*	*Bean common mosaic virus*	***Single-Stranded DNA Genomes***		
		Bean yellow mosaic virus	*Geminiviridae*	*Begomovirus*	*African cassava mosaic virus*
		Peanut mottle virus			*Bean golden mosaic virus*
		Soybean mosaic virus			*Cotton leaf curl Kokhran virus*
Secoviridae	*Comovirus*	*Bean pod mottle virus*			*Mungbean yellow mosaic India virus*
		Broad bean true mosaic virus			*Mungbean yellow mosaic virus*
		Cowpea mosaic virus			*Okra mottle virus*
		Cowpea severe mosaic virus			*Pepper golden mosaic virus*
		Quail pea mosaic virus			*Rhynchosia golden mosaic virus*
	Nepovirus	*Cycas necrotic stunt virus*			*Sida micrantha mosaic virus*
		Soybean severe stunt virus[b]			*Sida mottle virus*
		Tobacco ringspot virus			*Soybean blistering mosaic virus*
		Tomato ringspot virus			*Soybean chlorotic spot virus*
Tombusviridae	*Carmovirus*	*Bean mild mosaic virus*			*Soybean crinkle leaf virus*
		Blackgram mottle virus[b]			*Soybean mild mottle virus*
		Cowpea mottle virus			Soybean yellow mosaic virus[b]
		Soybean yellow mottle mosaic virus			
Virgaviridae	*Tobamovirus*	*Tobacco mosaic virus*			

[a] Genus is unclassified.
[b] Not classified as a distinct species by the International Committee on Taxonomy of Viruses (ICTV) (as of 2014).

by the ICTV are typically italicized, whereas names of viruses not yet recognized are not italicized. Much of the information can be accessed at the ICTV website (see Selected References below).

The viruses described in the following sections are grouped by genome classification and then organized alphabetically by family, genus, and species name.

Selected References

Boswell, K. F., and Gibbs, A. J., eds. 1983. Viruses of Legumes 1983—Descriptions and Keys from VIDE. Australian National University, Canberra.

Brunt, A., Crabtree, K., Dallwitz, M., Gibbs, A., and Watson, L. 1996. Viruses of Plants: Descriptions and Lists from the VIDE Database. CABI, Wallingford, Oxfordshire, U.K.

Edwardson, J. R., and Christie, R. G. 1991. CRC Handbook of Viruses Infecting Legumes. CRC Press, Boca Raton, FL.

Harrison, B. D., and Murant, A. F., eds. 1977. Descriptions of Plant Viruses, with Supplements. Commonwealth Mycological Institute and Association of Applied Biologists (CMI/AAB), Kew, U.K.

Hull, R. 2002. Matthews' Plant Virology. 4th ed. Academic Press, San Diego, CA.

Hull, R. 2009. Comparative Plant Virology. 2nd ed. Academic Press, San Diego, CA.

International Committee on Taxonomy of Viruses (ICTV). 2014. Virus Taxonomy. ICTV, Virology Division–IUMS. Available online at www.ictvonline.org

Khan, J., and Dijkstra, J., eds. 2006. Handbook of Plant Virology. Haworth Press, Binghamton, NY.

King, A. M. Q., Adams, M. J., Carstens, E. B., and Lefkowitz, E., eds. 2012. Virus Taxonomy: Ninth Report of the International Committee on Taxonomy of Viruses. Elsevier, Boston.

Walkey, D. G. A. 1991. Applied Plant Virology. 2nd ed. Chapman & Hall, London.

(Prepared by S. A. Tolin)

Single-Stranded, Positive-Sense RNA Genomes

Betaflexiviridae

Carlavirus

Cowpea mild mottle virus

Cowpea mild mottle virus (CpMMV) was first described in cowpea in Ghana in 1973 and then detected in soybean in the Ivory Coast in 1982. The virus also occurs in soybean in India, Indonesia, Malaysia, and Thailand, and it appears to be widely distributed in tropical countries, including the Fiji Islands. CpMMV has also been reported in Brazil, Egypt, Israel, and many other African nations.

CpMMV has a moderately wide host range that spans five plant families, but most of the hosts are legumes. CpMMV is currently assigned to the genus *Carlavirus* in the family *Betaflexiviridae*.

Symptoms and Signs

Field symptoms of *Cowpea mild mottle virus* in soybean can be striking, but the nature of the symptoms may depend on the cultivar. Vein clearing and leaf rolling (Fig. 166) have been observed in soybean cultivar Shirotsurunoko, whereas mosaic and vein necrosis have been observed in cultivar Toyosuzu. Reports from South America have associated CpMMV with foliar symptoms ranging from severe to mild (resembling those of *Soybean mosaic virus* [SMV]) and with stem necrosis. Symptoms in cowpea and yardlong bean (*Vigna unguiculata* subsp. *sesquipedalis*) are more striking than those in soybean and include bright mosaic patterns on foliage.

Fig. 166. Rolling of soybean leaves, caused by *Cowpea mild mottle virus*. (Courtesy S. A. Tolin; photographed by M. Iwaki)

Causal Agent

CpMMV has straight to flexuous, fragile, filamentous particles that are approximately 14 × 650 nm (diameter × length). The single-stranded, positive-sense RNA genome of approximately 7.6 kb is encapsidated in a coat protein of 32–33 kDa. In ultrathin sections of soybean leaves, featherlike and bundle inclusion bodies have been seen in the cytoplasm, which is characteristic of carlaviruses.

Disease Cycle and Epidemiology

Cowpea mild mottle virus is transmitted semipersistently by whiteflies (*Bemisia tabaci*), but nonpersistent transmission has also been reported. CpMMV is not transmitted by aphids, even though other carlaviruses are aphid transmitted in a nonpersistent manner. CpMMV was once reported to be seed transmitted at a rather high rate, but a reinvestigation revealed no evidence of seed transmissibility from soybean plants infected with CpMMV alone or infected with both CpMMV and SMV. However, for CpMMV in cowpea, the seed-transmission rate is as high as 40%.

Several reports of CpMMV in cowpea and in soybean have suggested that the virus is widely disseminated in tropical areas with high whitefly populations. In North America, distribution of the virus is currently limited to the U.S. state of Florida, where metagenome analysis of whiteflies has revealed a carlavirus-like CpMMV that infects legumes, including soybean.

Management

A recessive resistance gene (*rbc1*) for a CpMMV-like virus has been identified and mapped in soybean in Puerto Rico and Mexico.

Selected References

Almeida, A. M. R., Piuga, F. F., Marin, S. R., Kitajima, E. W., Gaspar, J. O., and Moraes, T. G. 2005. Detection and partial characterization of a carlavirus causing stem necrosis of soybeans in Brazil. Fitopatol. Bras. 30:191-194.

Brace, R. C., Fehr, W. R., and Graham, M. A. 2012. Inheritance and molecular mapping of an allele providing resistance to a cowpea mild mottle-like virus in soybean. Crop Sci. 52:2109-2114.

Brito, M., Fernández-Rodríguez, T., Garrido, M. J., Mejías, A., Ramano, M., and Marys, E. 2012. First report of *Cowpea mild mottle Carlavirus* on yardlong bean (*Vigna unguiculata* subsp. *sequipedalis*) in Venezuela. Viruses 4:3804-3811.

Brunt, A. A., and Kenten, R. H. 1973. Cowpea mild mottle, a newly recognized virus infecting cowpea (*Vigna unguiculata*) in Ghana. Ann. Appl. Biol. 74:67-74.

Laguna, I. G., Arneodo, J. D., Rodriguez-Pardina, P., and Kitajima, E. W. 2006. *Cowpea mild mottle virus* infecting soybean crops in North-Western Argentina. Fitopatol. Bras. 31:317.

Muniyappa, V., and Reddy, D. V. R. 1983. Transmission of Cowpea mild mottle virus by *Bemisia tabaci* in a nonpersistent manner. Plant Dis. 67:391-393.

Naidu, R. A., Gowda, S., Satyanarayana, T., Boyko, V., Reddy, A. S., Dawson, W. O., and Reddy, D. V. R. 1998. Evidence that whitefly-transmitted cowpea mild mottle virus belongs to the genus *Carlavirus*. Arch. Virol. 143:769-780.

Rosario, K., Capobianco, H., Ng, T. F. F., Breitbart, M., and Polston, J. E. 2014. RNA viral metagenome of whiteflies leads to the discovery and characterization of a whitefly-transmitted carlavirus in North America. PLoS One 9(1):e86748.

Tavasoli, M., Shahraeen, N., and Ghorbani, S. H. 2009. Serological and RT-PCR detection of Cowpea mild mottle carlavirus infecting soybean. J. Gen. Mol. Virol. 1:7-11.

Thouvenel, J. C., Monsarrat, A., and Fauquet, C. 1982. Isolation of Cowpea mild mosaic virus from soybeans in the Ivory Coast. Plant Dis. 66:336-337.

Yadav, M. K., Biswas, K. K., Lal, S. K., Baranwal, V. K., and Jain, R. K. 2013. A distinct strain of *Cowpea mild mottle virus* infecting soybean in India. J. Phytopath. 161:739-744.

Zanardo, L. G., Silva, F. N., Bicalho, A. A. C., Urquiza, G. P. C., Lima, A. T. M., Almeida, A. M. R., Zerbini, F. M., and Carvalho, C. M. 2014. Molecular and biological characterization of *Cowpea mild mottle virus* isolates infecting soybean in Brazil and evidence of recombination. Plant Pathol. 63:456-465.

(Prepared by S. A. Tolin)

Bromoviridae
Alfamovirus
Alfalfa mosaic virus

First reported in 1931 as causing mosaic in alfalfa in the United States, *Alfalfa mosaic virus* (AMV) has become distributed worldwide. AMV has a very wide natural host range, infecting plants from 22 dicotyledonous families, and it can be transmitted to hundreds of species in more than 50 families. Symptoms in infected plants may persist or disappear soon after infection.

AMV infection of soybean was first reported in 1960, and the virus can be prevalent in the U.S. Midwest. Since 2000, it has been identified at high incidences in the states of Nebraska and Wisconsin, and occasional infected plants have been observed in Illinois, North Dakota, and Virginia. Increases in AMV incidence in soybean and snap bean have paralleled increases in population and distribution of *Aphis glycines,* the soybean aphid, since its discovery in North America in 2000 (see Part II, the section Soybean Aphid). Yield loss estimates are not available.

Symptoms and Signs

In soybean, infection with *Alfalfa mosaic virus* can cause distortion, rugosity, and a bright-yellow mosaic pattern on leaves, along with stunting, reduced pod set, and seed coat mottling (Fig. 167). The nature of the symptoms depends on the virus strain and soybean cultivar. AMV-infected plants often show recovery; thus, some infected plants may be asymptomatic later in the growing season. AMV infection has been shown to reduce the numbers of nodules on infected plants compared with healthy plants, and an incidence of 30% or greater results in significant yield losses in soybean.

Infection of soybean plants by both AMV and *Soybean mosaic virus* (SMV) results in disease synergism, in which AMV symptoms are of greater severity and persist throughout the growing season.

Causal Agent

AMV virus particles are of four sizes—18 nm in diameter and 30, 34, 43, and 56 nm in length—and all are spherical to

Fig. 167. Yellowing and rugosity of soybean plants infected with *Alfalfa mosaic virus*. (Courtesy C. R. Grau)

bacilliform in shape. AMV has a tripartite, single-stranded RNA genome consisting of segments of 2.0, 2.6, and 3.6 kb in length. The three RNAs, along with a fourth subgenomic RNA (~0.9 kb) that encodes the coat protein, are encapsidated separately and account for the four sizes of virus particles.

The RNA genome of *Alfalfa mosaic virus* has been extensively studied as a model system. There are 82 AMV nucleotide sequence entries from various isolates deposited in GenBank.

Disease Cycle and Epidemiology

Alfalfa mosaic virus is transmitted by more than 15 species of aphids (including *A. glycines*) in a nonpersistent manner. AMV is also transmitted through alfalfa seeds and to a low extent through soybean seeds. In surveys conducted in Iran in 1999 and 2000, an overall incidence of 1.3% of AMV infection was reported in soybean. In Nebraska, AMV infections were detected in 52 and 56% of soybean fields surveyed in 2001 and 2002, respectively.

Management

Resistance to AMV infection has been identified in the Brazilian bean and soybean cultivars Pérola and Planalto, respectively, and in the U.S. cultivar Hood, which was a parent to both Pérola and Planalto. A dominant gene (*Rav1*) for resistance to AMV infection was identified and mapped from plant introduction (PI) 153282. The combination of AMV resistance genes from Hood and PI 153282 has the potential of providing broad-spectrum resistance to AMV in soybean.

Selected References

Allington, W. B., Moorhead, E. L., and Staples, R. 1960. Alfalfa mosaic virus in soybean. (Abstr.) Phytopathology 50:627.

Clark, A. J., and Perry, K. L. 2002. Transmissibility of field isolates of soybean viruses by *Aphis glycines*. Plant Dis. 86:1219-1222.

Giesler, L. J., and Ziems, A. D. 2006. Incidence of *Alfalfa mosaic virus, Bean pod mottle virus,* and *Soybean mosaic virus* in Nebraska soybean fields. Plant Health Progress. Plant Management Network. doi:10.1094/PHP-2006-0424-01-HM.

Golnaraghi, A. R., Shahraeen, N., Pourrahim, R., Farzadfar, Sh., and Ghasemi, A. 2004. Occurrence and relative incidence of viruses infecting soybeans in Iran. Plant Dis. 88:1069-1074.

Kopisch-Obuch, F. J., Koval, N. C., Mueller, E. M., Paine, C., Grau, C. R., and Diers, B. W. 2008. Inheritance of resistance to *Alfalfa mosaic virus* in soybean PI 153282. Crop Sci. 48:933-940.

Mueller, E. E., and Grau, C. R. 2007. Seasonal progression, symptom development, and yield effects of *Alfalfa mosaic virus* epidemics on soybean in Wisconsin. Plant Dis. 91:266-272.

Tu, J. C., and Tse, S. Y. 1976. Virus infectivity of Alfalfa mosaic virus in soybean as influenced by nodulation. J. Phytopathol. 85:170-175.

(Prepared by S. A. Tolin)

Bromovirus
Cowpea chlorotic mottle virus

Cowpea chlorotic mottle virus (CCMV) was first identified in soybean in 1967. There are multiple strains of this virus, and all can infect soybean. Most of the research on CCMV in soybean has been conducted with the soybean strain CCMV-S. In the United States, CCMV is found in Arkansas, Georgia, and South Carolina; a bean stipple strain occurs in Cuba, and a third strain occurs in Nigeria. Yield losses caused by CCMV-S can be as high as 37% when plants become infected in the seedling stage. Bean yellow stipple virus is considered a strain of CCMV, even though it was described earlier than CCMV.

Symptoms and Signs

In susceptible cultivars, the predominant symptoms caused by infection with CCMV-S are mosaic and stunting, which can reduce plant height up to 42%. Other leaf symptoms include the development of chlorotic patches (Fig. 168), as well as distortion and rugosity. Cultivars with intermediate resistance exhibit veinal necrosis, and resistant cultivars develop local lesions.

Plants infected with CCMV-S produce fewer seeds than healthy plants, and the seeds that are produced may have reduced total oil content and increased protein content. The changes in seed composition caused by CCMV-S are small, however, and would not be expected to reduce the value of the soybeans produced.

Causal Agent

Cowpea chlorotic mottle virus has a tripartite, single-stranded RNA genome that is packaged in three isometric particles with diameters of 26 nm. A single type of polypeptide

Fig. 168. Leaf chlorosis, caused by *Cowpea chlorotic mottle virus*. (Courtesy O. W. Barnett)

(180 protein subunits per particle) encapsidates four species of single-stranded, positive-sense RNA, which range from 1.0 to 3.2 kb in length. The genome consists of the three largest RNA molecules.

CCMV has a relatively wide host range and systemically infects mechanically inoculated plants from seven families. CCMV produces local lesions on resistant soybean cultivars and other hosts, and it produces veinal necrosis in soybean cultivars and other hosts with intermediate resistance.

Disease Cycle and Epidemiology

CCMV-S is sap transmissible, and there is no evidence of seed transmission. Two leaf-feeding beetles—*Cerotoma trifurcata* and *Diabrotica undecimpunctata howardi*—transmit the virus (see Part II, the section Bean Leaf Beetle). The bean yellow stipple strain is transmitted by another bean leaf beetle, *C. ruficornis*, and *D. balteata*.

Management

Cultivars and plant introductions (PIs) with the necrotic-local-lesion type of resistance (*Rcv1*) are common in soybean. In one test, 347 of 533 genotypes showed this type of response to CCMV infection. Another type of resistance shown to be effective under field conditions resulted in a reduced virus concentration and a lack of systemic virus movement.

Selected References

Bijaisoradat, M., and Kuhn, C. W. 1985. Nature of resistance in soybean to Cowpea chlorotic mottle virus. Phytopathology 75:351-355.

Boerma, H. R., Kuhn, C. W., and Harris, H. B. 1975. Inheritance of resistance to Cowpea chlorotic mottle virus (soybean strain) in soybeans. Crop Sci. 15:849-850.

Fulton, J. P., Gámez, R., and Scott, H. A. 1975. Cowpea chlorotic mottle and Bean yellow stipple viruses. Phytopathology 65:741-742.

Goodrick, B. J., Kuhn, C. W., and Boerma, H. R. 1991. Inheritance of nonnecrotic resistance to Cowpea chlorotic mottle virus in soybean. J. Hered. 82:512-514.

Harris, H. B., and Kuhn, C. W. 1971. Influence of Cowpea chlorotic mottle virus (soybean strain) on agronomic performance of soybeans. Crop Sci. 11:71-73.

Kuhn, C. W. 1968. Identification and specific infectivity of a soybean strain of Cowpea chlorotic mottle virus. Phytopathology 58:1441-1442.

Paguio, O. R., Boerma, H. R., and Kuhn, C. W. 1987. Disease resistance, virus concentration, and agronomic performance of soybean infected with Cowpea chlorotic mottle virus. Phytopathology 77:703-707.

Thottappilly, G., Seghal, O. P., and Rossel, H. W. 1993. Characteristics of a Cowpea chlorotic mottle virus isolate from Nigeria. Plant Dis. 77:60-63.

(Prepared by J. W. Demski and C. W. Kuhn; revised by S. A. Tolin and L. L. Domier)

Cucumovirus
Peanut stunt virus and *Cucumber mosaic virus* (Soybean stunt strain)

Peanut stunt virus (PSV) was originally described in the United States on peanut in 1964 and on soybean in 1977, and it has subsequently been found to occur in China, Japan, Russia, and several countries in Europe and Africa. PSV has been isolated from many natural hosts—including common bean, crownvetch (*Securigera varia*), lupine (*Lupinus* spp.), peanut, and white clover (*Trifolium repens*)—and is of greatest economic importance on these hosts. PSV sporadically infected soybean in an experimental nursery surrounded by PSV-infected clover in the state of Virginia, and it was also found to infect soybean naturally in Illinois, Indiana, and Kentucky. Viruses described in Germany as Robinia mosaic virus and Clover blotch virus are now considered strains of PSV.

The soybean stunt strain of *Cucumber mosaic virus* (CMV-SS) (formerly known as Soybean stunt virus) was described in 1958 as infecting soybean in Japan. It has also been isolated from soybean in China and Indonesia and from wild soybean in Russia. CMV exists worldwide as numerous strains in two subgroups and has an extremely broad host range, only a few of which are legumes. Except for CMV-SS, CMV has seldom been reported in soybean.

CCM-SS and PSV also reduce seed quality through seed mottling in Japan, but yield loss estimates are not available.

Symptoms and Signs

The symptoms produced by PSV and by CMV-SS in soybean are similar. Both viruses cause vein clearing followed by mild mottling and slight crinkling of systemically infected leaves (Fig. 169). Plants exhibit delayed maturity and produce mottled seed. Infection at an early growth stage results in greater symptom severity and yield loss than infection at a later growth stage.

Causal Agents

Both PSV and CMV-SS have icosahedral particles approximately 30 nm in diameter, single species of coat proteins of 20–25 kDa, and tripartite, positive-sense, single-stranded RNA genomes. Satellite RNAs may be associated with CMV or PSV and either enhance or attenuate symptoms.

Bean distortion mosaic virus, which has been isolated in Mexico, is thought to be a natural reassortant between genome segments of PSV and CMV. PSV is thought to be more diverse than CMV, because at least four genomically distinct groups of PSV are known. However, PSV has a more limited host range than most strains of CMV.

Disease Cycle and Epidemiology

Cucumoviruses are transmitted by a number of aphid species in a nonpersistent manner, as well as through seeds of specific hosts. CMV-SS is transmitted through soybean seeds at rates of 40–100%.

Since 2000, CMV has been recognized as a component of a virus disease complex in snap bean, which results from transmission of multiple viruses by *Aphis glycines,* the soybean aphid. Three legume-infecting strains of CMV have been isolated from diseased snap beans, but none is able to infect soybean.

Fig. 169. Mild mottling and crinkling of leaves, caused by *Peanut stunt virus,* in a field-grown plant inoculated at the V2 or V3 stage. (Courtesy S. A. Tolin)

The common Y strain of CMV does not infect soybean. However, chimeric RNAs of Y with two soybean strains have demonstrated swapping of domains of the 3a protein in RNA3, which enables either local or systemic infection of soybean.

Management

Resistance in soybean to two strains of PSV is associated with a single dominant gene (*Rps1*) from cultivar Hyuga that maps to chromosome 7 (linkage group M). Cultivar Harosoy, a source of the *Rsv3* gene for resistance to SMV, is also resistant to PSV. Resistance in soybean to CMV-SS has not been reported.

Selected References

Ghabrial, S. A., Pickard, C. M., and Stucky, R. E. 1977. Identification and distribution of virus diseases of soybean in Kentucky. Plant Dis. Rep. 61:690-694.

Hanada, K., and Tochihara, H. 1982. Some properties of an isolate of the soybean stunt strain of Cucumber mosaic virus. Phytopathology 72:761-764.

Hong, J. S., Masuta, C., Nakano, M., Abe, J., and Uyeda, I. 2003. Adaptation of Cucumber mosaic virus soybean strains (SSVs) to cultivated and wild soybeans. Theor. App. Genet. 107:49-53.

Hong, J. S., Ohnishi, S., Matsuta, C., Choi, J. K., and Ryu, K. H. 2007. Infection of soybean by *Cucumber mosaic virus* as determined by viral movement protein. Arch. Virol. 152:321-328.

Kiss, L., Balázs, E., and Salánki, K. 2009. Characterisation of black locust isolates of *Peanut stunt virus* (PSV) from the Pannon ecoregion show the frequent occurrence of the fourth taxonomic PSV subgroup. Eur. J. Plant Pathol. 125:671-677.

Milbrath, G. M., and Tolin, S. A. 1977. Identification, host range, and serology of peanut stunt virus isolated from soybean. Plant Dis. Rep. 61:637-640.

Roossinck, J. M. 2002. Evolutionary history of *Cucumber mosaic virus* deduced by phylogenetic analysis. J. Virol. 76:3382-3387.

Saruta, M., Takada, Y., Kikuchi, A., Yamada, T., Komatsu, K., Sayama, T., Ishimoto, M., and Okabe, A. 2012. Screening and genetic analysis of resistance to *Peanut stunt virus* in soybean: Identification of the putative *Rpsv1* resistance gene. Breed. Sci. 61:625-630.

Yan, L. Y., Xu, Z. Y., Goldbach, R., Kunrong, C., and Prins, M. 2005. Nucleotide sequence analysis of genomic RNAs of *Peanut stunt virus* Mi, the type strain representative of a novel PSV subgroup from China. Arch. Virol. 150:1203-1211.

(Prepared by S. A. Tolin)

Ilarvirus

Tobacco streak virus

Tobacco streak virus (TSV) is the type species of the genus *Ilarvirus,* which contains an additional 18 species, and it causes a wide variety of symptoms. TSV has been found to infect soybean in Brazil, India, and Iran, as well as the U.S. states of Illinois, Iowa, Kentucky, Oklahoma, and Wisconsin. Brazilian bud blight of soybean, which is caused by TSV, was first reported in Brazil in 1950, and the disease is common and destructive in the southeastern part of the country. In contrast, TSV isolates recovered from soybean in Illinois and Wisconsin have shown much less severe symptoms of disease.

TSV has a wide host range and infects plants from 17 families, including asparagus, cotton, dahlia, grape, papaya, pea, peanut, potato, raspberry, rose, tobacco, and tomato. Many strains of the virus are known, and cross-protection between strains is common.

Symptoms and Signs

Early reports from Brazil and Iowa described TSV symptoms as indistinguishable from the bud blight caused by *Tobacco ringspot virus,* whereas other reports did not observe pod and stem-terminus necrosis. In the latter case, soybean seedlings inoculated with TSV initially showed chlorosis and necrosis (Fig. 170), quickly recovered, and then displayed only very mild symptoms (Fig. 171); virus titers dropped dramatically as seeds matured.

Unlike infections from *Bean pod mottle virus* and *Soybean mosaic virus,* infection from TSV usually does not result in seed coat mottling.

Causal Agent

Tobacco streak virus has isometric particles 25–30 nm in diameter. The virus encapsidates four segments of positive-sense, single-stranded RNA that vary in size from 1.1 to 3.4 kb. The three largest RNAs are genomic but require either coat protein or the smallest (subgenomic) RNA for infectivity, similar to *Alfalfa mosaic virus.*

Disease Cycle and Epidemiology

Tobacco streak virus moves rapidly within a soybean plant—from the roots to the aerial parts and from the leaves to the shoots and the roots. The virus can be recovered from all parts of the plant except possibly the pollen. Virus concentrations increase in young, developing leaves and decrease as leaves mature.

Fig. 170. Young soybean plant displaying chlorosis and necrosis, caused by *Tobacco streak virus.* (Courtesy L. L. Domier)

Fig. 171. Soybean plant infected by *Tobacco streak virus,* showing severe chlorosis and necrosis on older leaves and few symptoms on newer leaves. (Courtesy L. L. Domier)

Indirect evidence suggests that TSV transmission is facilitated by multiple species of thrips, including *Caliothrips phaseoli, Frankliniella occidentalis, F. schultzei, F. tabaci, Megalurothrips usitatus, Microcephalothrips abdominalis, Scirtothrips dorsalis, Thrips parvispinus,* and *T. tabaci.* TSV is also seed transmitted and found in the embryos and seed coats of mature seeds. The extent of seed transmission depends on the virulence of the virus isolate, the susceptibility of the cultivar, and the time of infection. In early infected susceptible cultivars, seed transmission of up to 90% has been recorded.

Management

Resistance to *Tobacco streak virus* has been reported in at least one soybean cultivar (Tanner), but it has not been incorporated into adapted cultivars. Management measures should emphasize preventing the introduction of TSV-infected seeds into soybean fields. In areas in which the virus causes significant reductions in yield, the production of virus-free seed for planting may be useful; this has not been experimentally tested, however.

Selected References

Almeida, A. M. R., Filho, A. B., and Amorin, L. 1994. Disease progress of soybean bud blight and spatial pattern of diseased plants. J. Plant Dis. Prot. 101:386-392.

Almeida, A. M. R., Sakai, J., Hanada, K., Oliveira, T. G., Belintani, P., Kitajima, E. W., Souto, E. R., de Novaes, T. G., and Nora, P. S. 2005. Biological and molecular characterization of an isolate of *Tobacco streak virus* obtained from soybeans in Brazil. Fitopatol. Bras. 30:366-373.

Alvaro, M. R. A., and Corso, I. C. 1991. Effect of sowing time on the incidence of bud blight in soybean (*Glycine max* L. Merr.). J. Phytopathol. 132:251-257.

Fagbenle, H. H., and Ford, R. E. 1970. Tobacco streak virus isolated from soybeans, *Glycine max*. Phytopathology 60:814-820.

Ghanekar, A. M., and Schwenk, F. W. 1980. Comparison of Tobacco streak virus isolates from soybean and tobacco. J. Phytopathol. 97:148-155.

Hobbs, H. A., Jossey, S., Wang, Y., Hartman, G. L., and Domier, L. L. 2012. Diverse soybean accessions identified with temperature-sensitive resistance to *Tobacco streak virus*. Crop Sci. 52:738-744.

Jones, D. R. 2005. Plant viruses transmitted by thrips. Eur. J. Plant Pathol. 113:119-157.

Rabedeaux, P. F., Gaska, J. M., Kurtzweil, N. C., and Grau, C. R. 2005. Seasonal progression and agronomic impact of *Tobacco streak virus* on soybean in Wisconsin. Plant Dis. 89:391-396.

Sherwood, J. L., and Jackson, K. E. 1985. Tobacco streak virus in soybean in Oklahoma. Plant Dis. 69:727.

Wang, Y., Hobbs, H. A., Hill, C. B., Domier, L. L., Hartman, G. L., and Nelson, R. L. 2005. Evaluation of ancestral lines of US soybean cultivars for resistance to four soybean viruses. Crop Sci. 45:639-644.

(Prepared by J. W. Demski and C. W. Kuhn;
revised by S. A. Tolin and L. L. Domier)

Luteoviridae
Luteovirus
Soybean dwarf virus

Soybean dwarf virus (SbDV) was first described in Japan in 1952 and has subsequently been found on most continents worldwide. In Japan, the incidence and severity of soybean dwarf disease vary from year to year among locales, but the disease is economically important in the northern region. In years of high early season incidence (i.e., 50% of plants are infected), yield losses of 40% have been reported. The SbDV subterranean clover red-leaf (SCRL) strain has been reported to cause losses in Australia and New Zealand.

In addition to soybean, SbDV infects many genera and species within the Fabaceae, and natural infections in 29 species in six families have been reported. Depending on the virus strain and environmental conditions, symptoms may be obvious or latent. Legume hosts include Chinese milkvetch (*Astragalus sinicus*), fava bean, garden pea, garden vetch (*Vicia sativa*), kidney bean, lentil, and water medick (*Medicago littoralis*), along with a variety of clovers: alsike clover (*Trifolium hybridum*), barrelclover (*M. truncatula*), crimson clover (*T. incarnatum*), Egyptian clover (*T. alexandrinum*), red clover (*T. pratense*), strawberry clover (*T. fragiferum*), subterranean clover (*T. subterraneum*), suckling clover (*T. dubiumi*), and white clover (*T. repens*). A few nonleguminous hosts have been identified by enzyme-linked immunosorbent assay (ELISA)—including annual phlox, common beet, New Zealand spinach (*Tetragonia tetragonioides*), and spinach—and these identifications have been confirmed by back assays to soybean indicator cultivars Swift and Wayne.

Symptoms and Signs

Young soybean plants inoculated 1–3 weeks after emergence are severely affected by strains of SbDV. (See the following section, Causal Agents, for descriptions of the strains of SbDV.) Symptoms range from indiscernible to severe, depending on the cultivar and the strain. Infected plants are sometimes severely stunted because of the shortening of the internodes and the petioles. When leaflets mature, they are smaller than normal, curl downward, turn deep green, and become thick and brittle. At other times, infected plants are less stunted. However, their leaves have wavy margins and interveinal yellowing (especially older leaves), and they abort most florets (flowers). Greenhouse-grown plants infected with different strains of SbDV exhibit rugose leaves and display a mixture of symptoms (Figs. 172 and 173). Plants in the field can be infected with different strains.

Translocation of photosynthates from leaflets to pods is interrupted in infected plants, which results in reduced pod growth and seed number, depending on the age of the plant when infected. Diseased plants mature later than normal and remain green at harvest. Mottled seeds have not been observed from infected plants.

Fig. 172. Symptoms induced by the Japanese dwarfing (left) and severe-yellowing (right) strains of *Soybean dwarf virus* in soybean cultivar Williams 82 after inoculation with the viruliferous foxglove aphid (*Aulacorthum solani*). (Courtesy V. D. Damsteegt)

Causal Agents

Soybean dwarf virus consists of 26-nm isometric particles, each of which contains a single-stranded, positive-sense RNA and two capsid proteins. SbDV is difficult to purify because of low virus concentration (i.e., phloem limited). Purification requires freezing infected tissue in liquid nitrogen, pulverizing the tissue to a powder, and incubating it with cellulolytic enzymes.

SbDV is an excellent antigen, and several high-quality polyclonal and monoclonal antisera are available. Strain-specific monoclonal and polymerase chain reaction (PCR) primers have been developed for use in enzyme-linked immunosorbent assays (ELISAs), and various PCR platforms are useful for detection. SbDV strains are related to other members of the *Luteoviridae,* such as *Barley yellow dwarf virus, Bean leafroll virus,* and *Beet western yellows virus* but not *Potato leafroll virus.*

The causal luteovirus occurs as several distinct strains based on symptomatology in soybean, aphid relationships, physiochemical properties, and molecular composition. Historically, two strains have been identified based on symptoms expressed in soybean; both SbDV-D (dwarfing strain) and SbDV-Y (severe-yellowing strain) are monospecifically transmitted by the aphid *Aulacorthum solani.* The SCRL strain of SbDV, described in Australia, is also transmitted by *A. solani.* An SbDV isolate serologically related to the SCRL strain has been found in California, and other related isolates found worldwide are transmitted by *Acyrthosiphon pisum* and other aphids. The SCLR strain was placed in synonymy with the severe-yellowing strain in 1987. More recently, at least four unique strains of SbDV have been described in Japan based on their symptoms and principal aphid vectors: DP (dwarfing strain; vector *A. pisum*), DS (dwarfing strain; vector *A. solani*), YP (yellowing strain; vector *A. pisum*), and YS (yellowing strain; vector *A. solani*). Sequence analysis has indicated that the coat protein read-through product is tied to vector specificity.

Complete nucleotide sequences are available in GenBank for the genomic RNAs of the DP, DS, YP, and YS strains from Japan and 12 isolates from the United States.

Fig. 173. Soybean plant infected with a dwarfing isolate of *Soybean dwarf virus* (left) and an uninfected plant (right). (Courtesy L. L. Domier)

Disease Cycle and Epidemiology

Like all luteoviruses, *Soybean dwarf virus* is transmitted persistently by aphids and is localized in the phloem. Asian strains of SbDV are monospecifically transmitted by *A. solani* and *A. pisum,* whereas strains from the United States are nonspecifically transmitted by *A. pisum, A. solani, Myzus persicae,* and *Nearctaphis bakeri.* Aphid vectors retain SbDV through molts and require acquisition and inoculation access periods as long as 48 h from one or more of their many host species for efficient transmission of the virus.

No mechanical or seed transmission of SbDV has been demonstrated.

Management

Very few soybean cultivars have shown resistance or tolerance to *Soybean dwarf virus.* In Japan, certain early maturing cultivars have been found to provide some tolerance, and a highly resistant cultivar (Tsurukogane) was released in Hokkaido in 1984. More recently, virus-resistant soybean plants have been developed by transforming soybean with the sense-coat protein gene via microprojectile bombardment of soybean somatic embryos. An Indonesian cultivar (Wilis) with good resistance to SbDV has been identified, as well, but the inability of soybean to reach maturity and difficulties with seed production in northern latitudes limit the use of this cultivar.

Organophosphate insecticides have been used in Japan to control aphids that infect soybean crops. Application into the soil at planting time has been shown to be the most effective means of aphid control, although top-dressing is nearly equally effective. The use of insecticides does not prevent initial infection by viruliferous winged aphids that migrate into the fields, because they can infect plants before they die. However, using insecticides does provide satisfactory control of secondary virus spread.

Selected References

Abraham, A. D., Menzel, W., and Vetten, H. J. 2007. First report of *Soybean dwarf virus* (genus *Luteovirus*) infecting faba bean and clover in Germany. Plant Dis. 91:1059.

Banba, H., Tanimura, Y., and Matsukawa, I. 1986. Breeding for resistance to Soybean dwarf virus in soybeans. Trop. Agric. Res. 19:236-246.

Damsteegt, V. D., Hewings, A. D., and Sindermann, A. B. 1990. Soybean dwarf virus: Experimental host range, soybean germplasm reactions, and assessment of potential threat to U.S. soybean production. Plant Dis. 74:992-995.

Damsteegt, V. D., Stone, A. L., Kuhlmann, M., Gildow, F. E., Domier, L. L., Sherman, D. J., Tian, B., and Schneider, W. L. 2011. Acquisition and transmissibility of U.S. *Soybean dwarf virus* isolates by the soybean aphid, *Aphis glycines.* Plant Dis. 95:945-950.

Damsteegt, V. D., Stone, A. L., Russo, A. J., Luster, D. G., Gildow, F. E., and Smith, O. P. 1999. Identification, characterization, and relatedness of luteovirus isolates from forage legumes. Phytopathology 89:374-379.

Edwardson, J. R., and Christie, R. G. 1991. Luteoviruses. CRC Handbook of Viruses Infecting Legumes. CRC Press, Boca Raton, FL.

Harrison, B., Steinlage, T. A., Domier, L. L., and D'Arcy, C. J. 2005. Incidence of *Soybean dwarf virus* and identification of potential vectors in Illinois. Plant Dis. 89:28-32.

Hewings, A. D., Damsteegt, V. D., and Tolin, S. A. 1986. Purification and some properties of two strains of Soybean dwarf virus. Phytopathology 76:759-763.

Kellock, A. W. 1971. Red-leaf virus—A newly recognized virus disease of subterranean clover (*Trifolium subterraneum* L.). Aust. J. Agric. Res. 22:615-622.

Makkouk, K. M., Damsteegt, V., Johnstone, G. R., Katul, L., Lesemann, D.-E., and Kumari, S. G. 1997. Identification and some properties of soybean dwarf luteovirus affecting lentil in Syria. Phytopathol. Mediterr. 36:135-144.

Rathjen, J. P., Karageorgos, L. E., Habili, N., Waterhouse, P. M., and Symons, R. H. 1994. Soybean dwarf luteovirus contains the third variant genome type in the luteovirus group. Virology 198:671-679.
Tamada, T. 1973. Strains of Soybean dwarf virus. Ann. Phytopathol. Soc. Jpn. 39:27-34.
Terauchi, H., Kanematsu, S., Honda, K., Mikoshiba, Y., Ishiguro, K., and Hidaka, S. 2001. Comparison of complete nucleotide sequences of genomic RNAs of four Soybean dwarf virus strains that differ in their vector specificity and symptom production. Arch. Virol. 146:1885-1898.
Tougou, M., Yamagishi, N., Furutani, N., Shizukawa, Y., Takahata, Y., and Hidaka, S. 2007. Soybean dwarf virus-resistant transgenic soybeans with the sense coat protein gene. Plant Cell Rep. 26:1967-1975.
Yamashita, Y., Takeuchi, T., Ohnishi, S., Sasaki, J., and Tazawa, A. 2013. Fine mapping of the major Soybean dwarf virus resistance gene *Rsdv1* of the soybean cultivar 'Wilis.' Breed. Sci. 63:417-422.

(Prepared by V. Damsteegt)

Indonesian soybean dwarf virus

Indonesian soybean dwarf virus (ISDV) has been isolated from soybean in Indonesia and Thailand. (The genus for this virus is unassigned.)

Symptoms and Signs

Indonesian soybean dwarf virus induces dwarfing, along with shortened leaf petioles and internodes; infected plants are often dark green. The upper leaves are small and may be crinkled and curled, whereas the lower leaves are rugose and brittle (Fig. 174).

Causal Agent

Indonesian soybean dwarf virus has isometric particles 25–30 nm in diameter, and aggregates of particles occur in the vacuoles of phloem parenchyma cells. Particle morphology, mode of transmission, and tissue association suggest that ISDV is a member of the family *Luteoviridae,* but it has not been assigned to a genus (as of March 2015).

Fig. 174. Shortened leaf petioles and internodes and small, crinkled, curled upper leaves, caused by *Indonesian soybean dwarf virus.* (Courtesy S. A. Tolin; photographed by M. Iwaki)

ISDV has a moderate serological relationship with both *Soybean dwarf virus* (SbDV) and *Beet western yellows virus* (BWYV), making the classification of ISDV unclear in relation to other members of the *Luteoviridae* that infect legumes: namely, BWYV, *Bean leafroll virus, Groundnut rosette assistor virus,* and SbDV.

Disease Cycle and Epidemiology

Indonesian soybean dwarf virus is not mechanically transmissible, but it can be transmitted by the soybean aphid, *Aphis glycines,* in a persistent manner.

Management

Options for management are similar to those for *Soybean dwarf virus* in terms of controlling the aphid population to prevent secondary spread.

Selected Reference

Iwaki, M., Roechan, M., Hibino, H., Tochihara, H., and Tantera, D. M. 1980. A persistent aphidborne virus of soybean, Indonesian soybean dwarf virus. Plant Dis. 64:1027-1030.

(Prepared by S. A. Tolin)

Potyviridae

The family *Potyviridae* is the largest family of plant viruses, representing more than 20% of all known plant viruses. Genera within the family include those that contain the aphid-transmitted viruses (*Potyvirus* and *Macluravirus*), the mite-transmitted viruses (*Poacevirus, Rymovirus,* and *Tritimovirus*), the bipartite fungus-transmitted viruses (*Bymovirus*), and the whitefly-transmitted viruses (*Ipomovirus*). The vector of an eighth genus (*Brambyvirus*) is unknown.

Molecular analyses of the genomes of members of the genus *Potyvirus* have greatly advanced understanding of the relationships between the many potyviruses that have been found to naturally infect soybean and other legumes. Based on coat protein sequences, viruses have been placed within clusters related to *Bean common mosaic virus* (BCMV) and *Bean yellow mosaic virus* (BYMV). Azuki bean mosaic, blackeye cowpea mosaic, Dendrobium mosaic, and peanut stripe are considered strains of BCMV by the International Committee on Taxonomy of Viruses (ICTV). *Soybean mosaic virus* and *Watermelon mosaic virus* consistently group within the BCMV cluster but are distinct viruses. *Cowpea aphid-borne mosaic virus* is recognized by the ICTV, but South African passiflora is considered a strain of *Bean common mosaic necrosis virus* (BCMNV). BYMV, *Clover yellow vein virus,* and *Peanut mottle virus* are distinct but closely related viruses. Only the most highly characterized potyviruses that infect soybean are discussed in the following sections.

Selected References

Berger, P. H., Wyatt, S. D., Shiel, P. J., Silbernagel, M. J., Druffel, K., and Mink, G. I. 1997. Phylogenetic analysis of the *Potyviridae* with emphasis on legume-infecting potyviruses. Arch. Virol. 142:1979-1999.
International Committee on Taxonomy of Viruses (ICTV). 2014. Virus Taxonomy. ICTV, Virology Division–IUMS. Available online at www.ictvonline.org
King, A. M. Q., Adams, M. J., Carstens, E. B., and Lefkowitz, E. J. 2012. Virus Taxonomy: Classification and Nomenclature of Viruses. Ninth Report of the International Committee on Taxonomy of Viruses. Academic Press, London.

(Prepared by S. A. Tolin)

Potyvirus

Bean common mosaic virus (Peanut stripe strain)

The peanut stripe strain of *Bean common mosaic virus* (BCMV) was reported in 1982 infecting peanut in the United States, and it was reported infecting soybean in 1989. A virus producing a mild mottle in peanut was also found to infect soybean naturally in 1983 in Hubei Province, China. It reduced soybean yields by as much as 53% in greenhouse trials, even though leaf symptoms were mild and included primarily mottling. The virus appears to be the same as one from Taiwan, where peanut stripe strain of BCMV is as prevalent as *Soybean mosaic virus* (SMV), according to one report.

Symptoms and Signs

Symptoms of *Bean common mosaic virus* vary by soybean cultivar. For instance, only a mild, systemic mottle develops in cultivar Lawrence; a systemic mosaic develops in cultivar Elgin; local, chlorotic lesions followed by a systemic mottle develop in cultivar Coc Chun; and local, necrotic lesions followed by a systemic necrosis develop in cultivar Tulumayo-2.

Causal Agent

Bean common mosaic virus has flexuous rod particles approximately 12 × 750 nm (diameter × length). BCMV is serologically related to *Clover yellow vein virus* and SMV. The nucleotide sequence of the RNA of BCMV is closely related to *East Asian Passiflora virus,* SMV, *Watermelon mosaic virus,* and *Wisteria vein mosaic virus.*

Disease Cycle and Epidemiology

Bean common mosaic virus is seed transmitted in peanut but not in soybean. It is easily transmitted mechanically by inoculation and by some aphids. BCMV infects numerous legumes, including several clovers (arrowleaf [*Trifolium vesiculosum*], crimson [*T. incarnatum*], and subterranean [*T. subterraneum*]), blue and white lupine (*Lupinus angustifolius* and *L. albus,* respectively), peanut, and soybean. Lambsquarters (*Chenopodium amaranticolor*) is used as a local lesion host.

Management

Planting soybean more than 100 m from infected peanut will reduce the chance of infection.

Selected References

Demski, J. W., Reddy, D. V. R., Sowell, G., Jr., and Bays, D. 1984. Peanut stripe virus: A new seed borne potyvirus from China infecting groundnut (*Arachis hypogaea*). Ann. Appl. Biol. 105:495-501.

Demski, J. W., Reddy, D. V. R., Wongkaew, S., Kameya-Iwaki, M., Seleh, N., and Xu, Z. 1988. Letter to the editor: Naming of peanut stripe virus. Phytopathology 78:631-632.

Gillaspie, A. G., Jr., and Hopkins, M. S. 1991. Spread of peanut stripe virus from peanut to soybean and yield effects on soybean. Plant Dis. 75:1157-1159.

Green, S. K., and Lee, D. R. 1989. Occurrence of peanut stripe (PStV) on soybean in Taiwan—Effect on yield and screening for resistance. Trop. Pest Manage. 35:123-126.

Gunasinghe, U. B., Flasinski, S., Nelson, R. S., and Cassidy, B. G. 1994. Nucleotide sequence and genome organization of peanut stripe potyvirus. J. Gen. Virol. 75:2519-2526.

Warwick, D., and Demski, J. W. 1988. Susceptibility and resistance of soybean to peanut stripe virus. Plant Dis. 72:19-21.

Xu, Z., Yu, Z., Liu, J., and Barnett, O. W. 1983. A virus causing peanut mild mottle in Hubei Province, China. Plant Dis. 67:1029-1032.

(Prepared by S.A. Tolin)

Bean yellow mosaic virus

Bean yellow mosaic virus (BYMV) causes a mosaic and yellowing disease of soybean and occurs worldwide. In a survey of virus diseases in soybean in Iran, BYMV was detected in approximately 2% of soybean plants. Mixed infections of BYMV and *Soybean mosaic virus* (SMV) also occur. Reduced oil content and increased protein content have been recorded in seeds from BYMV-infected plants.

Symptoms and Signs

The initial symptoms of BYMV are essentially the same as those of SMV at its later stages. A conspicuous yellow mottling of the leaves is characteristic (Fig. 175); the yellow areas are scattered or produced in indefinite bands along the major veins. Rusty, necrotic spots appear within the yellow areas as leaves mature. Some strains of BYMV produce severe mottling and crinkling of leaves.

Causal Agent

Bean yellow mosaic virus has flexuous rod particles approximately 12 × 750 nm (diameter × length). The dilution endpoint is between 10^{-3} and 10^{-4}. BYMV has a single-stranded, positive-sense RNA genome that is 9.5 kb long.

Disease Cycle and Epidemiology

Bean yellow mosaic virus has a wide host range and is easily sap transmissible. More than 20 aphid species can transmit the virus in a nonpersistent manner. Common aphid vectors include *Acyrthosiphon pisum, Aphis fabae, Aphis glycines, Macrosiphum euphorbiae,* and *Myzus persicae.* Seed transmission in soybean has not been reported. In the United States, BYMV is frequently found in red clover (*Trifolium pratense*), which presumably serves as a reservoir for virus transmission to soybean.

Management

Little information is available about management strategies for BYMV. Resistance is known in other crops, such as bean and pea, but no soybean cultivars are reported to be resistant to BYMV. Aphid management will presumably provide some reduction in the incidence of BYMV.

Selected References

Conover, R. A. 1948. Studies of two viruses causing mosaic diseases of soybean. Phytopathology 38:724-735.

Fig. 175. Leaf chlorosis and puckering, symptoms similar to those of *Bean yellow mosaic virus,* caused by a geminivirus from Argentina. (Courtesy D. Ploper)

Golnaraghi, A. R., Shahraeen, N., Pourrahim, R., Farzadfar, Sh., and Ghasemi, A. 2004. Occurrence and relative incidence of viruses infecting soybeans in Iran. Plant Dis. 88:1069-1074.

Roechan, M., Iwaki, M., Nasir, S., and Tantera, D. 1978. Virus Diseases of Legume Plants in Indonesia. 3. Bean yellow mosaic virus. Contributions Central Research Institute for Agriculture, Bogor, Indonesia.

Wang, R. Y., Kritzman, A., Hershman, D. E., and Ghabrial, S. A. 2006. *Aphis glycines* as a vector of persistently and nonpersistently transmitted viruses and potential risks for soybean and other crops. Plant Dis. 90:920-926.

(Prepared by J. W. Demski and C. W. Kuhn; revised by S. A. Tolin and L. L. Domier)

Peanut mottle virus

Peanut mottle virus (PMV) causes leaf mottling in soybean. It was first found occurring naturally in 1971 in the U.S. state of Georgia. PMV is common in the southeastern states and in Oklahoma and can cause yield reductions of up to 60%. PMV has been reported in soybean in Argentina, Australia, China, Iran, and Thailand, and it occurs widely in peanut in many other countries.

PMV occurs in nature in several legume crops in addition to soybean: kidney bean and sieva bean (*Phaseolus lunatus*), arrowleaf clover (*Trifolium vesiculosum*) and subterranean clover (*T. subterraneum*), cowpea, garden pea, narrowleaf lupine (*Lupinus angustifolius*) and white lupine (*L. albus*), and peanut. The virus has also been isolated from weed hosts in nature, such as coffeeweed (*Senna occidentalis*), sicklepod (*S. obtusifolia;* syn. *Cassia obtusifolia*), woolly senna (*S. hirsuta* var. *leptocarpa;* syn. *C. leptocarpa*), and zarzabacoa comun (*Desmodium incanum*).

Symptoms and Signs

The first symptom of infection from *Peanut mottle virus* in soybean is the development of small, enlarging, chlorotic areas (Fig. 176). Dark-green islands are produced on young leaves. Later, yellow patches, line patterns, or ring patterns may develop on the third and fourth leaves formed after infection. A mosaic pattern is produced on older leaves. Leaves of infected plants may also pucker and curl down at the edges. Infected plants often produce mottled seeds.

Causal Agent

PMV particles are flexuous rods approximately 12 × 750 nm (diameter × length). Each contains one molecule of positive-sense, single-stranded RNA that is 9.7 kb in length. The RNA is encapsidated with a single type of capsid protein, and each subunit has an approximate molecular weight of 35 kDa.

Fig. 176. Chlorotic, puckered leaves, caused by *Peanut mottle virus*. (Courtesy S. A. Tolin)

Peanut mottle virus is moderately immunogenic and can be evaluated with standard serological tests. Immunosorbent electron microscopy and enzyme-linked immunosorbent assays (ELISAs) have demonstrated serological relationships between PMV and a few other potyviruses.

Variants of PMV have been found in bean, peanut, and soybean. Even though the variants cannot be differentiated serologically, they can be distinguished on the basis of host range and disease reactions. At least partial nucleotide sequence information is available for six strains of PMV.

Disease Cycle and Epidemiology

Peanut mottle virus is easily transmitted mechanically by inoculation to and from soybean plants, and it is more stable in basic buffers. To achieve transmission from some infected species (particularly peanut), antioxidants (e.g., sodium bisulfite, sodium diethyldithiocarbamate) should be added to the buffer. Seed transmission does not occur in soybean or naturally infected weed hosts. Several aphid species can readily transmit PMV in a nonpersistent manner, including *Aphis craccivora, A. gossypii, Hyperomyzus lactucae, Myzus persicae, Rhopalosiphum maidis,* and *R. padi.*

Peanut seeds infected with PMV appear to be the source of primary inoculum for infection of susceptible crops; this source has been established for cowpea, peanut, and soybean. With the presence of aphids, PMV can be spread within a field and move to nearby susceptible crops.

Management

Using resistant cultivars is the most effective means of reducing losses caused by PMV. Resistance in cultivars CNS and Dorman is conditioned by single dominant genes (*Rpv*1 and *Rpv*2, respectively), and a single recessive gene (*rpv*1) controls resistance in cultivar Peking. Cultivars Hutcheson and Young carry resistance to PMV and *Soybean mosaic virus* (SMV). Varieties with resistance to SMV are often also resistant to PMV because of the close linkage of the genes.

Studies have shown that very little or no infection develops in susceptible plants that are grown 100 m or more from PMV-infected peanut plants; thus, growers are advised to isolate soybean fields from peanut fields.

Selected References

Bays, D. C., Tolin, S. A., and Roane, C. W. 1986. Interactions of peanut mottle virus strains and soybean germplasm. Phytopathology 76:764-768.

Boerma, H. R., and Kuhn, C. W. 1976. Inheritance of resistance to peanut mottle virus in soybeans. Crop Sci. 16:533-534.

Golnaraghi, A. R., Shahraeen, N., Pourrahim, R., Farzadfar, Sh., and Ghasemi, A. 2004. Occurrence and relative incidence of viruses infecting soybeans in Iran. Plant Dis. 88:1069-1074.

Gore, M. A., Hayes, A. J., Jeong, S. C., Yue, Y. G., Buss, G. R., and Saghai Maroof, M. A. 2002. Mapping tightly linked genes controlling potyvirus infection at the *Rsv1* and *Rpv1* region in soybean. Genome 45:592-599.

Kuhn, C. W., Demski, J. W., and Harris, H. B. 1972. Peanut mottle virus in soybeans. Plant Dis. Rep. 56:146-147.

Shipe, E. R., Buss, G. R., and Tolin, S. A. 1979. A second gene for resistance to peanut mottle virus in soybean. Crop Sci. 19:656-658.

(Prepared by S. A. Tolin)

Soybean mosaic virus

Soybean mosaic disease, caused by *Soybean mosaic virus* (SMV), occurs in all the world's soybean production areas. The disease was first reported from the U.S. state of Connecticut in

1915 and then described around 1921. Reported yield reductions range from 8 to 35%, but losses as high as 94% have been documented.

Early season plant infections reduce pod set, increase seed coat mottling, and reduce seed size and weight, whereas late season infections have little effect on seed quality and yield. Additional effects of SMV include reduced seed size, oil content, and nodulation. SMV also adversely affects nitrogen fixation and can reduce seed germination. Seed quality may be reduced as a result of seed coat mottling. The increased occurrence of Phomopsis seed decay, resulting in poor seed quality and low vigor, has been associated with SMV-infected plants prior to growth stage R3 (see earlier in Part I, the section Phomopsis Seed Decay).

The most economically important host of *Soybean mosaic virus* is soybean. However, plant hosts infected by various isolates of SMV occur in six plant families, including more than 42 species in the Fabaceae.

Symptoms and Signs

Symptoms of *Soybean mosaic virus* vary with the host genotype, virus strain, plant age at infection, and environment. Most infected cultivars are slightly stunted and produce fewer pods that often contain fewer seeds. Trifoliolate leaves have a mosaic of light- and dark-green areas that may later become raised or blistered, particularly along the main veins, and chlorosis may develop between the dark-green areas. Leaf margins may be wavy or curl downward (Fig. 177). Leaf veins do not grow together, as is typical with herbicide injury. The primary leaves of some cultivars may show local, necrotic lesions, which may merge into veinal necrosis followed by yellowing and leaf abscission.

Some strains of SMV can induce severe stunting, systemic necrosis, leaf yellowing, petiole and stem necrosis, terminal necrosis, and defoliation, usually leading to death. Different strains often produce different symptoms on the same cultivar.

Temperature affects the development of rugosity. Rugosity is most severe in plants grown at approximately 18°C. Symptoms are less severe at 25°C and are largely masked at temperatures above 30°C. Temperature also influences the length of time between infection and symptom appearance, which ranges from 4 days at 30°C to 14 days at 19°C.

Fig. 177. Crinkled, downward-curving leaves, caused by *Soybean mosaic virus*. (Courtesy G. L. Hartman)

Depending on hilum color, seeds from infected plants may have a brown-and-black mottle (Fig. 178), which has been associated with suppression of posttranscriptional gene silencing of chalcone synthase by an SMV-encoded silencing suppressor protein. Seeds may be smaller and germination may be reduced compared with seeds from noninfected plants. Seed mottling does not indicate that SMV is present in seeds; plants not infected with SMV may also produce low levels of mottled seed, and not all seeds from virus-infected plants are mottled. The correlation of the percentage of infected seeds with mottling and the severity of mottling varies by year and by cultivar.

Synergism has been reported to occur in plants infected with SMV and either *Bean pod mottle* (BPMV) or *Alfalfa mosaic virus* (AMV). Yield reductions of plants infected with both SMV and BPMV may be as high as 66–86% in susceptible cultivars, compared with reductions of 8–25% in cultivars inoculated with SMV. Disease symptoms of plants infected with both SMV and BPMV may include severe dwarfing and foliar distortion, necrosis, and mottling. Doubly infected plants have fewer root nodules than singly infected plants. The ultimate effect on yield, seed mottling, and seed transmission of SMV is dependent on the growth stage at which soybean is infected by the virus, as well as the cultivar and the virus strain.

SMV-infected plant cells, particularly in vascular areas, contain intracellular inclusion bodies (characteristic of subdivisions III and IV), which look like pinwheels in cross section and bundles in longitudinal section. The length of the inclusion bodies ranges from 12 to 41 µm, and their number increases 2–3 weeks after inoculation. More inclusions occur in plants infected with a severe strain of the virus than a mild strain. Both SMV inclusion bodies and BPMV particles may occupy the same cell in a doubly infected plant.

Causal Agent

SMV is a flexuous rod approximately 14 × 750 nm (diameter × length). The virions consist of positive-sense, single-stranded RNA, which constitutes 5.3% of the particle. They have helical symmetry with a pitch of 33 Å, a central hole of radius 1.5 nm, and 8.8 subunits per turn of the viral helix. The radial location of the RNA is at 30–35 Å. Repeating monomeric coat protein subunits have a molecular weight of approximately 30,000 Da.

Complete nucleotide sequences of the viral RNA from several strains—consisting of 9,588 nucleotides, with a viral protein genome-linked (VPg) protein and a poly[A] tail at the 5′ end and the 3′ end, respectively—are available in GenBank. Nucleotide sequence data are also available for selected regions of several isolates. Infectious cDNA clones have been prepared

Fig. 178. Brown to nearly black mottling of seeds, caused by *Soybean mosaic virus*. (Courtesy S. A. Tolin)

from multiple strains. The positive-sense, single-stranded RNA encodes a single polyprotein precursor, which is cleaved into functional gene products by three virus-encoded proteases.

Numerous strains of *Soybean mosaic virus* have been identified based on reactions of a set of differential soybean cultivars. Reports of differentiation of SMV isolates using differential soybean cultivars have come from Brazil, China, Japan, Korea, and the United States. (Not all countries use the same differential set.) The large number of reported strains reflects the variability of the virus and most likely the lack of understanding of virus effectors and resultant resistance mechanisms. It is highly probable that strains exist that have not been reported.

SMV cross-reacts serologically with *Bean common mosaic virus, Bean yellow mosaic virus, Clover yellow vein virus, Lettuce mosaic virus,* and *Watermelon mosaic virus 2*, and highly specific antisera have been prepared. Serological differences among SMV isolates have been difficult to show using polyclonal antisera; nevertheless, the use of monoclonal antibodies has allowed serological differentiation of isolates characteristic of SMV strains. Coat protein trypsin peptide profiles of 14 strains or isolates have been found to be nearly identical and distinct from those of other potyviruses.

Disease Cycle and Epidemiology

Soybean mosaic virus is sap and graft transmissible, but it is not transmitted by dodder.

At least 32 aphid species, belonging to 15 different genera, transmit SMV in a nonpersistent manner. Virus isolates may show some vector specificity. Even though *Aphis glycines* can efficiently transmit SMV, the introduction of this soybean-colonizing aphid species into North America has not been documented as significantly increasing SMV incidence.

Plants infected from transmission through seed play an important role in SMV epidemiology, because they serve as sources of primary inoculum for the virus. Spread of the virus within and among fields is mostly aggregated from a point source, and secondary spread by aphids occurs at a moderately fast rate. In seeds, the virus remains infective for a long time, and viable virus can be recovered from seeds that no longer germinate. The level of seed transmission is dependent on the cultivar and the plant stage at infection; the incidence of seed transmission is higher in plants infected before the onset of flowering. In most commercial cultivars, seed transmission is less than 1%. However, no transmission occurs in some cultivars, and certain older cultivars can have transmission rates as high as 75%.

SMV moves both up and down in the plant and can be detected in all parts of a systemically infected plant, including the roots. Because of this, SMV has been used successfully as an expression vector.

Management

In the first field test of stable transgenic resistance against any pathogen of soybean, coat protein-mediated resistance was developed against *Soybean mosaic virus* by using a nonaphid-transmissible isolate to minimize the risk of spreading new pathogenic strains through recombination. This resistance was effective against numerous SMV strains. In a tactic intended to maximize the durability of resistance, the resistance could be successfully deployed in the field as rate-reducing resistance.

At least three naturally occurring independent loci—designated *Rsv1, Rsv3,* and *Rsv4*—have been identified and mapped for resistance to SMV. At the *Rsv1* locus, nine variants confer differential reactions to strains G1–G7; these variants have been identified as *Rsv1* from plant introduction (PI) 96983, *Rsv1-y* from York, *Rsv1-m* from Marshall, *Rsv1-h* from Suweon 97, *Rsv1-t* from Ogden, *Rsv1-k* from Kwanggyo, *Rsv1-r* from Raiden, *Rsv1-s* from LR1, and *Rsv1-n* from PI 507389. Five alleles at the *Rsv3* locus have been identified from OX686, L29, Harosoy, PI 399091, and PI 61947, and *Rsv4* has been identified from V94-5152 and PI 88788. Because of the high variability of SMV, use of single resistance genes is potentially dangerous; pyramiding available sources of resistance is recommended to achieve disease control.

Most studies have focused on *Rsv1*-mediated resistance. The *Rsv1* resistance mechanism operates against all major SMV strains, including the necrotic (N) isolate (SMV-N) of strain SMV-G2, but strain SMV-G7 overcomes *Rsv1* and induces a lethal systemic necrosis in PI 96983 (*Rsv1*). The mechanism of *Rsv1*-mediated resistance against SMV-N has been characterized as extreme resistance, in which no virus can be detected in an inoculated leaf. However, the SMV-N – *Rsv1* interaction has the potential to produce a restricted systemic hypersensitive response (HR) under certain conditions. The elicitors of *Rsv1*-mediated resistance have been mapped to the helper-component proteinase (HC-Pro) and P3 cistrons of SMV. The *Rsv1* locus consists of a cluster of related resistance genes that appear to recognize HC-Pro and P3 independently, which likely accounts for the fact that few naturally occurring strains overcome *Rsv1*-mediated resistance. Under certain conditions, however, SMV can mutate to evade *Rsv1*-mediated resistance. Moreover, other studies using different isolates of the same strains have reported that the cytoplasmic inclusion (CI) may also be an elicitor of *Rsv1* in these isolates.

In contrast to *Rsv1, Rsv3* confers resistance to SMV strains G5, G6, and G7. The resistance conferred by *Rsv3* is phenotypically different from that conferred by *Rsv1,* as it allows for a limited spread of avirulent SMV-G7 from the point of inoculation and does not induce a typical HR phenotype. The elicitors of *Rsv3*-mediated resistance have been mapped to the N- and the C-termini of the SMV cylindrical inclusion cistron.

Rsv4 resistance, which is effective against SMV strains G1–G7, is associated with a nonnecrotic reaction. Resistance-breaking isolates have been reported from Canada and South Korea. The elicitor for *Rsv4*-mediated resistance maps to P3. Single nucleotide mutations in P3 compromise resistance, leading to the appearance of new resistance-breaking isolates. Therefore, the use of soybean cultivars that carry this single resistance gene is likely dangerous.

Despite the development of transgenic resistance and the identification of naturally occurring resistance genes, resistance to *Soybean mosaic virus* has not been intentionally incorporated into most commercially marketed cultivars. Whether naturally occurring resistance genes exist in the cultivars used for commercial production is unknown. However, a 2005 study reported resistance to two SMV strains in 23 of 52 ancestral North American soybean lines. Of all the public registration articles cited since 2002, only 15 have reported resistance to SMV. Although the source of resistance is uncertain, it has been suggested to be one or more of the multiple alleles at the *Rsv1* locus. In a more extensive study using five SMV strains and 303 cultivars grown in the mid-South region of the United States, similar results revealed that the majority of lines tested did not carry resistance genes to SMV. However, 74 cultivars appeared to contain one of the *Rsv1* alleles, which was most likely *Rsv1-y*. Only a few cultivars appeared to carry *Rsv3* and *Rsv4*. Thus, at least in North America, SMV resistance may be more widespread than expected.

Insecticide applications to control potential aphid vectors of *Soybean mosaic virus* do not result in consistently successful virus disease management. Clearly, incorporating disease resistance (preferably using different sources) will provide the best means of controlling the disease caused by SMV. In addition, using SMV-free seed and timing planting to avoid high vector populations when plants are young are effective management practices for reducing yield losses induced by SMV. Serological seed-indexing techniques and/or grow-out tests can be used for virus detection in seed lots. Rogueing, although generally impractical in the field, may not be very effective because of the tendency for symptoms in soybean to be masked at temperatures above 30°C.

Selected References

Bowers, G. R., Jr., and Goodman, R. M. 1979. Soybean mosaic virus: Infection of soybean seed parts and seed transmission. Phytopathology 69:569-572.

Cho, E.-K., and Goodman, R. M. 1979. Strains of soybean mosaic virus: Classification based on virulence in resistant soybean cultivars. Phytopathology 69:467-470.

Chowda-Reddy, R. V., Sun, H., Chen, H., Poysa, V., Ling, H., Gijzen, M., and Wang, A. 2011. Mutations in the P3 protein of soybean mosaic virus G2 isolates determine virulence on *Rsv4*-genotype soybean. Mol. Plant-Microbe Interact. 24:37-43.

Chowda-Reddy, R. V., Sun, H., Hill, J. H., Poysa, V., and Wang, A. 2011. Simultaneous mutations in multi-viral proteins are required. PloS One 6:e28342.

Hajimorad, M. R., Eggenberger, A. L., and Hill, J. H. 2008. Adaptation of *Soybean mosaic virus* avirulent chimeras containing P3 sequences from virulent strains to *Rsv1*-genotype soybeans is mediated by mutations in HC-Pro. Mol. Plant-Microbe Interact. 21:937-946.

Hill, J. H., Alleman, R., Hogg, D. B., and Grau, C. R. 2001. First report of transmission of *Soybean mosaic virus* and *Alfalfa mosaic virus* by *Aphis glycines* in the New World. Plant Dis. 85:561.

Hill, J. H., Bailey, T. B., Benner, H. I., Tachibana, H., and Durand, D. P. 1987. Soybean mosaic virus: Effects of primary disease incidence on yield and seed quality. Plant Dis. 71:237-239.

Hill, J. H., Lucas, B. S., Benner, H. I., Tachibana, H., Hammond, R. B., and Pedigo, L. P. 1980. Factors associated with the epidemiology of soybean mosaic virus in Iowa. Phytopathology 70:536-540.

Irwin, M. E., and Goodman, R. M. 1981. Ecology and control of soybean mosaic virus. Pages 181-220 in: Plant Diseases and Their Vectors: Ecology and Epidemiology. K. Maramorosch and K. F. Harris, eds. Academic Press, New York.

Koning, G., TeKrony, D. M., Pfeiffer, T. W., and Ghabrial, S. A. 2001. Infection of soybean with *Soybean mosaic virus* increases susceptibility to *Phomopsis* spp. seed infection. Crop Sci. 41:1850-1856.

Li, K., Yang, Q. H., Zhi, H. J., and Gai, J. Y. 2010. Identification and distribution of *Soybean mosaic virus* strains in southern China. Plant Dis. 94:351-357.

Pedersen, P., Grau, C., Cullen, E., Koval, N., and Hill, J. H. 2007. Potential for integrated management of soybean virus disease. Plant Dis. 91:1255-1259.

Saghai Maroof, M. A., Jeong, S. C., Gunduz, I., Tucker, D. M., Buss, G. R., and Tolin, S. A. 2008. Pyramiding of *Soybean mosaic virus* resistance genes by marker-assisted selection. Crop Sci. 48:517-526.

Shakiba, E., Chen, P., Gergerich, R., Li, S., Dombek, D., and Shi, A. 2012. Reactions of commercial soybean cultivars from the mid South to *Soybean mosaic virus*. Crop Sci. 52:1990-1997.

Shakiba, E., Chen, P., Shi, A., Li, D., Dong, D., and Brye, K. 2012. Two novel alleles for resistance to *Soybean mosaic virus* in PI 399091 and PI 61947. Crop Sci. 52:2587-2594.

Shi, A., Chen, P., Li, D., Zheng, C., Zhang, B., and Hou, A. 2009. Pyramiding multiple genes for resistance to *Soybean mosaic virus* in soybean using molecular markers. Mol. Breed. 23:113-124.

Steinlage, T. A., Hill, J. H., and Nutter, F. W., Jr. 2002. Temporal and spatial spread of *Soybean mosaic virus* (SMV) in soybeans transformed with the coat protein gene of SMV. Phytopathology 92:478-486.

Takahashi, K., Tanaka, T., Wataru, I., and Tsuda, T. 1980. Studies on virus diseases and causal viruses of soybean in Japan. Bull. Tohoku Nat. Agric. Exp. Stn. 62:1-130.

Wang, L., Eggenberger, A., Hill, J., and Bogdanove, A. J. 2006. *Pseudomonas syringae* effector *avrB* confers soybean cultivar-specific avirulence on *Soybean mosaic virus* adapted for transgene expression but effector *avrPto* does not. Mol. Plant-Microbe Interact. 19:304-312.

Wang, Y., Hobbs, H. A., Hill, C. B., Domier, L. L., Hartman, G. L., and Nelson, R. L. 2005. Evaluation of ancestral lines of U.S. soybean cultivars for resistance to four soybean viruses. Crop Sci. 45:639-644.

Wen, R.-H., Khatabi, B., Ashfield, T., Saghai Maroof, M. A., and Hajimorad, M. R. 2013. The HC-Pro and P3 cistrons of an avirulent *Soybean mosaic virus* are recognized by different resistance genes at the complex *Rsv1* locus. Virology 26:203-215.

Xu, Z., Polston, J. E., and Goodman, R. M. 1986. Identification of soybean mosaic, southern bean mosaic, and tobacco ringspot viruses from soybean in the People's Republic of China. Ann. Appl. Biol. 108:51-57.

Yang, Y., Lin, J., Zheng, G., Zhang, M., and Zhi, H. 2014. Recombinant soybean mosaic virus is prevalent in Chinese soybean fields. Arch. Virol. 159:1793-1796.

Zhang, C., Hajimorad, M. R., Eggenberger, A. L., Tsang, S., Whitham, S., and Hill, J. H. 2009. Cytoplasmic inclusion cistron of *Soybean mosaic virus* serves as a virulence determinant on *Rsv3*-genotype soybean and a symptom determinant. Virology 391:240-248.

Zheng, C., Chen, P., and Gergerich, R. 2005. Characterization of resistance to *Soybean mosaic virus* in diverse soybean germplasm. Crop Sci. 45:2503-2509.

(Prepared by J. H. Hill)

Secoviridae

Comovirus

Bean pod mottle virus

Bean pod mottle virus (BPMV) is transmitted by multiple species of beetles and causes mottling of soybean leaves and discoloration of soybean seeds. BPMV was first noted on soybean in 1958 in the U.S. states of Arkansas, North Carolina, and Virginia, and it has since been reported as widespread in many of the soybean-growing areas in North America. Surveys of soybean fields in Arkansas, Illinois, Iowa, Kentucky, Nebraska, North Carolina, South Dakota, and Wisconsin have found the virus present at incidences ranging from 0 to 100%, depending on the abundance of vector beetles. BPMV was first reported in Canada in 2002 and in Iran in 2005.

Yield losses from BPMV range from 3 to 52% based on studies done in the United States. The maximum losses from BPMV infection occur when the plants are infected at the seedling stage.

Symptoms and Signs

Symptom development is most obvious and yields are lowest in plants infected with *Bean pod mottle virus* early in the growing season or during periods of rapid plant growth and cool conditions. Young leaves in the upper canopy exhibit green to yellow mottling (Fig. 179). Severe strains of BPMV may cause

Fig. 179. Green to yellow mottling of upper leaves, caused by *Bean pod mottle virus*. (Courtesy G. L. Hartman)

puckering and distortion of leaves. Under cool greenhouse conditions, leaves mechanically inoculated with some strains can develop chlorotic mottling. Symptoms are masked during periods of high temperatures and not observed on plants after pod set. Leaves of infected plants that are water stressed are less turgid than leaves of noninfected water-stressed plants, and under these conditions, symptom expression is more detrimental in the reproductive stages than in the vegetative stage, with pod formation being reduced by as much as 40%.

BPMV alone can cause necrosis of new terminal leaf growth. However, it does not cause the stem tip to curve or become extremely brittle, as it does in combination with *Soybean mosaic virus* (SMV) or as occurs with infection by *Tobacco ringspot virus, Tobacco streak virus,* or SMV singly. BPMV and SMV interact synergistically in mixed infections in soybean to cause more severe symptoms and greater yield losses than result from a single infection with either virus.

Infection with BPMV decreases pod formation and reduces seed size, weight, and number. It also predisposes soybean plants to infection with *Diaporthe longicolla,* a fungus that can infect soybean seed and reduce seed quality.

Causal Agent

Bean pod mottle virus has small, isometric particles approximately 28 nm in diameter and a bipartite, single-stranded RNA genome. Both RNA species are required for infection. Purified virus preparations contain three components—top, middle, and bottom—and the middle and bottom components are required for infectivity. The single-stranded RNAs of the middle and bottom components are 3.7 and 6.0 kb long, respectively. The top components consist of empty virus particles that do not contain nucleic acid.

BPMV is strongly immunogenic, and polyclonal antibodies are used routinely in enzyme-linked immunosorbent assay (ELISA) to detect the virus in soybean leaves, soybean seeds, and other herbaceous hosts, as well as in *Cerotoma trifurcata,* which is an important vector (see Part II, the section Bean Leaf Beetle). The concentration of BPMV in the leaves of infected soybean plants declines over time but can still be readily detected by ELISA and reverse-transcriptase polymerase chain reaction (RT-PCR) in field-collected soybean tissue. BPMV isolates have been divided into two subgroups based on nucleotide sequence variation and nucleic acid hybridization assays. Reassortants between the RNAs of the two subgroups typically produce more severe symptoms than viruses from either subgroup.

The RNA genome of BPMV was modified to silence the expression of soybean genes. Thus, it is a valuable tool for gene identification and functional genomics in soybean.

Life Cycle and Epidemiology

Bean pod mottle virus is readily sap transmissible and has been reported to be seed transmitted in soybean at very low levels. The host range of BPMV is largely restricted to soybean and other legumes.

Leaf-feeding beetles (order Coleoptera) of the families Chrysomelidae, Coccinellidae, and Meloidae transmit BPMV in nature. The most important vector in the field, *C. trifurcata,* transmits BPMV efficiently for several days after acquiring it from an infected plant. The incidence of soybean plants infected with BPMV often correlates with the population of *C. trifurcata.* In addition to *C. trifurcata,* beetles reported to be vectors of BPMV include *Colaspis brunnea, C. lata, Diabrotica balteata, D. undecimpunctata howardi, Epicauta vittata, Epilachna varivestis, Odontota horni,* and *Popillia japonica.* The relationship of BPMV with beetle vectors in general is typical of beetle-transmitted viruses: namely, the beetle acquires the virus at the initiation of feeding and can transmit it immediately, without a latent period.

Three sources of BPMV serve as the origins of primary inoculum in newly planted fields: virus-infected seeds, perennial weeds, and overwintering beetles. The perennial weeds showy tick trefoil (*Desmodium canadense*) and panicled tick trefoil (*D. paniculatum*) are naturally infected with BPMV. Because *C. trifurcata* can transmit BPMV from panicled tick trefoil to soybean, this species is purported to serve as a reservoir of the virus. Bean leaf beetles overwinter as adults in the borders of fields and along the edges of woods and may only rarely retain their ability to transmit BPMV during this dormant phase. When the adults become active in the spring, they are known to transmit the virus to newly emerging soybean plants.

Management

Resistance to BPMV has not been found in cultivated soybean, but it has been identified in the wild annual relative of soybean. Interspecific crosses have been produced between soybean and its wild annual relative to introduce immunity to BPMV into commercial cultivars. Differences in tolerance to BPMV infection have been reported, in which some soybean lines have lower yield losses than others when infected with BPMV. In addition, transgenic soybean plants that express the BPMV coat protein gene have been shown to resist infection from this virus.

Another recommendation for management of *Bean pod mottle virus* is to use herbicides to destroy alternative host plants. The use of insecticides to control beetle vectors has not always provided consistent protection from BPMV infection, but studies have shown that applying insecticides at the appropriate time will reduce the population of beetle vectors and provide a limited decrease in virus incidence. Seed treatments have been shown to reduce the population of overwintering adults but not necessarily the next generation of beetles.

Another management option is to alter planting dates. Typically, overwintering adults tend to congregate in early planted fields, so planting later may reduce the beetle population.

Selected References

Bradshaw, J. D., Rice, M. E., and Hill, J. H. 2008. Evaluation of management strategies for bean leaf beetles (Coleoptera: Chrysomelidae) and *Bean pod mottle virus* (Comoviridae) in soybean. J. Econ. Entomol. 101:1211-1227.

Ghabrial, S. A., and Schultz, F. J. 1983. Serological detection of bean pod mottle virus in bean leaf beetles. Phytopathology 73:480-483.

Giesler, L. J., Ghabrial, S. A., Hunt, T. E., and Hill, J. H. 2002. *Bean pod mottle virus:* A threat to U.S. soybean production. Plant Dis. 86:1280-1289.

Gu, H.-C., Clark, A. J., de Sá, P. B., Pfeiffer, T. W., Tolin, S., and Ghabrial, S. A. 2002. Diversity among isolates of *Bean pod mottle virus.* Phytopathology 92:446-452.

Hill, J. H., Koval, N. C., Gaska, J. M., and Grau, C. R. 2007. Identification of field tolerance to *Bean pod mottle* and *Soybean mosaic viruses* in soybean. Crop Sci. 47:212-218.

Hopkins, J. D., and Mueller, A. J. 1984. Effect of bean pod mottle virus on soybean yield. J. Econ. Entomol. 77:943-947.

Lin, M. T., and Hill, J. H. 1983. Bean pod mottle virus: Occurrence in Nebraska and seed transmission in soybeans. Plant Dis. 67:230-233.

Michelutti, R., Tu, J. C., Hunt, D. W. A., Gagnier, D., Anderson, T. R., and Welacky, T. W. 2002. First report of *Bean pod mottle virus* in soybean in Canada. Plant Dis. 86:330.

Patel, V. C., and Pitre, H. N. 1971. Transmission of bean pod mottle virus in soybean by the striped blister beetle, *Epicauta vittata.* Plant Dis. Rep. 55:628-629.

Pedersen, P., Grau, C., Cullen, E., Koval, N., and Hill, J. H. 2007. Potential for integrated management of soybean virus disease. Plant Dis. 91:1255-1259.

Ross, J. P. 1969. Effect of time and sequence of inoculation of soybeans with soybean mosaic and bean pod mottle viruses on yields and seed characters. Phytopathology 59:1404-1408.

Scott, H. A., van Scyoc, J. V., and van Scyoc, C. E. 1974. Reactions of *Glycine* spp. to bean pod mottle virus. Plant Dis. Rep. 58:191-192.
Shahraeen, N., Ghotbi, T., Salati, M., and Sahandi, A. 2005. First report of *Bean pod mottle virus* in soybean in Iran. Plant Dis. 89:775.
Srinivasa Reddy, M. S., Ghabrial, S. A., Redmond, C. T., Dinkins, R. D., and Collins, G. B. 2001. Resistance to *Bean pod mottle virus* in transgenic soybean lines expressing the capsid polyprotein. Phytopathology 91:831-838.
Zhang, C. Q., and Ghabrial, S. A. 2006. Development of *Bean pod mottle virus*-based vectors for stable protein expression and sequence-specific virus-induced gene silencing in soybean. Virology 344:401-411.
Zheng, C. M., Chen, P. Y., Hymowitz, T., and Gergerich, R. 2005. Evaluation of *Glycine* species for resistance to *Bean pod mottle virus*. Crop Prot. 24:49-56.
Ziems, A. D., Giesler, L. J., Graef, G. L., Redinbaugh, M. G., Vacha, J. L., Berry, S., Madden, L. V., and Dorrance, A. E. 2007. Response of soybean cultivars to *Bean pod mottle virus* infection. Plant Dis. 91:719-726.

(Prepared by R. C. Gergerich; revised by L. L. Domier)

Cowpea severe mosaic virus

Cowpea severe mosaic virus (CPSMV) has been isolated from soybean in Brazil, Puerto Rico, and Trinidad and in limited areas within the United States where cowpea is grown. In Brazil, CPSMV is highly prevalent in cowpea and thus an economic constraint to production; CSMV rarely seems to constrain soybean production, however.

Symptoms and Signs

Soybean plants infected with *Cowpea severe mosaic virus* display severe mosaic and stunting.

Causal Agent

CPSMV particles are isometric and approximately 25 nm in diameter. Isolates from different geographic regions are serologically related to each other and often weakly related to other comoviruses. CPSMV is most closely related to *Bean pod mottle virus*. CPSMV has two RNAs of 6.0 kb and 3.7 kb in length that are individually associated with capsids formed of 43 and 23 kDa proteins, respectively.

Disease Cycle and Epidemiology

Cowpea severe mosaic virus is transmitted in nature by numerous species of leaf-feeding chrysomelid beetles, especially *Cerotoma arcuata* in Brazil. The host range of CPSMV is restricted to legumes, yet it is wider than that of *Cowpea mosaic virus*. Cowpea is a diagnostic and propagation host, and lambsquarters (*Chenopodium amaranticolor*) is a local lesion host. CPSMV is seed transmitted in cowpea but not in soybean and is likely to occur only in fields grown near infected cowpea.

Management

Growers should avoid planting soybean near cowpea fields in which CPSMV is known to be a problem.

Selected References

Abreu, E. F. M., Tinoco, M. L. P., Andrade, E. D. and Aragão, F. J. L. 2012. Diversity among isolates of cowpea severe mosaic virus infecting cowpeas in northeastern Brazil. Genet. Mol. Res. 11:3146-3153.
Anjos, J. R. N., and Lin, M. T. 1984. Bud blight of soybeans caused by cowpea severe mosaic virus in central Brazil. Plant Dis. 68:405-407.
Bertacini, P. V., Almeida, A. M. R., Lima, J. A. A., and Chagas, C. M. 1998. Biological and physicochemical properties of cowpea severe mosaic comovirus isolated from soybean in the State of Paraná. Braz. Arch. Biol. Technol. 41:409-415.
Beserra, J. E., Jr., Andrade, E. C., Camarço, R. F. R. A., Nascimento, A. K. Q., and Lima, J. A. A. 2011. Sequence variability in the coat protein gene of *Cowpea severe mosaic virus* isolates from northeastern Brazil. Trop. Plant Pathol. 36:121-124.
Camarço, R. F. R. A., Nascimento, A. K., Andrade, E. C., and Lima, J. A. A. 2009. Biological, serological and molecular comparison between isolates of *Cowpea severe mosaic virus*. Trop. Plant Pathol. 34:239-244.
Cruz, A. R. R., and Aragão, F. J. L. 2013. RNAi-based enhanced resistance to *Cowpea severe mosaic virus* and *Cowpea aphid-borne mosaic virus* in transgenic cowpea. Plant Pathol. 63:831-837.
Lima, J. A. A., da Silva, A. K. F., Aragão, M. L., Ferreira, N. R. A., and Teófilo, E. M. 2011. Simple and multiple resistances to viruses in cowpea genotypes. Pesq. Agropec. Bras. 46:1432-1438.
Lin, M. T., Anjos, J. R. N., and Rios, G. P. 1980. Agrupamento sorológico de 14 isolados do "cowpea mosaic virus–Arkansas group" obtidos no Brasil central. (In Portuguese.) Fitopatol. Bras. 5:418-419.
McLaughlin, M. R., Thongmeearkom, P., Goodman, R. M., Milbrath, G. M., Ries, S. M., and Royse, D. J. 1978. Isolation and beetle transmission of cowpea mosaic virus (severe subgroup) from *Desmodium canescens* and soybeans in Illinois. Plant Dis. Rep. 62:1069-1073.

(Prepared by S. A. Tolin)

Other Comoviruses

Legumes are the primary hosts of 11 of the 15 members assigned to the genus *Comovirus*. These viruses occur globally, but individual species are somewhat geographically specific. Although *Bean pod mottle virus* (BPMV) and *Cowpea severe mosaic virus* (CPSMV) are often reported to naturally infect soybean in the United States and in Brazil, respectively, several other comoviruses can infect soybean and are potentially important pathogens (see Table 8).

Broad bean true mosaic virus (BBTMV) is severe in broad bean and pea and occurs in Europe and northern Africa. In Europe, BBTMV is commonly found together with *Broad bean stain virus* (BBSV) and *Red clover mottle virus* (RCMV) infecting clover. *Pea green mottle virus* (PGMV), from Czechoslovakia, is taxonomically similar to BBSV. *Cowpea mosaic virus* (CPMV), first described in 1924, clusters with BBTMV and is distinct from CPSMV. CPMV has been reported mainly in Iran, Kenya, Nigeria, and the Philippines, and despite its being seedborne in cowpea, it has not been reported from North America or South America.

Quail pea mosaic virus (QPMV) was first isolated from *Strophostyles helvula,* a wild legume in Arkansas, and it is found occasionally in soybean. It causes symptoms indistinguishable from those produced by BPMV, to which it is serologically related. Isolates of Bean curly dwarf mosaic virus from El Salvador and Costa Rica belong to a strain of QPMV.

Glycine mosaic virus (GMV) was isolated from wild *Glycine* species in Australia but has never been found in cultivated soybean. GMV produces vein yellowing and mosaic in inoculated soybean plants.

Pea mild mottle virus (PMMV) is from New Zealand and can also infect soybean.

Selected References

Petrzik, K., and Koloniuk, I. 2010. Emerging viruses in the genus *Comovirus*. Virus Genes 40:290-292.
Valverde, R. A., and Fulton, J. F. 1996. Comoviruses: Identification and diseases caused. Pages 17-33 in: The Plant Viruses: Polyhedral Virions and Bipartite RNA Genomes. B. D. Harrison and A. F. Murant, eds. Plenum Press, New York.

(Prepared by S. A. Tolin)

Nepovirus

Cycas necrotic stunt virus (Soybean mild mosaic strain)

The virus now known as *Cycas necrotic stunt virus* (CNSV) was first reported from Japan in 1974 and given the name "Soybean chlorotic spot virus." The name of the virus was subsequently changed to "Soybean mild mosaic virus" (SMMV). However, nucleotide sequence analysis has shown this virus to be a strain of CNSV, characterized from a gymnosperm. CNSV is a member of subgroup B of the genus *Nepovirus,* which includes *Beet ringspot virus*, *Grapevine chrome mosaic virus,* and *Tomato black ring virus.*

The soybean mild mosaic strain of CNSV has a wide host range. It infects more than 70 species of plants from 14 families.

Symptoms and Signs

Soybean plants infected with CNSV often display chlorotic spots and mild mottling of foliage.

Causal Agent

CNSV particles are icosahedral and 26–27 nm in diameter. Each particle individually encapsidates the two genomic single-stranded RNAs of 4.7 and 7.5 kb. No inclusions are evident in infected plants.

Disease Cycle and Epidemiology

The soybean mild mosaic strain of CNSV is transmitted mechanically by inoculation and through seed transmission at rates of 22–70%, depending on the cultivar. Nematode transmission is likely but has not been demonstrated.

Strains of CNSV isolated from sago palm (*Cycas revoluta*) did not infect soybean, whereas isolates from gladiola infected soybean and peanut systemically.

Management

Managing the nematode vectors and avoiding the planting of infected seeds may provide some protection against CNSV. No sources of resistance have been reported.

Selected References

Hanada, K., Inoue, T., Takahashi, K., and Iwanami, T. 2008. Soybean mild mosaic virus is a strain of *Cycas necrotic stunt virus.* Jpn. J. Phytopathol. 74:223.

Kusunoki, M., Hanada, K., Iwaki, M., Chang, M. U., Doi, Y., and Yora, K. 1986. Cycas necrotic stunt virus, a new member of nepoviruses found in *Cycas revoluta;* Host range, purification, serology and some properties. Ann. Phytopathol. Soc. Jpn. 52:302-311.

Takahashi, K., Tanaka, T., and Tsuda, Y. 1974. Soybean mild mosaic virus. Ann. Phytopathol. Soc. Jpn. 40:103.

(Prepared by S. A. Tolin and L. L. Domier)

Soybean severe stunt virus

Soybean severe stunt virus (SSSV) was first described in soybean in 1988 in the U.S. state of Delaware. The disease it causes—soybean severe stunt—may have been present in Delaware since 1964 on soybean plants exhibiting symptoms similar to those induced by the bud blight strain of *Tobacco ringspot virus* (TRSV). Severely stunted plants that have thick, dark-green leaves and produce little or no yield have been observed for many years in soybean plantings near Millsboro, Delaware, in coastal Sussex County. In fields in which the disease occurs, yield losses on individual plants can be 100%. The virus has not been confirmed outside Sussex County (as of March 2015).

SSSV, a putative nepovirus, is not recognized as a distinct virus species by the International Committee on Taxonomy of Viruses (ICTV).

Symptoms and Signs

Soybean plants infected with SSSV before epicotyl emergence in the field or via mechanical inoculation in vitro when roots are less than 5 cm long become stunted (Fig. 180). Symptoms occur on the first true leaves. Infected plants also have shortened internodes, resulting in severe stunting, as well as thickened, dark-green, mottled leaves; superficial stem lesions; and reduced numbers of flowers, pods, and viable seeds. A reduction in seedling emergence and stand establishment has been associated with SSSV both in the field and in greenhouse inoculation studies.

In the field, soybean severe stunt disease occurs in localized, circular, or elliptical areas that range from 10 m to several hundred meters in diameter. The disease is observed in many of the same fields year after year—particularly those that have been continuously cropped with susceptible soybean cultivars (e.g., Essex). In the most severe cases, the epicenter of the disease in the field has few if any soybean seedlings, and symptomatic plants can be found at the margins. Soybean severe stunt disease can be lethal when highly susceptible soybean cultivars are infected at an early age.

Causal Agent

Soybean severe stunt virus virions are isometric particles 29–30 nm in diameter, but virions are unstable and difficult to purify. Nucleic acid analyses have not been reported.

Virus-specific antisera are not available. In one-way serological tests, SSSV failed to react to antiserum to all the nepoviruses known to affect soybean and those occurring in the United States on other crops, including *Arabis mosaic virus, Grapevine fanleaf virus, Peach rosette mosaic virus, Strawberry latent ringspot virus,* TRSV, and *Tomato ringspot virus.* SSSV also failed to react to antiserum to other ilarviruses, including *Asparagus virus 2* and *Prunus necrotic ringspot virus.*

Disease Cycle and Epidemiology

SSSV is transmitted through soil in association with the dagger nematode *Xiphinema americanum.* The virus is not believed to be seed transmitted. SSSV has a limited host range and has been mechanically transmitted by inoculation to some soybean cultivars (including Essex), resulting in systemic infection.

Management

The occurrence of soybean severe stunt disease is most severe in highly susceptible cultivars planted from mid-May to

Fig. 180. Stunted growth of plants, associated with Soybean severe stunt virus. (Courtesy T. A. Evans)

early June, but the diseases rarely develops on soybean planted after July 1 in the United States. Cultivars Chesapeake, Cisne, Defiance, Delsoy 4710, and Stine 4650 (all from maturity group IV) have been shown resistant to SSSV and produced commercially acceptable yields. These results have been confirmed by an in vitro root inoculation assay.

Delsoy 4710 is the only SSSV-resistant cultivar that is also resistant to *Heterodera glycines,* the soybean cyst nematode, which is a common pest in areas in which this virus has been observed. Crop rotation with nonhosts, such as maize and sorghum, has reduced the dagger nematode population and the level of disease severity and is therefore the only cost effective means of managing SSSV.

Selected References

International Committee on Taxonomy of Viruses (ICTV). 2014. Virus Taxonomy. ICTV, Virology Division–IUMS. Available online at www.ictvonline.org

Mulrooney, R. P., Evans, T. A., and Taylor, R. W. 1997. Evaluation of selected soybean cultivars for resistance to soybean severe stunt virus, 1996. Biol. Cult. Tests 12:85.

Mulrooney, R. P., Evans, T. A., Taylor, R. W., and Uniatowski, R. 1997. Evaluation of selected soybean cultivars for resistance to soybean severe stunt virus, 1995. Biol. Cult. Tests 12:84.

Weldekidan, T., Evans, T. A., Carroll, R. B. and Mulrooney, R. P. 1992. Etiology of soybean severe stunt and some properties of the causal virus. Plant Dis. 76:747-750.

(Prepared by T. A. Evans; revised by L. L. Domier)

Tobacco ringspot virus

Tobacco ringspot virus (TRSV) is the type member of the genus *Nepovirus* in the *Comovirinae* subfamily of plant viruses. TRSV is in subgroup A of the nepoviruses, which includes *Arabis mosaic virus, Grapevine fanleaf virus,* and *Raspberry ringspot virus.* TRSV was first described in the United States in 1941 and has since been reported in Australia, Canada, China, and Russia.

TRSV infection can reduce soybean yields by 25–100%. In general, losses are greatest when young plants are infected or when seeds with a high percentage of TRSV are sown. Infected plants have lower yields, which results from reduced pod set and seed formation. Seeds from infected plants have increased total protein and decreased oil content.

Infection of soybean plants with both TRSV and *Soybean mosaic virus* (SMV) results in a synergism that produces disease symptoms more severe than those produced by either virus alone.

Symptoms and Signs

Plants infected with *Tobacco ringspot virus* at less than 5 weeks old become stunted, but the stunting is not evident when plants are grown at above 25°C in a greenhouse. The most striking symptom of TRSV infection is bud blight, indicated by the curving of the terminal bud to form a crook (Fig. 181). Later, other buds on the plant become brown, necrotic, and brittle, and they may fall off at the slightest touch. Adventitious leaf and floral buds may proliferate excessively.

The pith of stems and branches may develop a brown discoloration—first near the nodes and then throughout the stem. Brown streaks are occasionally observed on the petioles and large leaf veins. The petioles of the youngest trifoliolate leaves are often shortened and thickened and may be curved, distorting the shoot tips. Leaflets are dwarfed and tend to cup or roll, and the blades become rugose and bronzed. Ringspots may form on the leaves of some infected soybean genotypes (Fig. 182).

Pods are generally underdeveloped or aborted. Those that set before infection often develop dark blotches; these pods generally do not produce viable seeds and drop early.

Maturity is delayed in infected plants, which remain green and often stunted until harvested or killed by frost. TRSV infection significantly reduces root and nodule growth by *Rhizobium* spp. (Fig. 183). Nodulation is suppressed until plants

Fig. 181. Bud blight, caused by *Tobacco ringspot virus*. (Courtesy G. L. Hartman)

Fig. 182. Ringspots on leaves of soybean plants infected with *Tobacco ringspot virus*. (Courtesy L. L. Domier)

Fig. 183. Severe stunting of roots, caused by *Tobacco ringspot virus* (center and right), and healthy roots (left). (Courtesy O. W. Barnett)

are approximately 40 days old, and when it does begin, nodule fresh weight and leghemoglobin content are both reduced.

Causal Agent

The genome of *Tobacco ringspot virus* is bipartite and consists of two single-stranded, positive-sense, polyadenylated RNA molecules of 7.5 kb (RNA1) and 3.9 kb (RNA2); they are encapsidated separately in icosahedral particles 28 nm in diameter. Both types of particles are required for infection. The purified genomic RNA is infectious by mechanical inoculation.

Some strains of TRSV have satellite RNAs associated with their genomes. Satellite RNAs of TRSV either decrease or have little effect on the severity of symptoms caused by the virus. TRSV is moderately immunogenic. Serological tests have been used for diagnosis and to determine virus concentrations in infected plants.

Disease Cycle and Epidemiology

Tobacco ringspot virus causes systemic infection in susceptible cultivars, moving from infected leaves to stem tips and into the roots. Movement from the roots to the leaves is uncommon. Movement occurs most quickly at high temperatures, when plants grow more rapidly than normal, and movement from young leaves is greater than from maturing leaves. TRSV is easily sap transmissible. Circumstantial evidence suggests that root grafts from infected plants to healthy plants may transmit the virus.

No efficient arthropod vector of TRSV has been discovered. Nymphs of *Thrips tabaci* transmit TRSV to soybean at a low level of efficiency; they appear to retain the virus for at least 14 days after acquisition. The grasshopper *Melanoplus differentialis* is capable of transmission at a low rate (2–3%) after a single feeding. If feeding is extended to 30 sec, then no transmission occurs. Thus, transmission by grasshoppers in nature is questionable because of their voracious eating habits.

The dagger nematode *Xiphinema americanum* is also a vector of TRSV, but its efficiency in transmitting the virus to soybean is low. Even when it does occur, nematode transmission of TRSV to the roots of plants may be of no significance, because the infection generally remains confined to the roots.

Seed transmission is the most important mode of long-range dissemination and carryover of TRSV from season to season. Systemically infected plants often produce infected seeds, which give rise to diseased seedlings. The extent of seed transmission depends on when infection occurs; plants infected before bloom produce few or no seeds. Because all infected seeds come from plants infected before bloom, the maximum amount of viruliferous seeds possible is extremely small—barely enough to perpetuate the virus in a few widely scattered plants. The proportion of diseased seeds is greater when the time between infection and flowering is long. The locations of pods on the plant and of seeds in the pod does not have a significant effect on the frequency of seed infection. Less seed infection occurs during prolonged periods of cool temperatures. The virus remains viable in seeds for at least 5 years and is carried in the periplasm and the embryo.

TRSV induces local or systemic symptoms or both in a wide range of plants, and some symptomless carriers are known to act as reservoirs of the virus. In the absence of seedborne inoculum, the disease first appears at the edge of a soybean field and advances inward as the season progresses. The rate of spread depends on the crops and the weeds growing next to the field and probably on aerial vector populations, as well. More infections occur in fields next to pastures, and fewer infections occur in fields next to maize fields.

Management

A few soybean cultivars have been noted to have resistance to some strains of *Tobacco ringspot virus*—for instance, cultivar Young. In addition, TRSV resistance has been detected in soybean plant introductions (PIs) 92713 and 154194 and in PI 407287 of the annual ancestor of soybean (*Glycine soja*).

Virus-free soybean seeds should be planted in commercial fields. In addition, growers are advised to avoid fields infested with dagger nematodes or to treat them with an appropriate nematicide, if feasible and appropriate. Because the rate of spread of TRSV depends on the crops and weeds growing next to the soybean field and probably on arthropod vector populations, as well, growers should plant soybean next to fields of maize, rather than pastures.

Selected References

Allington, W. B. 1946. Bud blight of soybean caused by the tobacco ring-spot virus. Phytopathology 36:319-322.

Athow, K. L., and Laviolette, F. A. 1961. The relation of seed-transmitted tobacco ringspot virus to soybean yield. Phytopathology 51:341-342.

Bergeson, G. B., Athow, K. L., Laviolette, F. A., and Thomasine, S. M. 1964. Transmission, movement, and vector relationships of tobacco ringspot virus in soybean. Phytopathology 54:723-728.

Buzayan, J. M., Gerlach, W. L., Bruening, G., Keese, P., and Gould, A. R. 1986. Nucleotide sequence of satellite tobacco ringspot virus RNA and its relationship to multimeric forms. Virology 151:186-199.

Crittenden, H. W., Hastings, K. M., and Moore, G. M. 1966. Soybean losses caused by tobacco ringspot virus. Plant Dis. Rep. 50:910-913.

Fasoula, V. A., Harris, D. K., Bailey, M. A., Phillips, D. V., and Boerma, H. R. 2003. Identification, mapping, and confirmation of a soybean gene for bud blight resistance. Crop Sci. 43:1754-1759.

Gergerich, R., Asher, J. H., Jr., and Ramsdell, D. C. 1983. A comparison of some serological and biological properties of seven isolates of tobacco ringspot virus. J. Phytopathol. 107:289-300.

Messieha, M. 1969. Transmission of tobacco ringspot virus by thrips. Phytopathology 59:943-945.

Orellana, R. G. 1981. Resistance to bud blight in introductions from the germ plasm of wild soybean. Plant Dis. 65:594-595.

Tu, J. C. 1986. Strains of tobacco ringspot virus isolated from soybean (*Glycine max*) in southwestern Ontario (Canada). Can. J. Plant Sci. 66:491-498.

(Prepared by J. W. Demski and C. W. Kuhn; revised by G. L. Hartman and L. L. Domier)

Other Nepoviruses

Most nepoviruses have extremely wide host ranges and often occur in woody and perennial plants, although several also occur in alfalfa, crimson clover (*Trifolium incarnatum*), and white clover (*T. repens*). Nepoviruses that have been tested on soybean are usually capable of inducing symptoms resembling those caused by *Tobacco ringspot virus* (TRSV). Only *Tomato ringspot virus* (ToRSV) has been detected in cultivated soybean fields. Like TRSV, ToRSV is seedborne in soybean; it is apparently self-limiting, however, because it has been detected on only two occasions in the eastern United States.

Selected Reference

Murant, A. F., Jones, A. T., Martelli, G. P., and Stace-Smith, R. 1996. Nepoviruses: General properties, diseases, and virus identification. Pages 99-137 in: The Plant Viruses: Polyhedral Virions and Bipartite RNA Genomes. B. D. Harrison and A. F. Murant, eds. Plenum Press, New York.

(Prepared by S. A. Tolin)

Tombusviridae
Carmovirus
Soybean yellow mottle mosaic virus and Blackgram mottle virus

Blackgram mottle virus (BMoV) was first observed in 1968 on blackgram (*Vigna mungo*) in India, and in 1982, it was reported to cause a mild mottle disease of soybean in Thailand. BMoV is not recognized as a distinct species by the International Committee on Taxonomy of Viruses (ICTV).

Soybean yellow mottle mosaic virus (SYMMV) was first reported from soybean fields in South Korea in 2003. Six years later, it was reported from a research plot in the U.S. state of Mississippi.

Both BMoV and SYMMV have relatively narrow host ranges and infect plants in only a few plant families. The most systemic infections occur among plants in the Fabaceae.

Symptoms and Signs

Susceptible soybean genotypes infected with Blackgram mottle virus display a systemic mild mottle, and resistant genotypes display local, necrotic lesions. One report lists soybean as a diagnostically nonsusceptible host for BMoV.

Soybean yellow mottle mosaic virus induces veinal necrosis shortly after inoculation, followed by bright-yellow mosaic on systemic leaves. Stunting and reduced growth are evident in older leaves (Fig. 184). Symptoms become more obvious on the lower leaves as plants mature, and newly emerging leaves often show very mild symptoms.

Causal Agents

BMoV and SYMMV both consist of isometric particles approximately 28 nm in diameter; each particle has a single coat protein with a molecular weight of approximately 38 kDa. The genome of SYMMV consists of a positive-sense, single-stranded RNA molecule 4 kb in length.

BMoV is propagated in the bean cultivar Black Valentine and is serologically unrelated to three other beetle-transmitted viruses: *Bean mild mosaic virus* (BMMV), *Cowpea mottle virus* (CPMoV), and *Southern bean mosaic virus*. The nucleotide sequence of SYMMV is most closely related to that of CPMoV. BMMV and CPMoV both infect soybean.

Fig. 184. Light-yellow mosaic on leaves of cultivar Williams 82, caused by infection with *Soybean yellow mottle mosaic virus*. (Courtesy L. L. Domier)

Disease Cycle and Epidemiology

Both Blackgram mottle virus and *Soybean yellow mottle mosaic virus* are transmitted mechanically by inoculation. BMoV is also transmitted through seed and by three species of beetle: *Cerotoma trifurcata, Epilachna varivestis,* and *Monolepta signata*. Biological modes of transmission of SYMMV have not been characterized, but beetle vectors are likely.

Management

There are no reports of managing BMoV and SYMMV. However, reducing the impact of the beetle vectors and avoiding the planting of infected seed will likely reduce the incidences of these viruses.

Selected References

Honda, Y., Iwaki, M., Thongmeearkom, P., Deema, N., and Srithongchai, W. 1982. Blackgram mottle virus occurring on mungbean and soybean in Thailand. Jpn. Agric. Res. Q. 16:72-77.

International Committee on Taxonomy of Viruses (ICTV). 2014. Virus Taxonomy. ICTV, Virology Division–IUMS. Available online at www.ictvonline.org

Li, S., Moon, J. S., Lee, S. H., and Domier, L. L. 2009. First report of Soybean yellow mottle mosaic virus in soybean in North America. Plant Dis. 93:1214.

Nam, M., Kim, S. M., Domier, L. L., Koh, S., Moon, J. K., Choi, H. S., Kim, H. G., Moon, J. S., and Lee, S. H. 2009. Nucleotide sequence and genomic organization of a newly identified member of the genus *Carmovirus*, soybean yellow mottle mosaic virus, from soybean. Arch. Virol. 154:1679-1684.

Scott, H. A., and Phatak, H. C. 1979. Properties of blackgram mottle virus. Phytopathology 69:346-348.

(Prepared by J. W. Demski and C. W. Kuhn; revised by S. A. Tolin and L. L. Domier)

Virgaviridae
Tobamovirus
Tobacco mosaic virus

A strain of *Tobacco mosaic virus* (TMV) has been isolated from naturally infected soybean in Yugoslavia and has also been reported in Iran and Romania. The distribution and importance of the soybean strain (TMV-S) is unknown; this strain has not been detected in the United States.

Symptoms and Signs

TMV-S induces very mild vein clearing and mild chlorotic mosaic symptoms on soybean in the greenhouse.

Causal Agent

TMV-S particles are rigid rods, 300 nm in length; however, two or three sizes of smaller particles are present in leaves and purified preparations. Enzyme-linked immunosorbent assay (ELISA) with polyclonal antisera has shown that TMV-S is more closely related to the common strain of TMV than to the legume-infecting strains.

TMV-S causes symptoms similar to those caused by the common strain of TMV in common globe amaranth (*Gomphrena globosa*), jimsonweed (*Datura stramonium*), and tobacco. TMV-S also infects garden pea and red clover (*Trifolium pratense*) but not bean cultivars Bountiful, Pinto, Tendercrop, and Topcrop; cowpea cultivar Blackeye; or tomato cultivars Marglobe and Rutgers.

Disease Cycle and Epidemiology

TMV-S is mechanically transmitted but not seed transmitted in soybean.

Management

No strategies are reported for managing TMV-S, but applying sanitation practices and eliminating weedy hosts may reduce the incidence of the virus.

Selected References

McDaniel, L. L., Maratos, M. L., Goodman, J. E., and Tolin, S. A. 1995. Partial characterization of a soybean strain of tobacco mosaic virus. Plant Dis. 79:206-211.

Taraku, N., and Tolin, S. A. 1986. Properties of a strain of tobacco mosaic virus from soybean in Yugoslavia. Indian J. Virol. 2:188-200.

(Prepared by S. A. Tolin)

Single-Stranded, Ambisense RNA Genomes

Bunyaviridae
Tospovirus

Tospoviruses—including *Tomato spotted wilt virus* (TSWV) and provisionally named Soybean vein necrosis virus (SVNV)—are emerging viruses in many parts of the world. TSWV, which is transmitted by *Thrips setosus,* was first detected in soybean in South Africa in the early 1990s. TSWV was subsequently detected in surveys of soybean fields in Iran in 2004 and in the U.S. state of Georgia in 2006. In Iran and Georgia, incidences of TSWV infection in randomly collected plants ranged from 5.4 to 6.6%. Researchers have concluded that TSWV infections may go undiagnosed because the symptoms they cause are usually very mild.

The economic impact of tospovirus infections on soybean production has not been determined. (For additional yield loss information, see the following section, Soybean vein necrosis virus.)

Selected References

de Oliveira, A. S., Melo, F. L., Inoue-Nagata, A. K., Nagata, T., Kitajima, E. W., and Resende, R. O. 2012. Characterization of bean necrotic mosaic virus: A member of a novel evolutionary lineage within the genus *Tospovirus*. PloS One 7(6):e38634.

Golnaraghi, A. R., Shahraeen, N., Pourrahim, R., Farzadfar, Sh., and Ghasemi, A. 2004. Occurrence and relative incidence of viruses infecting soybeans in Iran. Plant Dis. 88:1069-1074.

Inoue, T., Sakurai, T., Murai, T., and Maeda, T. 2004. Specificity of accumulation and transmission of tomato spotted wilt virus (TSWV) in two genera, *Frankliniella* and *Thrips* (Thysanoptera: Thripidae). Bull. Entomol. Res. 94:501-507.

Nischwitz, C., Mullis, S. W., Gitaitis, R. D., and Csinos, A. S. 2006. First report of *Tomato spotted wilt virus* in soybean (*Glycine max*) in Georgia. Plant Dis. 90:524.

(Prepared by L. L. Domier and S. A. Tolin)

Soybean vein necrosis virus

Soybean vein necrosis virus (SVNV) was first reported in the U.S. state of Tennessee in 2008, and during the following years, it was reported in all soybean-producing areas of North America. The incidence of SVNV in commercial soybean fields and its economic importance are not yet known, although based on initial reports, the virus appears to be widespread in the midwestern soybean production area of the United States.

SVNV is not yet recognized as a distinct virus species by the International Committee on Taxonomy of Viruses (ICTV).

Symptoms and Signs

Symptoms of infection from Soybean vein necrosis virus begin with clearing along the main veins (Fig. 185) and progress to yellowing and then necrosis as the growing season advances (Fig. 186). Because of the absence of well-defined symptoms, distinguishing between noninfected and infected plants may be difficult in the early stage of symptom development. A mild infection causes thread-shaped vein clearing, whereas a severe infection produces purple or dark-brown lesions that expand to fill most of the leaf blade (Fig. 187). Disease symptoms are more evident higher in the canopy, presumably because the virus vector—the soybean thrips, *Neohydatothrips variabilis* (syn. *Sericothrips variabilis*)—prefers to feed on newly emerged leaves.

In the tobacco relative *Nicotiana benthamiana,* which is an experimental systemic host for the virus, SVNV initially in-

Fig. 185. Vein clearing and some interveinal necrosis—early or mild symptoms of Soybean vein necrosis virus. (Courtesy T. K. Herman)

Fig. 186. Yellow to brown patches on soybean leaves, caused by Soybean vein necrosis virus. (Courtesy T. K. Herman)

Fig. 187. Purple lesions on leaves, characteristic of Soybean vein necrosis virus. (Courtesy T. K. Herman)

duces ringspot symptoms followed by systemic chlorosis and rugosity.

Causal Agent

Like other tospoviruses, Soybean vein necrosis virus has a tripartite, ambisense RNA genome. It has low genetic variability and represents a new lineage in the genus—distinct from both the New and the Old World tospoviruses. SVNV shares less than a 60% amino acid sequence identity with the most closely related viruses and is identified primarily by sequence comparison.

SVNV forms a distinct new clade of tospoviruses with the recently described Bean necrotic mosaic virus, isolated from bean in Brazil. Both viruses are transmitted by the *Frankliniella* spp. of thrips, but common hosts have not yet been reported.

Disease Cycle and Epidemiology

SVNV is transmitted by *N. variabilis,* which is found in North America. A common weed found in soybean fields, ivyleaf morning-glory (*Ipomoea hederacea*), is a natural, symptomless host of the virus and may function as a virus reservoir. In addition to *N. benthamiana,* two other legume species—cowpea and mung bean (*Vigna radiata*)—have been experimentally verified as SVNV hosts.

SVNV infection can occur at any time during the growing season, but symptoms usually become most visible after flowering. Cool temperatures favor symptom development, and a mild winter followed by a warm, dry spring may promote vector proliferation. SVNV is not systemic in soybean, and the localized symptoms are believed to demonstrate a resistant reaction of the plant.

Management

Control of the soybean thrips, the natural vector of SVNV, and elimination of ivyleaf morning-glory and other potential hosts are recommended for reducing disease. Further experimentation is needed, however, to validate other management recommendations.

Selected References

International Committee on Taxonomy of Viruses (ICTV). 2014. Virus Taxonomy. ICTV, Virology Division–IUMS. Available online at www.ictvonline.org

Zhou, J., and Tzanetakis, I. E. 2013. Epidemiology of Soybean vein necrosis-associated virus. Phytopathology 103:966-971.

Zhou, J., Kantartzi, S. K., Wen, R.-H., Newman, M., Hajimorad, M. R., Rupe, J., and Tzanetakis, I. 2011. Molecular characterization of a new tospovirus infecting soybean. Virus Genes 43:289-295.

(Prepared by J. Zhou, L. L. Domier, S. Sabanadzovic, and I. E. Tzanetakis)

Double-Stranded DNA Genomes

Caulimoviridae
Caulimovirus and *Soymovirus*
Soybean chlorotic mottle virus and Soybean Putnam virus

The family *Caulimoviridae* currently consists of seven genera—among them, *Caulimovirus* and *Soymovirus. Soybean chlorotic mottle virus* (SbCMV) is the type species of the genus *Soymovirus,* which also includes a second legume-infecting species, *Peanut chlorotic streak virus,* and two other species. SbCMV was isolated from field-grown soybean with mosaic and stunting symptoms in Japan, but the virus has not been reported outside that country.

Soybean Putnam virus (SPuV) is distinct from SbCMV and more closely related to viruses in the *Caulimovirus* genus than in the *Soymovirus* genus. SPuV was discovered in 2010 in the U.S. state of Ohio during a survey of 24 counties for virus diseases of soybean conducted by high-throughput sequencing. The virus was subsequently detected in soybean field samples from Illinois and North Dakota.

Symptoms and Signs

Field-infected plants with *Soybean chlorotic mottle virus* display chlorotic mottling and mosaic on leaves. Plants are also stunted, and seed production is compromised. Under inoculated conditions in the greenhouse, infected soybean cultivars show vein clearing and chlorosis followed by leaf mottling and curling. Plants are stunted, have small leaves and shortened internodes, and produce vigorous lateral bud growth at later stages of infection.

The symptoms produced by Soybean Putnam virus have not been fully characterized.

Causal Agents

Particles of viruses in the family *Caulimoviridae* are spherical and approximately 50 nm in diameter; they contain circular, double-stranded DNA 7.8–8.2 kb in length. SbCMV and SPuV are considered pararetroviruses (like *Cauliflower mosaic virus*) but are distinct virus species. Inclusion bodies in soybean and bean tissues are frequently ovoid or elliptical and consist of an amorphous, vacuolated, electron-dense matrix.

Disease Cycle and Epidemiology

Soybean chlorotic mottle virus is transmitted mechanically by inoculation but not through seed. Research has shown that it is not transmissible by five aphid species, even though aphids are the expected vectors of caulimoviruses. SbCMV was detected in only a few samples in one field and produced symptoms similar to those produced by *Soybean mosaic virus* (SMV); symptoms of SMV and SbCMV may not be distinguishable.

Only four legumes—cowpea, hyacinth bean (*Lablab purpureus*), kidney bean, and soybean—have been shown (via mechanical inoculation) to be hosts of SbCMV. Hosts of Soybean Putnam virus have not been reported

Management

No specific strategies have been reported for managing these viruses.

Selected References

Fukuoka, H., Ogawa, T., Mitsuhara, I., Iwai, T., Isuzugawa, Y., Tagiri, A., Ugaki, M., Ohshima, M., Yano, H., Murai, N., Niwa, Y., Hibi, T., and Ohashi, Y. 2000. *Agrobacterium*-mediated transformation of monocot and dicot plants using the NCR promoter derived from soybean chlorotic mottle virus. Plant Cell Rep. 19:815-820.

Han, J. P., Domier, L. L., Dorrance, A., and Qu, F. 2012. Complete genome sequence of a novel pararetrovirus isolated from soybean. J. Virol. 86:9555.

Hibi, T., Iwaki, M., Saito, Y., Verver, J., and Goldbach, R. 1986. Double-stranded DNA of soybean chlorotic mottle virus. Ann. Phytopathol. Soc. Jpn. 52:785-792.

Iwaki, M., Isogawa, Y., Tsuzuki, H., and Honda, Y. 1984. Soybean chlorotic mottle, a new caulimovirus on soybean. Plant Dis. 68:1009-1011.

(Prepared by S. A. Tolin)

Nanoviridae

Nanovirus

Milk vetch dwarf virus

Milk vetch dwarf virus (MVDV) was initially classified as a luteovirus based on partial morphology and obligate transmission by aphids. However, it was later shown to be a member of an unusual group of viruses in the family *Nanoviridae,* genus *Nanovirus,* which have genomes consisting of several covalently closed, single-stranded circles of DNA. Other viruses in this genus include *Banana bunchy top virus, Faba bean necrotic yellows virus,* and *Subterranean clover stunt virus.*

MVDV occurs infrequently in soybean in Japan and has a narrow host range. It causes dwarfing along with leaf rolling and yellowing symptoms on soybean. MVDV is transmitted in a persistent manner by four species of aphids—*Aphis craccivora, Aphis gossypii, Acyrthosiphon pisum,* and *Acyrthosiphon solani*—but it is not mechanically transmissible.

Selected References

Inouye, T., Inouye, N., and Mitsuhata, K. 1968. Yellow dwarf of pea and broad-bean caused by milk vetch dwarf virus. Ann. Phytopathol. Soc. Jpn. 34:28-35.

Sano, Y., Wada, W., Hashimoto, Y., Matsumoto, T., and Kojima, M. 1998. Sequences of ten circular ssDNA components associated with the milk vetch dwarf virus genome. J. Gen. Virol. 79:3111-3118.

(Prepared by S. A. Tolin and L. L. Domier)

Single-Stranded DNA Genomes

Geminiviridae

Begomoviruses

Geminiviridae is the second-largest family of plant viruses and includes seven genera, three of which contain legume-infecting viruses: *Begomovirus, Curtovirus,* and *Mastrevirus*. The number of reported infections of soybean plants by begomoviruses has increased dramatically since 2000 for two main reasons: (1) the distribution of highly invasive and often insecticide-resistant biotypes of the principal vector of begomoviruses, the whitefly *Bemisia tabaci,* and (2) the development of polymerase chain reaction (PCR) detection techniques, which permit rapid and reliable identification of novel begomoviruses.

Begomoviruses that infect soybean in Asia include *Cotton leaf curl Kokhran virus, Mungbean yellow mosaic India virus, Mungbean yellow mosaic virus,* and *Soybean crinkle leaf virus.* These viruses have been isolated from naturally infected soybeans in Bangladesh, India, Pakistan, the Philippines, and Thailand. In Africa, the begomovirus *African cassava mosaic virus* has been identified in Nigeria. In North America, *Pepper golden mosaic virus* and *Rhynchosia golden mosaic virus* have been detected in Mexico. In South America, *Bean golden mosaic virus, Okra mottle virus, Sida micrantha mosaic virus,* and *Soybean chlorotic spot virus* have been detected in Brazil. In Argentina, *Sida mottle virus* is the most prevalent begomovirus in soybean, but *Soybean blistering mosaic virus* has also been detected.

Two additional viruses described for soybean are *Soybean mild mottle virus* and Soybean yellow mosaic virus. (The latter virus has not been approved as a species by the International Committee on Taxonomy of Viruses [ICTV].)

Symptoms and Signs

Mungbean yellow mosaic virus (MYMV) initially induces vein yellowing along the small veins in the leaves of systemically infected soybean. Bright-yellow patches develop over time.

In Mexico, begomovirus infections in soybean have been associated with stunted growth and leaf yellowing, curling, and crumpling (Fig. 188). A severe yellow mosaic develops at later stages of infection (Fig. 189).

In Thailand, *Soybean crinkle leaf virus* (SCLV) causes leaves to appear crinkled, curled, cupped, and twisted (Fig. 190), and

Fig. 188. Leaf yellowing and crumpling, caused by infection with a begomovirus. (Reprinted, by permission, from Méndez-Lozano et al., 2006)

veinal enations develop on the undersurfaces (Fig. 191). In the greenhouse, yellow netting of veins is a prominent symptom.

In Brazil, *Soybean chlorotic spot virus* produces chlorotic spots on leaves.

Causal Agents

Begomovirus particles are 18 × 30 nm (diameter × length). Most of the viruses in this genus have bipartite, single-stranded, circular DNA genomes (referred to as "DNA A" and "DNA B") that are approximately 2.5–2.7 kb in size. Loose aggregates of particles can be observed in phloem cell nuclei.

Fig. 189. Bright-yellow patches on leaves infected with *Mungbean yellow mosaic virus*. (Courtesy S. A. Tolin; photographed by S. Beniwal)

Fig. 190. Crinkled, curled, cupped, and twisted leaves, caused by *Soybean crinkle leaf virus*. (Courtesy S. A. Tolin; photographed by M. Iwaki)

Fig. 191. Veinal enations on the undersurface of a leaf, caused by *Soybean crinkle leaf virus*. (Courtesy S. A. Tolin; photographed by M. Iwaki)

Disease Cycle and Epidemiology

Begomoviruses are transmitted by whiteflies in a persistent manner and by grafting (but not mechanically), by aphids, and by seed. Host ranges vary among individual viruses. MYMV has an extensive host range within the Fabaceae but also includes rough cocklebur (*Xanthium strumarium*) and browntop millet (*Urochloa ramose;* syn. *Brachiaria ramosa*). SCLV causes systemic symptoms in soybean and in jimsonweed (*Datura stramonium*), kidney bean, petunia, sickle or wild senna (*Senna tora*), tomato, zinnia, and various species of tobacco (*Nicotiana clevelandii, N. debneyii, N. glutinosa,* and *N. tabacum*). SCLV is not transmissible to several other legumes, including clover, cowpea, and peanut.

In one study, the size and number of nodules on soybean cultivar Bragg was reduced 50% or more on MYMV-infected plants. In another study, SCLV infection caused yield reductions of 15–75% in eight soybean cultivars, decreasing with the increased age of plants at the time of inoculation.

Management

Few soybean cultivars have been identified as being resistant to *Mungbean yellow mosaic virus* and *Mungbean yellow mosaic India virus.*

Selected References

Alabi, O., Kumar, L. P., Mgbechi-Ezeri, J., and Naidu, R. 2010. Two new "legumoviruses" (genus *Begomovirus*) naturally infecting soybean in Nigeria. Arch. Virol. 155:643-656.

Coco, D., Calil, I. P., Brustolini, O. J., Santos, A. A., Inoue-Nagata, A. K., and Fontes, E. P. 2013. Soybean chlorotic spot virus, a novel begomovirus infecting soybean in Brazil. Arch. Virol. 158:457-462.

Dhingra, K. L., and Chenula, V. V. 1985. Effect of yellow mosaic on yield and nodulation of soybean. Indian Phytopathol. 38:248-251.

Fernandes, F. R., Cruz, A. R., Faria, J. C., Zerbini, F. M., and Aragão, F. J. 2009. Three distinct begomoviruses associated with soybean in central Brazil. Arch. Virol. 154:1567-1570.

Girish, K. R., and Usha, R. 2005. Molecular characterization of two soybean-infecting begomoviruses from India and evidence for recombination among legume-infecting begomoviruses from South-East Asia. Virus Res. 108:167-176.

Ilyas, M., Qazi, J., Mansoor, S., and Briddon, R. W. 2009. Molecular characterisation and infectivity of a "Legumovirus" (genus *Begomovirus*: family *Geminiviridae*) infecting the leguminous weed *Rhynchosia minima* in Pakistan. Virus Res. 145:279-284.

International Committee on Taxonomy of Viruses (ICTV). 2014. Virus Taxonomy. ICTV, Virology Division–IUMS. Available online at www.ictvonline.org

Iwaki, M., Thongmeearkom, P., Honda, Y., and Deema, N. 1983. Soybean crinkle leaf: A new whitefly-borne disease of soybean. Plant Dis. 67:546-548.

Méndez-Lozano, J., Quintero-Zamora, E., Barbosa-Jasso, M. P., Leyva-López, N. E., Garzón-Tiznado, J. A., and Argüello-Astorga, G. R. 2006. A begomovirus associated with leaf curling and chlorosis of soybean in Sinaloa, Mexico, is related to *Pepper golden mosaic virus*. Plant Dis. 90:109.

Mgbechi-Ezeri, J. U., Alabi, O. J., Naidu, R. A., and Lava Kumar, P. 2008. First report of the occurrence of *African cassava mosaic virus* in a mosaic disease of soybean in Nigeria. Plant Dis. 92:1709.

Raj, S. K., Khan, M. S., Snehi, S. K., Srivastava, S., and Singh, H. B. 2006. A yellow mosaic disease of soybean in northern India is caused by *Cotton leaf curl Kokhran virus*. Plant Dis. 90:975.

Rodríguez-Pardina, P. E., Zerbini, F. M., and Ducasse, D. A. 2006. Genetic diversity of begomoviruses infecting soybean, bean and associated weeds in northwestern Argentina. Fitopatol. Bras. 31:342-348.

Singh, B. B., and Malick, A. S. 1978. Inheritance of resistance to yellow mosaic in soybean. Indian J. Genet. Plant Breed. 38:258-261.

Varma, A., and Malathi, V. G. 2003. Emerging geminivirus problems: A serious threat to crop production. Ann. Appl. Biol. 142:145-164.

Yadav, R. K., Shukla, R. K., and Chattopadhyay, D. 2009. Soybean cultivar resistant to *Mungbean yellow mosaic India virus* infection induces viral RNA degradation earlier than the susceptible cultivar. Virus Res. 144:89-95.

(Prepared by S. A. Tolin and L. L. Domier)

Seed and Postharvest Pathology

The soybean seed industry is a multibillion-dollar business, and most of the soybean cultivars in production have been derived from highly regulated breeding programs that offer growers both genetically modified and nongenetically modified plants. The International Seed Testing Association (ISTA) provides international rules for seed health testing, including guidelines for comparative testing methods. The description of each method is a fully illustrated, self-contained package that includes the complete details of the test: media preparation, quality assurance information, pathogen identification, and test evaluation. This information is continuously updated as new methods are added following verification.

Many soybean pathogens that infect other parts of the soybean plant have a seed-infection phase. Not all of these pathogens are economically important, but many of them occur worldwide and are therefore significant. For pathogens that are seedborne, infected seeds not only provide a means of survival (often serving as a source of initial inoculum) but also a means of pathogen dissemination.

The pathogen invasion of seeds before harvest can reduce the yield and quality of the seed crop. The discoloration of seeds may limit the acceptance of seed lots, reduce the market grade, and lower the price the grower receives. And while some pathogens produce distinct symptoms on infected seeds, the absence of symptoms does not mean that seeds are pathogen free. More than one pathogen can infect the same seed, resulting in a variety of symptoms. In addition, multiple infections can increase seed susceptibility or preclude other pathogens from colonizing the same seed.

Descriptions of the symptoms on seeds caused by various pathogens appear in the sections on specific diseases throughout Part I of this book. Although most bacterial diseases are seedborne, distinctive names do not identify their seed-infection phases (except for Bacillus seed decay). Many fungal pathogens cause distinctive seed symptoms—for instance, purple seed stain, Phomopsis seed decay, and yeast spot—whereas others do not have distinctive names for the seed disease phases of infection. Viruses are often seedborne, as well, but symptoms of mottling, hilum bleeding, and other seed symptoms related to infection are not usually characteristic of any particular virus.

Selected References

Agarwal, V. K., and Sinclair, J. B. 1996. Principles of Seed Pathology. 2nd ed. CRC Press, Boca Raton, FL.

Barros, G. G., Oviedo, M. S., Ramirez, M. L., and Chulze, S. N. 2011. Safety aspect in soybean food and feed chains: Fungal and mycotoxins contamination. Pages 7-20 in: Soybean—Biochemistry, Chemistry, and Physiology. T.-B. Ng, ed. InTech, Rijeka, Croatia.

Bhattacharya, K., and Raha, S. 2002. Deteriorative changes of maize, groundnut and soybean seeds by fungi in storage. Mycopathologia 155:135-141.

Christensen, C. M., and Meronuck, R. A. 1986. Quality Maintenance in Stored Grains and Seeds. University of Minnesota Press, Minneapolis.

Christensen, C. M., Meronuck, R. A., Steele, J. A., and Behrens, J. C. 1973. Some morphological and chemical characteristics of binburned and fireburned soybeans. Trans. ASAE 18:899-901.

Esper, R. H., Gonçalez, E., Marques, M. O. M., Felicio, R. C., and Felicio, J. D. 2014. Potential of essential oils for protection of grains contaminated by aflatoxin produced by *Aspergillus flavus*. Front. Microbiol. doi:10.3389/fmicb.2014.00269.

Gupta, I. J., Schmitthenner, A. F., and McDonald, M. B. 1993. Effect of storage fungi on seed vigour of soybean. Seed Sci. Technol. 21:581-591.

International Seed Testing Association (ISTA). 2014. Seed health testing. Pages 7-1-7-3 in: International Rules for Seed Testing. ISTA, Bassersdorf, Switzerland.

Jacobsen, B. J., Harlin, K. S., Swanson, S. P., Lambert, R. J., Beasley, V. R., Sinclair, J. B., and Wei, L. S. 1995. Occurrence of fungi and mycotoxins associated with field mold damaged soybeans in the Midwest. Plant Dis. 79:86-89.

Jordan, E. G., Manandhar, J. B., Thapliyal, P. N., and Sinclair, J. B. 1988. Soybean seed quality of 16 cultivars and four maturity groups in Illinois. Plant Dis. 72:64-67.

Kedia, A., Prakash, B., Mishra, P. K., and Dubey, N. K. 2014. Antifungal and antiaflatoxigenic properties of *Cuminum cyminum* (L.) seed essential oil and its efficacy as a preservative in stored commodities. Int. J. Food Microbiol. 168-169:1-7.

Maude, R. B. 1996. Seedborne Diseases and Their Control: Principles and Practice. CABI, Wallingford, Oxfordshire, U.K.

McGee, D. C. 1992. Soybean Diseases: A Reference Source for Seed Technologists. American Phytopathological Society, St. Paul, MN.

Ploper, L. D., Abney, T. S., and Roy, K. W. 1992. Influence of soybean genotype on rate of seed maturation and its impact on seedborne fungi. Plant Dis. 76:287-292.

Sauer, D. B., ed. 1992. Storage of Cereal Grains and Their Products. 4th ed. American Association of Cereal Chemists International, St. Paul, MN.

Sinclair, J. B. 1992. Discoloration of soybean seeds—An indicator of quality. Plant Dis. 76:1087-1091.

Sinclair, J. B. 1993. Control of seedborne pathogens and diseases of soybean seeds and seedlings. Pestic. Sci. 37:15-19.

Sinclair, J. B. 1995. Reevaluation of grading standards and discounts for fungus damaged soybean seeds. J. Amer. Oil Chem. Soc. 72:1415-1419.

Soto-Arias, J. P., and Munkvold, G. P. 2011. Effects of virus infection on susceptibility of soybean plants to *Phomopsis longicolla*. Plant Dis. 95:530-536.

Walcott, R. R. 2003. Detection of seedborne pathogens. HortTechnology 13:40-47.

Yeh, C. C., and Sinclair, J. B. 1980. Effect of *Chaetomium cupreum* on seed germination and antagonism to other seedborne fungi of soybean. Plant Dis. 64:468-470.

(Prepared by M. M. Kulik and J. B. Sinclair;
revised by G. L. Hartman)

Detection of Seedborne Pathogens

Traditionally, seedborne bacteria and fungi are detected by placing surface-disinfected seeds on nutrient agar, potato dextrose agar (PDA), or a moist blotter. After a suitable period of incubation—usually, under long-wave, near-ultraviolet light to stimulate the sporulation of fungi—the seeds are examined individually under a dissecting microscope for the presence of characteristic fungal fruiting bodies. Examination of conidia under a bright-field compound microscope may also be necessary. Grow-out tests may be useful for detecting seedborne viruses, followed by serological techniques for confirmation.

Along with these traditional techniques, molecular assays are being used more commonly, especially those that involve multiplexing (see later in Part I, Molecular Methods for Pathogen Detection and Quantification). DNA-based detection systems offer a great deal of flexibility and often exhibit higher levels of sensitivity than conventional techniques.

(For additional information, see the Selected References provided with the introduction to the section Seed and Postharvest Pathology.)

(Prepared by M. M. Kulik and J. B. Sinclair; revised by G. L. Hartman)

Seedborne Bacteria

Bacteria that are seedborne in soybean include *Bacillus subtilis* (see the section Bacillus Seed Decay); *Curtobacterium flaccumfaciens* pv. *flaccumfaciens* (see Bacterial Wilt and Bacterial Tan Spot); *Pseudomonas savastanoi* pv. *glycinea* (see Bacterial Blight); *P. syringae* pv. *tabaci* (see Wildfire); *Ralstonia solanacearum* (see Bacterial Wilt and Bacterial Tan Spot); and *Xanthomonas axonopodis* pv. *glycines* (see Bacterial Pustule).

(For additional information, see the Selected References provided with the introduction to the section Seed and Postharvest Pathology.)

(Prepared by M. M. Kulik and J. B. Sinclair; revised by G. L. Hartman)

Seedborne Field Fungi

A number of fungi cause symptoms on soybean seeds, including the following: *Alternaria tenuissima* and *A. tenuis* (see the section Alternaria Leaf Spot, Pod Necrosis, and Seed Decay); *Cercospora sojina* (see Frogeye Leaf Spot); *Cercospora kikuchii* (see Cercospora Leaf Blight and Purple Seed Stain); *Chaetomium cupreum* (see the following section); *Colletotrichum* spp. (see Anthracnose); *Eremothecium ashbyi* and *E. coryli* (see Yeast Spot of Pods and Seeds); *Fusarium* spp. (see Fusarium Seed Rot and *Fusarium* Species Associated with Seed); *Macrophomina phaseolina* (see Charcoal Rot); *Peronospora manshurica* (see Downy Mildew); and *Phomopsis* and *Diaporthe* species (see Phomopsis Seed Decay; Pod and Stem Blight; Stem Canker). Some diseases, such as Alternaria pod necrosis and seed decay, are associated with pod injury caused by the bean leaf beetle or another insect (see Part II, Arthropod Pests).

Root rot fungi—such as *Fusarium* spp., *M. phaseolina,* and *Rhizoctonia solani*—and stem-infecting fungi—such as *Sclerotinia sclerotiorum* and *Athelia rolfsii*—can also infect seeds with propagules from infested soil and when pods come in contact with infested plant parts. Infection rates for *Fusarium* spp. and *M. phaseolina* have reached 50%, but infection rates for *R. solani, S. sclerotiorum,* and *A. rolfsii* are often 10% or less.

Chaetomium cupreum

Chaetomium spp. and *C. cupreum,* in particular, have been associated with soybean seeds in Brazil and the United States. *C. cupreum* produces a compound that inhibits seed germination and the growth of other fungi associated with soybean seeds. This fungus also produces a copper-colored pigment on agar, which can be confused with that produced by *Cercospora kikuchii* (see the section Cercospora Leaf Blight and Purple Seed Stain).

Other Seedborne Field Fungi

Numerous other genera of fungi have been reported to be seedborne in soybean, including species of *Ascochyta, Cladosporium, Corynespora, Curvularia, Drechslera, Epicoccum, Helminthosporium, Macrophoma, Microdochium, Monilia, Nigrospora, Paecilomyces, Pestalotia, Phialophora, Phoma, Phyllosticta, Phytophthora, Rhizoctonia, Rhizopus, Sclerotium, Septoria, Stemphylium, Thielaviopsis,* and *Verticillium.* Some are recognized as pathogens of soybean, but further evidence is needed to ascertain their importance in affecting seed quality.

(For additional information, see the Selected References provided with the introduction to the section Seed and Postharvest Pathology.)

(Prepared by M. M. Kulik and J. B. Sinclair; revised by G. L. Hartman)

Storage Fungi

Deterioration of soybean seeds caused by fungi in the field following a weather-delayed harvest or during storage is a serious economic problem for both processors and producers. During storage, there is a gradual decrease of most field fungi with a simultaneous increase in storage fungi, which is often accompanied by a reduction in seed germination and vigor. Seed deterioration includes a decrease in carbohydrates (both soluble and insoluble), a decrease in protein content, and a decrease in oil content plus an increase in fatty acids. Deterioration caused by fungi results in increased damage discounts against the producer because of the reduced meal and oil quality and the reduced storability of soybean products. In some situations, mycotoxins produced by *Aspergillus flavus, A. parasiticus, Fusarium graminearum, Alternaria* spp., and *Penicillium* spp. can cause problems for animals and humans that consume soybean and its products.

Although losses caused by storage fungi are sporadic and difficult to assess, they are estimated to exceed $20 million a year in the United States. This estimate does not include the reductions in prices that result from the damage discounts assessed when mold-damaged soybean seeds are marketed. Price reductions are frequently in the range of 30–50% and may approach $100 million a year in the United States.

Causal Agents

More than 150 fungal species have been associated with stored soybean seeds. Serious economic losses frequently occur in storage as a result of deterioration caused by several species of *Aspergillus, Curvularia, Fusarium, Mucor, Paecilomyces, Penicillium, Rhizopus,* and *Scopulariopsis,* as well as by bacteria and yeasts.

Symptoms and Signs

Storage fungi produce both heat and water from respiration and are capable of raising temperatures higher than 76°C. At 55°C, hotspots are usually detectable in the grain mass, and seeds become clumped and turn dark brown. As a result of exothermic and oxidative chemical reactions, heating to ignition can occur when sufficient oxygen is available. At temperatures below the level of ignition, soybean seeds can appear charred and blackened—a condition known as "bin burning." Bin-burned seeds contain excessive amounts of free fatty acids, which cause off flavors in soybean oil and oil products and dramatically reduce the shelf lives of these products. Seeds heated to ignition are blackened and cavities form in the cotyledons, as observed in cross section. Blackened seeds lose little or no weight when artificially heated to 220°C for 3–6 h.

Disease Cycle and Epidemiology

Storage losses occur when moisture is present at a level sufficient to promote the growth of mold. Storage fungi of the species *Aspergillus* and *Penicillium* are especially adapted for growth in seeds with a moisture content greater than 12–13%

(i.e., in equilibrium with relative humidity of 70–75%). Generally, the growth of *A. restrictus* produces little detectable heating. However, *A. glaucus* becomes active as the moisture level increases, producing temperature increases of 10–15°C plus moisture, particularly as temperature and moisture levels reach the optimal ranges for this species. At a moisture content greater than 14–15%, the growth of *A. candidus* and of *A. flavus* rapidly raises the temperature to 55°C. At a moisture content greater than 20%, these fungi are commonly succeeded by species of *Fusarium, Mucor,* and *Rhizopus* and by thermophilic bacteria and yeasts. Each fungus has a distinct optimal environment for growth, and the presence or absence of a given fungus can provide evidence of the moisture, temperature, and degree of deterioration of a given seed lot.

The respiration of fungi increases the temperature and the available moisture in the grain mass and allows a more diverse fungal community to grow under these favorable conditions. Losses can still occur if the average grain moisture content is less than 12% in the following situations:

1. If some seeds in storage have a moisture content greater than 12%, then the fungi growing on them will produce heat and moisture, increasing the likelihood of fungal growth on adjacent seeds.
2. Weed seeds harvested with soybean seeds have a moisture content high enough to support fungal growth. Weed seeds frequently congregate in cone-shaped, compacted layers that accumulate in the grain mass under the loading spout. These layers of fine material—including broken soybeans, as well as weed seeds—are known as "spout lines." Hotspots often develop within spout lines.
3. Ice harvested with soybean seeds eventually becomes free water.
4. Moisture transfer from the air can occur when the air and the grain are at different temperatures; unequal temperatures cause convection currents, resulting in condensation in cool areas. This phenomenon often occurs during the fall, when warm grain is placed in a storage bin and the air becomes cooler, and more frequently following winter storage in temperate climates, when the grain mass is cold and then exposed to warm, moist air during unloading and transfer. Damage resulting from moisture transfer is frequently found in the upper layers and near the center of the bin.

Management

Seed moisture should not exceed 12%. If the temperature is less than 10°C, storage at 13–14% moisture is safe for up to 6 months. Moisture-measuring equipment should be regularly calibrated. Also, the wettest seeds (not those with average moisture) should be used to determine storability. High-moisture seeds adjacent to low-moisture seeds may lose moisture over time, but the average moisture of a grain lot does not provide enough information to evaluate the potential storage mold activity.

Soybean seed lots should be cleaned to remove broken seeds and weed seeds. Cleaning not only removes potentially high-moisture weed seeds but also helps to eliminate spout lines, which slow air movement in the storage bin. Aeration via fans of the sizes common in most storage structures can be used to regulate temperature. Greater airflow is necessary to affect the equilibrium relative humidity in the grain mass.

The temperature of the grain mass should be maintained within 5–10°C of the average monthly outdoor temperature. The grain mass temperature should be constantly monitored with preplaced thermocouple cables or temperature probes. Any sign of a temperature increase signals advanced storage decay, and immediate action should be taken to cool the grain or to move it before a greater loss occurs. A temperature increase of more than 1–3°C indicates that advanced spoilage is under way. Grain is an excellent insulator, and a temperature of 40–50°C may be undetectable as little as 0.3–1.0 m away.

Insects that are associated with stored grain should be managed, because their activities can raise both grain moisture and temperature levels. The use of essential oils, such as mentrasto and oregano, has been suggested to control aflatoxin production in seeds.

Selected References

Barros, G. G., Oviedo, M. S., Ramirez, M. L., and Chulze, S. N. 2011. Safety aspect in soybean food and feed chains: Fungal and mycotoxins contamination. Pages 7-20 in: Soybean—Biochemistry, Chemistry, and Physiology. T.-B. Ng, ed. InTech, Rijeka, Croatia.

Bhattacharya, K., and Raha, S. 2002. Deteriorative changes of maize, groundnut and soybean seeds by fungi in storage. Mycopathologia 155:135-141.

Christensen, C. M., and Meronuck, R. A. 1986. Quality Maintenance in Stored Grains and Seeds. University of Minnesota Press, Minneapolis.

Christensen, C. M., Meronuck, R. A., Steele, J. A., and Behrens, J. C. 1973. Some morphological and chemical characteristics of bin-burned and fireburned soybeans. Trans. ASAE 18:899-901.

Esper, R. H., Gonçalez, E., Marques, M. O. M., Felicio, R. C., and Felicio, J. D. 2014. Potential of essential oils for protection of grains contaminated by aflatoxin produced by *Aspergillus flavus*. Front. Microbiol. doi:10.3389/fmicb.2014.00269.

Gupta, I. J., Schmitthenner, A. F., and McDonald, M. B. 1993. Effect of storage fungi on seed vigour of soybean. Seed Sci. Technol. 21:581-591.

International Seed Testing Association (ISTA). 2014. Seed health testing. Pages 7-1-7-3 in: International Rules for Seed Testing. ISTA, Bassersdorf, Switzerland.

Jacobsen, B. J., Harlin, K. S., Swanson, S. P., Lambert, R. J., Beasley, V. R., Sinclair, J. B., and Wei, L. S. 1995. Occurrence of fungi and mycotoxins associated with field mold damaged soybeans in the Midwest. Plant Dis. 79:86-89.

Kedia, A., Prakash, B., Mishra, P. K., and Dubey, N. K. 2014. Antifungal and antiaflatoxigenic properties of *Cuminum cyminum* (L.) seed essential oil and its efficacy as a preservative in stored commodities. Int. J. Food Microbiol. 168-169:1-7.

Sauer, D. B., ed. 1992. Storage of Cereal Grains and Their Products. 4th ed. American Association of Cereal Chemists International, St. Paul, MN.

Sinclair, J. B. 1992. Discoloration of soybean seeds—An indicator of quality. Plant Dis. 76:1087-1091.

Sinclair, J. B. 1995. Reevaluation of grading standards and discounts for fungus damaged soybean seeds. J. Amer. Oil Chem. Soc. 72:1415-1419.

(Prepared by M. M. Kulik and J. B. Sinclair; revised by G. L. Hartman)

Molecular Methods for Pathogen Detection and Quantification

Ongoing interest in the availability of convenient, inexpensive, fast, sensitive, and accurate techniques for detecting and/or quantifying the presence of soybean pathogens has resulted in the increased usage of molecular tools for both research and diagnostic purposes. The method of extracting a molecular target (usually, DNA or RNA) for detection depends entirely on the nature of the sample, because pathogens can be present in foliar, woody, root, and reproductive tissues; in soil and in irrigation water; in arthropod vectors; on mechanical equipment; as airborne propagules; and even in association with people. Consequently, different interfering agents can be present and act to inhibit or obscure detection of a pathogen. Automated DNA extraction is possible in some cases, but throughput may be limited by the investigator's ability to prepare the input samples. Thus, an assay used successfully to quantify a soybean pathogen in foliar samples may require modification when used to detect the same pathogen in roots or soil—for example, cases in which washing or sieving, drying, milling, and subsampling may be necessary. Some methods developed for pathogens on different hosts can be used with soybean with little or no modification, as long as the isolates are not delimited by host range.

A selection of molecular assays used for detecting pathogens relevant to soybean is provided in Table 9.

Bacterial Pathogens

Because of how quickly many bacterial pathogens can be cultured in vitro and how easily colony-forming units (CFUs) can be counted, the use of metabolic panels, selective media, and immunological methods remains widespread. Molecular methods for bacterial pathogen detection are generally based on polymerase chain reaction (PCR), followed by electrophoretic analysis of the amplification products. The majority of plant-pathogenic bacteria of soybean can be identified collectively via multiplex PCR using a battery of specific primers. Standard PCR is relatively inexpensive and straightforward but also low throughput and semiquantitative at best.

An isothermal amplification method (LAMP) has been described that allows screening for some mollicutes within hours. This method has potential for use in the field.

Fungal and Oomycete Pathogens

Molecular detection of plant-pathogenic fungi is generally PCR based but with some variations—for example, sequential amplification with nested primers, used for detecting the Sclerotinia stem rot pathogen (*Sclerotinia sclerotiorum*), and amplification of an intersimple sequence repeat (ISSR) marker, used for detecting the target spot pathogen (*Corynespora cassiicola*). In addition, random amplified polymorphic DNA (RAPD) markers have been useful in identifying isolates of *Fusarium oxysporum* and led to sequence-characterized amplified region (SCAR) assays. Like standard PCR, ISSR and RAPD assays are fairly straightforward and inexpensive but quantitative and low throughput.

The target sequence for PCR amplification is often located in the nuclear genes for ribosomal RNAs and the affiliated internal transcribed spacers (ITSs) and intergenic sequences (IGSs). The multicopy nature of the gene cluster increases sensitivity considerably, as compared with a single-copy gene. The rRNA large and small subunit genes offer sequence conservation, which can increase the inclusivity of an assay, whereas the ITS and IGS sequences offer greater sequence diversity, which can provide increased specificity.

Unfortunately, it is not always possible to identify a locus that simultaneously includes all the isolates of a particular fungus and excludes all the nonpathogenic related species. Extensive validation may be important in some but not all cases. Sometimes, the geospatial distribution of pathogenic species can be exploited to overcome such a problem. For example, in the case of sudden death syndrome (caused by several *Fusarium* spp.), an assay may only need to specifically detect *F. virguliforme* for it to be useful in the United States, because the other causative species do not occur outside South America. Other loci have been useful, such as the cytochrome oxidase gene for the anthracnose pathogen (*Colletotrichum* species) and gypsi-like retroelements for detecting the Phytophthora root rot pathogen (*Phytophthora sojae*).

As real-time, quantitative thermocycling (qPCR) instruments became available, they were widely adopted for detecting and quantifying fungi and oomycetes. Although this technology has a relatively high capital cost, using it permits absolute and relative quantification of a target, accommodates the need for low or medium throughput (96-well), and permits ready adaptation to a high-throughput production context. Robotic pipettors can prepare, seal, thermocycle, and evaluate 384-well reaction plates, potentially completing thousands of reactions a day. Portable qPCR instruments are also available for remote deployment, extending the benefits of this technology to mobile and field diagnostics (although with modest throughput).

In one form of qPCR, an intercalating dye (i.e., DNA dye binding) is present that increasingly fluoresces as amplification products accumulate and is measured in situ by the incorporated fluorometer in real time, producing estimates of DNA quantity. Dye-binding assays have been reported for both the Phytophthora root rot and the charcoal rot pathogens (*P. sojae* and *Macrophomina phaseolina,* respectively). In another form of qPCR, oligonucleotide probes are utilized that produce fluorescence by virtue of sequence-specific binding to regions between the two PCR primers. Several types of probes can be designed; the most common is the fluorogenic 5′-exonuclease linear hydrolysis probe, as reported for the pathogens of sudden death syndrome and brown stem rot (*F. virguliforme* and *Cadophora gregata,* respectively). qPCR can be extremely sensitive, as well. In some cases—for example, with the soybean rust pathogen (*Phakopsora pachyrhizi*)—the presence of a single dikaryotic urediniospore in the original sample can be detected. This technique is also useful in detecting latent pathogens that are not causing any visual symptoms in tissues.

Nematode Pathogens

Nematologists have also found the use of molecular assays to be advantageous. PCR, SCAR, and qPCR assays are available for detection of *Heterodera glycines* and *Meloidogyne* spp., and qPCR assays have been reported for both *Rotylenchulus*

TABLE 9. Selected Molecular Assays Reported for Detecting Soybean Pathogens[a]

Bacterial or Mollicute Disease(s)	**Target(s) for Detection**	**Assay Type[b]**	**Reference**
Bacterial pustule	*Xanthomonas axonopodis* pv. *glycines*	PCR	Llop et al., 1999
Bacterial tan spot	*Curtobacterium flaccumfaciens* pv. *flaccumfaciens*	PCR	Tegli et al., 2002
Phyllody	Phytoplasma	LAMP	Tomlinson et al., 2010
Multiple diseases	Multiple bacteria	PCR	Lee et al., 2013
Fungal or Oomycete Disease(s)	**Target(s) for Detection**	**Assay Type(s)[b]**	**Reference(s)**
Alternaria leaf spot and pod necrosis	*Alternaria alternata*	PCR	Johnson et al., 2000
Anthracnose	*Colletotrichum* spp.	PCR	Chen et al., 2006
		qPCR-dye	Yang et al., 2015
Brown stem rot	*Cadophora gregata*	qPCR-probe	Malvick and Impullitti, 2007
Cercospora leaf blight and purple seed stain	*Cercospora kikuchii*	qPCR-probe	Chanda et al., 2014
Charcoal rot	*Macrophomina phaseolina*	DNA spot blot and PCR	Babu et al., 2007
		qPCR-dye and qPCR-probe	Babu et al., 2011
Fusarium blight, wilt, and root rot	*Fusarium oxysporum*	RAPD and SCAR-PCR	Alves-Santos et al., 2002
Phomopsis seed decay	*Diaporthe longicolla*	qPCR-probe	Zhang et al., 1999
Phytophthora root rot	*Phytophthora sojae*	PCR	Wang et al., 2006
		PCR and qPCR-dye	Bienapfl et al., 2011
Pod and stem blight	*Diaporthe phaseolorum* var. *sojae*	qPCR-probe	Zhang et al., 1999
Pythium damping-off and root rot	*Pythium aphanidermatum* and	PCR	Asano et al., 2010
	Globisporangium ultimum	PCR	Kageyama et al., 1997
Rhizoctonia foliar blight, damping-off, and root rot	*Rhizoctonia solani*	PCR	Grosch et al., 2007
			Matsumoto, 2002
Rust	*Phakopsora pachyrhizi* and *P. meibomiae*	qPCR-probe	Frederick et al., 2002
		Nested PCR and qPCR-probe	Barnes et al., 2009
Sclerotinia stem rot	*Sclerotinia sclerotiorum*	PCR	Hirschhäuser and Fröhlich, 2007
		Nested PCR	Qin et al., 2011
Stem canker	*Diaporthe phaseolorum* var. *caulivora* and *D. phaseolorum* var. *meridionalis*	qPCR-probe	Zhang et al., 1999
Sudden death syndrome	*Fusarium virguliforme*[c]	qPCR-probe	Gao et al., 2004; Li et al., 2008; Mbofung et al., 2011; Westphal et al., 2014
Target spot	*Corynespora cassiicola*	ISSR	Qi et al., 2009a
		Nested PCR	Qi et al., 2009b
Thielaviopsis root rot	*Thielaviopsis basicola*	PCR	Zhao et al., 2009
		qPCR-probe	Huang and Kang, 2010
Nematode Disease	**Target(s) for Detection**	**Assay Type[b]**	**Reference**
Lance nematode	*Hoplolaimus columbus, H. galeatus,* and *H. magnistylus*	PCR	Bae et al., 2009
Lesion nematode	*Pratylenchus* spp.	qPCR-dye	Berry et al., 2008
Reniform nematode	*Rotylenchulus reniformis*	qPCR-probe	Showmaker et al., 2011
		qPCR-probe	Sayler et al., 2012
Root-knot nematode	*Meloidogyne arenaria, M. incognita,* and *M. javanica*	SCAR-PCR	Zijlstra et al., 2000
		qPCR-dye	Berry et al., 2008
Soybean cyst nematode	*Heterodera glycines*	PCR	Subbotin et al., 2001
		qPCR-probe	Lambert et al., 2005
		SCAR-PCR	Ou et al., 2008
Viral Disease	**Target for Detection**	**Assay Type(s)[b]**	**Reference**
Alfalfa mosaic	*Alfalfa mosaic virus*	RT-PCR	Bariana et al., 1994
		IC-RT-PCR	Naghavi et al., 2008
		MRLP	Wang et al., 2011
Bean pod mottle	*Bean pod mottle virus*	LAMP	Wei et al., 2012
Bean yellow mosaic	*Bean yellow mosaic virus*	RT-PCR	Vunsh et al., 1990
Cowpea mild mottle	*Cowpea mild mottle virus*	RT-PCR	Tavasoli et al., 2009
Cucumber mosaic (Soybean stunt)	*Cucumber mosaic virus*	RT-PCR	De Blas et al., 1994
		IC-RT-PCR	Yu et al., 2005
		MRLP	Wang et al., 2011
Peanut mottle	*Peanut mottle virus*	Northern blot	Dietzgen et al., 1994
		IC-RT-PCR	Gillaspie et al., 2000
Peanut stripe	*Peanut stripe virus*	Northern blot	Dietzgen et al., 1994
Soybean dwarf	*Soybean dwarf virus*	RT-PCR and Dotblot	Yamagishi et al., 2008
Soybean mosaic	*Soybean mosaic virus*	RT-PCR	Omunyin et al., 1996
Tobacco ringspot	*Tobacco ringspot virus*	MRLP	Wang et al., 2011
Tobacco streak	*Tobacco streak virus*	RT-PCR	Bhat et al., 2002
		IC-RT-PCR	Sarovar et al., 2010

[a] Some assays listed were developed using pathogen species isolated from non-soybean hosts.

[b] DNA spot blot = hybridization assay; Dotblot and Northern blot = nucleic acid blots; ISSR = intersimple sequence repeat; LAMP = loop-mediated isothermal amplification; MRLP = microarray-based multiplex RT-ligase; PCR = polymerase chain reaction; RAPD = random amplified polymorphic DNA; RT-PCR = reverse transcription PCR; IC-RT-PCR = immunocapture reverse transcription PCR; SCAR-PCR = sequence-characterized amplified region PCR; qPCR-dye = intercalating dye; qPCR-probe = fluorogenic 5′-exonuclease linear hydrolysis probe.

[c] Sudden death syndrome is caused by additional *Fusarium* spp. in South America.

reniformis and *Pratylenchus* spp. Once again, the target loci may vary. rDNA sequences (*R. reniformis, H. glycines,* and *Meloidogyne* spp.), the chorismate mutase gene (*Heterodera* spp.), a SCAR region (*H. glycines*), and the β-tubulin gene (*R. reniformis*) have all been used.

Viral Pathogens

Nucleic acid blots with hybridization probes of varying types quickly proved to be valuable tools in viral diagnostics, as the small genomes of viral pathogens became well characterized and in some cases fully cloned. Such blots permit only rough estimates of relative abundance, but they are relatively straightforward and inexpensive (if the analyst has the resources to purify the viral genome from host material and access to the necessary radio-labeled or enzyme-linked probe).

In the 1990s, the use of reverse transcription- (RT-) PCR assays became common. RT-PCR provides high specificity in detection, and the electrophoreogram offers a semiquantitative result at a modest cost. Around the year 2000, RT-qPCR began proving useful for many viral pathogens, and since then, variants such as immunocapture (IC-RT-qPCR) and microarray-based multiplex RT-ligase (MRLP) PCR have emerged. Although these technologies offer improved quantification and have the capacity to detect many different viruses, they have significant capital costs and complexity of use. An RT-LAMP system is also being used in cases in which a qualitative answer at a lower cost suffices.

Method Selection

When selecting a molecular assay, the analyst must be cognizant of the validation methods described by the original publication about the assay and whether it was used in the context of discovery research versus production. Questions to consider include the following:

- Is the assay inclusive enough to detect all relevant pathogen isolates and species?
- Is the assay exclusive enough to avoid false positives?
- Is the assay sensitive enough to detect the target organism in the quantity of sample being collected and with respect to the possible bulking of the sample?
- Is the assay robust enough to tolerate both automated and manual processing?
- What is the lower limit of detection of the assay, and is it adequate to detect pathogens at the minimum impact threshold?
- How are type 1 and type 2 errors detected and minimized, and what controls are needed?

With reference to qPCR, specifically, the MIQE (minimum information for publication of quantitative real-time PCR experiments) guidelines address many of these questions.

When considering a qPCR approach, the choice among technologies may not be straightforward. Probe-based methods offer high specificity and the potential to differentiate a single nucleotide polymorphism (SNP), the ability to multiplex several reactions in each tube (enabling quantification of multiple pathogens at once), and the potential to include an exogenous control reaction. Probe-based assays are also less susceptible to interference by off-target or primer-dimer amplification, as compared with dye-binding assays, which generate a positive signal from any amplicon, intended or not. In contrast, singleplex dye-binding assays are fairly straightforward to design and do not require the purchase of expensive fluorescently labeled (and quenched) custom oligonucleotides. Furthermore, a dye-binding assay can often be readily adapted from an established standard PCR method using the identical primers, and it may perform well even when the target is large (e.g., 500 base pairs [bp]), because the fluorescent signal is produced in proportion to the mass (not the stoichiometry) of the amplified product. Thus, a standard PCR assay that has already been proven against a wide variety of target and nontarget species might be adapted with neither the need for nor the expense of revalidation. Some dye-binding assays use a dissociation curve analysis to determine the melting temperatures of amplification products to differentiate among target species.

Molecular genetic assays function rather independently of the host species. Even so, it is important to consider the precise target of the assay (e.g., an SNP or another identifying genomic sequence), because that trait may be present only in host-specific strains of the pathogen. An assay useful to a worker focusing on maize, for example, may fail to detect soybean isolates; this is especially true for the rapidly evolving genomes of viruses. Conversely, an assay that has been demonstrated to exclude nonphytopathogenic isolates in sugar beet may fail to exclude such isolates in soybean. Finally, it is important to note that none of these assays provides insight as to the viable status of what is being detected. DNA can be easily detected from both a vigorously infected root system and an herbarium specimen that has been dead for a century. For instance, the presence of a picogram of soybean rust DNA in an atmospheric spore trap reveals only that sporuliferous material was present upwind—not whether the spores represent a threat to crops. Despite this drawback, molecular quantification offers specific and quantifiable data on pathogen presence.

The use of high-throughput DNA sequencing in combination with metagenomic analysis has made it feasible to screen many hundreds or thousands of plant, soil, water, and seed samples for the presence of all known viral species in a quantitative manner. Similarly, such samples can be screened for all eubacteria and archaebacteria, all fungi, and all eukaryotes for which the requisite rDNA sequences are publicly available. Metagenomic sequencing produces large volumes of raw data (many billions of nucleotide bases), requiring a bioinformaticist for analysis, and methods of normalizing the metagenomes between samples may require scrutiny. Yet it is likely that core-sequencing facilities will soon automate the process toward a "DNA in, ID out" model, which would be more readily accessible to a busy pathologist or diagnostician. Thus, efficiency may be limited only by the practicality of bulking samples and extracting nucleic acids.

Ultimately, the imperatives and the context (research versus diagnostic) of the individual researcher, diagnostician, business, or agency or the science itself will dictate the most appropriate approach. In some cases, the most appropriate approach may be to rely on the expertise of a skilled pathologist who can visually identify and rate, for example, soybean rust, frogeye leaf spot, soybean cyst nematodes, and aphids within minutes. Serological tests can be speedy and are likely to remain valuable. Finally, because some disorders are abiotic and because there is no molecular substitute on the horizon for the prudent exercise of completing Koch's postulates, using molecular methods to detect the DNA from a plant pathogen must not be conflated with diagnosing the cause of a disease. Nevertheless, the molecular techniques available to researchers and diagnosticians are powerful and present new opportunities in soybean pathology.

Selected References

Adams, I. P., Glover, R. H., Monger, W. A., Mumford, R., Jackeviciene, E., Navalinskiene, M., Samuitiene, M., and Bonham, N. 2009. Next-generation sequencing and metagenomic analysis: A universal diagnostic tool in plant virology. Mol. Plant Pathol. 10:537-545.

Atkins, S. D., and Clark, I. M. 2004. Fungal molecular diagnostics: A minireview. J. Appl. Genet. 45:3-15.

Alves-Santos, F. M., Ramos, B., García-Sánchez, M. A., Eslava, A. P., and Díaz-Mínguez, J. M. 2002. A DNA-based procedure for in planta detection of *Fusarium oxysporum* f. sp. *phaseoli*. Phytopathology 92:237-244.

Asano, T., Senda, M., Suga, H., and Kageyama, K. 2010. Development of multiplex PCR to detect five *Pythium* species related to turfgrass diseases. J. Phytopathol. 158:609-615.

Babu, B. K., Mesapogu, S., Sharma, A., Somasani, S. R., and Arora, D. K. 2011. Quantitative real-time PCR assay for rapid detection of plant and human pathogenic *Macrophomina phaseolina* from field and environmental samples. Mycologia 103:466-473.

Babu, B. K., Saxena, A. K., Srivastava, A. K., and Arora, D. K. 2007. Identification and detection of *Macrophomina phaseolina* by using species-specific oligonucleotide primers and probe. Mycologia 99:797-803.

Bae, C.-H., Robbins, R. T., and Szalanski, A. L. 2009. Molecular identification of some *Hoplolaimus* species from the USA based on duplex PCR, multiplex PCR and PCR-RFLP analysis. Nematology 11:471-480.

Bariana, H. S., Shannon, A. L., Chu, P. W. G., and Waterhouse, P. M. 1994. Detection of five seedborne legume viruses in one sensitive multiplex polymerase chain reaction test. Phytopathology 84:1201-1205.

Barnes, C. W., Szabo, L. J., and Bowersox, V. C. 2009. Identifying and quantifying *Phakopsora pachyrhizi* spores in rain. Phytopathology 99:328-338.

Berry, S. D., Fargette, M., Spaull, V. W., Morand, S., and Cadet, P. 2008. Detection and quantification of root-knot nematode (*Meloidogyne javanica*), lesion nematode (*Pratylenchus zeae*) and dagger nematode (*Xiphinema elongatum*) parasites of sugarcane using real-time PCR. Mol. Cell. Probe 22:168-176.

Bhat, A. I., Jain, R. K., and Ramiah, M. 2002. Detection of *Tobacco streak virus* from sunflower and other crops by reverse transcription polymerase chain reaction. Indian Phytopathol. 55:216-218.

Bienapfl, J. C., Malvick, D. K., and Percich, J. A. 2011. Specific molecular detection of *Phytophthora sojae* using conventional and real-time PCR. Fungal Biol. 115:733-740.

Blok, V. C. 2005. Achievements in and future prospects for molecular diagnostics of plant-parasitic nematodes. Can. J. Plant Pathol. 27:176-185.

Bustin, S. A., Benes, V., Garson, J. A., Hellemans, J., Huggett, J., Kubista, M., Mueller, R., Nolan, T., Pfaffl, M. W., Shipley, G. L., Vandesompele, J., and Wittwer, C. T. 2009. The MIQE Guidelines: Minimum information for publication of quantitative real-time PCR experiments. Clin. Chem. 55:611-622.

Chalupová, J., Raus, M., Sedlářová, M., and Šebela, M. 2014. Identification of fungal microorganisms by MALDI-TOF mass spectrometry. Biotech. Adv. 32:230-241.

Chanda, A. K., Ward, N. A., Robertson, C. L., Chen, Z.-Y., and Schneider, R. W. 2014. Development of a quantitative polymerase chain reaction detection protocol for *Cercospora kikuchii* in soybean leaves and its use for documenting latent infection as affected by fungicide applications. Phytopathology 104:1118-1124.

Chen, L. S., Chu, C., Liu, C. D., Chen, R. S., and Tsay, J. G. 2006. PCR-based detection and differentiation of anthracnose pathogens, *Colletotrichum gloeosporioides* and *C. truncatum*, from vegetable soybean in Taiwan. J. Phytopathol. 154:654-662.

De Blas, C., Borja, M. J., Saiz, M., and Romero, J. 1994. Broad spectrum detection of cucumber mosaic virus (CMV) using the polymerase chain reaction. J. Phytopathol. 141:323-329.

Dietzgen, R. G., Zeyong, X., and Teycheney, P.-Y. 1994. Digoxigenin-labeled cRNA probes for the detection of two potyviruses infecting peanut (*Arachis hypogaea*). Plant Dis. 78:708-711.

Frederick, R. D., Snyder, C. L., Peterson, G. L., and Bonde, M. R. 2002. Polymerase chain reaction assays for the detection and discrimination of the soybean rust pathogens *Phakopsora pachyrhizi* and *P. meibomiae*. Phytopathology 92:217-227.

Gao, X., Jackson, T. A., Lambert, K. N., Li, S., Hartman, G. L., and Niblack, T. L. 2004. Detection and quantification of *Fusarium solani* f. sp. *glycines* in soybean roots with real-time quantitative polymerase chain reaction. Plant Dis. 88:1372-1380.

Gillaspie, A. G., Jr., Pittman, R. N., Pinnow, D. L., and Cassidy, B. G. 2000. Sensitive method for testing peanut seed lots for *Peanut stripe* and *Peanut mottle viruses* by immunocapture-reverse transcription-polymerase chain reaction. Plant Dis. 84:559-561.

Grosch, R., Schneider, J. H. M., Peth, A., Waschke, A., Franken, A., Kofoet, P., and Jabaji-Hare, S. H. 2007. Development of a specific PCR assay for the detection of *Rhizoctonia solani* AG 1-IB using SCAR primers. J. Appl. Microbiol. 102:806-819.

Hartman, G. L., and Haudenshield, J. S. 2009. Movement of *Phakopsora pachyrhizi* (soybean rust) spores by non-conventional means. Eur. J. Plant Pathol. 123:225-228.

Haudenshield, J. S., and Hartman, G. L. 2011. Exogenous controls increase negative call veracity in multiplexed, quantitative PCR assays for *Phakopsora pachyrhizi*. Plant Dis. 95:343-352.

Hirschhäuser, S., and Fröhlich, J. 2007. Multiplex PCR for species discrimination of *Sclerotiniaceae* by novel laccase introns. Int. J. Food Microbiol. 118:151-157.

Huang, J., and Kang, Z. 2010. Detection of *Thielaviopsis basicola* in soil with real-time quantitative PCR assays. Microbiol. Res. 165:411-417.

Johnson, R. D., Johnson, L., Kohmoto, K., Otani, H., Lane, C. R., and Kodama, M. 2000. A polymerase chain reaction-based method to specifically detect *Alternaria alternata* apple pathotype (*A. mali*), the causal agent of Alternaria blotch of apple. Phytopathology 90:973-976.

Kageyama, K., Ohyama, A., and Hyakumachi, M. 1997. Detection of *Pythium ultimum* using polymerase chain reaction with species-specific primers. Plant Dis. 81:1155-1160.

Lambert, K. N., Bekal, S., Domier, L. L., Niblack, T. L., Noel, G. R., and Smyth, C. A. 2005. Selection of *Heterodera glycines Chorismate Mutase-1* alleles on nematode-resistant soybean. Mol. Plant-Microbe Interact. 18:593-601.

Lee, Y.-H., Kim, N.-G., Yoon, Y.-N., Lim, S.-T., Kim, H.-T., Yun, H.-T., Baek, I.-Y., and Lee, Y.-K. 2013. Multiplex PCR assay for the simultaneous detection of major pathogenic bacteria in soybean. Korean J. Crop Sci. 58:142-148.

Li, S., Hartman, G. L., Domier, L. L., and Boykin, D. 2008. Quantification of *Fusarium solani* f. sp. *glycines* isolates in soybean roots by colony-forming unit assays and real-time quantitative PCR. Theor. Appl. Genet. 117:343-352.

Llop, P., Caruso, P., Cubero, J., Morente, C., and López, M. M. 1999. A simple extraction procedure for efficient routine detection of pathogenic bacteria in plant material by polymerase chain reaction. J. Microbiol. Methods 37:23-31.

Malvick, D. K., and Impullitti, A. E. 2007. Detection and quantification of *Phialophora gregata* in soybean and soil samples with a quantitative, real-time PCR assay. Plant Dis. 91:736-742.

Matsumoto, M. 2002. Trials of direct detection and identification of *Rhizoctonia solani* AG 1 and AG 2 subgroups using specifically primed PCR analysis. Mycoscience 43:185-189.

Mbofung, G. C. Y., Fessehaie, A., Bhattacharyya, M. K., and Leandro, L. F. S. 2011. A new TaqMan real-time polymerase chain reaction assay for quantification of *Fusarium virguliforme* in soil. Plant Dis. 95:1420-1426.

Naghavi, A., Habibi, M. K., and Firouzabadi, F. N. 2008. Detection and identification of some soybean viral mosaic viruses, using molecular techniques in Lorestan province, south west of Iran. Asian J. Plant Sci. 7:557-562.

Omunyin, M. E., Hill, J. H., and Miller, W. A. 1996. Use of unique RNA sequence-specific oligonucleotide primers for RT-PCR to detect and differentiate soybean mosaic virus strains. Plant Dis. 80:1170-1174.

Ou, S., Peng, D., Liu, X., Li, Y., and Moens, M. 2008. Identification of *Heterodera glycines* using PCR with sequence characterised amplified region (SCAR) primers. Nematology 10:397-403.

Powers, T. 2004. Nematode molecular diagnostics: From bands to barcodes. Annu. Rev. Phytopathol. 42:367-383.

Qi, Y. X., Xie, Y., Zhang, X., Pu, J., Zhang, H., Huang, S., and Zhang, H. 2009a. Molecular and pathogenic variation identified among isolates of *Corynespora cassiicola*. Mol. Biotechnol. 41:145-151.

Qi, Y. X., Zhang, X., Pu, J. J., Xie, Y. X., Zhang, H. Q., Huang, S. L., Li, S. L., and Zhang, H. 2009b. Nested PCR assay for detection of Corynespora leaf fall disease caused by *Corynespora cassiicola*. Australas. Plant Pathol. 38:141-148.

Qin, L., Fu, Y., Xie, J., Cheng, J., Jiang, D., Li, G., and Huang, J. 2011. A nested-PCR method for rapid detection of *Sclerotinia sclerotiorum* on petals of oilseed rape (*Brassica napus*). Plant Pathol. 60:271-277.

Sarovar, B., Prasad, Y. S., and Sai Gopal, D. V. R. 2010. Detection of *Tobacco streak virus* by immunocapture-reverse transcriptase-polymerase chain reaction and molecular variability analysis of a part of RNA3 of sunflower, gherkin, and pumpkin from Andhra Pradesh, India. ScienceAsia 36:194-198.

Sayler, R. J., Walker, C., Goggin, F., Agudelo, P., and Kirkpatrick, T. 2012. Conventional PCR detection and real-time PCR quantification of reniform nematodes. Plant Dis. 96:1757-1762.

Schaad, N. W., and Frederick, R. D. 2002. Real-time PCR and its application for rapid plant disease diagnostics. Can. J. Plant Pathol. 24:250-258.

Schena, L., Nigro, F., Ippolito, A., and Gallitelli, D. 2004. Real-time quantitative PCR: A new technology to detect and study phytopathogenic and antagonistic fungi. Eur. J. Plant Pathol. 110:893-908.

Song, J., Jun, N., Li, S., Kim, H., and Hartman, G. L. 2007. Development of PCR assay using species-specific primers for *Phytophthora sojae* based on the DNA sequence of its transposable element. Phytopathology 97:S110.

Showmaker, K., Lawrence, G. W., Lu, S., Balbalian, C., and Klink, V. P. 2011. Quantitative field testing *Rotylenchulus reniformis* DNA from metagenomic samples isolated directly from soil. PloS One 6:e28954.

Subbotin, S. A., Peng, D., and Moens, M. 2001. A rapid method for the identification of the soybean cyst nematode *Heterodera glycines* using duplex PCR. Nematology 3:365-371.

Tavasoli, M., Shahraeen, N., and Ghorbani, S. H. 2009. Serological and RT-PCR detection of *Cowpea mild mottle* carlavirus infecting soybean. J. Gen. Mol. Virol. 1:7-11.

Tegli, S., Sereni, A., and Surico, G. 2002. PCR-based assay for the detection of *Curtobacterium flaccumfaciens* pv. *flaccumfaciens* in bean seeds. Lett. Appl. Microbiol. 35:331-337.

Tomlinson, J. A., Boonham, N., and Dickinson, M. 2010. Development and evaluation of a one-hour DNA extraction and loop-mediated isothermal amplification assay for rapid detection of phytoplasmas. Plant Pathol. 59:465-471.

Vunsh, R., Rosner, A., and Stein, A. 1990. The use of the polymerase chain reaction (PCR) for the detection of bean yellow mosaic virus in gladiolus. Ann. Appl. Biol. 117:561-569.

Wang, P., Guo, Y., Cheng, J., Dong, Q., Ding, X., Guo, J., and Jiang, Y. 2011. Development of multiplex reverse transcription-ligase detection reaction-polymerase chain reaction (MRLP) mediated universal DNA microarray for diagnostic platform. Biosens. Bioelectron. 26:3719-3724.

Wang, Y., Zhang, W., Wang, Y., and Zheng, X. 2006. Rapid and sensitive detection of *Phytophthora sojae* in soil and infected soybeans by species-specific polymerase chain reaction assays. Phytopathology 96:1315-1321.

Wei, Q.-W., Yu, C., Zhang, S.-Y., Yag, C.-Y., Miriam, K., Zhang, W.-N., Dou, D.-L., and Tao, X.-R. 2012. One-step detection of *Bean pod mottle virus* in soybean seeds by the reverse-transcription loop-mediated isothermal amplification. Virol. J. 9:187-193.

Westphal, A., Li, C., Xing, L., McKay, A., and Malvick, D. 2014. Contributions of *Fusarium virguliforme* and *Heterodera glycines* to the disease complex of sudden death syndrome of soybean. PloS One 9:e99529.

Yamagishi, N., Terauchi, H., and Hidaka, S. 2008. Detection and strain differentiation of *Soybean dwarf virus* by nucleic acid techniques. Pages 77-92 in: Techniques in Diagnosis of Plant Viruses. Studium Press, Houston, TX.

Yang, H.-C., Haudenshield, J. S., and Hartman, G. L. 2015. Multiplex real-time PCR detection and differentiation of *Colletotrichum* species infecting soybean. Plant Dis. doi:10.1094/PDIS-11-14-1189-RE.

Yu, C., Wu, J., and Zhou, X. 2005. Detection and subgrouping of *Cucumber mosaic virus* isolates by TAS-ELISA and immunocapture RT-PCR. J. Virol. Methods 123:155-161.

Zhang, A. W., Hartman, G. L., Curio-Penny, B., Pedersen, W. L., and Becker, K. B. 1999. Molecular detection of *Diaporthe phaseolorum* and *Phomopsis longicolla* from soybean seeds. Phytopathology 89:796-804.

Zhao, Y.-Q, Zhang, C.-L., Zhang, W., Li, L.-N., and Zhang, G.-M. 2009. Molecular detection of *Thielaviopsis basicola* by PCR assay. Acta Phytopathol. Sin. 39:23-29.

Zijlstra, C., Donkers-Venne, D. T. H. M., and Fargette, M. 2000. Identification of *Meloidogyne incognita, M. javanica* and *M. arenaria* using sequence characterised amplified region (SCAR) based PCR assays. Nematology 2:847-853.

(Prepared by J. S. Haudenshield and G. L. Hartman)

Part II. Arthropod Pests

Insects and Their Management

Insects are capable of limiting soybean production throughout the world, and failing to manage them when economically damaging populations have developed can result in significant yield losses. Although soybean can harbor more than 700 species of phytophagous insects, only a few species pose significant threats to soybean production each year.

Insect pests of soybean can be categorized into feeding guilds, which are based on the types of injuries caused and the parts of the plant on which pests feed:

- Defoliators
- Sap feeders
- Stem feeders
- Root, nodule, and planted-seed feeders
- Fruit (flower, pod, and seed) feeders

These categories are not mutually exclusive, because some pests can be placed in more than one category. Nonetheless, the categories define the pests by the most identifiable types of injuries they cause.

The economic importance of the guilds and the different species within them vary annually and geographically in the United States. Local variations in species and the intensity of outbreaks depend primarily on crop maturity and weather conditions. For the purpose of defining insect pests in the United States, the following regions are characterized:

- **Midwest:** Illinois, Indiana, Iowa, Michigan, Minnesota, Ohio, and Wisconsin, plus most of Missouri
- **Great Plains:** Kansas, Nebraska, North Dakota, and South Dakota
- **Delta/Mid-South/Gulf Coast:** Alabama, Arkansas, Kentucky, Louisiana, Mississippi, Tennessee, and Texas, plus the "boot-heel" region of Missouri
- **Atlantic:** Delaware, Florida, Georgia, Maryland, North Carolina, South Carolina, and Virginia

Survey results from research and extension entomologists in the major soybean-producing states in the United States in 2009 and 2014 are presented in Table 10. In completing the survey, the entomologists listed the most economically significant insect pests in their respective states and ranked the order of importance of the feeding guilds within their states. For the United States as a whole, the results are as follows:

- The most economically important defoliators are the bean leaf beetle (*Cerotoma trifurcata*); grasshoppers (*Melanoplus* spp.); the green cloverworm (*Plathypena scabra*); the Japanese beetle (*Popillia japonica*); the soybean looper (*Chrysodeixis includens*); and the velvetbean caterpillar (*Anticarsia gemmatalis*).

TABLE 10. Arthropod Pests of Soybean, by Order of Importance Within Five Feeding Guilds and Four Regions of the United States[a]

Feeding Guild	Midwest	Great Plains	Delta/Mid-South/Gulf Coast	Atlantic
Defoliators	Bean leaf beetle Twospotted spider mite Japanese beetle Grasshoppers	Bean leaf beetle Grasshoppers Green cloverworm Twospotted spider mite	Loopers Green cloverworm Velvetbean caterpillar Bean leaf beetle Twospotted spider mite	Bean leaf beetle Grasshoppers Loopers Velvetbean caterpillar Japanese beetle
Sap feeders	Soybean aphid Potato leafhopper Twospotted spider mite	Soybean aphid Twospotted spider mite Potato leafhopper	Kudzu bug Thrips Soybean aphid	Kudzu bug Twospotted spider mite
Stem feeders	Dectes stem borer	Dectes stem borer	Threecornered alfalfa hopper Dectes stem borer Lesser cornstalk borer	Dectes stem borer Threecornered alfalfa hopper Lesser cornstalk borer
Root, nodule, and planted-seed feeders	Seedcorn maggot White grubs	Seedcorn maggot White grubs	Wireworms Colaspis grubs	Colaspis grubs White grubs
Fruit (flower, pod, and seed) feeders	Bean leaf beetle Grasshoppers Stink bugs	Bean leaf beetle Stink bugs Corn earworm	Stink bugs Corn earworm Bean leaf beetle	Stink bugs Corn earworm Bean leaf beetle

[a] Based on rankings of entomologists provided in survey results from 2009 and 2014. Information for this table was contributed from the following individuals: Jack Bacheler (North Carolina State University); Wayne Bailey (University of Missouri); Larry Buschman (Kansas State University); Angus Catchot (Mississippi State University); Eileen Cullen (University of Wisconsin); Jeremy Greene (Clemson University); Ron Hammond (Ohio State University); Ames Herbert (Virginia Polytechnic Institute and State University); Tom Hunt (University of Nebraska); Doug Johnson (University of Kentucky); Janet Knodel (North Dakota State University); Christian Krupke (Purdue University); Rogers Leonard (Louisiana State University); Gus Lorenz (University of Arkansas); Bob McPherson (University of Georgia); John Obermeyer (Purdue University); Matt O'Neal (Iowa State University); Ken Ostlie (University of Minnesota); Bruce Potter (University of Minnesota); Dave Ragsdale (University of Minnesota); Tim Reed (Auburn University); Dominic Reisig (North Carolina State University); Ron Smith (Auburn University); Kevin Steffey (Dow AgroSciences); Scott Stewart (University of Tennessee); Kelley Tilmon (South Dakota State University); Mo Way (Texas A & M University); and Joanne Whalen (University of Delaware).

- The most important sap feeders are the soybean aphid (*Aphis glycines*), which was discovered in North America in 2000, and the kudzu bug (*Megacopta cribraria*), which was discovered in the U.S. state of Georgia in 2009.
- The most important pod and seed feeders are the bean leaf beetle; the corn earworm (*Helicoverpa zea*), which is sometimes called the "soybean podworm"; several species of stink bugs, including the brown stink bug (*Euschistus servus*), the green stink bug (*Chinavia hilaris*), the redbanded stink bug (*Piezodorus guildinii*), the southern green stink bug (*Nezara viridula*), and the brown marmorated stink bug (*Halyomorpha halys*); and the tobacco budworm (*Heliothis verescens*).
- The most important stem feeders are the Dectes stem borer (*Dectes texanus*) and the threecornered alfalfa hopper (*Spissistilus festinus*).
- Insects that feed on the belowground parts of soybean plants (roots, nodules, planted seeds) are considered relatively minor pests throughout the United States; the most commonly identified belowground pests are colaspis grubs (primarily the grape colaspis [*Colaspis brunnea*]); the seedcorn maggot (*Delia platura*); white grubs (primarily *Phyllophaga* spp.); and wireworms (*Melanotus* spp., *Agriotes mancus, Limonius dubitans*).
- Although not an insect, the twospotted spider mite (*Tetranychus urticae*) is considered a major sap feeder/defoliator of soybean, especially when hot, dry conditions prevail.

As stated previously, entomologists' rankings of insect pests varied by region and among states within regions. Overall, the feeding guilds were ranked as follows:

- **Midwest:** (1) sap feeders; (2) defoliators; (3) fruit (flower, pod, and seed) feeders; (4) root, nodule, and planted-seed feeders; (5) stem feeders
- **Great Plains:** (1 and 2) sap feeders and stem feeders (varies across states); (3) defoliators; (4) fruit (flower, pod, and seed) feeders; (5) root, nodule, and planted-seed feeders
- **Delta/Mid-South/Gulf Coast:** (1) fruit (flower, pod, and seed) feeders; (2) defoliators; (3) sap feeders; (4) stem feeders; (5) root, nodule, and planted-seed feeders
- **Atlantic:** (1) fruit (flower, pod, and seed) feeders; (2) sap feeders; (3) defoliators; (4) stem feeders; (5) root, nodule, and planted-seed feeders

In the Midwest and the Great Plains, the soybean aphid (sap feeder) is the most economically damaging insect pest, followed by defoliators, including the beetle complex (primarily, the bean leaf beetle and the Japanese beetle) and grasshoppers. In the Delta/Mid-South/Gulf Coast states and the Atlantic states, stink bugs and caterpillars are the most significant complexes of insect pests. The Dectes stem borer is the most important stem feeder in the Midwest, the Great Plains, and the Atlantic states, whereas the threecornered alfalfa hopper is most important in the Delta/Mid-South/Gulf Coast states. The primary pod-feeding threats are the bean leaf beetle in the Midwest and the Great Plains states and stink bugs and corn earworms in the Delta/Mid-South/Gulf Coast states and the Atlantic states. Because of the rapid spread of the kudzu bug from Georgia to Delaware and Maryland, it is considered the primary sap feeder in the Atlantic states.

Details about other insect pests that are occasional threats to soybean production (Table 11) can be found in other publications, including those provided in the Selected References for this section. An overview of insect pest management is provided in Part IV of this compendium, and it offers strategies that can be used for managing all insect pests of soybean. Regional and local sources of information will provide specific details about insects and their management in soybean.

TABLE 11. Other Arthropod Pests Within Four Feeding Guilds Known to Cause Occasional Damage to Soybean in the United States[a]

Feeding Guild	Common Name (Scientific Name)
Defoliators	Alfalfa caterpillar (*Colias eurytheme*) Armyworms: – Beet (*Spodoptera exigua*) – Fall (*Spodoptera frugiperda*) – Yellowstriped (*Spodoptera ornithogalli*) Blister beetles (Multiple species; most in genus *Epicauta*) Mexican bean beetle (*Epilachna varivestis*) Northern corn rootworm—adult (*Diabrotica barberi*) Painted lady caterpillar (*Vanessa cardui*) Saltmarsh caterpillar (*Estigmene acrea*) Spotted cucumber beetle—larva (*Diabrotica undecimpunctata howardi*)[b] Webworms (Multiple species) Western corn rootworm—adult (*Diabrotica virgifera virgifera*) Woollybear caterpillars (Multiple species)
Sap feeders	Whiteflies (Multiple species)
Stem feeders	Cutworms (Multiple species) Pea leaf weevil (*Sitona lineatus*)
Root, nodule, and planted-seed feeders	Asiatic garden beetle (*Maladera castanea*) Mealybugs (Multiple species) Pea leaf weevil (*Sitona lineatus*) Southern corn rootworm—larva (*Diabrotica undecimpunctata howardi*)[b] Soybean nodule fly (*Rivellia quadrifasciata*) Whitefringed beetle (*Graphognathus* spp.)

[a] Based on reports of entomologists provided in survey results from 2009 and 2014. Information for this table was contributed from the following individuals: Jack Bacheler (North Carolina State University); Wayne Bailey (University of Missouri); Larry Buschman (Kansas State University); Angus Catchot (Mississippi State University); Eileen Cullen (University of Wisconsin); Jeremy Greene (Clemson University); Ron Hammond (Ohio State University); Ames Herbert (Virginia Polytechnic Institute and State University); Tom Hunt (University of Nebraska); Doug Johnson (University of Kentucky); Janet Knodel (North Dakota State University); Christian Krupke (Purdue University); Rogers Leonard (Louisiana State University); Gus Lorenz (University of Arkansas); Bob McPherson (University of Georgia); John Obermeyer (Purdue University); Matt O'Neal (Iowa State University); Ken Ostlie (University of Minnesota); Bruce Potter (University of Minnesota); Dave Ragsdale (University of Minnesota); Tim Reed (Auburn University); Dominic Reisig (North Carolina State University); Ron Smith (Auburn University); Kevin Steffey (Dow AgroSciences); Scott Stewart (University of Tennessee); Kelley Tilmon (South Dakota State University); Mo Way (Texas A & M University); and Joanne Whalen (University of Delaware).

[b] Common names for the adult and larval stages of *Diabrotica undecimpunctata howardi* differ: adult = spotted cucumber beetle; larva = southern corn rootworm.

Selected References

Hammond, R. B., Higgins, R. A., Mack, T. P., Pedigo, L. P., and Bechinski, E. J. 1991. Soybean pest management. Pages 341-474 in: Handbook of Pest Management in Agriculture. 2nd ed. D. Pimentel, ed. CRC Press, Boca Raton, FL.

Higley, L. G., and Boethel, D. J., eds. 1994. Handbook of Soybean Insect Pests. Entomological Society of America, Lanham, MD.

(Prepared by K. L. Steffey)

Bean Leaf Beetle

The bean leaf beetle (*Cerotoma trifurcata*) is a consistent presence in soybean fields throughout North America, and occasionally, the population density of this insect reaches a level that requires management. In addition to being a pest of soybean, this native insect feeds on a number of other leguminous plants, including crops such as cowpea and green bean.

Description

The bean leaf beetle is approximately 6 mm long; it is light tan or red-orange and often has a pattern of black spots on its back. There is also a distinctive black, triangular pattern on the center of its back, where the wing covers meet the thorax (Fig. 192). The female has a blackened frons (i.e., face), and the male has a light-tan frons; the male also has a brushlike pad on its fore tarsi.

Ellipsoid eggs are approximately 1 mm long and orange (Fig. 193); they are often laid singly. Larvae are white and have dark head capsules and terminal segments; they grow to a maximum length of approximately 1 cm.

Life Cycle

Two or three generations of bean leaf beetles are produced a year in North America. Overwintering adult beetles become active during April or May and often begin feeding and mating in crops of alfalfa and other perennial legumes. As soon as soybean seedlings emerge from the soil, the beetles move into the fields to feed on the young plants and to lay eggs (as many as 300 per female). Although egg-laying behaviors depend on soil characteristics and temperatures, 90% of eggs in soybean fields are laid in the top 8 cm of the soil, near the bases of plants.

Larvae develop through three stages while feeding on the internal content of soybean nodules (Fig. 194) and possibly on root hairs, as well. Pupation occurs within the soil, and the entire egg–adult life cycle lasts approximately 30 days (depending on the temperature). In regions at northern latitudes, only the earliest of the first-generation beetles are reproductive; beetles that emerge subsequently must overwinter before laying eggs. In regions at southern latitudes, there are three generations of beetles per year. The beetles continue to feed on soybean plants until they move to overwintering sites during the late summer and autumn. Overwintering sites include forested areas and soybean residue. Adults diapause in their overwintering habitats from November to February in southern regions and from August to April in northern regions.

Damage

Adult bean leaf beetles feed on soybean from plant emergence until seed maturation, causing both direct and indirect damage. The primary damage is defoliation, which is recognized as round holes (Fig. 195). One adult can consume approximately 2 cm^2 of foliage per day. Early defoliation of cotyledons can kill seedlings, requiring replanting if the number of plants is reduced significantly (Fig. 196). Feeding during the vegetative growth stages is concentrated on the tender upper foliage. During the reproductive stages, adults feed on green soybean pods, removing their outer layers (Fig. 197). Because of this feeding, the seeds are exposed to secondary pathogens, particularly *Alternaria* species, which results in reduced yield and quality. Bean leaf beetles also feed on the pod peduncle, causing the clipped pod to drop from the plant.

In addition to causing direct feeding damage, adult bean leaf beetles transmit *Bean pod mottle virus, Cowpea mosaic virus, Cowpea chlorotic mottle virus,* and *Southern bean mosaic virus*

Fig. 192. Bean leaf beetle adult (*Cerotoma trifurcata*). (Courtesy J. G. Lundgren)

Fig. 193. Bean leaf beetle egg (*Cerotoma trifurcata*). (Courtesy J. G. Lundgren)

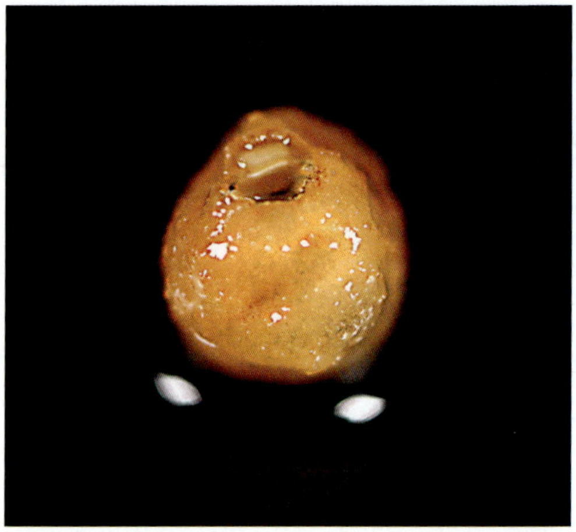

Fig. 194. Soybean nodule damaged by feeding of bean leaf beetle larvae (*Cerotoma trifurcata*). (Courtesy J. G. Lundgren)

Fig. 195. Roundish holes, indicating feeding by the bean leaf beetle (*Cerotoma trifurcata*). (Courtesy F. R. Musser)

(see Part I, Diseases Caused by Viruses). Beetle-transmitted viruses may reduce the soybean yield even when the density of the beetle population is low, potentially lowering the economic injury level for this insect. The impact of virus transmission on soybean productivity is inconsistent, however, depending on the proximity of the virus source and the age of the plant when transmission occurs. Injury from viral infection is greatest when infection occurs during the early growth stages.

Feeding by bean leaf beetle larvae may cause reduced nitrogen fixation and physiological changes to the soybean plant. The impact of larval feeding on yield has not been measured experimentally, but larval feeding may contribute to yield losses, particularly when coupled with other plant stress factors.

Management

Management of the bean leaf beetle should focus on minimizing the damage caused by adult feeding. A cultural control used to minimize the population of bean leaf beetles in northern areas is trap cropping, in which 5–10% of the field is planted 2–3 weeks earlier than the rest of the field. Bean leaf beetles migrate to the early planted trap crop, where they can be treated with an insecticide, if necessary; treating only a small part of the field reduces insecticide costs. A related approach is to plant soybean later than usual to avoid most first-generation beetles. Seeds treated with neonicotinoid insecticides provide effective protection from the bean leaf beetle for 3–4 weeks, shielding soybean plants during the vulnerable seedling stage. However, given the sporadic nature of the bean leaf beetle, prophylactic seed treatments are seldom warranted without knowledge of the pest's local history. Later in the season, the primary management option is application of foliar insecticides when the number of bean leaf beetles exceeds established treatment thresholds.

Treatment thresholds from VC to V2 are from two to five beetles per plant. Defoliation thresholds are approximately 30–40% defoliation and 15–25% defoliation during the vegetative stages and the reproductive stages, respectively. However, a variety of insects can defoliate soybean, including bean leaf beetles, numerous caterpillars, and several grasshoppers. Consequently, sampling to identify the primary defoliating species is required to select the appropriate control method. Treatment thresholds during pod feeding are based on injury (10% of pods showing feeding damage) or beetle density (two beetles per sweep with a 38-cm sweep net or 40 beetles/m^2).

Fig. 196. Injury to soybean seedlings, caused by the bean leaf beetle (*Cerotoma trifurcata*). (Courtesy A. Catchot)

Fig. 197. Bean leaf beetle (*Cerotoma trifurcata*) feeding and causing damage to a soybean pod. (Courtesy A. Catchot)

Selected References

Hadi, B. A. R., Bradshaw, J. D., Rice, M. E., and Hill, J. H. 2012. Bean leaf beetle (Coleoptera: Chrysomelidae) and bean pod mottle virus in soybean: Biology, ecology, and management. J. Integr. Pest Manage. 3:B1-B7.

Kogan, M., Waldbauer, G. P., Boiteau, G., and Eastman, C. E. 1980. Sampling bean leaf beetles on soybean. Pages 201-236 in: Sampling Methods in Soybean Entomology. M. Kogan and D. C. Herzog, eds. Springer, New York.

Krell, R. K., Pedigo, L. P., Hill, J. H., and Rice, M. E. 2004. Bean leaf beetle (Coleoptera: Chrysomelidae) management for reduction of bean pod mottle virus. J. Econ. Entomol. 97:192-202.

Lundgren, J. G., and Riedell, W. E. 2008. Soybean nitrogen relations and root characteristics after *Cerotoma trifurcata* (Coleoptera: Chrysomelidae) larval feeding injury. J. Entomol. Sci. 43:107-116.

Pedigo, L. P. 1994. Bean leaf beetle. Pages 42-44 in: Handbook of Soybean Insect Pests. L. G. Higley and D. J. Boethel, eds. Entomological Society of America, Lanham, MD.

(Prepared by J. G. Lundgren and F. R. Musser)

Caterpillars

Pod-Feeding Caterpillar: Corn Earworm

The corn earworm (*Helicoverpa zea*) is a primary pest of soybean throughout much of the eastern United States, extending from Delaware–Maryland through the Mississippi Delta, eastern Texas, and southern Louisiana and through much of Arkansas, eastern Kansas, and southeastern Missouri. The tobacco budworm (*Heliothis verescens*) is also found in the same general areas but further south than the corn earworm. The feeding patterns of the two insects are the same.

Description. A caterpillar (i.e., Lepidoptera larva) can be green, yellow, brown, or pink and have light and/or dark stripes along the length of its body. It has four pairs of abdominal prolegs and a light-green- to amber-colored head capsule (never dark brown), and it is covered with tiny, dark hairs or spinules (Fig. 198).

Life Cycle. In most areas, the corn earworm overwinters in the pupal stage, emerges as an adult (i.e., moth) during April or May, and progresses through three or four generations during summer to early autumn. Before the corn earworm moves into soybean, it completes one generation in field corn, where it develop in the ears, but this generation has been known to feed on soybean pods and potentially other hosts. Although the introduction of corn containing *Bacillus thuringiensis* toxins has decreased the earworm's survival in this crop, infestations in soybean are still common, in part because of the availability of non-*B. thuringiensis* corn.

Moths deposit eggs on soybean leaves and stems, and the eggs hatch into caterpillars in 2–3 days. The caterpillars develop through six stages (instars) and reach full size in approximately 2 weeks.

Damage. Young caterpillars (less than 31 mm long) typically feed on flowers and leaves. Older instars also feed on leaves, occasionally causing extensive defoliation on preblooming soybean plants. In the absence of other leaf feeders, defoliation by the corn earworm rarely causes economically significant damage.

Medium and large caterpillars cause the greatest damage to soybean by feeding directly on pods and seeds (Fig. 198). One caterpillar can consume 15–20 flat pods (before seed enlargement) or 6–10 older pods. The last two instars (fifth and sixth) do most of the pod feeding. Research during the late 1980s in the U.S. state of North Carolina showed that a single large caterpillar per row foot in a full-season soybean field reduced yield by an average of 28.7 t/ha.

Management. Many extension entomologists in the United States provide advisory information about the current corn earworm population and moth activity throughout the season based on pheromone-baited and black-light trapping systems. Most states have also developed economic thresholds for caterpillars, along with sampling procedures and insecticide application recommendations, based on local research. These recommendations can vary across a given region, so local extension personnel should be consulted.

Defoliating Caterpillars

Soybean plants can be attacked by several defoliating caterpillars during the vegetative and reproductive stages of development. The most common species are the green cloverworm (*Plathypena scabra*), the velvetbean caterpillar (*Anticarsia gemmatalis*), the soybean looper (*Chrysodeixis includens*), the cabbage looper (*Trichoplusia ni*), the garden webworm (*Achyra rantalis*), and the saltmarsh caterpillar (*Estigmene acrea*).

Descriptions. Larvae of the green cloverworm, the soybean and cabbage loopers, and the velvetbean caterpillar are typically smooth skinned and light green and have a series of pale, longitudinal stripes along the top of the abdomen. The color may vary depending on the population density, the time of the season, and the local field environment. The diagnostic key used to separate these species in the field focuses on the number of abdominal prolegs. Both the cabbage looper and the soybean looper have two pairs of prolegs, the green cloverworm has three pairs (Fig. 199), and the velvetbean caterpillar has four pairs. Soybean looper larvae frequently have dark thoracic legs, whereas those of cabbage looper larvae are almost always green.

Garden webworm larvae are green and have black spots on all body segments; they also have four pair of prolegs. These larvae are easily distinguished from larvae of other caterpillars by the silken webbing they produce to bind leaves, where they normally reside while feeding. When disturbed, the garden webworm backs away from the disturbance.

The saltmarsh caterpillar has four pairs of prolegs. It can be distinguished easily from the other caterpillars, however, because its body is almost entirely covered with long, yellow or black hairs.

Life Cycles. The life cycles of the green cloverworm, the velvetbean caterpillar, and the soybean and cabbage loopers are somewhat similar, particularly in relation to soybean field infestation. The complete life cycle for each species usually takes 25–30 days during the summer, and each produces three or four generations annually. All of these caterpillars exhibit some level of migration, and overwintering populations survive on legume plants in locations that do not experience freezing temperatures (e.g., southern United States, Caribbean basin, Central America). They may be detected on soybean early in the production season but are more commonly pests at the reproductive stage of the crop.

The garden webworm usually appears on vegetative soybean during middle to late June in the U.S. South and during late June and July in the Midwest, migrating from other plants—including weeds such as pigweed (*Amaranthus* spp.)—and localized infestations of soybean fields may be found in close proximity to this weed.

The saltmarsh caterpillar moth deposits clusters of eggs (400 to more than 1,000 per cluster), whereas the other species in this group deposit eggs singly. Late-stage saltmarsh caterpillars also actively disperse and may be found crawling over soil or roads, searching for suitable host plants, such as soybean. Infestations of saltmarsh caterpillars are often limited to the margins of fields, but insects might be found in low numbers throughout a given field.

Damage. Small defoliating caterpillars may cause a windowpane type of symptom on upper leaf surfaces; this damage is considered to be minor. Other typical damage includes large holes in leaves, skeletonization of entire leaves, and defoliation throughout the plant profile. The destruction of leaf tissue reduces the plant's ability for photosynthesis—an injury that may indirectly reduce pod and seed numbers and seed size. However, a soybean plant can tolerate a significant loss of foliage—35% prebloom and 15–20% during bloom and postbloom—before its yield is reduced.

Management. Action thresholds for the initiation of insecticide treatment to control defoliating caterpillars have been established in most U.S. states. Fields should be scouted at least weekly to measure infestation levels with a sweep net or shake sheet and to make visual observations and assess levels of defoliation. Numerous insecticides provide effective control, but achieving natural control with native predators, parasitoids, and insect pathogens should be considered before deciding to apply an insecticide.

Fig. 198. Corn earworm (*Helicoverpa zea*) feeding on and causing injury to a soybean pod. (Courtesy G. Lorenz)

Fig. 199. Green cloverworm larva (*Plathypena scabra*). (Courtesy G. Lorenz)

Armyworms

Several species of armyworm can be found in soybean during the growing season—primarily, the yellowstriped armyworm (*Spodoptera ornithogalli*), the fall armyworm (*S. frugiperda*), and the beet armyworm (*S. exigua*). The yellowstriped armyworm may occur in the southern United States during early April, whereas the fall armyworm and the beet armyworm usually occur later in the season (although the grass strain of the fall armyworm can occur around flowering). The fall armyworm and the beet armyworm are native to tropical and subtropical areas of Florida, Texas, and southern Louisiana. Both armyworms disperse by adult migration throughout much of the southern and midwestern United States, usually later in the growing season.

Descriptions. Armyworm caterpillars have four pairs of prolegs. The yellowstriped armyworm is usually dark brown to black and has a prominent yellow stripe along each side of its body. In addition, a black spot can usually be found on each side of the first abdominal segment, just behind the first pair of true legs.

The fall armyworm larva varies in color from tan to green and has black bumps with dark-black hairs on its body (Fig. 200). On the upper surface of the eighth (last) abdominal segment, there are four distinct black spots arranged in a square. This larva also has a light-colored, inverted "Y" on the front of its head. A mature beet armyworm larva is usually green and has prominent dark, lateral stripes. Unlike the fall armyworm larva, the beet armyworm larva does not have stout, black hairs on its body. Rather, the beet armyworm larva has a black spot on each side of its body on the second thoracic segment, just above the middle pair of true legs.

Life Cycles. The life cycles of these armyworms are similar. The entire life cycle takes 24–36 days, depending on environmental conditions. In the United States, the number of generations ranges from one to six a year, depending on geography and latitude.

All three species lay their eggs in clusters, usually on the undersides of leaves, and the females cover the eggs with scales from their bodies. All three species pupate in the soil. After the eggs hatch, the larvae are initially found in clusters but quickly disperse by spinning silken threads and "ballooning" via the wind.

Damage. The yellowstriped armyworm rarely occurs on soybean seedlings in sufficient numbers to cause economically significant damage. Plants usually recover from the injuries caused by these armyworms with no yield loss, and control is seldom necessary.

Some infestations of the fall armyworm occur on soybean seedlings during the early spring in numbers large enough to cause significant yield losses, usually in double-cropped or late-planted crops, in which grass becomes established and fall armyworms first infest the grass. When herbicides are used to control the grass, the fall armyworm moves from the dying grass onto the soybean plants. Additionally, a population of the fall armyworm may occur late in the season from the direct deposit of eggs on soybean and not build up to damaging levels.

Occasionally, a large population of the beet armyworm develops late in the season. Heavy infestations of larvae feeding on foliage, blooms, and pods can cause significant yield losses.

Management. Action thresholds are based on percentages of defoliation with larvae present and vary slightly from state to state. Scouting fields weekly is critical to detect potentially threatening populations before defoliation levels exceed economically significant damage levels.

Predators, parasitoids, and insect pathogens play important roles in regulating populations of armyworms. Disregarding the value of these natural enemies by making unneeded insecticide applications can destroy this complex and make pests even more problematic. Many beet armyworm and fall armyworm populations have developed resistance to pyrethroid insecticides and may be difficult to control.

Fig. 200. Fall armyworm larva (*Spodoptera frugiperda*). (Courtesy G. Lorenz)

Selected References

Herbert, A. 2000. Corn Earworm Economic Threshold Calculator. Virginia Tech University, Tidewater Agricultural Research and Extension Center. Available online at www.ipm.vt.edu/cew

Herbert, D. A., Jr. 2009. Essentials of a good soybean insect pest management program. Sect. 4-59 in: Pest Management Guide, Field Crops: 2015. Virginia Tech University, Virginia Cooperative Extension. Available online at http://pubs.ext.vt.edu/456/456-016/456-016-15-Field-Crops.pdf

(Prepared by G. Lorenz, A. Herbert, and B. R. Leonard)

Dectes Stem Borer

The Dectes stem borer (*Dectes texanus*) is native to North America and was first noted as a pest of soybean in 1968 in the U.S. state of North Carolina and in the southeastern "boot-heel" region of Missouri. Since then, this insect has been reported as a significant pest in isolated areas of Arkansas, Kansas, Kentucky, Missouri, and Tennessee. It is also a significant pest of sunflower. Other hosts include cocklebur (*Xanthium strumarian*), common and giant or great ragweed (*Ambrosia artemisiifolia* and *A. trifida,* respectively), cowpea, crested anoda (*Anoda cristata*), and wild sunflower (*Helianthus annuus*).

There are many common names for *D. texanus*. In addition to "Dectes stem borer," other common names are "long-horned beetle," "soybean stem borer," "sunflower stem borer," and "sunflower stem girdler."

Description

The Dectes stem borer is a small (6–13 mm), gray, flat-faced, long-horned beetle. Its antennae are longer than its body and are marked with black and gray rings (Fig. 201). The larvae are legless, up to 13 mm long, and creamy white to yellow with brown, incomplete head capsules (Fig. 202). The larvae also have accordion-like body segments, and their bodies taper toward their posteriors.

Life Cycle

Most adult Dectes stem borers emerge from mid-June to early October; the peak emergence period runs from mid-July through early August. Adults can live nearly 60 days in captivity, but their field survival time is likely much shorter. Adults feed on the epidermises of tender, newly emerged stems and leaves for approximately10 days. They mate after 10–12 days and begin oviposition 3–12 days after mating. The female lays up to 50 eggs, and each egg is laid individually in the pith of a tender leaf petiole.

Larvae hatch in 3–5 days and feed on the pith inside leaf petioles until the late-second- to third-instar stage; then, they bore down from the leaves into the main stems. A larva tunnels in the main stem until the plant matures. As the plant dries down, the larva tunnels toward the soil, where it overwinters inside a chamber at the base of the plant. Before sealing off the overwintering chamber with frass and shredded plant material, the larva may cut the plant from inside approximately 5–10 cm above the soil line (i.e., "girdling").

During the spring, the larva creates an exit hole for its emergence as an adult and then pupates inside the stem. The pupal stage lasts 10–15 days. An adult remains in the stem 1–2 days (until its wings harden) before emerging.

Damage

The presence of Dectes stem borer larvae in soybean plants may not be detected unless the stems are dissected to reveal tunneling (Fig. 202). However, the first symptom of damage from the Dectes stem borer is the presence of wilted leaves in the upper canopy. The leaves of infested petioles wilt and then drop, exposing the entry hole where the larva tunneled into the main stem. The entry hole is the second symptom. Several larvae may enter the plant's main stem from the same petiole or from different petioles, but because the larvae are cannibalistic, only one larva typically survives per plant.

Fig. 201. Dectes stem borer adult (*Dectes texanus*). (Courtesy K. V. Tindall)

Fig. 202. Dectes stem borer larva (*Dectes texanus*) tunneling in a soybean stem. (Courtesy K. Fothergill)

During the autumn, the larva tunnels to the base of the plant to develop an overwintering chamber. It then girdles the plant approximately 5–10 cm above the soil line. Girdled plants are likely to lodge (i.e., fall over), especially during high winds and severe storms and when dew or rain make plants top heavy. Lodged plants are difficult to harvest, because they can bunch up and clog the machinery. Harvest losses resulting from lodging can be as great as 40%. Physiological yield reductions of approximately 10% are associated with tunneling.

Management

Management options for the Dectes stem borer are limited and not universally effective. Cultural control practices have been widely recommended, but some of them are not compatible with current farming practices. The most useful practice is to harvest early. Infested fields should be identified and harvested promptly once they reach maturity.

Crop rotation may be effective in managing the Dectes stem borer when the soybean acreage is limited (isolated fields), but rotation is less effective in areas with significant acreage, because the insect can easily find nearby soybean fields each year. Destruction of stubble by tillage can also be effective, but this practice is not acceptable in areas with residue coverage requirements. Moreover, the insect moves readily from reduced-tillage fields to tilled fields.

Commercially available insecticides have not been effective for controlling the Dectes stem borer, because they target the adult stage and there is a lengthy infestation period. Biological control agents have been identified—primarily, parasitic wasps in the family Pteromalidae—but their impact on the Dectes stem borer population has not been determined.

Soybeans resistant to the Dectes stem borer have been identified, but resistant cultivars are not commercially available.

Selected References

Buschman L. L., and Sloderbeck P. E. 2010. Pest status and distribution of the stem borer, *Dectes texanus*, in Kansas. J. Insect Sci. 10:198.

Campbell, W. V., and Van Duyn, J. W. 1977. Cultural and chemical control of *Dectes texanus* on soybeans. J. Econ. Entomol. 70:256-258.

Falter, J. M. 1969. *Dectes* sp. (Coleoptera: Cerambycidae): A unique and potentially important pest of soybeans. J. Elisha Mitchell Sci. Soc. 85:123.

Hatchett, J. H., Daugherty, D. M., Robbins, J. C., Barry, R. M., and Houser, E. C. 1975. Biology in Missouri of *Dectes texanus*, a new pest of soybeans. Ann. Entomol. Soc. Am. 68:209-213.

Hatchett, J. H., Jackson, R. D., and Barry, R. M. 1973. Rearing a weed cerambycid, *Dectes texanus,* on artificial medium, with notes on biology. Ann. Entomol. Soc. Am. 66:519-522.

Michaud, J. P. 2009. Soybean Insects: Dectes Stem Borer, *Dectes texanus*. Kansas State University, Department of Entomology, Manhattan. Available online at http://entomology.k-state.edu/extension/insect-information/crop-pests/soybeans/sbsb/

Michaud, J. P., and Grant, A. K. 2005. The biology and behavior of the longhorned beetle, *Dectes texanus,* on sunflower and soybean. J. Insect Sci. 5:25.

Michaud, J. P., and Grant, A. K. 2010. Variation in fitness of the longhorned beetle, *Dectes texanus*, as a function of host plant. J. Insect Sci. 10:206.

Piper, G. L. 1978. Biology and immature stages of *Dectes sayi* Dillon and Dillon (Coleoptera: Cerambycidae). Coleopt. Bull. 32:299-306.

Richardson, L. G. 1975. Resistance of soybeans to a stem borer, *Dectes texanus texanus* LeConte. Ph.D. thesis. North Carolina State University, Raleigh.

Sloderbeck P. E., and Buschman L. L. 2011. Aerial insecticide treatments for management of Dectes stem borer, *Dectes texanus*, in soybean. J. Insect Sci. 11:49.

Tindall, K. V., Stewart, S., Musser, F., Lorenz, G., Bailey, W., House, J., Henry, R., Hastings, D., Wallace, M., and Fothergill, K. 2010. Distribution of the long-horned beetle, *Dectes texanus*, in soybeans of Missouri, western Tennessee, Mississippi, and Arkansas. J. Insect Sci. 10:178.

(Prepared by K. V. Tindall and L. L. Buschman)

Kudzu Bug

The insect commonly known as the "kudzu bug" gets its name from its strong affinity for kudzu as its preferred wild host. Both invasive species (insect and weed) are indigenous to the Eastern Hemisphere and are commonly found in China, India, Japan, and other countries in that part of the world.

Also known as the "bean plataspid" (*Megacopta cribraria*) (Hemiptera: Plataspidae), this insect has been an economically important pest of soybean and a tremendous nuisance pest for U.S. homeowners since its discovery in 2009 in northeastern Georgia. By the end of 2013, it had spread to 12 additional states, plus the District of Columbia (Fig. 203). South Carolina was the first state to report all of its counties being infested with the kudzu bug (confirmed on July 15, 2011). Undoubtedly, the prevailing weather patterns that move across the southeastern United States from west to east assisted in distribution of the insect in that direction and thereby explain its slower movement to the west.

Description

Adult kudzu bugs are approximately the same size as adult lady beetles. They are small, almost square (approximately 0.6 cm long), and light brown with an olive-green hue and dark specks (Fig. 204). The eggs are light tan and deposited in a slanting position in two-row masses (Fig. 205). Each barrel-shaped egg has a row of spines around the operculum (i.e., the "lid" opening for the emerging nymph).

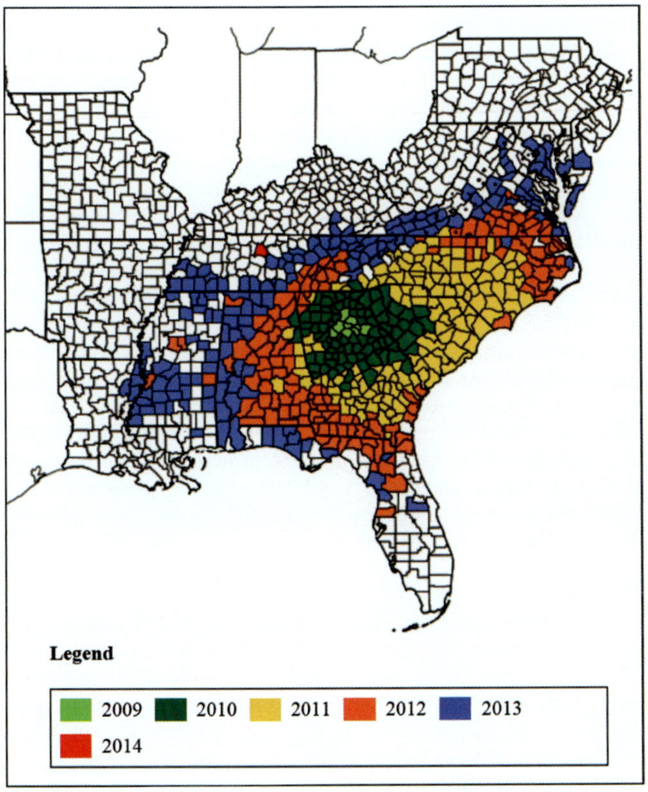

Fig. 203. Distribution of the kudzu bug (*Megacopta cribraria*) in the United States. (Courtesy J. K. Greene)

The immatures (i.e., nymphs) that hatch from eggs are orange and remain in close proximity to the egg mass for a short time before dispersing (Fig. 206). As the nymphs grow, they take on a pale-green color and a very hairy appearance, particularly late in the immature stages (Fig. 207). When large numbers of adults and/or nymphs are present, a distinct odor can be detected; it can be quite strong, often to the point of being useful in detecting the presence of the insect before visually confirming it.

Life Cycle

The kudzu bug has a life cycle similar to the life cycles of stink bugs and other hemipterans. It goes through a simple, gradual metamorphosis; the major stages of the life cycle are eggs, nymphs, and adults.

The eggs are deposited in masses, typically on the undersides of leaves. The nymphs usually hatch from eggs within 1–2 weeks. There are five immature stages (instars) for the developing nymphs, and molts separate the stages. Development

Fig. 204. Dorsal view of an adult kudzu bug (*Megacopta cribraria*). (Courtesy J. K. Greene)

Fig. 205. Egg mass deposited by a kudzu bug (*Megacopta cribraria*) on a soybean leaf. (Courtesy J. K. Greene)

of the first through the fifth instar takes 6–8 weeks, depending on the temperature. After the fifth instar completes its molt, the winged adult emerges.

Adults are reproductively capable within days and seek mates very quickly, starting the entire life cycle again. Adults overwinter and can live for months, depending on the time of year.

Damage

The kudzu bug is a phytophagous stem feeder that prefers species of legumes. Like other vectors of plant pathogens (e.g., aphids, bean leaf beetles, leafhoppers, thrips), the kudzu bug has piercing–sucking mouthparts that can penetrate plant stems and petioles during probing. Kudzu bugs have not been found to be vectors of specific soybean diseases, although they can be associated with delayed plant maturity. Infestations of soybean fields typically begin around the edges or borders.

Management

Using insecticides to control the kudzu bug in soybean is highly effective, as the insect appears to be susceptible to a variety of insecticide chemistries. Many of the insecticides labeled and recommended for use on other insect pests of soybean (e.g., stink bugs)—including the pyrethroids and organophosphates, among others—are effective in controlling the kudzu bug. The pyrethroid insecticides, as well as products premixed with a pyrethroid, offer the best initial and residual control of kudzu bugs in soybean. The recommendation is to interrupt development of each generation in soybean with an insecticide application targeting the immature stage of the insect. Treatment thresholds and sampling plans for kudzu bugs are being refined, but the recommendation is to apply an insecticide when immatures (i.e., nymphs) are detected in sweep-net samples at approximately one nymph per sweep.

As an alternative to sweep-net sampling, visual inspections of insect density down in the canopy may suffice. If immature kudzu bugs are easily and repeatedly found on plants' petioles and main stems using this method of canopy observation, then treatment is likely warranted.

Fig. 206. Newly hatched kudzu bug (*Megacopta cribraria*) immatures in close proximity to the egg mass. (Courtesy J. K. Greene)

Fig. 207. Kudzu bug nymphs (*Megacopta cribraria*) on soybean stems. (Courtesy J. K. Greene)

Selected References

Eger, J. E., Jr., Ames, L. M., Suiter, D. R., Jenkins, T. M., Rider, D. A., and Halbert, S. E. 2010. Occurrence of the Old World bug *Megacopta cribraria* (F.) (Heteroptera: Plataspidae) in Georgia: A serious home invader and potential legume pest. Insecta Mundi 121:1-11.

Gardner, W. A., Peeler, H. B., La Forest, J., Roberts, P. M., Sparks, A. N., Greene, J. K., Reisig, D., Suiter, D. R., Bacheler, J. S., Kidd, K., Ray, C. H., Hu, X. P., Kemerait, R. C., Scocco, E. A., Eger, J. E., Ruberson, J. R., Sikora, E. J., Herbert, D. A., Jr., Campana, C., Halbert, S., Stewart, S. D., Buntin, G. D., Toews, M. D., and Bargeron, C. T. 2013. Confirmed distribution and occurrence of *Megacopta cribraria* (F.) (Hemiptera: Heteroptera: Plataspidae) in the southeastern United States. J. Entomol. Sci. 48:118-127.

Seiter, N. J., Reay-Jones, F. P. F., and Greene, J. K. 2013. Within-field spatial distribution of *Megacopta cribraria* (Hemiptera: Plataspidae) in soybeans. Environ. Entomol. 42:1363-1374.

Suiter, D. R., Eger, J. E., Kemerait, R. C., All, J. N., Roberts, P. M., Gardner, W. A., Greene, J. K., Ames, L. M., Buntin, G. D., Jenkins, T. M., and Douce, G. K. 2010. Discovery and distribution of *Megacopta cribraria* (Hemiptera: Heteroptera: Plataspidae) in northeast Georgia. J. Integr. Pest Manage. 1:1-4.

Zhang, Y., Hanula, J. L., and Horn, S. 2012. The biology and preliminary host range of *Megacopta cribraria* (Heteroptera: Plataspidae) and its impact on kudzu growth. Environ. Entomol. 41:40-50.

(Prepared by J. K. Greene)

Soybean Aphid

The soybean aphid (*Aphis glycines*) is native to Asia. It was first detected in the United States in 2000 and quickly spread across 22 states and three provinces of Canada. The soybean aphid is a pest predominantly in regions at middle to high latitudes in the United States and is seldom found south of Kansas, Kentucky, Missouri, and Virginia. In the southern United States, the soybean aphid is sometimes confused with the cotton aphid (*Aphis gossypii*), which may occur on plants at the same time.

The soybean aphid is a significant pest of soybean in the north-central United States, causing an estimated annual yield loss of $2.4 billion or more if left untreated. Before the arrival of the soybean aphid, a very small amount of insecticide was used on soybean in the north-central region, but following widespread infestations of this insect, soybean growers dramatically increased the use of insecticides. Significant field-based research has demonstrated that integrated pest management (IPM) strategies can limit yield losses while preventing unnecessary input costs and environmental damage (see Management section).

Description

The soybean aphid is small (1.6 mm long), soft bodied, yellow to pale green, and generally shaped like a lemon drop (Fig. 208). Dark-tipped cornicles ("tailpipes") are found on the abdomen of the adult. During the late summer, some aphids may be much smaller and paler in appearance. A winged aphid has a black head and thorax (Fig. 209).

Soybean aphids may occur in low densities of only a few insects per plant, or they may form large, persistent colonies of several hundreds or thousands per plant.

Life Cycle

Like many aphid species, the soybean aphid has a complex life cycle that involves different physical forms (morphs) and two host plant species. Through most of their life cycle, soybean aphids reproduce asexually. On soybean, the aphids are female and give live birth to all female offspring, which are themselves only days away from reproducing.

These aspects of the insect's biology enable the soybean aphid to increase its population very rapidly when conditions are favorable. In general, the soybean aphid reproduces at a faster rate in temperatures of 25–28°C. Consistently high temperatures can result in aphid mortality or reduced population growth. Reproduction is also greater on plants grown in potassium-deficient soils.

The soybean aphid overwinters as eggs on buckthorn (*Rhamnus* spp.), a shrub common in shelterbelts and woods throughout the north-central United States. Buckthorn is a critical part of the insect's life cycle; without it, the soybean aphid cannot survive the winter. Soybean fields with buckthorn nearby are often the first to be colonized by aphids during the spring and should be scouted early.

Aphid eggs hatch on buckthorn after overwintering, usually in April. Upon hatching, the population increases and produces all female winged morphs, which disperse from the buckthorn to colonize the nearby soybean—the summer host of the aphid. The winged females bear wingless offspring that bear their own wingless offspring within days. Consequently, the soybean aphid can complete many generations during a single summer. When conditions become locally crowded, many of the wingless aphids give birth to offspring that develop into winged morphs, which then disperse to colonize other soybean fields.

At the end of the growing season, the aphids produce both male and female winged morphs (the only sexual generation), which migrate back to the buckthorn to mate. The females lay eggs on buckthorn, completing the life cycle.

Damage

Soybean aphids have piercing–sucking mouthparts and feed on phloem sap. A heavy infestation can stunt soybean plants' growth and development, causing a significant yield loss. A large infestation can also negatively affect the seed quality and size, the pod number, the plant height, and photosynthesis. Heavily infested plants are stunted and may be covered with dark, sooty mold, which grows on the sugary "honeydew" excreted by the aphids. When a large outbreak of thousands of aphids per plant occurs, the insects' feeding can reduce yield by 40% or more.

Like many aphids, soybean aphids can transmit plant viruses. They can transmit both *Soybean mosaic virus* and *Alfalfa mosaic virus* to soybean and other transiently visited crops (see Part I, the section Diseases Caused by Viruses). Efforts to manage these viruses by targeting the soybean aphid have not reduced their incidences, in part because several other aphids also transmit these viruses to soybean.

Management

Damage caused by phloem-feeding insects may go undetected without close visual inspection, and feeding injury may become apparent only after large, yield-reducing populations have developed. Therefore, scouting is an important component of soybean aphid management. Aphid population levels can vary greatly from field to field, even on the same farm. The population often remains small throughout the season and the cost of treatment is not justified. In other instances, an initially small aphid population reaches a damaging level quickly. Scouts are advised to check fields weekly from late June through late August.

During the early to mid-reproductive stages of soybean (R1–R5), the threshold for treatment is when the population exceeds 250 aphids per plant and 80% of plants are infested. Recognizing this economic threshold allows sufficient time to arrange for treatment before the aphid population exceeds the economic injury level. Application of a foliar insecticide is recommended within 3–7 days after the population reaches the economic threshold. When soybean plants reach the R6 growth stage (i.e., full seed, when a pod on one of the four top nodes has green seeds that fill the pod to capacity), the yield loss is limited and insecticide applications may not provide a return on investment, regardless of the aphid density. Applying insecticides only when the number of soybean aphids exceeds the economic threshold is more cost effective than making preventive applications based solely on the plants' growth stage (R3), irrespective of aphid density. Sampling protocols such as "speed scouting" have been developed to aid in the rapid assessment of a soybean aphid population relative to the economic threshold.

Insecticides from multiple classes (e.g., organophosphates, pyrethroids, and neonicotinoids) are effective for control of the soybean aphid. Seed-applied insecticides (i.e., seed treatments), such as thiamethoxam and imidacloprid, can limit growth of the aphid population early in the season. Yet these insecticides lack the residual activity needed to suppress population growth during late July and August, when damaging populations typically occur. In the absence of other early season insect pests (e.g., bean leaf beetles) at levels that may justify use of a seed treatment, seed treatments alone or in combination with a foliar insecticide are not likely cost effective for soybean aphid control, based on available evidence.

Fig. 208. Soybean aphids (*Aphis glycines*). Note the female in the upper-left corner giving live birth to another female. (Courtesy R. Scott)

Fig. 209. Black-headed, winged soybean aphid (*Aphis glycines*), along with wingless aphids. (Courtesy R. Scott)

Fig. 210. Soybean plant (right) heavily infested with hundreds of soybean aphids (*Aphis glycines*), compared with a plant (left) that is aphid free because it contains the *Rag1* aphid-resistance gene. (Courtesy C. B. Hill)

Scouting protocols for soybean aphids include walking a broad "U" or "X" pattern through a field and examining a total of 20–30 plants throughout the field. Aphids can occur in hotspots, but treatment decisions should be based on a broad sample of randomly selected plants. During July and early August, soybean aphids are found most often on the newest leaves and on the undersides of leaves. Late in the season, they are often concentrated in the lower canopy.

Field studies have revealed a diverse community of natural enemies (e.g., lady beetles, minute pirate bugs, and insect-killing fungi) that can help suppress soybean aphid colonization and population growth. In the absence of predators, population growth is significantly faster (two to seven times) than in the presence of predators. Prophylactic applications of broad-spectrum insecticides or fungicides may exacerbate aphid infestations or cause secondary outbreaks of other pests (e.g., spider mites) by removing the natural enemies that often keep aphid and other insect populations in check.

Host plant resistance is a promising management tactic for soybean aphids. Several genes have been found in soybean germplasm that provide resistance to the soybean aphid; they are called *Rag* genes (for "resistance to *Aphis glycines*"). Beginning in 2009, the first aphid-resistant lines containing the *Rag1* gene were made commercially available in North America (Fig. 210). Growing soybean cultivars with *Rag1* may reduce the aphid population significantly but not entirely; applying an insecticide may be required to reach the optimal yield. Since the discovery of *Rag1,* additional *Rag* genes have been identified and are being incorporated into soybean lines. Pyramid lines containing more than one resistance gene show particular promise for protecting yields.

Selected References

Clark, A. J., and Perry, K. L. 2002. Transmissibility of field isolates of soybean viruses by *Aphis glycines*. Plant Dis. 86:1219-1222.

Hesler, L. S., Chiozza, M. V., O'Neal, M. E., MacIntosh, G. C., Tilmon, K. J., Chandrasena, D. I., Tinsley, N. A., Cianzio, S. R., Costamagna, A. C., Cullen, E., Di Fonzo, C. D., Potter, B. D., Steffey, K., and Koehler, K. J. 2013. Performance and prospects of *Rag* genes for management of soybean aphid. Entomol. Exp. Appl. 147:201-216.

Hill, C. B., Chirumamilla, A., and Hartman, G. L. 2012. Resistance and virulence in the soybean-*Aphis glycines* interaction. Euphytica 186:635-646.

Hodgson, E. W., McCornack, B. P., Koch, K. A., Ragsdale, D. W., Johnson, K. D., O'Neal, M. E., Cullen, E. M., Kraiss, H. J., Di Fonzo, C. D., and Behnken, L. M. 2007. Field validation of speed scouting for soybean aphid. Crop Management. doi:10.1094/CM-2007-0511-01-RS.

Hodgson, E. W., McCornack, B. P., Tilmon, K., and Knodel, J. J. 2012. Management recommendations for soybean aphid (Hemiptera: Aphididae) in the United States. J. Integr. Pest Manage. 3:1-10.

Johnson, K. D., O'Neal, M. E., Ragsdale, D. W., Di Fonzo, C. D., Swinton, S. M., Dixon, P. M., Potter, B. D., Hodgson, E. W., and Costamagna, A. C. 2009. Probability of economical management of soybean aphid (Hemiptera: Aphididae) in North America. J. Econ. Entomol. 102:2101-2108.

Ragsdale, D. W., Landis, D. A., Brodeur, J., Heimpel, G. E., and Desneux, N. 2011. Ecology and management of the soybean aphid in North America. Annu. Rev. Entomol. 56:375-399.

Ragsdale, D. W., McCornack, B. P., Venette, R. C., Potter, B. D., MacRae, I. V., Hodgson, E. W., O'Neal, M. E., Johnson, K. D., O'Neil, R. J., Di Fonzo, C. D., Hunt, T. E., Glogoza, P., and Cullen, E. M. 2007. Economic threshold for soybean aphid (Hemiptera: Aphididae). J. Econ. Entomol. 100:1258-1267.

Tilmon, K. J., Hodgson, E. W., O'Neal, M. E., and Ragsdale, D. W. 2011. Biology of the soybean aphid, *Aphis glycines* (Hemiptera: Aphididae), in the U.S. J. Integr. Pest Manage. 2:1-7.

(Prepared by K. J. Tilmon and M. E. O'Neal)

Stink Bugs

Stink bugs are one of the most important groups of pestiferous insects in soybean. Stink bug feeding not only causes losses in yield, quality, and oil content, but it also reduces germination and delays plant maturity. Numerous species of stink bugs can be encountered in soybean, but only a few are regularly of economic importance: namely, the green stink bug (*Chinavia hilaris*); the southern green stink bug (*Nezara viridula*); the brown stink bug (*Euschistus servus*); and the redbanded stink bug (*Piezodorus guildinii*).

The green and the brown stink bugs are distributed across all soybean-producing areas of the United States, whereas the southern green and the redbanded stink bugs occur primarily in the southernmost region of the country. The southern green and the redbanded stink bugs are the most reproductively fecund species and are thus often the principal species when problems occur with stink bugs in the southern United States. However, all four species typically occur in a complex and are significant annual problems responsible for limiting soybean yields in the South.

Descriptions

The four species of stink bug have distinct characteristics. In the adult stage, the brown stink bug is the only species that is brown (Fig. 211); the other three species are green. Of the green species, the green and the redbanded stink bugs have noticeable ventral abdominal spines. The ventral (i.e., abdominal) spine on the redbanded stink bug is very long (Fig. 212), extending between the bases of the middle pair of legs, and the ventral spine on the green stink bug is short and pointed near the bases of the hind legs (Fig. 213). Antennae of the southern green stink bug have red stripes (Fig. 214), whereas antennae of the green stink bug have black stripes (Fig. 215).

During the immature stages of development, the nymphs of these species also have distinguishable characteristics. The brown stink bug nymph is yellow-tan to green-brown (Fig. 216). The nymph of the southern green stink bug has spots on the dorsal surface of the abdomen (Fig. 217), whereas the nymph of the green stink bug has lateral stripes (Fig. 218). The redbanded stink bug nymph has several compact, black bands on the dorsum confined near the midline of the abdomen (Fig. 219).

Life Cycle

All species of stink bugs go through simple, gradual metamorphosis, in which eggs, nymphs, and adults are the major stages of the life cycle. Eggs are deposited in masses (Figs. 220

Fig. 211. Brown stink bug adult (*Euschistus servus*). (Courtesy S. D. Stewart)

Fig. 212. Underside of a redbanded stink bug adult (*Piezodorus guildinii*). Note the long ventral spine. (Courtesy A. Catchot)

Fig. 213. Underside of a green stink bug adult (*Chinavia hilaris*). Note the short, pointed ventral spine. (Courtesy S. Bundy)

Fig. 214. Southern green stink bug adult (*Nezara viridula*), which has red-striped antennae. (Courtesy J. K. Greene)

Fig. 215. Green stink bug adult (*Chinavia hilaris*), which has black-striped antennae. (Courtesy S. D. Stewart)

Fig. 216. Brown stink bug nymph (*Euschistus servus*). (Courtesy S. D. Stewart)

and 221), typically on the undersides of leaves, and the number of eggs per mass varies by species.

The nymphs usually hatch from eggs within a week (Fig. 222). There are five immature stages (instars) for the developing nymphs, and molts separate these stages. Development of the first through the fifth instar takes several weeks, depending on the temperature. After the fifth instar completes its molt, the winged adult emerges.

Adults are reproductively capable within days and seek mates very quickly, starting the entire life cycle again. Adults overwinter and can live for months, depending on the time of year.

Damage

Phytophagous species of stink bugs are fruit and seed feeders, and species important to soybean damage the seeds in developing pods. Stink bugs have piercing–sucking mouthparts, similar to other vectors that penetrate plant stems and pod hulls when probing; thus, they have the potential to transmit microorganisms directly to the plant or to create feeding injury to pods and seeds, which facilitates the entry of microorganisms.

Stink bugs have not been found to be vectors of specific diseases of soybean. However, they have been associated with delayed maturity, possibly through the loss of or injury to pods,

Fig. 217. Southern green stink bug nymph (*Nezara viridula*). (Courtesy S. D. Stewart)

Fig. 218. Green stink bug nymph (*Chinavia hilaris*). (Courtesy S. D. Stewart)

Fig. 219. Redbanded stink bug nymph (left) and adult (right) (*Piezodorus guildinii*). (Courtesy A. Catchot)

Fig. 220. Egg mass deposited by a southern green stink bug (*Nezara viridula*) on the underside of a leaf. (Courtesy S. D. Stewart)

Fig. 221. Egg mass deposited by a redbanded stink bug (*Piezodorus guildinii*) on a soybean pod. (Courtesy S. D. Stewart)

Fig. 222. Green stink bug nymphs (*Chinavia hilaris*) hatching from an egg mass. (Courtesy S. D. Stewart)

and with obscure maladies such as green stem (or green bean) syndrome (see Part III, the section Green Stem Disorder).

Management

Management approaches for stink bugs in soybean are primarily reactive, and few or no preventive measures are available. Soybean growers rely almost exclusively on insecticides to control stink bugs when the insect population reaches the threshold. The treatment or action threshold for stink bugs in soybean—commonly assessed via a sweep net or drop cloth—varies from state to state, depending on the sampling technique, the development stage of the crop, and whether the crop is being grown for seed production. For sampling with a sweep net, the general threshold for all regions is five to nine bugs per 25 sweeps (or two to four per 10 sweeps). For sampling with a drop cloth, almost all states have consistent recommendations for treatment when the population exceeds one stink bug per foot of row.

When the number of stink bugs exceeds the treatment threshold, the choice of insecticide is dependent on the species present and whether it is part of a greater pest complex attacking the crop at the same time. Pyrethroid insecticides are often used, because they provide broad-spectrum control of numerous caterpillars, selected species of stink bugs, and other insect groups. Organophosphate insecticides and combinations of insecticide chemistries have been used to decrease populations of the more difficult-to-control species, such as the brown and the redbanded stink bugs. Fungicides and insecticides are often applied together in a tank mix to minimize operational activities in the field. Populations of disease-causing organisms and insect pests often overlap at least once during the growing season (e.g., growth stage R3), providing an opportune time for intervention with chemical control and allowing for one trip across the field with a pesticide combination. However, insecticides can negatively affect natural enemies, leading to pest resurgence and problems with other soybean pests.

Selected References

Boethel, D. J., Russin, J. S., Wier, A. T., Layton, M. B., Mink, J. S., and Boyd, M. L. 2000. Delayed maturity associated with southern green stink bug (Heteroptera: Pentatomidae) injury at various soybean phenological stages. J. Econ. Entomol. 93:707-712.

Jensen, R. L., and Newsom, L. D. 1972. Effect of stink bug damaged soybean seeds on germination, emergence and yield. J. Econ. Entomol. 65:261-264.

Nyval, R. F. 1999. Field Crop Diseases. 3rd ed. Iowa State University Press, Ames.

Panizzi, A. R., and Slansky, F., Jr. 1985. Review of phytophagous pentatomids (Hemiptera: Pentatomidae) associated with soybean in the Americas. Fla. Entomol. 68:184-214.

Todd, J. W., and Turnipseed, S. G. 1974. Effects of southern green stink bug damage on yield and quality of soybeans. J. Econ. Entomol. 67:421-426.

Tolin, S. A., and Lacy, G. H. 2003. Viral, bacterial and phytoplasmal diseases of soybean. Pages 765-819 in: Soybean Monograph. R. Boerma and J. Specht, eds. American Society of Agronomy, Crop Science Society of America, Soil Science Society of America, Madison, WI.

(Prepared by J. K. Greene and J. A. Davis)

Threecornered Alfalfa Hopper

The threecornered alfalfa hopper (*Spissistilus festinus*) is native to the southern United States and Mexico. It feeds on a wide variety of cultivated and uncultivated plants, including clover, cocklebur (*Xanthium strumarian*), cotton, dock (*Rumex* spp.), legume crops, and wild geranium (*Geranium maculatum*). This insect is found in most fields in the U.S. South, but its population only occasionally reaches a level high enough to justify treatment.

Description

Both immatures (i.e., nymphs) and adults have piercing–sucking mouthparts. Adults are green, wedge shaped, and approximately 6 mm long (Fig. 223). Nymphs are also wedge shaped and have spines (i.e., projections) along the tops of their bodies (Fig. 224). Small nymphs are nearly transparent; older nymphs are typically green but sometimes brown.

Life Cycle

Overwintering adults become active during the spring and feed on alternative hosts, including clovers, dock, vetch (*Vicia* spp.), and wild geranium. Two generations of insects a year usually occur in soybean.

Eggs are laid singly or in small clusters and are embedded into the plant tissue—generally, at the main stem nodes or in the terminal of the plant. Fecundity varies considerably, ranging from 38 to 223 eggs per female. Nymphs complete five instars before molting to the adult stage.

Damage

Both nymphs and adults cause injury by puncturing the plant with their beaklike mouthparts and feeding on phloem. They commonly feed by making a series of punctures around the stems, lateral braches, and leaf petioles. This girdling causes

Fig. 223. Threecornered alfalfa hopper adult (*Spissistilus festinus*). (Courtesy S. D. Stewart)

Fig. 224. Threecornered alfalfa hopper nymph (*Spissistilus festinus*). (Courtesy S. D. Stewart)

swollen calluses to develop on the stems and lateral branches (Fig. 225). Girdling on the main stem of a plant that is less than 25 cm tall may cause lodging and death, but the lodging may not occur until later in the season. The upper stems, lateral branches, and leaf petioles may be girdled on larger plants.

Management

Threecornered alfalfa hoppers are more common in reduced tillage systems than other types of systems. The largest populations are often observed in late-planted fields and in fields of soybean planted after wheat. Lodging caused by feeding on small plants is often first noticed following high winds, after plants have developed a pod load and long after the girdling has occurred. However, smaller plants will often snap at the girdle when bent over or when people walk through the field (Fig. 226). Fields with reduced stands are more prone to yield losses, because girdled plants are more likely to lodge and adjacent plants are less likely to compensate for stand loss.

Fig. 225. Girdling of a soybean stem, caused by feeding of the threecornered alfalfa hopper (*Spissistilus festinus*). (Courtesy S. D. Stewart)

Fig. 226. Snapped soybean plant, caused by feeding of the threecornered alfalfa hopper (*Spissistilus festinus*). (Courtesy S. D. Stewart)

Sampling for threecornered alfalfa hoppers in seedling soybean is difficult and usually involves the visual examination of plants. Some states recommend treatment when 10% of seedling plants are infested or 50% of seedling plants are girdled. For plants that are setting pods, a treatment threshold of one or two hoppers per sweep is sometimes suggested. However, research has indicated that injuries on larger plants have minimal effects on yield.

At-planting insecticide applications and insecticide seed treatments can provide good control of early season infestations. Nonetheless, damage in small fields and along field edges may still occur.

Selected Reference

Mueller, A. J., and Sparks, A. N., Jr. 1994. Three cornered alfalfa hopper. Pages 91-92 in: Handbook of Soybean Insect Pests. L. G. Higley and D. J. Boethel, eds. Entomological Society of America, Lanham, MD.

(Prepared by S. D. Stewart and A. Catchot)

Twospotted Spider Mite

The twospotted spider mite (*Tetranychus urticae*) is a phytophagous pest with worldwide distribution. It injures shrubs, trees, and dozens of field and greenhouse crops, including soybean. Although the twospotted spider mite can damage soybean anywhere in the United States, it is best described as an occasional pest in most regions. Even so, in fields in which outbreaks occur, damage can be severe. Severe outbreaks generally occur in association with hot, dry conditions. In the mid-Atlantic region of the United States, the twospotted spider mite is ranked as a major pest of soybean.

Description

The twospotted spider mite is an arachnid in the family Tetranychidae. It has five stages of development: egg, larva, protonymph, deutonymph, and adult. Eggs are round, very small (0.1 mm), and clear to light yellow or tan. Larvae are also small and have six legs. Adults (0.35 mm) and both nymph stages have eight legs and two dark spots on the abdomen—hence, the insect's common name (Fig. 227). Adults and nymphs range in color from yellow to green, although diapausing and dormant mites become orange during the cool months.

Fig. 227. Twospotted spider mite adult (*Tetranychus urticae*) on a soybean leaf. (Courtesy M. L. Pawlowski)

Generally, all life stages are found together on the undersides of leaves. Heavy infestations often result in webbing.

Life Cycle

The twospotted spider mite overwinters as an adult female in a dormant state under clods of soil, in crevices of tree bark, in leaf litter, and in clumps of grass. In the southern United States, some of the mite population may remain active throughout the winter. Females become active during the early spring and begin laying eggs in fields or on the grasses at the edges of fields. The twospotted spider mite exhibits a type of reproduction in which unfertilized eggs produce males and fertilized eggs produce females or males at an approximate ratio of 3 to 1.

Depending on the air temperature, eggs hatch in 3–10 days. After adults become active, the population can complete a generation every 7–21 days. In outbreak years, five to 10 generations can be completed within a growing season.

Twospotted spider mites are dispersed passively by wind and can crawl from adjacent weeds and cultivated crops to soybean. As the population grows, young females disperse from infested plants by producing fine webs that catch the wind and move them to new areas. In general, economically damaging populations occur in the middle to late parts of the season.

Damage

The twospotted spider mite injures soybean while feeding by puncturing the mesophyll cells on the undersides of leaves. This damage can reduce transpiration, nutrient transport, and photosynthetic rates, and it causes stippling (i.e., white or yellow spots) that is visible on the upper and lower surfaces of leaves (Figs. 228 and 229). Extensive stippling causes leaves to appear gray-green and then yellow to bronze; after this change in color, the leaves drop. Feeding damage can cause reduced pod set, fewer seeds in pods, and smaller beans. Severe injury may cause plant death.

Mites initially colonize the lower leaves of soybean, and as their population grows, they also colonize the upper portions of plants. Outbreaks often begin around the field margins (Fig. 230) but may proceed across the field within a few days. When early drought causes the mites to begin feeding early in the summer, infestations can be found on field edges and throughout a field at the same time. Infested areas within a field likely extend well beyond the areas in which damage is first noticed.

Management

Sampling for early signs of stippling at the bases of leaves should begin at field borders, and entire plants should be examined for symptoms. Tapping symptomatic soybean leaves over a white sheet of paper or cardboard will dislodge the mites onto the paper or cardboard, verifying their presence.

Various sets of decision-making guidelines are available. However, the following guidelines, which are used in the state of Ohio, provide a system that can be used to make treatment decisions:

- **Observation:** Mites are barely detected on the undersides of leaves in dry locations or on the edges of fields. Damage is barely detected.
  Recommendation: This is not an economically significant population; do nothing.
- **Observation:** Mites are easily detected on the undersides of leaves along the edges of a field and perhaps on leaves in dry areas throughout a field. Most of the foliage is green, but yellow stippling caused by mite feeding is detectable on the upper sides of leaves, and mites are feeding on the undersides of leaves.
  Recommendation: This is not an economically significant population, but closely monitor the situation.
- **Observation:** Many plants are infested with mites. Varying degrees of stippling are present, and some speckling and discoloration are apparent, as well. Injured plants may be limited to the edges of fields but perhaps also found throughout a field. The edges of fields might display signs of injury.

Fig. 229. Stippling on the underside of a soybean leaf, caused by feeding of the twospotted spider mite (*Tetranychus urticae*), with active mites. (Courtesy R. B. Hammond)

Fig. 228. Stippling on the upper side of a soybean leaf, caused by feeding of the twospotted spider mite (*Tetranychus urticae*). (Courtesy R. B. Hammond)

Fig. 230. Damage along the edge of a soybean field, caused by an outbreak of the twospotted spider mite (*Tetranychus urticae*). (Courtesy R. B. Hammond)

Recommendation: An economically significant population is developing; thus, treatment is warranted. Consider treating the entire field if mites are prevalent throughout it.

- **Observation:** All the plants in an area are heavily infested, include those along the edges of a field and within a field. Plants are discolored and have wilted leaves, and these symptoms are usually obvious from a distance. Severe damage is occurring.
  Recommendation: This is an economically significant population; treatment is warranted and will help reduce a yield loss.
- **Observation:** There is an extremely high population density of twospotted spider mites, and much of the field is discolored and stunted. Many plants are dying or already dead.
  Recommendation: This is an economically significant population; treatment will be beneficial only if new growth is expected after late summer rain.[1]

When temperatures are cooler than 29°C and the relative humidity is greater than 90%, the spider mite population can be suppressed with fungal pathogens. However, in many situations, this option is ignored until after the treatment threshold has been exceeded and an economically significant yield loss has already occurred.

In the mid-Atlantic United States, planting group V soybean cyst-tolerant cultivars may provide some degree of suppression of spider mites. These cultivars are late maturing, however, and may not produce yields comparable to those of other maturity groups; moreover, these cultivars do not fit into many growers' production systems.

Although chemical controls are available, options are limited for soybean. Also, efficacy varies among products. A treatment is generally most effective when applied before an infestation becomes severe, after which multiple applications are often needed. Local extension recommendations should be considered before applying an insecticide, because some insecticides may cause the mite population to flare up.

Selected References

Hammond, R. B., Michel, A., and Eisley, J. B. 2014. Two-spotted spider mite on soybean. Ohio State University Extension, Agriculture and Natural Resources Fact Sheet, Publ. ENT-24-15. Available online at http://ohioline.osu.edu/ent-fact/pdf/ENT_24_14.pdf

Ostlie, K., and Potter, B. 2013. Managing two-spotted spider mites on soybeans in Minnesota. University of Minnesota Extension, Soybean Production. Available online at www.extension.umn.edu/agriculture/soybean/pest/managing-two-spotted-spider-mites-on-soybeans/

(Prepared by J. F. Smith and J. Whalen)

Insect Management Strategies

Traditionally, soybean growers have dealt with a diverse complex of insect pests. Although soybean pest diversity and pressure may differ among the major soybean-growing areas of the United States, the strategies for insect management are similar. Soybean growers throughout the country minimize economic losses from soybean insect pests by using multiple tactics that are ecologically based and that maximize the use of natural control factors. These strategies can be organized into two broad categories: preventive and curative.

Preventive Strategies

Preventive strategies focus on manipulating the agroecosystem to prevent or reduce the likelihood that insect populations will colonize and reproduce. Applying these strategies reduces the potential for pest populations to develop to economically significant levels.

Heritable factors that reduce insect injury to plants (referred to as "host plant resistance") can provide substantial preventive control. Soybean-breeding programs have focused on reducing yield losses to insects by incorporating resistance genes into commercial cultivars. Although insect-resistant cultivars have not been widely planted, advances in plant sciences have allowed for the identification, cloning, and sequencing of specific plant-resistance factors and thus the greater application of biotechnological approaches in preventing insect damage.

Cultural practices are inexpensive and easy to employ and effective as preventive strategies. Rotating crops isolates insects from their hosts, thereby reducing populations. Practicing timely tillage and weed control eliminates host plants, interfering with reproduction and overwintering. Early planting can disrupt the crop–pest synchrony and enable the crop to escape high pest densities. For instance, to reduce late-season caterpillar pressure, growers in the southern United States have adopted an early soybean production system. Early canopy closure and pod set reduce insect oviposition by reducing host attractiveness. However, this early system has significantly increased stink bug populations, as these early planted cultivars produce pods earlier in the season. Although not widely adopted, trap cropping (e.g., planting an attractive host to concentrate insect pests in a single area, which is then treated with an insecticide) has proven effective for controlling beetles and stink bugs.

Insecticide seed treatments (i.e., insecticides applied directly to the seed) have become a widely used preventive strategy. Based on current adoption rates and market trends, so-called naked seed (i.e., without fungicide or insecticide) is less likely to be available in the future. Insecticide seed treatments prevent injury from some early season pests for approximately 30 days. However, these products should be recommended for use only in areas in which heavy insect pressure occurs annually.

Curative Strategies

Curative strategies focus on monitoring insect pest populations throughout the growing season and controlling pests when their populations reach levels that will limit soybean production. Monitoring pest population levels can be achieved by trapping for some insects and by scouting fields for all insect pests. Various techniques are used for pest population monitoring, including whole-plant counts (i.e., of numbers of insects per plant), traps (e.g., water, pan, suction, black light), sticky cards, sweep nets, and shake sheets.

Insecticide application is the strategy of choice when population levels of insect pests reach or exceed economic thresholds. These thresholds are often established on the basis of empirical studies and meant to be responsive, not static. However, most economic thresholds do not consider the market value of soybean or the cost of inputs.

Selected References

Baur, M. E., Boethel, D. J., Boyd, M. L., Bowers, G. R., Way, M. O., Heatherly, L. G., Rabb, J., and Ashlock, L. 2000. Arthropod populations in early soybean production systems in the mid-south. Environ. Entomol. 29:312-328.

Boethel, D. J. 1999. Assessment of soybean germplasm for multiple insect resistance. Pages 101-129 in: Global Plant Genetic Resources for Insect-Resistant Crops. S. L. Clement and S. S. Quisenberry, eds. CRC Press, Boca Raton, FL.

Funderburk, J. E., McPherson, R., and Buntin, D. 1999. Soybean insect management. Pages 273-290 in: Soybean Production in the Midsouth. L. G. Heatherly and H. F. Hodges, eds. CRC Press, Boca Raton, FL.

Hill, C. B., Kim, K., Crull, L., Diers, B. W., and Hartman, G. L. 2009. Inheritance of resistance to the soybean aphid in soybean PI 200538. Crop Sci. 49:1193-1200.

(Prepared by J. A. Davis)

---

[1] Guidelines adapted, by permission of The Ohio State University Extension Entomology Department, from Hammond et al. (2014).

Part III. Noninfectious and Stress Disorders

Noninfectious and stress disorders of soybean include injuries related to weather and climate, soil nutrients, air pollutants, excesses of pesticides, and mechanical and other types of damage. The severity and type of injury varies with the plant growth stage, the plant part involved, and the timing of the causal event or condition. Symptoms of these disorders are often confused with those caused by infectious agents. Moreover, stress conditions can predispose soybean plants to attack by infectious agents.

Weather Factors Associated with Stress

Weather is often the driving force behind many abiotic stresses, because extremes in temperature and rainfall provide specific conditions for stresses that affect soybean. With ongoing changes in climate, more extreme variations in global surface temperatures are expected to lead to changes in rainfall patterns—either more or less water, depending on the location. These changes will, in turn, likely produce additional abiotic stresses.

Crusting and Compaction

Following a heavy rain event and/or flooding, a hard crust may form on the soil and prevent germinated seeds from emerging, especially with fine-textured soils (Fig. 231). The soybean seedlings that do emerge will have thick stems, and in severe cases, their hypocotyl arches will be cracked or broken. Rotary or hand hoeing to break up the surface crust may aid in emergence, but in some cases, replanting may be necessary.

Fig. 231. Germination of soybean plants inhibited by soil crusting. (Courtesy W. O. Scott)

Frost

Frost damage to soybean plants is common in temperate growing areas. Often, only the upper part of the plant is killed or damaged (Fig. 232), and regrowth soon appears at one of the unaffected nodes. If the new growth is strong and vigorous and if few plants are severely damaged, then replanting may not be needed. Some frost damage—especially in areas with temperate climates midway or late in the season, before seed fill is complete—can cause complete loss of the crop.

Hail

Leaves injured by hail are ragged and torn, and large areas of tissues may be beaten away. Stems may be cut off or broken, or they may develop dark, sunken areas (Fig. 233). Most of the

Fig. 232. Soybean seedlings killed by frost. (Courtesy W. O. Scott)

Fig. 233. Stems of a soybean plant broken by hail. (Courtesy W. O. Scott)

damage is usually on the side of the plant facing the prevailing winds of the storm. Pathogenic fungi may colonize hail-damaged areas and produce stem-girdling cankers or weaken the stems. The decision to replant depends on the extent of damage, the stage of plant development, the regrowth that appears, and, in many situations, whether the crop was insured.

Heat Canker

Heat canker occurs at temperatures greater than 35°C and causes girdling of the hypocotyl at or just above the soil line. Affected areas vary in length from 0.25 to 1.0 mm, and affected tissue appears reddish brown and shriveled. The hypocotyl tissue directly above or below the constriction appears normal. An affected plant may continue to grow for 1–2 days, elevating the constricted area several millimeters above the soil line. Within a few days, however, the entire plant will shrivel, wilt, and die.

Lightning

Lightning kills soybean in the field in circular spots (up to 15 m in diameter), which have wide borders of damaged plants (Fig. 234). The lower parts of plant stems may be blackened, and many dead leaves may still be attached. The pith of a plant struck by lightning appears "cooked" or blanched.

Lightning damage is sometimes mistaken for a parasitic disease, such as a type of root rot. Lightning damage can be easily distinguished, however, by the sudden death of both soybean plants and weeds, the clearly defined margins of the affected area, the lack of any signs of a pathogen, and the fact that the affected area does not expand.

Fig. 234. Soybean plants killed by a lightning strike in the field. (Courtesy J. C. Rupe)

Fig. 235. Severe leaf discoloration, caused by sunburn. (Courtesy W. O. Scott)

Sunburn

Minor sunburn damage appears as small (1–2 cm), interveinal, brick-red spots on both the upper and lower leaf surfaces. In severe cases, the discoloration spreads over and along the leaf veins, as well (Fig. 235). The spots later develop brownish centers, which may crack. Sunburn damage on petioles and stems appears as elongated, brick-red lesions, which may be confused with lesions from bacterial infections. On pods, brown spots appear, spread, and are often colonized later by saprophytic fungi, such as species of *Alternaria* and *Penicillium*.

Water Stress

Soybean plants will die if they are grown in fields that are flooded or submerged for extended periods. How long plants can survive under such conditions depends on the temperature and whether the water is still or moving. If soybean plants are not submerged, they can generally withstand flooding relatively well. Some cultivars are more flood tolerant than others. Root injury or death results from an exhaustion of the available oxygen (i.e., anoxia) in the soil environment.

Soybean plants growing in low, wet areas may die from Phytophthora rot, rather than water damage (see Part I, the section Phytophthora Root and Stem Rot). Herbicide injury may also be more likely to develop in wet soils.

Water stress can be caused by drought, as well. In fact, throughout the soybean-growing world, drought is probably the most important yield-limiting factor affecting the crop. Symptoms of drought include stunting, poor vigor, and chlorosis of the interveinal areas of leaves, which changes to a general chlorosis with interveinal necrosis in severe situations. Plants in advanced moisture stress may lose many of their leaves.

Yield is affected only slightly by drought before growth stages R1–R3 if watering becomes sufficient later in production. However, drought that occurs during growth stages R4–R5 can cause irreversible yield losses. In research trials conducted in the U.S. state of Alabama, involving 35 experiments over 7 years, rainfall in a 10-day period centered at growth stage R3 was responsible for 40% of the variation in yield. Although drought occurring in later growth stages is less significant, it can also have strong effects on yield.

Damage to roots caused by fungal or nematode pathogens can increase drought sensitivity in soybean plants. Similarly, pathogens that cause blockage of the plant's vascular system or disrupt it by the formation of cankers or galls can increase drought stress. Many of the symptoms observed on plants with diseases related to these conditions are caused by water stress more than any other aspects of the diseases. For example, the increased occurrence and severity of charcoal rot is associated with drought.

As drought conditions become more severe, soil nutrient solutions become more concentrated. This change in composition alters the availability of nutrients such as iron and nitrogen, causing symptoms of nitrogen deficiency and iron chlorosis.

Managing water-stress conditions can be as simple as providing proper field drainage under wet conditions (e.g., underground tiling for removing excess water) and applying irrigation under dry conditions. In addition, efforts have been made to identify and breed plants to survive longer in both waterlogged and drought conditions. Specifically, traits related to root depth, water use efficiency, nitrogen fixation, and leaf wilting are important for developing drought-tolerant soybean lines. Some progress has also been made in identifying stress-related genes that may increase soybean plants' tolerance to abiotic stresses.

Selected References

de Paiva Rolla, A. A., de Fátima Corrêa Carvalho, J., Fuganti-Pagliarini, R., Engels, C., Do Rio, A., Marin, S. R., de Oliveira, M. C., Beneventi, M. A., Marcelino-Guimarães, F. C., Farias, J. R., Neumaier, N., Nakashima, K., Yamaguchi-Shinozaki, K., and

Nepomuceno, A. L. 2013. Phenotyping soybean plants transformed with rd29A:AtDREB1A for drought tolerance in the greenhouse and field. Transgenic Res. 23:75-87.
Hoeft, R. G., Nafziger, E. D., Johnson, R. R., and Aldrich, S. R. 2000. Modern Corn and Soybean Production. McSp, Champaign, IL.
Khatoon, A., Rehman, S., Hiraga, S., Makino, T., and Komatsu, S. 2012. Organ-specific proteomics analysis for response mechanism in soybean seedlings under flooding stress. J. Proteomics 75:5706-5723.
Kůdela, V. 2011. On the need for revision of some names of plant health malfunctions and their categorization. Plant Protect. Sci. 47:133-148.
Lee, G. J., Boerma, H. R., Villagarcia, M. R., Zhou, X., Carter, T. E., Jr., Li, Z., and Gibbs, M. O. 2004. A major QTL conditioning salt tolerance in S-100 soybean and descendent cultivars. Theor. Appl. Genet. 109:1610-1619.
Mittler, R. 2006. Abiotic stress, the field environment and stress combination. Trends Plant Sci. 11:15-19.
Mohammadi, P. P., Moieni, A., Hiraga, S., and Komatsu, S. 2012. Organ-specific proteomic analysis of drought-stressed soybean seedlings. J. Proteomics 75:1906-1923.
Nakashima, K., Yamaguchi-Shinozaki, K., and Shinozaki, K. 2014. The transcriptional regulatory network in the drought response and its crosstalk in abiotic stress responses including drought, cold, and heat. Front. Plant Sci. doi:10.3389/fpls.2014.00170.
Raper, C. C., Jr., and Kramer, P. J. 1987. Stress physiology. Pages 590-641 in: Soybeans: Improvement, Production, and Uses. 2nd ed. J. R. Wilcox, ed. American Society of Agronomy, Madison, WI.
Sobhanian, H., Razavizadeh, R., Nanjo, Y., Ehsanpour, A. A., Jazii, F. R., Motamed, N., and Komatsu, S. 2010. Proteome analysis of soybean leaves, hypocotyls and roots under salt stress. Proteome Sci. 8:1-15.
Valliyodan, B., and Nguyen, H. T. 2006. Understanding regulatory networks and engineering for enhanced drought tolerance in plants. Curr. Opin. Plant Biol. 9:1-7.

(Prepared by L. E. Sweets and J. B. Sinclair;
revised by G. L. Hartman)

Nutrient Deficiencies and Toxicities

Nutrient deficiencies occur most commonly in sandy, weathered, and highly organic soils that are exposed to abundant rainfall or heavy irrigation. Aeration, moisture, soil pH, temperature, and microbial activity also affect the availability of nutrients (i.e., elements). Changing the soil pH to slightly acidic (pH 6.0–6.5) may alter the availability of many elements and reduce the need for direct application of required elements in fertilizers.

Symptoms of nutrient deficiencies and toxicities may be confused with symptoms of infectious diseases. Identifying a deficiency or toxicity requires having complete information about previous soil treatments. In addition, a soil or plant analysis may be needed for positive diagnosis. Samples should be taken from both affected and unaffected areas.

Aluminum

Aluminum toxicity is uncommon in soybean but may occur if the soil pH falls below 5, causing the development of ionic forms of aluminum that are toxic to plants. These ionic forms inhibit root elongation by destroying the cell structure of the root apex and thus affecting water and nutrient uptake by the roots. The most recognized symptom of aluminum toxicity is the inhibition of root elongation, which, if severe, causes abnormal or stunted growth of aboveground plant parts, as well.

Boron

Soybean is relatively tolerant of boron deficiency, which occurs in alkaline and strongly acidic soils in areas of low rainfall and in soils with high or very low levels of organic matter. Symptoms of this deficiency include shortened internodes and yellowing or reddening of the upper leaves (Fig. 236). In addition, flowers fail to develop, particularly in dry weather. Symptoms of boron deficiency may be confused with those of calcium or potassium deficiency and with leafhopper injury.

An excess of boron—especially near turn rows and in sandy soils—is extremely toxic to soybean plants, particularly seedlings. Affected plants are stunted. Toxicity symptoms appear only on the leaves (usually, those that are just opening). The pubescence at the leaf tip becomes brown, which is followed by progressive browning of the entire leaf (Fig. 237). The leaf continues to develop, but eventually, the dead tissue at the tip and along the leaf margins prevents the blade from expanding. As a result, the leaf puckers or crinkles and its edges cup up or down. Older leaves may fall prematurely.

Calcium

Calcium deficiency is most common in acidic, sandy soils that are low in organic matter. Plants grown in calcium-deficient soils may display symptoms of nitrogen deficiency. The emergence of the primary leaves is delayed, and when they do emerge, they are cup shaped (Fig. 238). The terminal buds of primary leaves become necrotic, and other parts become chlorotic. The tissues between veins tend to be ridged. The terminal buds wither, and the petioles fall. Later, the primary leaves become flaccid and drop prematurely.

Plants deficient in calcium are poorly nodulated and more vulnerable to infection by microorganisms that cause damping-off. Calcium deficiency also affects the uptake of magnesium and potassium by plants.

Fig. 236. Soybean plant with leaf marginal necrosis, necrotic leaf spots, and meristem tip dieback, caused by boron deficiency. (Courtesy T. K. Herman)

Fig. 237. Chlorosis and necrosis of leaf margins with some interveinal necrosis, caused by boron toxicity. (Courtesy T. K. Herman)

Calcium deficiency can be corrected by liming the soil to pH 6 or above. Adding calcium does not neutralize acidic soils; a basic anion is required. However, most liming materials that occur in nature contain basic anion compounds.

Chloride

Symptoms of chloride toxicity range from slight leaf chlorosis to plant death, depending on the foliar concentration of the ion (Figs. 239 and 240). Toxic levels may occur in soils that receive large applications of muriate of potash or that are irrigated repeatedly with water containing chloride salts. Poorly drained soils are more likely to develop chloride toxicity than well-drained soils.

The chloride concentration in the soil changes as the moisture level changes. Rainfall can reduce the concentration near the soil surface, but as the soil dries, chloride moves toward the surface. Any variation in topography of the soil surface can produce variable chloride concentrations at the surface. For instance, ridges and beds will have much higher concentrations when dry than the areas between them. This high variability and the rapid movement of chloride make it difficult to precisely determine toxic soil levels.

The reaction of soybean cultivars to chloride is controlled by a single gene (*ncl*). When a potentially toxic level of chloride is present in the soil, homozygous recessive genotypes transport large amounts of chloride to plant stems and leaves. These genotypes are referred to as "chloride includers." Foliar chlorosis has been noted on leaves containing 8,000–10,000 µg per gram of chloride. A foliar concentration of approximately 15,000 µg per gram of chloride is associated with increased chlorosis and the beginning of necrosis at leaf margins. Plant death occurs when the concentration reaches or exceeds 20,000 µg per gram of chloride. Genotypes with the dominant allele (*Ncl*) transport relatively small amounts of chloride to the upper part of the plant and are referred to as "chloride excluders." These plants retain high concentrations of chloride in their roots. Foliar symptoms of chloride toxicity have not been seen in production fields with high chloride concentrations when excluder cultivars are grown. However, yields of excluder cultivars are lower when grown in soils that contain high amounts of chloride than in soils that do not.

Planting soybean cultivars that are chloride excluders is an effective means of avoiding losses from chloride toxicity. In some cases, southern blight and sudden death syndrome may be more serious in plants grown in soils high in chloride. Growers should plant cultivars that have resistance to sudden death syndrome and that are chloride excluders to minimize yield losses from both problems.

Fig. 238. Chlorotic leaves with cupped margins, caused by calcium deficiency. (Courtesy M. C. Shurtleff)

Copper

Research on copper deficiency in soybean is limited. Although the deficiency reduces yields, plants may not show symptoms. Copper deficiency is likely to occur in plants grown in alkaline peat and muck soils (pH 7–8), inorganic soils high in organic matter, and acidic, highly leached, sandy soils. The deficiency can be detected using soil and plant analyses, and it can be corrected by applying copper sulfate or chelated copper compounds to the soil or by changing the soil pH to slightly acidic (pH 6.0–6.5).

Iron

The initial foliar symptom of iron deficiency is a yellowing of the interveinal areas on young leaves (Fig. 241). Later, even the veins may become chlorotic, and finally, the whole leaf turns ivory to white. Brown, necrotic spots may occur near leaf margins, and overall plant growth is repressed. Chlorosis caused by iron deficiency is often referred to as "lime-induced chlorosis." Chlorotic symptoms may also appear when plants do not use the available iron effectively and large amounts of it accumulate in the leaves. Iron deficiency is often severe in high-pH soils (usually, well above pH 7), in which iron salts are converted to the ferric form, which is unavailable to soybean plants.

Factors that increase the severity of iron deficiency include well-aerated, alkaline soils that are low in organic matter; unusually low or high temperatures; and intense sunlight. Excess levels of calcium, copper, manganese, molybdenum, phosphorus, vanadium, and zinc induce iron chlorosis by inhibiting root absorption of this nutrient or interfering with its utilization. A high concentration of manganese in poorly aerated soil may also inhibit iron absorption and utilization.

Fig. 239. Plants displaying chlorosis and necrosis of the upper leaves, caused by chloride toxicity. (Courtesy D. J. Widick)

Fig. 240. Necrosis of leaf margins, caused by chloride toxicity. (Courtesy D. J. Widick)

Soybean cultivars differ in their tolerance of iron deficiency. At marginal deficiency levels, some cultivars show complete chlorosis, whereas others are not affected.

Approaches to managing iron chlorosis are still being investigated. Foliar application of an aqueous solution of 0.5–1.0% ferric sulfate is a possible control.

Magnesium

Magnesium is required for chlorophyll formation, because chlorophyll is structured around a magnesium atom. A deficiency of magnesium inhibits nitrogen fixation by nodule-forming bacteria.

In the early stages of soybean growth, magnesium deficiency is recognized by the development of pale-green to yellow interveinal tissues on leaf blades. Leaves later become deep yellow, except at the bases. The lower leaves display symptoms first; later, rusty specks and necrotic blotches may appear between the veins and around the edges of the middle and upper leaves. Magnesium deficiency gives older plants the appearance of early maturity. Leaf margins curve down, and a gradual yellowing develops from the margins inward; it is followed by a bronzing of the entire leaf.

Magnesium deficiency can occur after the heavy application of ammonia or potassium to the soil and is most prevalent in deep, sandy, acidic soils. Applying dolomitic limestone corrects the deficiency. If lime is not needed, sulfate of potash magnesia can be used to supply magnesium, potassium, and sulfur. Applying magnesium may increase the oil content and improve the growth and yield of soybean.

Manganese

Soybean is sensitive to manganese deficiency, which commonly occurs in plants in well-drained alkaline and neutral soils (e.g., muck and peat soils). Interveinal areas of leaves become light green to white; the veins remain green (Fig. 242). These symptoms resemble the early symptoms of iron deficiency, except that the veins of manganese-deficient leaves remain green and stand out prominently. As manganese deficiency becomes more severe, necrotic brown spots develop. Leaves may drop prematurely.

Symptoms are more pronounced in cool weather; when temperatures rise, affected plants generally outgrow symptoms of mild deficiency. Symptoms of manganese deficiency frequently appear if the iron concentration is increased.

Applying a fertilizer that contains manganese to the soil can easily treat the deficiency. Aerial application of this nutrient to young deficient plants can increase yields. The element is best applied as a spray of manganese sulfate when plants are 15–25 cm tall; one application is usually sufficient. Another effective means of correcting the deficiency is to mix manganese salts with acid-forming fertilizers (e.g., superphosphate) and then apply the mixture in bands beneath rows of plants; this method is especially recommended for soils high in organic matter. The availability of manganese is effectively increased by combining it with phosphorus and potash; with nitrogen, phosphorus, and potash; or with nitrogen and potash. Manganese must be applied yearly, because alkaline soils bind it in an unavailable form.

Manganese toxicity appears in plants growing in soils that are below pH 5 and that contain abundant iron–manganese pebbles, and it is associated with high levels of soil moisture and management practices that increase soil acidity. Plants are stunted, and leaves are crinkled and cup downward (Fig. 243). The youngest leaves are affected first. Yellowing between leaf veins may or may not be observed. Manganese toxicity can be corrected by liming the soil to pH 6 or slightly higher.

Molybdenum

Molybdenum is required for nitrogen metabolism in leaf tissues and for nitrogen fixation (see Part IV, the section Beneficial

Fig. 241. Yellowing of the interveinal areas on young leaves, caused by iron deficiency. (Courtesy N. R. Usherwood)

Fig. 242. Light-green to chlorotic interveinal areas on leaves, caused by manganese deficiency. (Courtesy E. C. Corrigan)

Fig. 243. Stunted plants with crinkled, cupped leaves, caused by manganese toxicity. (Courtesy N. R. Usherwood)

Bacteria). Molybdenum deficiency occurs in highly acidic soils that are strongly weathered and leached and in soils in which the element occurs in an unusable form. Yields are greatly reduced by molybdenum deficiency. Plant growth, number of pods, number of seeds per pod, seed size, nodulation, and total nitrogen and protein contents of seeds are all reduced.

Symptoms of molybdenum deficiency in soybean resemble those of nitrogen deficiency and are probably caused indirectly by reduced nitrogen utilization, rather than directly by lack of molybdenum. Leaves are pale green or yellow, necrotic, and twisted; the necrosis is largely confined to the midribs, interveinal areas, and margins.

This deficiency can also arise when seeds from plants grown in molybdenum-deficient soil are planted. Seeds with low molybdenum content may produce plants that display deficiency symptoms when grown in soils with a marginal supply of this nutrient.

Molybdenum deficiency can be corrected by treating seeds with sodium molybdate or ammonium molybdate. Applying foliar sprays containing these salts at the prebloom stage is also effective. In many soils, the deficiency can be remedied by liming the soil to pH 6.0–6.3.

Nitrogen

Nitrogen-deficient plants initially become pale green, and later, the leaves turn distinctly and uniformly yellow (Fig. 244). Symptoms appear first on the basal leaves and quickly spread to the upper parts. Plants eventually defoliate and are often spindly and stunted. Analyzing the leaves for nitrogen, uprooting the plant to check for nodule formation, and analyzing the soil to determine its pH and calcium content can diagnose a nitrogen deficiency.

Soybean normally does not need nitrogen fertilization, because the requirement for this nutrient is met by nodulating bacteria and by available soil nitrogen (see Part I, the section Ineffective Biological Nitrogen Fixation). The deficiency occurs when soil nitrogen is low, soil moisture is limited, and the molybdenum level is deficient. When soybean is grown for the first time, seed inoculation with nodulating bacteria is required.

Occasionally, symptoms of nitrogen deficiency are caused by root-rotting organisms, such as the soybean cyst nematode (*Heterodera glycines*) or another nematode, or by molybdenum deficiency—all of which interfere with nodule development. In addition, symptoms may appear early in the season because of slow nodulation, but this condition is usually temporary.

Phosphorus

Soybean requires relatively large amounts of phosphorus, especially at pod set. Phosphorus is required for normal nitrogen fixation. Symptoms of phosphorus deficiency include small, spindly leaflets. Leaves also turn dark green or bluish green, giving the impression that plants are quite healthy, and leaf blades curl up and appear pointed. Blooming and maturity are delayed.

Phosphorous is easily fixed by clay minerals (which are abundant in acidic soils), including various iron oxides and kaolinite. These minerals render phosphorus unavailable for root uptake. Phosphorus deficiency is more pronounced in acidic soils.

Diagnosis involves conducting leaf analyses for phosphorus content. Soil analyses for available phosphorus and soil pH are also good indicators of the availability of the nutrient. In acidic soils, phosphorus exists in a form not available to soybean roots. Phosphorus uptake is reduced in cool, wet soils.

A phosphorous deficiency can be corrected by applying fertilizers that contain phosphorus. Doing so is profitable when the availability of the nutrient in the soil (as determined by a soil test) is low and other nutrients are in adequate supply. The fertilizer should be well mixed into the soil, because phosphorous moves slowly through most soils. Application of phosphorus increases the nitrogen-fixing capacity of nodulating bacteria. The presence of mycorrhizal fungi associated with soybean roots may also help overcome a phosphorus deficiency (see Part IV, Beneficial Fungi).

An elevated phosphorus level in the soil may induce iron and zinc deficiencies, if levels of those elements are low or nearly low.

Potassium

Soybean requires large amounts of potassium. This nutrient is important for all aspects of plant growth and influences the plant's nutritional balance. Potassium is also involved in the uptake of calcium and magnesium. The maximum need for potassium occurs during periods of rapid vegetative growth. Plants deficient in this nutrient tend to have weak stems and are more susceptible to some plant pathogens.

Symptoms of potassium deficiency are well defined. In the early stages of growth, an irregular, yellow mottling appears around the margins of leaflets. These yellow areas coalesce to form more or less continuous, irregular, yellow borders (Fig. 245). This symptom is often followed by necrosis of the chlorotic areas and downward cupping of leaf margins. The dead tissues then drop away, giving leaves a ragged appearance. Chlorosis and necrosis may spread inward to include half or more of a leaflet. The basal portions remain green. Symptoms appear first on older leaves. A severe potassium deficiency tends to produce wrinkled and misshapen seeds.

This deficiency, as determined by a soil test, can be corrected by applying fertilizers that contain potassium. Fertilizing will increase yield by increasing the number of pods; the number, size, and weight of seed; and nodulation.

Plants infected by the soybean cyst nematode (*Heterodera glycines*) often show symptoms of potassium deficiency, but a nutrient deficiency is not the actual cause. Rather, the diseased

Fig. 244. Pale-green to yellow discoloration of soybean plants, caused by nitrogen deficiency. (Courtesy L. F. Welch)

Fig. 245. Chlorosis of leaf margins, caused by potassium deficiency. (Courtesy J. C. Rupe)

plants are unable to pick up sufficient potassium as a result of the root feeding and damage caused by the nematode (see Part I, the section Diseases Caused by Nematodes). Injury from leafhoppers feeding on cultivars with glabrous leaves often mimics potassium deficiency, as well.

Sulfur

The main symptom of sulfur deficiency is the development of small, yellow-green leaves. In addition, stems are thin, hard, and elongated. Foliar symptoms resemble those produced by deficiencies of other elements (e.g., nitrogen and phosphorus), but stem elongation is characteristic of sulfur deficiency.

Sulfur deficiency results in the accumulation of carbohydrates, nitrate, and amino acids that do not contain sulfur. The synthesis of sulfur-containing amino acids is interrupted; thus, protein synthesis is reduced and proteolysis is increased. Cell walls thicken from the accumulation of hemicellulose and starch.

The availability of sulfur depends on the rate at which it is released from organic matter, which is in turn influenced by the kinds of plant residues in the soil, as well as soil moisture and pH. Sulfur deficiency can be corrected by applying fertilizers that contain sulfur to the soil.

Zinc

Soybean yields are reduced considerably in zinc-deficient soils. Deficient plants have stunted stems and leaves with chlorotic interveinal areas (Fig. 246). Later, entire leaves turn yellow or light green. Lower leaves may turn brown or gray and drop early. Few flowers form, and the pods that set are abnormal and mature slowly. Yield reduction results primarily from the reduced number of seeds that form. Soybean cultivars vary greatly in their tolerance for zinc deficiency and toxicity.

Zinc deficiency is most likely to occur in strongly weathered, coarse-textured soils that are alkaline (with free lime); in eroded soils; and in soils excessively fertilized with phosphorus and low in organic matter. Adding lime and phosphorus may reduce zinc availability and thus cause a deficiency of this nutrient.

Zinc deficiency can be corrected by applying a zinc salt—most commonly, zinc sulfate or zinc oxide—to the foliage or the soil. The salt may be broadcast and worked in or placed in a band at least 5 cm to the side and 5 cm below the seed. A single broadcast application is generally sufficient for 2–4 years. Seed treatment with zinc salts is also helpful.

Soybean is sensitive to zinc toxicity, particularly in fields near industrial plants that emit large amounts of zinc or after heavy applications of zinc salts (especially in acidic soils). Toxicity can be corrected by liming soil to pH 6.2–6.5.

Fig. 246. Chlorosis of interveinal leaf areas, caused by zinc deficiency. (Courtesy K. Onki)

Selected References

Liang, Q., Cheng, X., Mei, M., Yan, X., and Liao, H. 2010. QTL analysis of root traits as related to phosphorus efficiency in soybean. Ann. Bot. 106:223-234.

Parker, M. B., Gaines, T. P., and Gascho, G. J. 1986. Sensitivity of soybean cultivars to soil chloride. Ga. Agric. Exp. Stn. Res. Bull. 347.

Rupe, J. C., Widick, J. D., Sabbe, W. E., Robbins, R. T., and Becton, C. B. 2000. Effect of chloride and soybean cultivar on yield and the development of sudden death syndrome, soybean cyst nematode, and southern blight. Plant Dis. 84:669-674.

Sinclair, J. B. 1993. Soybeans. Pages 99-103 in: Nutrient Deficiencies and Toxicities in Crop Plants. W. F. Bennett, ed. American Phytopathological Society, St. Paul, MN.

Wang, X., Yan, X., and Liao, H. 2010. Genetic improvement for phosphorus efficiency in soybean: A radical approach. Ann. Bot. 106:215-222.

Yang, J., and Blanchar, R. W. 1993. Differentiating chloride susceptibility in soybean cultivars. Agron. J. 85:880-885.

Zheng, S. J. 2010. Crop production on acidic soils: Overcoming aluminum toxicity. Ann. Bot. 106:183-184.

(Prepared by D. J. Widick, J. C. Rupe, and J. B. Sinclair; revised by G. L. Hartman)

Green Stem Disorder

Green stem disorder (also called "green stem" and "green stem syndrome") was first reported in Kansas in 1974. Since then, the disorder has become widely distributed throughout all soybean-growing regions of North America. It is the most common type of delayed maturity found in the main U.S. soybean production region. Because the cause of green stem disorder is unknown, it is referred to as a "disorder" rather than a "disease."

Numerous other factors are associated with the delayed or incomplete maturation of soybean plants and plant parts. Some of the known causes of delayed maturity are described in the sections of this book on stink bugs (see Part II, the section Stink Bugs) and viruses (see Part I, the section Diseases Caused by Viruses). The symptoms associated with these kinds of delayed maturity often result in lower yields and/or poor seed quality.

In the main soybean production region of North America, plants with other forms of delayed maturity are encountered much less frequently than plants with symptoms of green stem disorder. Green stem disorder is distinguished from other types of delayed maturity by the normal ripening of pods and seeds.

Symptoms and Signs

Green stem is a disorder of soybean that causes the stems to remain immature and moist after the pods and the seeds have fully ripened. Normally, the stems mature and dry down along with the pods and the seeds. Thus, the key diagnostic feature is the presence of mature pods and seeds on moist, immature, pliable stems (Fig. 247). This feature distinguishes green stem disorder from general delayed maturity and other causes of plants remaining green at harvest time.

Stem color may range from yellow to light green, green, and even purple. Sometimes, leaves and petioles in various degrees of senescence may remain attached to plants with symptoms of green stem disorder.

The numbers of pods and seeds are not affected by the disorder. Only the normal maturation of stem tissue appears to be affected. Plants affected with the disorder may be scattered, occur in patches, or found throughout large areas in a field.

Another interesting feature of green stem disorder is that the stems of affected plants appear to be relatively free of fungal stem diseases, such as charcoal rot (*Macrophomina phaseolina*), and diseases caused by the *Diaporthe–Phomopsis* complex.

There is no evidence that green stem disorder negatively affects soybean yields. On the contrary, higher yields have been associated with higher incidence of the disorder in some instances.

Epidemiology

Soybean producers worldwide have long recognized consistent differences among soybean cultivars for sensitivity to green stem disorder. Most soybean cultivars are insensitive to the disorder, with less than 10% of plants developing symptoms. Among sensitive cultivars, however, more than 90% of plants may develop the disorder.

Bean pod mottle virus (BPMV) was once thought to be the main cause of green stem disorder, but research has shown that the disorder occurs independently of BPMV infection. *Soybean mosaic virus* and other viruses that delay plant maturity have also been considered in relation to the disorder.

In the southern United States and in South America, a disorder known as "green bean syndrome" is caused by stinkbug feeding. This disorder is different from green stem disorder, however, because the maturity of the whole plant is delayed, including the pods and the seeds.

Other factors that can delay the maturity of the whole plant include haploidy (i.e., a plant has only one set of chromosomes, instead of two), male or female sterility, and late plant emergence. Haploid and sterile plants remain green because they continue to attempt to produce seed, even though they are unable to do so. Artificial depodding of soybean plants has also been shown to delay plant maturity, but there is no evidence that loss of pods is associated with green stem disorder. The soybean mutant genotype *d1d1d2d2* gives soybean plants the "stay-green" phenotype, in which leaves remain green longer and senesce later than normal. It is not known, however, if green stem disorder is associated with this genotype.

The late application of herbicides has also been associated with delayed maturity of soybean plants. Moreover, fungicide application—especially the application of strobilurin fungicides during midseason—has been demonstrated to increase the incidence of green stem disorder, particularly in sensitive genotypes. This finding supports the notions that fungal infection is associated with the lack of symptoms of green stem disorder and that soybean plants with the disorder are relatively free of disease, compared with asymptomatic plants. In addition, strobilurin fungicides may cause a delayed maturity that is associated with a greening effect, in which leaf senescence is commonly delayed.

Management

Green stem disorder makes harvesting soybean significantly more difficult, because affected plants are much harder to cut. As a result, the ground speed of the combine must be slowed to help the header knives cut the tough, pliable green stems, while the engine and cylinder speed must be kept high to reduce the risk of thresher clogging. Working in this manner reduces fuel efficiency in the combining operation. Furthermore, moisture can be transferred from affected stems to the seed during the threshing process, reducing seed grade and storability.

Until the cause or causes of green stem disorder are determined, no specific control strategies can be recommended. Given the consistent differences among soybean cultivars in sensitivity to green stem disorder, growers should select cultivars that are less prone to it. In fact, choosing cultivars that are insensitive to the disorder is the only means of management available to growers. Growers should keep track of the history of green stem disorder in the cultivars they plant in their fields to identify sensitive and insensitive cultivars.

Fig. 247. Soybean plants with and without green stem disorder. (Courtesy C. B. Hill)

Selected References

Boethel, D. J., Russin, J. R., Wier, A. T., Layton, M. B., Mink, J. S., and Boyd, M. C. 2000. Delayed maturity associated with southern stink bug (Heteroptera: Pentatomidae) injury at various soybean phenological stages. J. Econ. Entomol. 93:707-712.

Egli, D. B., and Bruening, W. P. 2006. Depodding causes green-stem syndrome in soybean. Crop Manag. doi:10.1094/CM-2006-0104-01-RS.

Hill, C. B., Bowen, C. R., and Hartman, G. L. 2013. Effect of fungicide application on soybean green stem disorder. Plant Dis. 97:1212-1220.

Hill, C. B., Hartman, G. L., Esgar, R., and Hobbs, H. A. 2006. Field evaluation of green stem disorder in soybean cultivars. Crop Sci. 46:879-885.

Hobbs, H. A., Hill, C. B., Grau, C. R., Koval, N. C., Wang, Y., Pedersen, W. L., Domier, L. L., and Hartman, G. L. 2006. Green stem disorder of soybean. Plant Dis. 90:513-518.

Leonard, B. R., Boquet, D. J., Padgett, B., Davis, J. A., Schneider, R. W., Griffin, J. L., Valverde, R. A., and Levy, R. J., Jr. 2011. Soybean green plant malady: Contributing factors and mitigation. La. Agric. 54:32-34.

Lustosa, P. R., Zanuncio, J. C., Leite, G. L. D., and Picanco, M. 1999. Quality of seeds and senescence of soybean genotypes under two levels of bug infestation (*Pentatomidae*). Pesqui. Agropecu. Bras. 34:1347-1351.

Padgett, B., Schneider, R., and Whitam, K. 2003. Foliar-applied fungicides in soybean disease management. La. Agric. 46:7-9.

Schwenk, F. W., and Nickell, C. D. 1980. Soybean green stem caused by bean pod mottle virus. Plant Dis. 64:863-865.

(Prepared by C. B. Hill and G. L. Hartman)

Herbicide Injury

Many of the herbicides used to control weeds in soybean have the potential to injure the crop. Additionally, soybean can be injured by herbicides not labeled for use in soybean but to which plants are exposed via herbicide carryover, drift from applications made to adjacent fields, herbicide residues becoming dislodged from spray application equipment, and mixing and application errors. Determining the reason or reasons for an observed soybean injury can be challenging, because more than one factor may be involved.

Growth abnormalities in soybean can be caused by corn herbicides, including 2,4-Dichlorophenoxyacetic acid (2,4-D) (Figs. 248 and 249) and dicamba (Fig. 250), which can drift to soybean fields. Residual carryover from cyopyralid can cause abnormal growth, as well. Symptoms of some diseases, such as those caused by viruses and sudden death syndrome, can be confused with symptoms of herbicide injury. Injury symptoms caused by linuron (Fig. 251), metribuzin (Fig. 252), and trifluralin (Fig. 253) may include seedling dieback, scorching of leaf tips, and general chlorosis and stunting. On occasion, glyphosate has been used on plants that are glyphosate intolerant (Fig. 254).

More information on the effects of herbicides on soybean is presented in Table 12.

Soybean Cultivar Sensitivity

Soybean cultivars vary in sensitivity to certain herbicides. Whereas there are some differences in sensitivity to post-emergence herbicides, most cultivar sensitivity differences are expressed to soil-residual herbicides, such as flumioxazin, metribuzin, and sulfentrazone.

Symptoms of injury caused by soil-applied herbicides are often noticed shortly after soybean emergence. In severe cases, plants may die following emergence of the cotyledons. Other symptoms include necrotic lesions on the soybean hypocotyl very near the soil surface, as well as reddish spots or lesions on the cotyledons. Lesions on the hypocotyl may not always kill a young soybean plant, but they often create an area of weakened tissue, which may cause the plant to snap off during rain or high wind.

Fig. 248. Malformation of leaves, caused by 2,4-Dichlorophenoxyacetic acid (2,4-D) drift and 4-(2,4-Dichlorophenoxy)butyric acid (2,4-DB) direct injury. (Courtesy E. L. Knake)

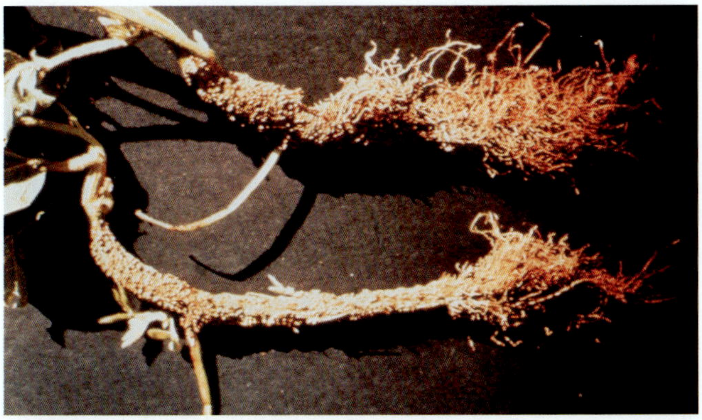

Fig. 249. Malformation of roots and lower stems, caused by drift of 2,4-Dichlorophenoxyacetic acid (2,4-D). (Courtesy E. L. Knake)

Fig. 250. Soybean plants with dicamba injury, displaying cup and leaf malformation on newer leaves. (Courtesy E. L. Knake)

Fig. 251. Leaf scorching, caused by exposure to linuron. (Courtesy E. L. Knake)

Fig. 252. Plant death caused by metribuzin injury, resulting from use of the herbicide with an organophosphate insecticide. (Courtesy D. E. Kuhlman)

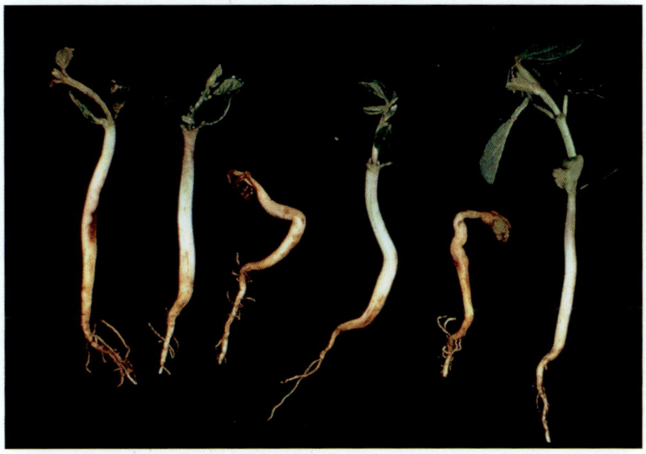

Fig. 253. Swelling of hypocotyls, caused by trifluralin exposure. (Courtesy E. L. Knake)

Environmental Variables

Environmental factors greatly influence the severity of soybean injuries caused by both soil-applied and postemergence herbicides. For instance, high temperatures and levels of relative humidity can enhance the absorption of postemergence herbicides, whereas adequate soil moisture and low relative humidity can enhance the uptake of soil-applied herbicides.

In addition to enhancing herbicide uptake, environmentally induced crop stresses often exacerbate soybean injuries from herbicides. Cool, wet soil conditions are examples of stresses induced by the environment. In general, herbicide selectivity arises from the soybean plant's ability to rapidly metabolize the herbicide to a nonphytotoxic form before it causes much visible injury. Soybean plants growing under favorable conditions are able to metabolize the herbicide well before injury symptoms are expressed. However, when plants are under stress (which may be caused by a variety of factors), their ability to metabolize the herbicide can be sufficiently reduced that injury symptoms develop.

Herbicides

Herbicides labeled for use in soybean vary in their inherent potential to injure plants. A spray additive applied with a postemergence herbicide can often enhance the crop's response. Postemergence herbicides that do not extensively translocate within plants following absorption can cause general stunting, as well as foliar chlorosis and necrosis. Foliar damage is usually confined to the leaves directly exposed to the herbicide. Postemergence herbicides that translocate within plants following absorption can cause stunting along with general chlorosis on both existing leaves and leaves that emerge after herbicide application.

Soybean injury from herbicides applied in crop is not always an accurate predictor of yield response, because soybean plants generally have a tremendous ability to recover. However, injury that occurs late in the season is more likely to adversely affect yield.

Herbicide Carryover

Herbicides vary in their persistence in the soil. Some have very little persistence, whereas others can last for several months. The soil properties that influence herbicide persistence can be broadly categorized as chemical, physical, and microbial.

Soil pH can impact how much and how quickly certain herbicides are degraded via a chemical process known as "hydrolysis." Hydrolysis occurs more rapidly under acidic conditions than under neutral or basic conditions.

The physical properties of a soil are related to the relative amounts of sand, silt, clay, and organic matter it contains. Soils with high amounts of clay and organic matter have a greater potential for herbicide carryover than coarse-textured soils and soils with less organic matter.

The rate of herbicide degradation by soil microorganisms is often directly related to prevailing environmental and soil conditions. Temperature and moisture are both factors that can greatly influence soil microbial populations. The activity levels of most soil microorganisms are higher under warm soil temperatures than under cool soil temperatures. This partially explains why herbicide degradation tends to slow down as soil temperatures drop during the weeks after harvest. Minimal herbicide degradation by soil microorganisms occurs when soil temperatures fall below approximately 4°C and/or when soil moisture is limited.

Fig. 254. Glyphosate damage on glyphosate-intolerant plants. (Courtesy G. L. Hartman)

Selected References

Egan, J. F., and Mortensen, D. A. 2012. Quantifying vapor drift of dicamba herbicides applied to soybean. Environ. Toxicol. Chem. 31:1023-1031.

Egan, J. F., Barlow, K. M., and Mortensen, D. A. 2014. A meta-analysis on the effects of 2,4-D and dicamba drift on soybean and cotton. Weed Sci. 62:193-206.

Huang, H., Reddy, K. N., Thomson, S. J., and Yao, H. 2015. Assessment of soybean injury from glyphosate using airborne multispectral remote sensing. Pest Manag. Sci. 71:545-552.

Smith, D., Davis, V., and Conley, S. 2013. Is That *Soybean mosaic virus* or Herbicide Injury? University of Wisconsin–Extension. Available online at http://fyi.uwex.edu/fieldcroppathology/files/2013/08/SMVHerb.pdf

Taylor-Lovell, S., Wax, L. M., and Nelson, R. L. 2001. Phytoxic response and yield of soybean (*Glycine max*) varieties treated with sulfentrazone or flumioxazin. Weed Tech. 15:95-102.

(Prepared by A. Hager)

Fungicide and Insecticide Injury

When fungicides and insecticides are used according to label directions, they rarely induce phytotoxicity to soybean. However, certain combinations of soil-applied insecticides and herbicides can injure soybean seedlings, and certain foliar-applied fungicides and insecticides can cause phytotoxicity under some environmental conditions. Like the injuries caused by herbicides, the phytotoxicity caused by fungicides and insecticides is usually short lived, and soybean plants typically outgrow injuries with little or no subsequent yield loss.

Fungicides applied to soybean rarely cause crop damage or yield losses. In some soybean cultivars, applications of tebuconazole were found to produce foliar symptoms similar to those of brown stem rot and sudden death syndrome.

Soybean may be injured if the herbicide metribuzin is used with a soil-applied organophosphate insecticide. Several organophosphates—including chlorpyrifos, diazinon, ethoprop, and phorate—are labeled for use on soybean to prevent insects and nematodes from attacking planted seeds and underground parts of plants. Using any of these insecticides with metribuzin can induce injury symptoms such as seedling dieback, chlorosis, and necrosis, especially along leaf margins. If the injury is severe, the entire leaf may become chlorotic or necrotic, and soybean stands may be thin or spotty and uneven. Evidence suggests that organophosphates may enhance the activity of metribuzin within the soybean plant, even when metribuzin is applied at the labeled rate. This type of injury occurs most commonly in soybean grown from poor-quality seed or when seedlings are stressed by environmental conditions. Injury can be prevented by not applying soil-applied organophosphates and metribuzin in the same field.

Labels of soil-applied organophosphates also state that they should not be applied in the seed furrow or allowed

TABLE 12. Herbicides (by Mode of Action), Symptoms of Injury, and Management Strategies for Soybean

Amino Acid Synthesis Inhibitors	Symptoms	Management Strategies	Other Remarks
Acetolactate Synthase (ALS) Inhibitors Imidazolinones Sulfonylureas Triazolopyrimidines	Plants are stunted and have shortened internodes. Terminal leaves are golden, and purple to red venation appears on the undersides of leaves.	Apply accurately, and select the appropriate spray additive according to label recommendations.	Applying these products when air temperatures and humidity levels are high can enhance injury.

Cell Membrane Disruptors	Symptoms	Management Strategies	Other Remarks
Protoporphyrinogen Oxidase (PPO) Inhibitors *Foliar Applied:* Acifluorfen Flumiclorac Fluthiacet Fomesafen Lactofen	Chlorosis and necrosis of leaf tissue directly contacted by the spray solution.	Consult product labels to select the appropriate spray additives. Tank mixing with other postemergence herbicides can increase the potential for injury.	The likelihood of injury tends to increase when these products are applied with oil-based spray additives. These products are contact herbicides. Injury can be exacerbated by high temperatures, humidity levels, and crop stress levels.
Soil Applied: Flumioxazin Saflufenacil Sulfentrazone	Lesions develop on the hypocotyls and the cotyledons and are often surrounded by red or orange tissue. Plants may fail to emerge, and those that do emerge often have stunted growth. The margins of true leaves are sometimes wrinkled.	Use the appropriate application rate according to label recommendations based on soil texture and organic matter content. Avoid overlapping of treatments.	Sensitivity can vary by cultivar. Precipitation soon after soybean emergence can increase the potential for injury.

Growth Regulators	Symptoms	Management Strategies	Other Remarks
TIR1 Auxin Receptors *Benzoic Acid:* Dicamba	Young leaves are cupped and crinkled; they often display parallel venation and are sometimes greatly reduced in size. Internodes on the upper portion of the plant can be shortened. Symptoms resemble those caused by *Soybean mosaic virus.*	Avoid spraying in cornfields and pastures when the wind is blowing toward soybean fields. Thoroughly clean application equipment after spraying to avoid applying residues during application of a postemergence soybean herbicide. Follow label restrictions when using prior to soybean planting.	Plants are usually exposed through spray drift from applications to nearby fields and through residues dislodged from spray equipment. This product can be applied prior to soybean planting but can cause severe injury if label guidelines are not followed.
Phenoxy: 2,4-Dichlorophenoxyacetic acid (2,4-D)	Young leaves are thickened, crinkled, and paddle or strap shaped and have parallel veins. Injury is often observed in a diminishing pattern, according to spray operations and drift.	Avoid spraying near soybean fields, especially when the wind is blowing toward them. Increase the average droplet size to help reduce spray drift.	Injury usually occurs along roadsides, along fencerows, and next to other fields that have been sprayed. Symptoms are sometimes confused with those of *Soybean mosaic virus* and other viral diseases.
4-(2,4-Dichlorophenoxy) butyric acid (2,4-DB)	Plants may have curved or cracked stems, crinkled leaves, and callused tissue on the lower stems. Adventitious roots might form along the stem.	Apply at the recommended rate, and avoid overlapping of applications. Directed spray applications are safer than broadcast applications. Do not spray more than the lower one-third of the plant.	2,4-DB is used in established soybean to control cocklebur and some other weeds. Injury is more likely when applications are made during hot and humid conditions.

Photosynthesis Inhibitors[a]	Symptoms	Management Strategies	Other Remarks
Photosystem II Inhibitors *Benzothiadiazole:* Bentazon	The initial symptom is bronzing of the leaf tissue in direct contact with the spray solution; this tissue eventually becomes chlorotic and necrotic.	Use the correct spray additive, based on the weed spectrum.	Injury tends to become more severe when applications include oil-based spray additives and coincide with high temperatures and humidity.
Triazine: Atrazine Simazine	Carryover causes progressive yellow mottling of leaves, beginning along the margins; in severe cases, entire plants turn yellow to brown and die. Interveinal chlorosis and necrosis often develop. Symptoms are similar to those caused by linuron and metribuzin exposure.	Do not exceed the recommended application rate. Do not overlap or apply late in the growing season.	Carryover injury occurs when soybean follows a crop to which atrazine or simazine was applied during the previous season. Injury is most prevalent after a cool, dry year and in plants growing in high-pH soils.
Triazinone: Metribuzin	Injury symptoms are difficult to distinguish from those caused by linuron and triazine herbicide carryover.	Adjust the rate for soil type, and apply accurately. Avoid overlapping of treatments. Use other herbicides on fields with variable soils. Do not apply on sandy soils.	Cultivars can vary in sensitivity. Even at the recommended application rate, metribuzin can injure soybean if used with a soil-applied organophosphate insecticide.
Urea: Linuron	Plants are chlorotic, and seedlings may die back. Leaf margins and tips appear scorched and are yellow to brown; in severe cases, entire leaves turn yellow to brown.	Adjust the rate for soil type, and apply accurately. Avoid overlapping of treatments. Use other herbicides on fields with variable soils. Do not apply on sandy soils.	Injury might be mistaken for seedling blight. Injury occurs most often on plants grown in sandy soils low in organic matter.

(continued)

[a] Products in this category have several different binding mechanisms.
[b] Encapsulated formulations.

TABLE 12. *(continued)*

Pigment Inhibitors	Symptoms	Management Strategies	Other Remarks
4-Hydroxyphenyl-Pyruvate-Dioxygenase (HPPD) Inhibitors Mesotrione Tembotrione Topramezone	Symptoms caused by persistent soil residues include stunted growth and gradual yellowing of upper leaves. Injury caused by spray drift includes whitening that begins at the margins of leaflets that emerge following spraying.	Avoid overlapping applications of soil-applied herbicides. Do not make postemergence applications when the wind is blowing toward nearby soybean fields.	Carryover is favored by dry soil conditions during the season of application. Injury severity caused by drift depends on the dose received and the soybean growth stage.
Seedling Root Growth Inhibitors	**Symptoms**	**Management Strategies**	**Other Remarks**
Microtubule Inhibitors Pendimethalin Trifluralin	Plants have stunted growth and are slow to emerge. The hypocotyls are enlarged, and the secondary roots are limited, thick, and stubby. Leaves may be small and crinkled. Callused tissue may form on the stem at the soil surface. The apical meristem can be damaged if treated soil is splashed into the meristem region by precipitation.	Adjust the application rate for the soil type and organic matter content. Apply accurately and uniformly. Incorporate evenly for uniform distribution. Do not exceed the recommended rate.	Thickened hypocotyls may also be produced when crusting or compacted soil interferes with seedling emergence, but in these cases, the root systems will be normal. Injury may increase plants' susceptibility to root rot. Plants under stress from disease, deep planting, or cold, wet weather are often more susceptible to injury from these herbicides. Injury may be more likely in fields in which overlapping has occurred and/or the application rate was not adjusted for the soil texture and organic matter content.
Seedling Shoot Growth Inhibitors	**Symptoms**	**Management Strategies**	**Other Remarks**
Long-Chain Fatty Acid Inhibitors Acetochlor[b] Alachlor Dimethenamid-P *S*-metolachlor Pyroxasulfone	Young leaves can be crinkled. Leaflets have shortened midribs and are heart shaped; their tips appear indented.	Use the correct formulation and rate. Apply carefully, and avoid overlapping treatments.	Symptoms of injury are usually temporary and rarely result in the loss of plant vigor or yield. Injury is more likely when the soil is cool and wet.

to contact seeds. Insecticides applied incorrectly or during periods of cool, wet weather (during which soybean growth slows down) can cause seedling injury. Under these conditions, seedling emergence may be delayed and plants may be stunted. Some insecticides can cause leaf necrosis if they are applied to wet, tender soybean foliage or during periods of high humidity.

Many insecticide labels prohibit application in combination with specific foliar-applied herbicides. Conversely, certain herbicide labels specify a time interval that should elapse either before or after herbicide application and before an insecticide is applied.

(Prepared by K. L. Steffey and A. Hager)

Air Pollutant Injury

Many airborne chemicals are known to injure soybean leaves. The symptoms are diverse and often mimic those caused by other factors. Thus, the precise cause of foliar symptoms is usually difficult to determine.

The injuries caused by most air pollutants are infrequent and limited to areas near the source of emission. For example, pesticides can be considered air pollutants in cases of injury caused by the drift of compounds such as carbaryl (Fig. 255) and dicofol (Fig. 256) from nearby fields. Injuries from chlorine (Fig. 257), hydrogen chloride, and ammonia (Fig. 258) can occur after accidental spills. Recurring injuries can affect soybean plants near industrial sources of fluorides and sulfur dioxide, but recent control strategies involving scrubbers and tall stacks have greatly decreased emissions.

Ozone is considered the most important phytotoxic air pollutant, because it occurs regularly on a regional basis and at levels injurious to many crops, including soybean (Fig. 259). Over the years, open-top field chambers and free-air concentration enrichment have been used to investigate phytotoxic air pollutants such as ozone to determine more precisely how they affect various soybean genotypes. Also addressed in most of these studies were the effects of elevated carbon dioxide levels on soybean and other crop plants. In at least some cases, an elevated level of carbon dioxide alone or with ozone was shown to reduce the severity of downy mildew but to increase the severity of brown spot. More studies are needed in these controlled

Fig. 255. Leaf flecking, caused by carbaryl exposure. (Courtesy R. K. Howell)

free-air concentration enrichments systems with inoculations of pathogens to determine more precisely the effects of air pollutants on soybean diseases. (The effects of ozone exposure are discussed further in a following section.)

Ammonia

Soybean leaves that are almost fully expanded are the most sensitive to ammonia. The nature of the injury ranges from local, reddish-brown lesions to broad, cream to brown bands between veins (see Fig. 258). The source of the pollutant must be known to distinguish ammonia damage from that caused by other pollutants.

Fluoride

The most common symptom of fluoride injury to soybean, as well as most other crops, is marginal chlorosis and necrosis of older leaves. Identification of a nearby fluoride source and foliar analysis for fluoride content are both required for diagnosis.

Ozone

Most of the ozone in rural areas is formed by complex photochemical reactions involving nitrogen oxides and hydrocarbons transported from major urban areas. The daily exposure level is determined by the concentrations of hydrocarbons and nitrogen oxides in the air along with the amount of sunshine. Seasonal exposure consists of days when ozone levels are relatively low or near average, punctuated by days when levels are high. Because of wind patterns, concentrations of ozone in rural areas that are downwind of urban centers may be higher than concentrations in urban areas.

Soybean leaves that have recently reached full expansion are more sensitive to ozone than younger leaves. Growing conditions that favor leaf gas exchange also favor ozone uptake by the leaves. Symptoms of ozone injury may include fairly discrete upper-surface flecking, which produces white to tan necrotic areas less than 1 mm in diameter, and upper-surface stippling, which produces dark areas less than 1–2 mm in diameter (see Fig. 259). More commonly, however, the only foliar symptoms are increased levels of chlorosis, necrosis, and abscission—symptoms that are virtually indistinguishable from those of normal senescence. Field research estimates of soybean yield losses caused by ozone can be as high as 23%, depending on the cultivar, season, and location.

Because ozone is one of the leading pollutants detrimental to soybean and different soybean genotypes exhibit different tolerance to ozone damage, efficient methods for high-throughput plant phenotyping through leaf optical properties such as remote sensing have been used to assess soybean plants with ozone tolerance. Even so, no commercial soybean cultivars with tolerance to ozone are available.

Sulfur Dioxide

The sources of sulfur dioxide pollution—utilities, factories, and engines that burn high-sulfur fuel or smelt high-sulfur ore—are usually nearby. Symptoms of injury from sulfur dioxide exposure include chlorosis and the development of irregular, cream-colored to light-brown, necrotic areas between leaf veins. As with exposure to ozone, repeated exposure to sulfur dioxide in low concentrations may accelerate the senescence of leaves. Figure 260 shows soybean plants with unusually severe symptoms from an experimental exposure to sulfur dioxide. Yields may be reduced if soybean is exposed to large amounts of sulfur dioxide after flowering.

Fig. 256. Leaf chlorosis and bleaching, caused by dicofol exposure. (Courtesy R. K. Howell)

Fig. 258. Interveinal bleaching of leaves, caused by exposure to ammonia. (Courtesy R. K. Howell)

Fig. 257. Leaf chlorosis and necrosis, caused by chlorine exposure. (Courtesy J. B. Sinclair)

Fig. 259. Leaf bronzing and necrosis, caused by ozone exposure. (Courtesy R. K. Howell)

Fig. 260. Leaf bleaching, caused by exposure to sulfur dioxide. (Courtesy R. K. Howell)

Injury that resembles that caused by sulfur dioxide can be caused by dicofol, a miticide. Dicofol usually causes injury to fully expanded soybean leaves; dead areas appear between veins and at leaf margins. Knowledge of what chemicals have been applied to soybean is essential for a proper diagnosis of injury.

Selected References

Ainsworth, E. A., Rogers A., and Leakey, A. D. B. 2008. Targets for crop biotechnology in a future high-CO_2 and high-O_3 world. Plant Physiol. 147:13-19

Burkey, K. O., and Carter, T. E., Jr. 2009. Foliar resistance to ozone injury in the genetic base of U.S. and Canadian soybean and prediction of resistance in descendent cultivars using coefficient of parentage. Field Crop Res. 111:207-217.

Eastburn, D., Degennaro, M., De Lucia, E., Dermody, O., and McElrone, A. 2010. Elevated atmospheric carbon dioxide and ozone alter soybean diseases at SoyFACE. GCB Bioenergy 16:320-330.

Heagle, A. S. 1989. Ozone and crop yield. Annu. Rev. Phytopathol. 27:397-423.

Heagle, A. S., Heck, W. W., Rawlings, J. O., and Philbeck, R. B. 1983. Effects of chronic doses of ozone and sulfur dioxide on injury and yield of soybeans in open-top field chambers. Crop Sci. 23:1184-1191.

Krupa, S., McGrath, M. T., Andersen, C. P., Booker, F. L., Burkey, K. O., Chappelka, A. H., Chevone, B. I., Pell, E. J., and Zilinskas, B. A. 2001. Ambient ozone and plant health. Plant Dis. 85:4-17.

Noggle, J. C., Meagher, J. F., and Jones, H. C. 1986. Sulfur in the atmosphere and its effects on plant growth. Pages 251-278 in: Sulfur in Agriculture. M. A. Tabatabai, ed. Agron. Ser. No. 27. American Society of Agronomy, Crop Science Society of America, and Soil Science Society of America, Madison, WI.

(Prepared by A. S. Heagle; revised by G. L. Hartman)

Part IV. Soybean Disease and Pest Management Strategies

Losses of yield and quality from diseases and pests can be reduced by implementing a comprehensive program that incorporates the principles of pest management, including preventive, preemptive, and remedial strategies. Disease and pest management comprises a decision matrix that considers the context of suitable agronomic (i.e., best-management) practices, pesticide use regulations, state and federal land regulations, production logistics, product marketing, and economics. In addition, disease and pest management often utilizes economic injury levels, economic thresholds, scouting, recordkeeping, and pest–crop mapping.

As management programs become more refined, it will be increasingly important to identify pathogens and pests to their specific types—namely, biotypes, groups, pathotypes, and races that overcome the deployment of specific genes for resistance. The use of computer programs that use location, identification, and real-time reporting of occurrences and risks for outbreaks of diseases and pests provides a faster and more appropriate management response and will likely continue to improve with technological advances in sampling and identification.

Selected References

Copping, L. G., Green, M. B., and Rees, R. T., eds. 1992. Pest Management in Soybean. Elsevier Applied Science, London.

Frisbie, R. E., and Adkisson, P. L., eds. 1985. Integrated Pest Management on Major Agricultural Systems. Texas A&M University, College Station, TX.

Fry, W. E. 1982. Principles of Plant Disease Management. Academic Press. New York.

Hartman, G. L., and Hill, C. B. 2010. Diseases of soybean and their management. Pages 276-279 in: Soybean: Botany, Production, and Uses. G. Singh, ed. CABI, Wallingford, Oxfordshire, U.K.

Jacobsen, B. 2001. Disease management. Pages 351-356 in: Encyclopedia of Plant Pathology. O. C. Maloy and T. D. Murray, eds. Wiley, New York.

Maloy, O. C. 2005. Plant Disease Management. Plant Health Instructor. doi:10.1094/PHI-I-2005-0202-01.

Ngugi, H. K., Esker, P. D., and Scherm, H. 2011. Meta-analysis to determine the effects of plant disease management measures: Review and case studies on soybean and apple. Phytopathology 101:31-41.

O'Neal, M., and Johnson, K. 2010. Insect pests of soybean and their management. Pages 300-324 in: Soybean: Botany, Production, and Uses. G. Singh, ed. CABI, Wallingford, Oxfordshire, U.K.

(Prepared by G. L. Hartman)

Biological Enhancements

Biological enhancements are used to reduce the impact of diseases and pests, often by augmenting naturally occurring pest control systems. These enhancements can be applied directly to the seed or the soil. In addition, amending the soil with specific plant materials and following certain cropping patterns may increase levels of beneficial organisms while directly or indirectly reducing the effects of diseases and pests.

Treating seed with rhizosphere-competent bacteria has been known for years to provide a degree of protection to the plant, either directly or indirectly, and this practice has been used to reduce Rhizoctonia and Phytophthora root rots. A similar scenario occurs with selected fungal species that protect the plant directly or indirectly by inducing host resistance or by antagonizing pathogens such as bacteria, fungi, and nematodes. Viruses, too, have been shown to reduce fungal pathogen activity and may serve a useful function in managing other pathogenic organisms, including bacteria and nematodes.

The engineering of microbes (including viruses) to infect plant pathogens has great potential, and researchers continue to explore the genomics of host plants and their pathogens. There are many examples of how genetic engineering (e.g., protoplast fusion) has increased the potential of biocontrol agents. In the future, it may be common to manipulate opportunistic pathogens to be more specific to their hosts, so that they produce specific antipathogen and/or antipest molecules that will control the targeted pathogen or pest. In addition, there are numerous controlling pests with biological control agents. Many of these agents occur naturally, such as insect predators, parasites, and entomophagous fungi.

Beneficial Bacteria

Among the many plant growth-promoting bacteria are those that nodulate soybean roots to form a symbiotic relationship (usually, *Bradyrhizobium diazoefficiens*) that is beneficial to the plant by fixing or trapping atmospheric nitrogen, which is then converted into an organic form in the host. Nodules develop after direct penetration of root hairs or epidermal cells by the bacteria.

Nodules vary in size and shape, depending on many factors, including the strain of bacterium, the time and place of infection, and the soil conditions. Generally, nodules range from 1 mm to 2–3 cm in diameter; they may be cylindrical, round, or irregularly shaped and rough or smooth. A mature nodule consists of an epidermal layer, a cortical layer, and a central region, which contains bacteroids (i.e., membrane-enclosed bacteria). The cortical tissues contain vascular elements, which are not in contact with the tissues of the host root. A meristematic region develops around the nodule or at the tip. The colors of nodules vary from whitish to cream to brown, depending on the nodule's age and the soil conditions. Nodules are pink internally when actively fixing nitrogen from the atmosphere to ammonium.

The nodulating bacteria are gram negative, aerobic, polar or subpolar flagellate, and generally rod shaped ($1.2–3.0 \times 0.5–0.9$ μm), although they can also be irregularly shaped or club shaped. Isolates or strains show crop preferences and even cultivar preferences within a given crop. The bacteria survive in tissues of susceptible hosts and in the soil. Their survival depends on soil conditions and whether the particular host or cultivar is planted continuously in the field.

Plant growth-promoting rhizobacteria can also establish a nonsymbiotic association that can affect the plants' nutrition, as well as soil properties and other soil microbes. Bacteria are commonly isolated from the soybean rhizosphere, including species of *Azospirillum, Bacillus,* and *Pseudomonas.*

Beneficial Fungi

Among the many species of fungi, more are beneficial to plants than attack plants. Beneficial fungi include those that are associated with plants as decomposers of organic matter (saprotrophs) and those that live with plants, such as endophytes and mycorrhizae. Other fungi benefit plants by interacting and attacking pathogens and pests, such as those that are entomophagous parasites and hyperparasites (pathogens of fungi).

A fungal endophyte is an organism that lives part of its life inside the plant without causing visual symptoms. Many of the fungi that are pathogens of soybean have a phase in the life cycle in which they do not cause any visual symptoms (often referred to as "latent infections"). Other endophytes live in soybean but do not normally have a pathogenic phase. Fungi belonging to numerous species—including *Alternaria, Ampelomyces, Chaetomium, Cladosporium, Davidella, Diaporthe, Epicoccum, Fusarium, Phoma, Plectosphaerella,* and *Verticillium*—may play a role in protecting soybean from other pathogens and pests and may contribute to inducing resistance.

Mycorrhizal fungi are frequently the most abundant fungi in soil. Arbuscular mycorrhizae (AM) fungi are a group of ubiquitous soil organisms that form mutualistic symbioses with most terrestrial plants. Generally, this root–fungal relationship is beneficial to plants, but under certain conditions (e.g., a high level of phosphorous in the soil), mycorrhizal development may be pathogenic. Clearly, the majority of mycorrhizal associations benefit plants by enhancing the acquisition of nutrients via the fungus to the plant. The fungus partner also benefits from the association via the translocation of carbohydrates from the plant.

Mycorrhizal fungi are important in the uptake of phosphorous, especially in soils that have low to moderate levels of this element. Carbon fixation may be enhanced through increased photosynthetic rates, and nitrogen nutrition may be improved by increased uptake of ammonium or by enhanced nitrogen fixation in legumes. Moreover, mycorrhizal fungi increase the uptake of several minor elements. Additional benefits of AM fungi colonization to host plants include improved drought tolerance, increased disease tolerance, and biogeochemical cycling of nutrients. The hyphae of AM fungi serve valuable functions in reducing soil erosion through the physical entanglement of soil particles and the production of a protein on the surface of hyphae that is involved in forming water-stable aggregates.

Reproduction of AM fungi occurs asexually via spores. In addition, other structures are formed during the AM fungi life cycle, such as extra- and intraradical hyphae, arbuscules, auxiliary cells, and spores. Extraradical hyphae have several important functions, one of which is the acquisition of nutrients from the soil. Following acquisition, the nutrients are translocated to the plant via intraradical hyphae. As the roots continue to grow, extraradical hyphae grow along the root surfaces and usually enter the cortical cells by forming appressoria. Once hyphae enter the cortex, extensive proliferation of intraradical hyphal growth occurs between cells. A protein called "glomalin" produced on the surfaces of the hyphae of isolates representing all genera of AM fungi may be involved with water-stable aggregate formation. Arbuscules form on hyphal tips of intraradical hyphae in the root cortex. They are composed of fine, dichotomously branched hyphae that form an interface between the arbuscular cell and the host plasmalemma. Arbuscules are the sites of nutrient exchanges between the host and the mycorrhizae.

Most spores of AM fungi are produced asexually and form singly on a sporogenous hypha, in clusters, or without sporocarps. The mode of spore formation differs among families and genera. Some *Glomus* spp. form high numbers of spores within the roots, causing the root cortex to rupture. Spore germination of *Glomus* spp. occurs with the formation of a germ tube that grows through the subtending hypha. In *Gigaspora* spp., one or more germ tubes grow directly through the spore wall, but in *Acaulospora, Entrophospora,* and *Scutellospora* spp., a germination shield is formed and germ tube initials grow through the shield.

AM fungi are generally not host specific. Yet different species of AM fungi and isolates of the same species from different geographic locations have different effects on the growth of the same soybean cultivar. Numerous studies involving AM fungi–plant interactions have been conducted in the greenhouse under controlled conditions. Studies involving these fungi under field conditions are limited, however, because of the complexity associated with the soil environment and its impact on the plant host and its interactions with AM fungi.

Certain components of agricultural management systems have demonstrated the effects on AM fungi spore populations. For example, inclusion of cover crops, reduced tillage methods, and choice of crop sequences influence AM fungi colonization and spore populations. In general, soybean growers do not follow any particular management scheme based on recommendation to increase AM fungi populations.

Selected References

Brodeur, J. 2012. Host specificity in biological control: Insights from opportunistic pathogens. Evol. Appl. 5:470-480.

Carling, D. E., Roncadori, R. W., and Hussey, R. S. 1989. Interactions of vesicular-arbuscular mycorrhizal fungi, root-knot nematode, and phosphorous fertilization on soybean. Plant Dis. 73:730-733.

Carris, L., Little, C., and Stiles, C. 2012. Introduction to Fungi. Plant Health Instructor. doi:10.1094/PHI-I-2012-0426-1001.

de Souza Leite, T., Cnossen-Fassoni, A., Pereira, O. L., Mizubuti, E. S. G., de Araújo, E. F., and de Queiroz, M. V. 2013. Novel and highly diverse fungal endophytes in soybean revealed by the consortium of two different techniques. J. Microbiol. 51:56-69.

Douds, D. D., Jr., Janke, R. R., and Peters, S. E. 1993. VAM fungus spore populations and colonization of roots of maize and soybean under conventional and low-input sustainable agriculture. Agric. Ecosyst. Environ. 43:325-335.

Higo, M., Isobe, K., Drijber, R. A., Kondo, T., Yamaguchi, M., Takeyama, S., Suzuki, Y., Niijima, D., Matsuda, Y., and Ishii, R. 2014. Impact of a 5-year winter cover crop rotational system on the molecular diversity of arbuscular mycorrhizal fungi colonizing roots of subsequent soybean. Biol. Fertil. Soils. doi:10.1007/s00374-014-0912-0.

Higo, M., Isobe, K., Yamaguchi, M., Drijber, R. A., Jeske, E. S., and Ishii, R. 2013. Diversity and vertical distribution of indigenous arbuscular mycorrhizal fungi under two soybean rotational systems. Biol. Fertil. Soils 49:1085-1096.

Impullitti, A., and Malvick, D. 2013. Fungal endophyte diversity in soybean. J. App. Microbiol. 114:1500-1506.

Johnson, N. C., Copland, P. J., Crookston, R. K., and Pfleger, F. L. 1992. Mycorrhizae: A possible explanation for yield decline associated with continuous cropping of corn and soybean. Agron. J. 84:378-390.

Johnson, N. C., Pfleger, F. L., Crookston, R. K., Simmons, S. R., and Copland, P. J. 1991. Vesicular-arbuscular mycorrhizas respond to corn and soybean cropping history. New Phytol. 117:657-663.

Miransari, M. 2014. Plant, mycorrhizal fungi, and bacterial network. Pages 315-325 in: Plant Signaling: Understanding the Molecular Crosstalk. K. R. Hakeem, R. U. Rehman, and I. Tahir, eds. Springer, New Delhi, India.

Rodriguez, R., White, J., Jr., Arnold, A., and Redman, R. 2009. Fungal endophytes: Diversity and functional roles. New Phytol. 182:314-330.

(Prepared by F. L. Pfleger and J. B. Sinclair; revised by G. L. Hartman)

Cultural Practices

Exclusion and eradication are two major approaches to disease and pest management, and they can range from passing legislation to applying specific agronomic practices.

Exclusion is useful in situations in which a pathogen or pest does not exist, cannot survive between crops, or is not present in an area and will not likely be introduced as airborne or soilborne inoculum. Legislative embargoes, inspections, and quarantines are widely used to prevent the introduction of exotic plant pathogens, weeds, and arthropods into new areas. These legislative actions limit the spread of pathogens and pests that do not occur worldwide but pose significant risks to soybean production in currently noninfested areas. Knowledge of the epidemiology, life cycle, and distribution of the pathogen and the risk it poses for production in a particular environment is needed for legislative embargoes, inspections, and quarantines to be effective.

Eradication includes methods that eliminate pathogens and pests. It may occur at the local level or across a regional production area. For the most part, complete eradication has not been possible, except in a few specific cases in localized areas. Many of the cultural practices commonly in use reduce levels of inocula or reduce the effectiveness of inocula. These practices include rotating crops, plowing or tilling to destroy crop residues where pathogens overseason, using pathogen-free seeds, ensuring timely irrigation, and selecting appropriate seeds and cultivars.

Agronomic Practices

Many important agronomic factors must be considered in managing diseases and pests—among them, soil compaction, drainage, and fertility; cultivar selection; seed treatments; tillage used in preparing a seedbed; type of seeding equipment used; row spacing; seeding density; pesticide use; irrigation; planting date; and other influences on pathogen and pest severity. For example, both an excessive seeding rate and a high plant density contribute to increased damage from charcoal rot, Septoria leaf spot, Sclerotinia stem rot, Sclerotium blight, and diseases caused by *Pythium* and *Rhizoctonia* spp. To manage these diseases, the seeding rate may need to be adjusted based on seed quality, soil fertility and moisture, cultural factors, row spacing, and the growth habit of the cultivar. Certain soil amendments (e.g., silicon) have been shown to decrease soybean rust, and adding plant materials of acacia, brassica, neem, and seaweed to the soil has been shown to reduce propagules of the soybean cyst nematode (*Heterodera glycines*) while increasing the number of soybean pods produced. Disease suppression also occurs through the use of cover crops, which can act as biofumigants and inducers of host resistance and also stimulate beneficial microbes that aid in root health or act directly on soilborne pathogens. Several reports have shown a reduction in the population of the soybean cyst nematode after growing a winter crop of annual ryegrass.

Crop Rotation. The effectiveness of crop rotation often depends on how the pathogen survives—whether via long-lived survival structures (such as sclerotia) or other hosts (such as the viruses associated with insect vectors). When nonhost crops are planted for one or more seasons in rotation with soybean, pathogens and pests either die out or their populations are reduced. Growing members of the grass family (e.g., barley, maize, oats, rye, sorghum, and wheat) is often particularly beneficial, since these plants are hosts of only a few soybean pathogens. In addition, these plants can provide benefits in accelerating population levels of beneficial microbes, which can directly or indirectly benefit soybean growth. For example, rye has been shown to provide microbial benefits to subsequent soybean crops.

In some instances, rotation with resistant cultivars can also reduce inoculum buildup and be effective in managing disease. Examples include rotation with resistant cultivars to lower the population of the soybean cyst nematode and the fungus that causes brown stem rot (*Cadophora gregata*) and rotation with cotton or grain sorghum to reduce the fungus that causes charcoal rot (*Macrophomina phaseolina*). The economics of rotation over several years should be considered, not just for the value generated in a given year but also for the long-term benefits.

There are numerous examples of yield increases in soybean after rotation with nonlegumes, even in cases in which pathogens do not appear to be limited. These increases may have resulted from suppression of disease, improved weed control, improved soil tilth (i.e., physical condition of the soil), or other factors. In Brazil, long rotations have been used to limit the occurrence of soybean rust by not growing soybean in the offseason, which causes a reduction of fungal spores for the main season crop. Less is known about intercropping soybean with other crops, primarily because this practice is uncommon. However, in one case, a reduction in red crown rot was reported when maize was intercropped with soybean.

Fertility. Optimal fertility—particularly adequate phosphorous and potash nutrition—is important in managing bacterial blight and pustule; charcoal rot; Fusarium blight, pod, and stem blight; Rhizoctonia root rot; and soybean cyst nematode. Maintaining the optimal soil pH (6.2–7.0) is important for root nodulation. Acidic soils accentuate the damage from Sclerotium blight. Suboptimal soil pH, which reduces nutrient availability, increases the damage from a number of pathogens.

Planting. Seedbed conditions that favor rapid seed germination and emergence minimize the damage caused by pre- and postemergence damping-off. Soybean seeds germinate when soil temperatures are higher than 12–14°C and seed moisture exceeds 50%. When these conditions are met (e.g., when soybean is planted early), a fungicide seed treatment is suggested, even if high-quality seed is used. If conditions are favorable for germination and emergence, the need for seed treatment depends on the seed quality (see Harvest and Seed Quality, later in this section).

The planting date can also influence the development of diseases, other than those that affect seedlings. Early planting tends to minimize the damage from charcoal rot, stem canker, soybean cyst nematode, and soybean mosaic, because it minimizes the likelihood that soybean plants will be in the critical growth stages when conditions favor infection. However, early planting can accentuate the damage caused by some diseases, including pod and stem blight and sudden death syndrome, when conditions in the middle to late summer favor their development.

Planting in narrow rows increases the severity of downy mildew and Sclerotinia stem rot but has little effect on the development of Septoria brown spot. When soybeans are planted early, the rapid canopy closure that occurs in narrow rows may lower the soil temperature, reduce the severity of charcoal rot, and suppress damaging weed populations.

The soybean growth habit also affects disease severity. Dwarf and semidwarf determinate cultivars suffer greater yield losses from Septoria brown spot and pod and stem blight than nondeterminate cultivars.

Tillage. Destruction of crop residues by tillage and other practices can be an effective means of reducing overseasonal inocula of most foliage- and stem-infecting bacteria and fungi that survive locally. Most these pathogens survive between crops on or in infested residue, and when this residue is forced to decay following tillage, the pathogens cannot compete with the soil microorganisms and die out. Tillage can also influence soil compaction and thereby drainage. Systems that reduce compaction can reduce the incidence and severity of Phytophthora root rot and sudden death syndrome. In many areas, tillage—especially deep tillage—is not desirable or even legal because of the concern over soil erosion.

Water Management. Water management is critical in managing root diseases. Saturated soils create favorable conditions for infection by species of *Phytophthora* and *Pythium,* and soil-water deficits contribute significantly to losses caused by brown stem rot, charcoal rot, and nematodes. Allowing seedbeds to become excessively dry contributes to seed decay by species of *Aspergillus* and *Penicillium.* These fungi decay seeds with a moisture content far less than the 50% moisture required for germination.

Irrigation water management can be important in reducing diseases such as anthracnose, bacterial blight, pod and stem blight, Septoria brown spot, and Sclerotinia stem rot. Furrow irrigation reduces pod and seed infections in many areas, whereas excessive sprinkler irrigation during pod fill and maturation increases these infections. In areas in which Sclerotinia stem rot is a threat, avoiding irrigation during flowering will help reduce the incidence of floral infection. Irrigation to maintain adequate moisture can reduce damage from the brown stem rot and charcoal rot pathogens and the soybean cyst nematode. In managing brown stem rot, adequate moisture is particularly important during growth stages R6–R8.

Weed Management. Weeds growing in and around production fields can serve as hosts of bacterial, fungal, nematode, and viral pathogens. Weed management and crop rotation can significantly reduce the effects of the soybean cyst nematode and root-knot nematodes (*Meloidogyne* spp.). Uncultivated perennial plants growing around fields are often hosts of the viruses that cause bean pod mottle, bean yellow mosaic, soybean mosaic, and tobacco ringspot.

Weeds growing in soybean fields also compete with soybean plants for space, nutrients, and water, thereby increasing the likelihood of stress and stress-related diseases and disorders. The selection of herbicides for weed management can also influence disease development by other pesticides. For example, applying the herbicide chloramben can increase the severity of Thielaviopsis root rot. Some herbicides may increase or even decrease the severity of some diseases. For example, the herbicide alachlor influences the effectiveness of nematicides in the control of soybean cyst nematode, and the herbicide lactofen has been used by some growers to suppress white mold.

Harvest and Seed Quality. The timely harvest of soybean is necessary to keep certain fungal pathogens from infecting seeds. Green stem disorder and other maladies that delay harvest can be detrimental to seed quality. In some cases, an herbicide desiccant is used to diminish the effects of delayed harvest, thus keeping seeds from prolonged environmental exposure.

A number of tests can be used to assess soybean seed quality, and in most situations, accurate assessment requires using several of them. In temperate environments, indexing seed lots by means of a standard warm-germination test, a cold-soil test, an accelerated-aging test, and an assessment of seedborne *Phomopsis* spp. provides an excellent evaluation of seed quality (see Part I, Phomopsis Seed Decay).

Physical and mechanical damage can be assessed using several methods. One method is to use a tetrazolium chloride test, which identifies cracks and dead or damaged areas in seed tissues. A second method is to soak seeds in a bleach (0.5% sodium hypochlorite) solution for 1 h; doing so identifies seeds with cracked or damaged coats. Mechanical damage is a serious problem in harvesting and handling seeds with a moisture content of less than 10%.

Seed cleaning is an important step in producing a high-quality seed lot. Gravity tables and spiral cleaners remove a large percentage of poor-quality seeds; small, lightweight seeds; fungal sclerotia; and soil peds. Removal of soil peds can be significant in reducing the spread of the soybean cyst nematode and the sclerotia of various fungi, including *Sclerotinia sclerotiorum.*

Selected References

Adee, F. A., Oplinger, E. S., and Grau, C. R. 1994. Tillage, rotation sequence, and cultivar influences on brown stem rot and soybean yield. J. Prod. Agric. 7:341-347.

Agrwal, V. K., and Sinclair, J. B. 1996. Principles of Seed Pathology. 2nd ed. CRC Press, Boca Raton, FL.

Anderson, I. C., and Caieney, J. W. G. 2004. Diversity and ecology of soil fungal communities: Increased understanding through the application of molecular techniques. Environ. Microbiol. 68:769-779.

Auwal, H. M., Wang, H., Li, L., Kakar, K.-U., and Zheng, J. 2014. Utilization of plant materials for control of soybean cysts nematode. Acta. Agric. Scand. Sect. B Soil Plant Sci. 64:1-6.

Baird, R. E., Mullinix, B. G., Peery, A. B., Minton, N. A., and Parker, M. B. 1987. Root-knot nematode management and yield of soybean as affected by winter cover crops, tillage systems, nematicides. J. Nematol. 19:38-43.

Bowman, J. E., Hartman, G. L., McClary, R. D., Sinclair, J. B., Hummel, J. W., and Wax, L. M. 1986. Effects of weed control and row spacing in conventional tillage, reduced tillage, and nontillage on soybean seed quality. Plant Dis. 70:673-676.

Cook, R. J., and Baker, K. F. 1983. The Nature and Practice of Biological Control of Plant Pathogens. American Phytopathological Society, St. Paul, MN.

Creech, J. E., Westphal, A., Ferris, V. R., Faghihi, J., Vyn, T. J., Santini, J. B., and Johnson W. G. 2008. Influence of winter annual weed management and crop rotation on soybean cyst nematode (*Heterodera glycines*) and winter annual weeds. Weed Sci. 56:103-111.

Curl, E. A. 1963. Control of plant diseases by crop rotation. Bot. Rev. 29:413-479.

Dhingra, O. D., and Sinclair, J. B. 1994. Basic Plant Pathology Methods. 2nd ed. CRC Press, Boca Raton, FL.

Gao, X., Wu, M., Xu, R., Wang, X., and Pan, R. 2014 Root interactions in a maize/soybean intercropping system control soybean soilborne disease, red crown rot. PLoS One 9:e95031.

Lemes, E. M., Mackowaiak, C. L., Blount, A., Marois, J. J., Wright, D. L., Coelho, L., and Datnoff, L. E. 2011. Effects of silicon applications on soybean rust development under greenhouse and field conditions. Plant Dis. 95:317-324.

Riga, E., Topp, E., Potter, J., Welacky, T., Anderson, T., and Tenuta, A. 2001. The impact of plant residues on the soybean cyst nematode, *Heterodera glycines.* Can. J. Plant Pathol. 23:168-173.

Rothrock, C. S., Hobbs, T. W., and Phillips, D. V. 1985. Effects of tillage and cropping system on incidence and severity of southern stem canker of soybean. Phytopathology 75:1156-1159.

Sumner, D. R., Doupnik, B., Jr., and Boosalis, M. G. 1981. Effect of reduced tillage and multiple cropping on plant diseases. Annu. Rev. Phytopathol. 19:167-187.

(Prepared by J. B. Sinclair; revised by G. L. Hartman)

Host Plant Resistance

The discovery and use of sources of disease and pest resistance have, over time, produced incremental yield gains and yield stability in soybean. The use of pathogen- and pest-resistant cultivars dates back to 1951, when a specific gene was discovered to control the inheritance of resistance to bacterial pustule. For many decades following, the improvement of disease and pest resistance in soybean was achieved primarily through public programs operated by government agencies and land-grant universities. A complete list of known resistance sources in soybean was reported in 1980, and it included a list of resistance sources to many common pathogens.

More up-to-date public information is available online in the U.S. Department of Agriculture–Agricultural Research Service (USDA–ARS) Germplasm Resources Information

Network, commonly known as "GRIN." In addition, thousands of molecular markers associated with disease and pest resistance are located online at SoyBase and the Soybean Breeder's Toolbox. Performing genomewide association studies is now possible because of completion of the entire sequence of the soybean genome. Also, the availability of 50,000 single-nucleotide polymorphic markers in each of the soybean accessions in GRIN will allow for connecting genotypes–phenotypes to rapidly locate resistance genes for disease and pest resistance.

Commercially grown soybean cultivars in the three largest soybean-producing countries (Argentina, Brazil, and the United States) are now developed and supplied by private industry, mostly with specialty traits such as glyphosate tolerance. Few publicly developed soybean cultivars are still available. Some public programs evaluate these privately developed cultivars for pathogen and pest resistance and provide unbiased reports. Based on the information provided on the websites of commercial seed companies, many commercial cultivars are advertised to have resistance to at least one or more pathogens and pests—often including bacterial blight, bacterial pustule, frogeye leaf spot, Phytophthora root rot, Sclerotinia stem rot, stem canker, soybean aphid, soybean cyst nematode, and sudden death syndrome. For a number of diseases and pests of soybean, either no resistance has been found or no commercial soybean cultivars are advertised to have resistance.

Resistant cultivars have generally been effective in management programs that address the primary pathogens and pests of soybean. Even so, these privately developed cultivars are often short lived in terms of market life (3 years or less), and newer lines in development are not always evaluated thoroughly for resistance to all the common pathogens and pests because of the need to get them to market. In addition, pathogens and pests may develop variants in biotypes, groups, pathotypes, and races. Single-gene resistance is often a problem in cultivars used for controlling soybean cyst nematode, downy mildew, frogeye leaf spot, Phytophthora root rot, soybean rust, and soybean aphids, because numerous isolates are known to cause susceptible reactions on specific resistance genes.

(More specific information on using resistance as a management tool is given in the discussions of individual diseases and pests in Parts I and II of this book.)

Selected References

Bastien, M., Sonah, H., and Belzile, F. 2014. Genome wide association mapping of resistance in soybean with a genotyping-by-sequencing approach. Plant Genome 7:1-13.

Chawla, S., Bowen, C. R., Slaminko, T. L., Hobbs, H. A., and Hartman, G. L. 2013. A public program to evaluate commercial soybean cultivars for pathogen and pest resistance. Plant Dis. 97:568-578.

Germplasm Resources Information Network (GRIN). 2013. U.S. Department of Agriculture, Agriculture Research Service. Available online at www.ars-grin.gov

Hartwig, E., and Lehman, S. 1951. Inheritance of resistance to the bacterial pustule disease of soybeans. J. Agron. 43:226-229.

Korte, A., and Farlow, A. 2013. The advantages and limitations of trait analysis with GWAS: A review. Plant Methods 9:29.

SoyBase and the Soybean Breeder's Toolbox. n.d. U.S. Department of Agriculture and Iowa State University. Available online at www.soybase.org

Tisseli, O., Sinclair, J. B., and Hymowitz, T. 1980. Sources of Resistance to Selected Fungal, Bacterial, Viral, and Nematode Diseases of Soybean. INTSOY Ser. No. 18. University of Illinois at Urbana–Champaign, College of Agriculture.

Vidic, M., Dordevic, V., Petrovic, K., and Miladinovic, J. 2013. Review of soybean resistance to pathogens. Ratar. Povrt. 50:52-61.

Wilcox, J. 2001. Sixty years of improvement in publicly developed elite soybean lines. Crop Sci. 41:1711-1716.

(Prepared by G. L. Hartman)

Pesticides

Using pesticides to reduce the development and effects of pathogens and pests is an effective management strategy in certain situations. Fungicides and nematicides can be used to treat seed, soil, and foliage, and insecticides can be used to control insect pests and vectors. The use of both foliar- and seed-applied fungicides on soybean has increased dramatically over the years.

Foliar Fungicides

Foliage-applied fungicides are used to improve seed quality and to protect yields through the management of specific foliar, pod, and stem diseases. When weather conditions favor disease development during growth stages R1–R5, outbreaks of fungal diseases can reduce yields and seed quality. In general, applying fungicides at growth stages R2–R5 provides maximum yield responses and improves seed quality by reducing fungal infections. The use of foliar fungicides has expanded both because of the development of soybean rust around the world and because of the increase in soybean market prices.

Although many acres of soybean are treated with fungicides in the United States and worldwide on a calendar basis (not according to the principles of integrated pest management [IPM]), the decision to apply a fungicide should be based on several elements: namely, disease risk factors, disease presence, developmental stage of the crop, and current and forecasted weather conditions. Before applying a foliar fungicide, growers should consider the potential for economic benefit, which is based on thresholds (if known), the susceptibility of the cultivar, the time of application, and the possible seed-quality enhancements.

Canopy penetration and minimal loss to drift are critical to the success of a foliar fungicide application. Both aerial and ground application methods are routinely used. Ground applications made during the reproductive growth stages may reduce yields because of wheel traffic through the field; nevertheless, many growers own their own spray equipment, so ground applications are common in many areas. Fungicides are typically applied only once during the season, if at all. However, additional applications are used to achieve acceptable control of certain fungal diseases, especially in areas with warm, humid climates. Factors that influence the number of applications include the stage of crop development when the disease first appears, the incidence and severity of foliar fungal diseases (as determined by crop scouting), the current and forecasted weather conditions, the yield potential of the crop, the market price of soybeans, and the cost of application.

Both protective and locally systemic fungicides have been registered for use on soybean (Table 13). Traditionally, systemic fungicides included products such as benomyl and thiophanate–methyl, both of which are in the methyl benzimidazole carbamate (MBC) chemical group. Benomyl is no longer available, however, and most new products belong to one of these three fungicide classes:

1. quinone outside inhibitors (QoIs), which include the strobilurin fungicides
2. demethylation inhibitors (DMIs), which include the triazole fungicides
3. succinate-dehydrogenase inhibitors (SDHIs)

Most fungicides can reduce the impact of fungal pathogens and/or improve seed quality in medium to high disease environments. Systemic fungicides perform better than protective fungicides (e.g., chlorothalonil and mancozeb), but they are more expensive.

The timing of foliar-applied fungicides is critical to maximizing their effectiveness. If the target disease is Sclerotinia stem rot, for instance, then making applications during flowering is critical. For most other foliar diseases, protecting the plant during pod fill is critical. The onset of disease also influences application timing. For example, early detection of soybean rust is critical to successful management of the disease. Fungicides must be applied in the early stages of a soybean rust epidemic to be effective. Control of all diseases may be severely compromised if applications are made well before infection occurs in the field or after the disease has become well established. Generally, modern foliar fungicides protect leaf tissues for 14–21 days following application.

Certain soil-applied fungicides are effective for managing root rot diseases, such as Phytophthora root rot, Pythium root rot, and Rhizoctonia root rot. Effective programs that use soil-applied fungicides to manage Phytophthora root rot also require an integrated approach that uses resistant cultivars and other cultural practices. Soil fungicides are generally applied in-furrow or in bands over rows.

Seed Treatments

Fungicide seed treatments have been used for decades, but the percentage of soybean seed treated has greatly increased since 2000. Seed treatments are generally not needed in areas in which high-quality seed is planted at an appropriate seeding rate under conditions that favor rapid germination and emergence. Also, seed treatments are not useful in situations in which seed quality has been compromised by mechanical damage or physiological factors.

Seed treatments may be useful, however, in fields with a history of a certain soilborne pathogen (e.g., *Phytophthora sojae, Pythium* spp.) and in situations in which seed quality has been

TABLE 13. Efficacy Ratings of Foliar Fungicides for Control of Foliar Soybean Diseases[a,b]

		Disease								
Classification	**Active Ingredient (%)[c]**	**Aerial Blight and Web Blight**	**Anthracnose**	**Brown Spot**	**Cercospora Leaf Blight[d]**	**Frogeye Leaf Spot[e]**	**Pod and Stem Blight**	**Soybean Rust**	**Sclerotinia Stem Rot[f]**	**Harvest Restriction[g]**
Quinone Outside Inhibitor (QoI) Strobilurins (Group 11)	Azoxystrobin (22.9%)	VG	VG	G	F	P	U	G-VG	P	14 days
	Fluoxastrobin (40.3%)	VG	G	G	F	P	U	U	NL	R5, 30 days
	Picoxystrobin (22.5%)	VG	G	G	F	P	U	G	G	14 days
	Pyraclostrobin (23.6%)	VG	VG	G	F	P	U	VG	NL	21 days
Demethylation Inhibitor (DMI) Triazoles (Group 3)	Cyproconazole (8.9%)	U	U	VG	U	F	U	VG	NL	30 days
	Flutriafol (11.8%)	U	VG	VG	F	VG	U	VG-E	F	21 days
	Propiconazole (41.8%)	P	VG	G	NL	F	NL	VG	NL	R5
	Prothioconazole (41.0%)	NL	NL	NL	NL	G-VG	NL	VG	F	21 days
	Tetraconazole (20.5%)	NL	VG	VG	F	G	U	VG-E	F	R5
Methyl Benzimidazole Carbamate (MBC) Thiophanates (Group 1)	Thiophanate-methyl (70%)	U	U	U	F	VG	U	G	F	21 days
SDHI Carboximides (Group 7)	Boscalid (70%)	U	NL	VG	U	P	NL	NL	G-VG	21 days
Mixed Mode of Action	Azoxystrobin (18.2%) Difenoconazole (11.4%)	U	U	U	U	VG	U	VG	NL	14 days
	Azoxystrobin (7.0%) Propiconazole (11.7%)	U	U	G	U	G	U	VG	NL	21 days
	Azoxystrobin (13.5%) Propiconazole (11.7%)	E	VG	G	F	VG	U	VG	NL	R6
	Fluoxastrobin (18.0%) Tebuconazole (25.0%)	U	F	VG	P-F	F	U	U	NL	30 days
	Pyraclostrobin (28.58%) Fluxapyroxad (14.3%)	E	VG	E	F	VG	U	VG	P-F	21 days
	Trifloxystrobin (32.3%) Prothioconazole (10.8%)	VG	VG	VG	F	VG	U	VG	NL	21 days

[a] The North Central Regional Committee on Soybean Diseases and the Regional Committee for Soybean Rust Pathology (NCERA-212 and NCERA-208, respectively) have developed this information on the efficacy of foliar fungicides for control of major foliar soybean diseases in the United States. The efficacy rating for each fungicide was determined by field testing the product over multiple years and at multiple locations by members of the committees. Efficacy ratings are based on the level of disease control achieved by the product and do not necessarily reflect yield increases obtained from product application. Efficacy depends on proper application timing, rate, and method to achieve optimum effectiveness of the fungicide, as determined by labeled instructions and overall level of disease in the field at the time of application. Differences in efficacy among products were determined by making direct comparisons among products in field tests and are based on single applications of the labeled rates (as listed in the table), unless otherwise noted. This table includes products that are systemic fungicides; it is not intended to identify all labeled products, and new products are generated annually. Products may have specific use restrictions; growers should check with local and state authorities for information about their proper use.

[b] Efficacy ratings: NR = Not Recommended; P = Poor; F = Fair; G = Good; VG = Very Good; E = Excellent; NL = Not Labeled for use against this disease; U = Unknown efficacy or insufficient data to rank product efficacy.

[c] Multiple fungicides are labeled exclusively for soybean rust, powdery mildew, and Alternaria leaf spot, including tebuconazole (multiple products) and myclobutanil. Contact fungicides, such as chlorothalonil, may also be labeled for use.

[d] Efficacy in treating Cercospora leaf blight relies on accurate application timing, and the standard R3 application timing may not provide adequate disease control. Fungicide efficacy may improve with earlier or later applications. Fungicides with a solo or mixed QoI or MBC mode of action may not be effective in areas in which QoI and MBC resistance has been detected in the fungal population that causes Cercospora leaf blight.

[e] Fungicides with a solo or mixed QoI mode of action may be effective in areas in which QoI resistance has not been detected in the fungal population that causes frogeye leaf spot.

[f] Efficacy in treating Sclerotinia stem rot is based on an R1–R2 application. Lower efficacy is obtained at an R3 application or if disease symptoms are already present at the time of application.

[g] Harvest restrictions are listed for soybean harvested for grain. Restrictions may vary for other types of soybean (e.g., edamame, etc.) and soybean for other uses (e.g., forage, fodder). R5 is the growth stage when seeds begin to develop at one of four upper-most nodes with a completely unrolled leaf. R6 is the growth stage when pods contain full-sized green beans at one of four upper-most nodes with a completely unrolled leaf.

reduced by a fungal disease (e.g., pod and stem blight, purple seed stain). Seed treatments may also be useful when seeds must be planted in seedbeds that favor delayed germination (i.e., cold, wet soils; compacted fields; reduced or no-till fields) and when low seeding rates are used. Under these circumstances, a seed treatment may effectively reduce losses from both seed-borne fungi and fungi that cause seedling pre- and postemergence damping-off. Many commercial seed-treatment facilities provide at least two fungicide modes of action that are effective against fungi (e.g., *Diaporthe, Fusarium,* and *Rhizoctonia* spp.) and oomycetes (e.g., *Pythium* and *Phytophthora* spp.).

Seed treatments are available in several forms—dusts, flowable suspensions, liquids, and wettable powders—and can be applied commercially or by growers on farms. Commercial seed treatments are more common for soybean because of the precision required to treat each seed uniformly with each active ingredient. Commercial seed-treatment facilities may use a direct-injection method to coat or to pellet the seed with the treatment, which is more efficient than the simple seed-dressing process done on a farm or with slurry seed-treatment equipment. Coating or pelleting the seed can improve seed-treatment adherence and seed uniformity, which aids in planting.

On-farm seed treatments are also done in the planter or drill box prior to planting. Another common on-farm method is to add a liquid fungicide seed treatment into the base of the auger used to fill the drill box. Although these methods are easy and convenient, they may not ensure accurate coverage of the seed. Also, when a machine is used for treating seed, it is important to avoid causing mechanical damage. Such damage can be particularly severe in the handling of seeds with a moisture content of less than 10–11%. The risk of seed injury when using a machine to treat seeds can be offset by the greater degree of fungicide distribution on seeds.

The use of soybean seed treatments may not be economical in all situations. Field history, predicted planting date, and production practices should all be considered in making the decision to use a seed treatment. Environmental conditions at and before planting will greatly influence the development of soilborne diseases and consequently the likelihood that a seed treatment will be profitable. The additional cost of a seed treatment must be considered against these factors to determine the probability of breaking even following the gain in yield that might result. Often, growers consider seed treatments as "insurance": In the event they are needed, they will be useful, and if they are not needed, then their relatively low cost can be justified.

Inoculants and Seed-Treatment Fungicides

The effectiveness of nitrogen-fixing bacteria coated on seeds may be reduced if the inoculant is applied to fungicide-treated seeds in advance of planting. Some fungicides may adversely affect bacterial inoculants and limit the conditions under which the inoculants can be used effectively with fungicides. These fungicides include copper-based compounds, benzimidazoles, and dithiocarbamates. The viability and nodulation of Rhizobium inoculants may be reduced if these fungicides are applied more than 10–12 h in advance to seeds treated with some fungicides. The impact of other fungicides, such as triazoles and strobilurins, on nitrogen-fixing bacterial inoculants is not clear; growers should check the compatibility of inoculants and seed-treatment fungicides if they are used together.

Nematode Management Products

Nematode management products are not solely applied to the soil nor are they exclusively nematicides, directly affecting the nematodes. There are both foliar-applied and seed-treatment products, and there are both nematode repellants and compounds that affect host plant defenses. For the management of nematodes on soybean, there are three categories of products: soil-applied systemic nematicides; protectant seed treatments; and foliar-applied products.

Only a few soil-applied, systemic nematicides are available for soybean. These products are applied when the soybean crop is planted. They affect the functioning of the nematode nervous system, resulting in the paralysis and death of the worm. The use of such products has declined greatly, primarily because of their high cost and inconsistent efficacy, as well as concerns over possible negative environmental and health effects.

Protectant seed treatments are a new category of nematode management products, and their economic value may vary from year to year. These products are applied to the seed as a seed treatment, and they function using several different active ingredients or modes of action, including a bacterium, a bacterial fermentation compound, and a harpin protein that elicits changes in natural plant defenses. Many of these products are described as providing early season protection of roots from nematode parasitism.

A few nematode management products are applied to the soybean canopy. These products are applied to the growing soybean crop weeks after planting. One product is a harpin protein that affects natural plant defenses, and another is a compound that is translocated to the roots and disrupts the nematode nervous system.

Product names, formulations, application rates, and label restrictions can change annually and vary by geographical region. Growers should consult local agrichemical professionals to learn what options exist. Also, the economic benefit of using any of these products will be determined by the severity of yield loss caused by the nematode parasitism, the efficacy of the product used, and the value of the soybean crop. All of these factors should be considered when deciding whether to use a nematode management product.

Insecticides

Foliar-applied and seed-treatment insecticides are available, but they are generally ineffective in managing aphid-vectored viruses of soybean and rarely used to control thrips-transmitted viruses (e.g., *Tobacco ringspot virus*) and beetle-transmitted viruses (e.g., *Bean pod mottle virus*). Application of foliar-applied insecticides may be considered to reduce the fungal infections that follow pod feeding by insects, such as the bean leaf beetle (*Cerotoma trifurcata*), grasshoppers (several species), and stinkbugs (various species).

In deciding whether to use an insecticide to reduce arthropod damage, growers should take into account the added damage caused by opportunistic fungi. This is particularly true with stinkbugs that transmit the yeast spot fungus and with the bean leaf beetle, whose feeding causes pod injuries that provide entry sites for several seed-infecting fungi.

Selected References

Esker, P. D., and Conley, S. P. 2012. Probability of yield response and breaking even for soybean seed treatments. Crop Sci. 52:351-359.

Ferriss, R. S., Stuckey, R. E., Gleason, M. L., and Siegel, M. R. 1987. Effects of seed quality, seed treatment, soil source, and initial soil moisture on soybean seedling performance. Phytopathology 77:140-148.

Hall, R., and Xue, A. G. 1995. Effectiveness of fungicidal seed treatments applied to smooth and shriveled soybean seeds contaminated by *Diaporthe phaseolorum*. Phytoprotection 76:47-56.

Mueller, D. S., Dorrance, A. E., Derksen, R. C., Ozkan, E., Kurle, J. E., Grau, C. R., Gaska, J. M., Hartman, G. L., Bradley, C. A., and Pedersen, W. L. 2002. Efficacy of fungicides on *Sclerotinia sclerotiorum* and their potential for control of Sclerotinia stem rot on soybean. Plant Dis. 86:26-31.

Mueller, T. A., Miles, M. R., Morel, W., Marois, J. J., Wright, D. L., Kemerait, R. C., Levy, C., and Hartman, G. L. 2009. Effect of fungicide and timing of application on soybean rust severity and yield. Plant Dis. 93:243-248.

Sinclair, J. B. 1993. Phomopsis seed decay of soybeans—A prototype for studying seed disease. Plant Dis. 77:329-334.

(Prepared by D. S. Mueller, K. A. Wise, G. L. Tylka, and N. Dufault)

Appendix I

Diseases of Soybean

BACTERIAL DISEASES

Bacillus seed decay
Bacillus subtilis (Ehrenberg) Cohn
Bacterial blight
Pseudomonas savastanoi pv. *glycinea* (Coerper) Gardan et al.
(syn. *P. syringae* pv. *glycinea* [Coerper] Young et al.)
Bacterial pustule
Xanthomonas axonopodis pv. *glycines* (Nakano) Vauterin et al.
Bacterial tan spot
Curtobacterium flaccumfaciens pv. *flaccumfaciens* (Hedges) Collins & Jones
Bacterial wilt
Curtobacterium flaccumfaciens pv. *flaccumfaciens* (Hedges) Collins & Jones
Ralstonia solanacearum (E. F. Smith) Yabuuchi et al.
(syn. *Pseudomonas solanacearum* [Smith] Smith)
Wildfire
Pseudomonas syringae pv. *tabaci* (Wolf & Forster) Young et al.

Other Bacteria Known to Infect Soybean

Burkholderia andropogonis (E. F. Smith) Gillis et al.
Curtobacterium plantarum Dunleavy
Erwinia amylovora (Burrill) Winslow et al.
E. carotovora subsp. *atroseptica* (van Hall) Dye
Pseudomonas savastanoi pv. *phaseolicola* (Burkholder) Gardan et al.
P. syringae pv. *syringae* van Hall
Xanthomonas campestris pv. *cannabis* Severin

FUNGAL DISEASES

Foliage, Upper Stems, Pods, and Seeds

Alternaria leaf spot, pod necrosis, and seed decay
Alternaria alternata (Fr.:Fr.) Keissl.
A. atrans W. Gibson
A. tenuissima (Nees & T. Nees:Fr.) Wiltshire
Anthracnose
Colletotrichum chlorophyti S. Chandra & Tandon
C. coccodes (Wallr.) S. Hughes
C. gloeosporioides (Penz.) Penz. & Sacc.
(syn. *Glomerella cingulata* [Stoneman] Spauld. & H. Schrenk)
C. graminicola (Ces.) G. W. Wilson
(syn. *G. graminicola* Politis)
C. incanum H. C. Yang et al.
C. truncatum (Schwein.) Andrus & W. D. Moore
Glomerella glycines Lehman & F. A. Wolf
Aristastoma leaf spot and yellowing
Aristastoma camarographioides B. Sutton
A. guttulosum B. Sutton
Black leaf blight
Arkoola nigra J. Walker & Stovold
Brown spot (Septoria leaf spot)
Septoria glycines Hemmi
Cercospora leaf blight and purple seed stain
Cercospora kikuchii (T. Matsumoto & Tomoy.) M. W. Gardner
Choanephora leaf blight
Choanephora cucurbitarum (Berk. & Ravenel) Thaxt.
C. infundibulifera (Curr.) Sacc.
Downy mildew
Peronospora manshurica (Naumov) Syd.
Drechslera blight
Bipolaris glycines (S. Naray. & Durairaj) B. A. Khasanov
(syn. *Drechslera glycines* S. Naray. & Durairaj)
Exserohilum rostratum (Drechsler) K. J. Leonard & Suggs
(syn. *Setosphaeria rostrata* K. J. Leonard)
Frogeye leaf spot (Cercospora leaf spot)
Cercospora sojina Hara
Fusarium seed rot and *Fusarium* spp. associated with seeds
Fusarium acuminatum Ellis & Everh.
(syn. *Gibberella acuminata* Wollenw.)
F. decemcellulare C. Brick
(syn. *Albonectria rigidiuscula* [Berk. & Broome] Rossman & Samuels)
F. equiseti (Corda) Sacc.
(syn. *G. intricans* Wollenw.)
F. graminearum Schwabe
(syn. *G. zeae* [Schwein.:Fr.] Petch)
F. incarnatum (Desm.) Sacc.
(syn. *F. semitectum* Berk. & Ravenel)
F. lateritium Nees.:Fr.
(syn. *G. baccata* [Wallr.] Sacc.)
F. oxysporum Schltdl.:Fr.
F. solani (Mart.) Sacc.
(syn. *Haemanectria haematococca* [Berk. & Broome] Samuels & Nirenberg)
F. sporotrichioides Sherb.
F. tricinctum (Corda) Sacc.
(syn. *G. tricincta* El-Gholl et al.)
F. verticillioides (Sacc.) Nirenberg
(syn. *G. moniliformis* Wineland)
Leptosphaerulina leaf spot
Leptosphaerulina trifolii (Rostr.) Petr.
Phomopsis seed decay
Diaporthe longicolla (Hobbs) J. M. Santos
(syn. *Phomopsis longicolla* Hobbs)
Phyllosticta leaf spot (Phyllosticta leaf blight)
Pleosphaerulina sojicola Miura
(syn. *Phyllosticta sojicola* C. Massal.)

Pod and stem blight
Diaporthe sojae Lehman
(syn. *Phomopsis phaseoli* [Desm.] Sacc.)
Powdery mildew
Erysiphe diffusa (Cooke & Peck) U. Braun & S. Takam.
E. glycines F. L. Tai
Red leaf blotch (Dactuliophora leaf spot, Pyrenochaeta leaf blotch, Pyrenochaeta leaf spot)
Coniothyrium glycines (R. B. Stewart) Verkley & Gruyter
(syns. *Dactuliochaeta glycines* [R. B. Stewart] G. L. Hartm. & J. B. Sinclair; *Dactuliophora glycines* C. L. Leakey; *Phoma glycinicola* Gruyter & Boerema nom. illegit; *Pyrenochaeta glycines* R. B. Stewart)
Rhizoctonia aerial blight and web blight (Rhizoctonia foliar blight, Rhizoctonia web blight)
Rhizoctonia solani J. G. Kühn
(syn. *Thanatephorus cucumeris* [A. B. Frank] Donk)
Rust
Phakopsora meibomiae (Arthur) Arthur
(syn. *Malupa vignae* [Bres.] Y. Ono et al.)
P. pachyrhizi Syd.
(syn. *M. sojae* [Henn.] Y. Ono et al.)
Scab
Sphaceloma glycines Kurata & Kuribayashi
Sclerotinia stem rot (white mold)
Sclerotinia minor Jagger
S. sclerotiorum (Lib.) de Bary
Stemphylium leaf blight
Pleospora herbarum (Pers.:Fr.) Rabenh.
(syn. *Stemphylium herbarum* E. G. Simmons)
S. botryosum Wallr.
(syn. *P. tarda* E. G. Simmons)
S. vesicarium (Wallr.) E. G. Simmons
(syn. *P. allii* [Rabenh.] Ces. & De Not.)
Target spot
Corynespora cassiicola (Berk. & M. A. Curtis) C. T. Wei
Yeast spot of pods and seeds
Eremothecium ashbyi Guilliermond.
E. coryli (Peglion) Kurtzman

Fungal Diseases of Lower Stems and Roots
Brown stem rot
Cadophora gregata (Allington & D. W. Chamb.) T. C. Harr. & McNew
(syn. *Phialophora gregata* [Allington & D. W. Chamb.] W. Gams)
Charcoal rot
Macrophomina phaseolina (Tassi) Goid.
Fusarium blight or wilt
Fusarium oxysporum Schltdl.:Fr.
Fusarium pod and collar rot
Fusarium incarnatum (Desm.) Sacc.
(syns. *F. pallidoroseum* [Cooke] Sacc.; *F. semitectum* Berk. & Ravenel)
Fusarium root rot
Fusarium acuminatum Ellis & Everh.
(syn. *Gibberella acuminata* Wollenw.)
F. avenaceum (Fr.:Fr.) Sacc.
(syn. *G. avenacea* R. J. Cook)
F. commune K. Skovgaard et al.
F. equiseti (Corda) Sacc.
(syn. *G. intricans* Wollenw.)
F. graminearum Schwabe
F. heterosporum Nees & T. Nees:Fr.
(syn. *G. cyanea* [Sollm.] Wollenw.)
F. oxysporum Schltdl.:Fr.
F. proliferatum (Matsush.) Nirenberg ex Gerlach & Nirenberg
(syn. *G. intermedia* [Kuhlman] Samuels et al.)
F. redolens Wollenw.
F. solani (Mart.) Sacc.
F. sporotrichioides Sherb.
F. tricinctum (Corda) Sacc.
(syn. *G. tricincta* El-Gholl et al.)
F. verticillioides (Sacc.) Nirenberg
(syns. *F. moniliforme* J. Sheld.; *G. moniliformis* Wineland)
Phytophthora root and stem rot
Phytophthora sansomeana E. M. Hansen
P. sojae Kaufm. & Gerd.
Pythium seed rot, damping-off, and root rot
Globisporangium irregulare (Buisman) Uzuhashi et al.
(syn. *Pythium irregulare* Buisman)
G. sylvaticum (W. A. Camp. & F. F. Hendrix) Uzuhasi et al.
(syn. *P. sylvaticum* W. A. Camp. & F. F. Hendrix)
G. ultimum (Trow) Uzuhashi et al.
(syn. *P. ultimum* Trow)
P. aphanidermatum (Edson) Fitzp.
P. dissotocum Drechsler
P. inflatum V. D. Matthews
P. myriotylum Drechsler
P. torulosum Coker & F. Patt.
Red crown rot (Cylindrocladium root rot, black root rot, Calonectria root rot)
Calonectria ilicicola Boedijn & Reitsma
(syn. *Cylindrocladium parasiticum* Crous et al.)
Calonectria morganii Crous et al.
(syn. *Cylindrocladium scoparium* Morg.)
Rhizoctonia damping-off and root rot
Rhizoctonia solani J. G. Kühn
(syn. *Thanatephorus cucumeris* [A. B. Frank)] Donk)
Sclerotium blight (southern blight, southern stem rot, white mold)
Athelia rolfsii (Curzi) Tu & Kimbr.
(syn. *Sclerotium rolfsii* Sacc.)
Stachybotrys root rot
Stachybotrys chartarum (Ehrenb.) S. Hughes
Stem canker
Diaporthe phaseolorum var. *caulivora* Athow & Caldwell
(syn. *D. caulivora* [Athow & Caldwell] J. M. Santos et al.)
D. phaseolorum var. *meridionalis* F. A. Fernández
(syn. *D. aspalathi* E. Jansen et al.)
Sudden death syndrome
Fusarium brasiliense T. Aoki & O'Donnell
F. crassistipitatum Scandiani et al.
F. tucumaniae T. Aoki et al.
F. virguliforme O'Donnell & T. Aoki
(syn. *F. solani* f. sp. *glycines*)
Thielaviopsis root rot (black root rot)
Thielaviopsis basicola (Berk. & Broome) Ferraris

Storage Fungi
Alternaria spp.
Aspergillus spp.
A. candidus Link:Fr.
A. flavus Link:Fr.
A. glaucus Link:Fr.
A. parasiticus Speare
A. restrictus G. Sm.
Curvularia spp.
Fusarium spp.
F. graminearum Schwabe
Mucor spp.
Paecilomyces spp.
Penicillium spp.
Rhizopus spp.
Scopulariopsis spp.

Other Fungi and Oomycetes Reported in Soybean
Ascochyta sojaecola Abramov
Bipolaris sorokiniana (Sacc.) Shoemaker
Chaetomium spp.
C. cupreum L. M. Ames

Cladosporium tenuissimum Cooke
Epicoccum spp.
Exserohilum rostratum (Drechsler) K. J. Leonard & Suggs
Gaeumannomyces graminis (Sacc.) Arx & D. L. Olivier
Gliocladium roseum Bainier
Helminthosporium spp.
Macrophoma spp.
Marssonina sojaecola Mashkina et al.
Microdochium spp.
Monilia spp.
Mycoleptodiscus terrestris (Gerd.) Ostaz.
Mycosphaerella phaseolorum Siemaszko
Neocosmospora vasinfecta E. F. Sm.
(syn. *Acremonium* sp.)
Nigrospora spp.
Olpidium viciae Kusano
Pestalotia spp.
Boeremia exigua var. *exigua* (Desm.) Aveskamp et al.
(syn. *Phoma exigua* Desm.)
Phymatotrichopsis omnivora (Shear) Hennebert
Plectosphaerella cucumerina (Lindf.) W. Gams
Pleosphaerulina sojicola Miura
(syn. *Phyllosticta glycineum* Tehon & E. Y. Daniels)
Setosphaeria glycinea L. S. Srivast. et al.
Stauronematopsis sojae (Uecker & Kulik) Abbas et al.
(syn. *Pseudorobillarda sojae* Uecker & Kulik)
Verticillium nigrescens Pethybr.

MOLLICUTE DISEASES

Bud proliferation
Buds proliferate from leaf axils on plant; plant maturity is delayed
Machismo
First symptoms occur after flowering or pod formation; pods are rigid, flat, and empty or corrugated and leaflike; buds proliferate from leaf axils anywhere on plant, causing witches'-broom
Phyllody and witches'-broom
Phylloid disorder of floral organs; shoot proliferation and reduced leaflets; resembles machismo
Phytoplasmas
Bushy or stunted plants; purple streaks and patches on stems; marginal necrosis or yellow blotches on leaves; bud proliferation; reduction of pod numbers at nodes; dark-bluish foliage

NEMATODES (PARASITIC)

Dagger nematode:
Xiphinema americanum Cobb
Lance nematodes:
Hoplolaimus spp.
H. columbus Sher
H. galeatus (Cobb) Thorne
H. magnistylus Robbins
H. stephanus Sher
Lesion nematodes:
Pratylenchus spp.
P. agilis Thorne & Malek
P. alleni Ferris
P. brachyurus (Godfrey) Filip. & Sch. Stek.
P. coffeae (Zimmerman) Filip. & Sch. Stek.
P. crenatus Loof
P. hexincisus Taylor & Jenkins
P. neglectus (Rencsh) Filip. & Sch. Stek.
P. penetrans (Cobb) Filip. & Sch. Stek.
P. pratensis (de Man) Filip.
P. scribneri Steiner
P. sefaenis Fortuner
P. teres Khan & Singh
P. thornei Sher & Allen
P. vulnus Allen & Jensen
P. zeae Graham
Reniform nematode:
Rotylenchulus reniformis Linford & Oliveira
Root-knot nematodes:
Meloidogyne spp.
M. arenaria (Neal) Chitwood
M. hapla Chitwood
M. incognita (Kofoid & White) Chitwood
M. javanica (Treub) Chitwood
Soybean cyst nematode:
Heterodera glycines Ichinohe
Sting nematodes:
Belonolaimus spp.
B. longicaudatus Rau

NONINFECTIOUS DISEASES, DISORDERS, AND DAMAGE

Anoxia
Root injury or death caused by extended flooding or submergence in water
Crusting and compaction of soil
Thickened seedling stems; broken or cracked hypocotyl arches; occurs after heavy rain or flooding
Drought damage
Stunting, poor vigor, and chlorosis of interveinal areas of leaves, which changes to general chlorosis with interveinal necrosis in severe situations
Frost damage
Upper plant damage or death; new growth at lower nodes
Green stem disorder
Immature, still-green stems after ripening of pods and seeds
Hail damage
Ragged and torn leaves; broken or cut stems with dark, sunken areas
Heat canker
Girdling of the hypocotyl at or just above the soil line; affected tissue is reddish brown and shriveled; entire plant dies within days; occurs at temperatures above 35°C
Lightning damage
Circular areas of damage within the field; blackened lower stems and dead leaves; blanched or "cooked" pith
Sunburn
Brick-red spots on leaves that develop brownish centers; brick-red lesions on petioles and stems

VIRUS DISEASES

African cassava mosaic
Genus *Begomovirus; African cassava mosaic virus* (ACMV)
Alfalfa mosaic
Genus *Alfamovirus; Alfalfa mosaic virus* (AMV)
Bean common mosaic
Genus *Potyvirus; Bean common mosaic virus* (BCMV), peanut stripe strain, Azuki bean mosaic strain
Bean common mosaic necrosis
Genus *Potyvirus; Bean common mosaic necrosis virus* (BCMNV), blackeye cowpea mosaic strain
Bean golden mosaic
Genus *Begomovirus; Bean golden mosaic virus* (BGMV)
Bean leafroll
Genus *Luteovirus; Bean leafroll virus* (BLRV)
Bean mild mosaic
Genus *Carmovirus; Bean mild mosaic virus* (BMMV)
Bean pod mottle
Genus *Comovirus; Bean pod mottle virus* (BPMV)
Bean yellow mosaic
Genus *Potyvirus; Bean yellow mosaic virus* (BYMV)

Beet western yellows
Genus *Polerovirus; Beet western yellows virus* (BWYV)
Blackgram mottle
Blackgram mottle virus (BMoV)*
Broad bean true mosaic
Genus *Comovirus; Broad bean true mosaic virus* (BBTMV)
Cotton leaf curl
Genus *Begomovirus; Cotton leaf curl Kokhran virus* (CLCuKoV)
Cowpea chlorotic mottle
Genus *Bromovirus; Cowpea chlorotic mottle virus* (CCMV), bean yellow stipple strain, soybean strain (CCMV-S)
Cowpea mild mottle
Genus *Carlavirus; Cowpea mild mottle virus* (CpMMV)
Cowpea mosaic
Genus *Comovirus; Cowpea mosaic virus* (CPMV)
Cowpea mottle
Genus *Carmovirus; Cowpea mottle virus* (CPMoV)
Cowpea severe mosaic
Genus *Comovirus; Cowpea severe mosaic virus* (CPSMV)
Cucumber mosaic (Soybean stunt)
Genus *Cucumovirus; Cucumber mosaic virus* (CMV), soybean stunt strain (CMV-SS)
(syn. Soybean stunt virus)
Cycas necrotic stunt
Genus *Nepovirus; Cycas necrotic stunt virus* (CNSV), soybean mild mosaic strain
(syns. Soybean chlorotic spot virus; Soybean mild mosaic virus [SMMV])
Euphorbia mosaic
Genus *Begomovirus; Euphorbia mosaic virus* (EuMV)
Glycine mosaic
Genus *Comovirus; Glycine mosaic virus* (GMV)
Indonesian soybean dwarf
Genus unassigned; *Indonesian soybean dwarf virus* (ISDV)
Milk vetch dwarf
Genus *Nanovirus; Milk vetch dwarf virus* (MVDV)
Mungbean yellow mosaic
Genus *Begomovirus; Mungbean yellow mosaic India virus* (MYMIV)
Genus *Begomovirus; Mungbean yellow mosaic virus* (MYMV)
Okra mottle
Genus *Begomovirus; Okra mottle virus* (OMoV)
Pea green mottle
Genus *Comovirus; Pea green mottle virus* (PGMV)
Pea mild mottle
Genus *Comovirus; Pea mild mottle virus* (PMMV)
Peanut mottle
Genus *Potyvirus; Peanut mottle virus* (PMV)
Peanut stunt
Genus *Cucumovirus; Peanut stunt virus* (PSV)

---

* Not classified as a distinct species by the International Committee on Taxonomy of Viruses (ICTV) (as of 2014) (www.ictvonline.org).

Pepper golden mosaic
Genus *Begomovirus; Pepper golden mosaic virus* (PepGMV)
Quail pea mosaic
Genus *Comovirus; Quail pea mosaic virus* (QPMV)
Rhynchosia golden mosaic
Genus *Begomovirus; Rhynchosia golden mosaic virus* (RhGMV)
Sida micrantha mosaic
Genus *Begomovirus; Sida micrantha mosaic virus* (SimMV)
Sida mottle
Genus *Begomovirus; Sida mottle virus* (SiMoV)
Southern bean mosaic
Genus *Sobemovirus; Southern bean mosaic virus* (SBMV)
Soybean blistering mosaic
Genus *Begomovirus; Soybean blistering mosaic virus* (SbBMV)
Soybean chlorotic mottle
Genus *Soymovirus; Soybean chlorotic mottle virus* (SbCMV)
Soybean chlorotic spot
Genus *Begomovirus; Soybean chlorotic spot virus* (SoCSV)
Soybean crinkle leaf
Genus *Begomovirus; Soybean crinkle leaf virus* (SCLV)
Soybean dwarf
Genus *Luteovirus; Soybean dwarf virus* (SbDV)
Soybean mild mottle
Genus *Begomovirus; Soybean mild mottle virus* (SbMMV)
Soybean mosaic
Genus *Potyvirus; Soybean mosaic virus* (SMV)
Soybean Putnam
Soybean Putnam virus (SPuV)*
Soybean severe stunt
Soybean severe stunt virus (SSSV)*
Soybean vein necrosis
Soybean vein necrosis virus (SVNV)*
Soybean yellow mosaic
Soybean yellow mosaic virus (SYMV)*
Soybean yellow mottle mosaic
Genus *Carmovirus; Soybean yellow mottle mosaic virus* (SYMMV)
Tobacco mosaic
Genus *Tobamovirus; Tobacco mosaic virus* (TMV), soybean strain (TMV-S)
Tobacco ringspot
Genus *Nepovirus; Tobacco ringspot virus* (TRSV)
Tobacco streak
Genus *Ilarvirus; Tobacco streak virus* (TSV)
Tomato ringspot
Genus *Nepovirus; Tomato ringspot virus* (ToRSV)
Tomato spotted wilt
Genus *Tospovirus; Tomato spotted wilt virus* (TSWV)

Appendix II

Pests of Soybean

Bean leaf beetle
Cerotoma trifurcata (Forster)
Beet armyworm
Spodoptera exigua Hübner
Brown stink bug
Euschistus servus (Say)
Cabbage looper
Trichoplusia ni Hübner
Corn earworm
Helicoverpa zea Boddie
Dectes stem borer
Dectes texanus LeConte
Fall armyworm
Spodoptera frugiperda Smith & Abbot
Garden webworm
Achyra rantalis Guenée
Green cloverworm
Plathypena scabra Fabricius
Green stink bug
Chinavia hilaris (Say)
Kudzu bug (bean plataspid)
Megacopta cribraria (Fabricius)
Redbanded stink bug
Piezodorus guildinii (Westwood)
Saltmarsh caterpillar
Estigmene acrea Drury
Southern green stink bug
Nezara viridula (L.)
Soybean aphid
Aphis glycines Matsumura
Soybean looper
Chrysodeixis includens (Walker)
Soybean thrips
Neohydatothrips variabilis
(syn. *Sericothrips variabilis*)
Threecornered alfalfa hopper
Spissistilus festinus (Say)
Tobacco budworm
Heliothis virescens (Fabricius)
Twospotted spider mite
Tetranychus urticae Koch
Velvetbean caterpillar
Anticarsia gemmatalis Hübner
Yellowstriped armyworm
Spodoptera ornithogalli (Guenée)

Glossary

Å—active
a.i.—active ingredient
C—Celsius or centigrade
cm—centimeter (1 cm = 0.01 m = 0.3937 in.)
Da—Dalton
g—gram (1 g = 0.03527 oz)
h—hour
ha—hectare (1 ha = 2.471 acres)
kg—kilogram (1 kg = 2.205 lb)
kDa—kilodalton
L—liter (1 L = 1.057 quarts liquid [U.S.])
m—meter (1 m = 39.37 in.)
mg—milligram (1 mg = 0.001 g)
min—minute
ml—milliliter (1 ml = 0.001 L)
mm—millimeter (1 mm = 0.001 m = 0.03937 in.)
μg—microgram (1 μg = 10^{-6} g)
μm—micrometer (1 μg = 10^{-6} m)
nm—nanometer (1 nm = 10^{-9} m)
ppm—parts per million
sec—second

abiotic—pertaining to the absence of life, as diseases not caused by living organisms
abscise—to separate from a plant, as leaves, flowers, and fruits do when they fall
abscission—the shedding of leaves or other plant parts as the result of physical weakness in a specialized layer of cells (the abscission layer), which develops at the base of the structure
acervulus (pl. acervuli)—an erumpent, cushionlike fruiting body bearing conidiophores, conidia, and sometimes setae
acid—having a pH less than 7
acid precipitation—precipitation of low pH because of the presence of nitric and sulfuric acid formed by the combination of air pollutants (NO_x and SO_2) with water
acid rain—*see* acid precipitation
acropetal—upward from the base to the apex of a shoot of a plant (*see* basipetal); in fungi, producing spores in succession in the direction of the apex, so that the apical spore is the youngest
acute—with reference to symptoms, developing suddenly (*see* chronic)
adventitious—arising from other than the usual place, as roots from a stem rather than branches of a root
aeciospore—a dikaryotic spore produced in the aecium of a rust fungus; in heteroecious rust fungi, the spore stage that infects the alternate host
aecium (pl. aecia; adj. aecial)—the fruiting body of a rust fungus in which the first dikaryotic spores (aeciospores) are produced
aerial—occurring in air
aerobic—living only in the presence of oxygen
aflatoxin—a chemical by-product of *Aspergillus flavus* and *A. parasiticus* that is harmful to humans and other animals
agar—a jellylike material derived from algae and used to solidify liquid culture media; a medium containing agar
alate—winged (*see* apterous)
albino (n. albinism)—white or light colored; having a marked deficiency in pigmentation
alkaline—having basic (nonacidic) properties; having a pH greater than 7
allele—any of one or more alternative forms of a gene
allelopathy (adj. allelopathic)—the ability of one species to inhibit or prevent the growth of another species through the production of one or more toxic substances
alternate host—one of two kinds of plants on which a parasitic fungus (e.g., a rust fungus) must develop to complete its life cycle
alternative host—a plant other than the main host that can be colonized by a parasite but that is not required for completion of the developmental cycle of the parasite
anaerobic—living in the absence of oxygen
anamorph (adj. anamorphic)—the imperfect state or asexual form in the life cycles of certain fungi, producing asexual spores (e.g., conidia) or no spores (*see* holomorph; teleomorph)
anastomosis (pl. anastomoses)—the fusion of branches of the same or different structures (e.g., hyphae) to make a network
annual—a plant that completes its life cycle and dies within one year (*see* biennial; perennial)
antagonist—an organism or substance that counteracts or limits the action of another organism or substance
anterior—situated toward the front or head (*see* posterior)
anther—the pollen-bearing portion of a flower
antheridium (pl. antheridia)—the male sexual organ (male gametangium) of certain fungi
anthracnose—disease caused by acervuli-forming fungi (order Melanconiales); characterized by sunken lesions and necrosis
antibiotic—any of various chemical compounds produced by microorganisms and killing or inhibiting the growth of other living organisms
antibody—a protein formed in the blood of warm-blooded animals in response to the presence of an antigen
antigen—any foreign chemical (normally, a protein) that induces antibody formation in warm-blooded animals
antiserum (pl. antisera)—blood serum containing an antibody
apex (pl. apices; adj. apical)—the tip of a root or shoot; contains the apical meristem
aphid—any of numerous species of small, sucking insects of the family Aphididae (order Homoptera) that produce honeydew and injure plants when present in large populations
apothecium (pl. apothecia)—open, cuplike or saucerlike, ascus-bearing fruiting body (ascocarp) of ascomycetous fungi; often supported on a stalk
appressed—closely flattened down or pressed against a surface (syn. adpressed)
appressorium (pl. appressoria)—a swollen, flattened portion of a fungal filament that adheres to the surface of a higher plant, providing anchorage for invasion by the fungus
apterous—wingless (*see* alate)
arbuscule (adj. arbuscular)—a branched haustorial structure of certain endomycorrhizal fungi; formed within living cells of the root
arthropod—any member of the phylum Arthropoda, which consists of animals with articulated bodies and limbs, including insects, arachnids, and crustaceans
ascocarp—a sexual fruiting body (e.g., apothecium, ascostroma, cleistothecium, perithecium, pseudothecium) of ascomycetous fungi, producing asci and ascospores (syn. ascoma)
ascogenous—pertaining to ascus-producing hyphae
ascogonium (pl. ascogonia)—a specialized cell that gives rise to hyphae that produce asci
ascoma (pl. ascomata)—*see* ascocarp

ascomycetes (adj. ascomycetous)—a group of fungi in the phylum Ascomycota (kingdom Fungi) characterized by the production of sexual spores (ascopores) within an ascus
ascospore—a sexual spore borne in an ascus
ascostroma (pl. ascostromata)—a fruiting body of ascomycetous fungi containing bitunicate (double-walled) asci in locules (cavities); usually dark and containing multiple locules but sometimes containing only a single locule (*see* pseudothecium)
ascus (pl. **asci**)—a saclike structure containing ascospores (typically eight) and usually borne in a fruiting body
aseptate—lacking cross-walls (septa); nonseptate; coenocytic
asexual—vegetative; lacking sex organs, gametes, or sexual spores (as in the imperfect or anamorphic stage of a fungus)
asexual reproduction—any type of reproduction not involving the union of gametes and meiosis
asexual stage—the imperfect (anamorphic) stage of certain fungi
attenuate—to narrow; to weaken; to decrease in virulence or pathogenicity
autoecious—with reference to rust fungi, producing all spore forms on one species of host plant (*see* heteroecious)
autotroph—any of various organisms (e.g., plants capable of photosynthesis) that synthesize their nutritive substances from inorganic molecules
avirulent—nonpathogenic; unable to cause disease (*see* virulent)
auxin—a class of plant hormones
axenic—culture in the absence of living bacteria or other organisms; pure culture
axil—the angle formed by a leaf petiole and the stem
axillary—pertaining to or placed within an axil

bacilliform—shaped like a short rod with rounded ends
backcross—to cross (mate) an offspring with one of its parents
bactericide—a chemical or physical agent that kills bacteria
bacterium (pl. bacteria)—a prokaryotic, microscopic, single-celled organism having a cell wall and increasing by binary fission
basidiocarp—the sexual fruiting body of fungi in the group basidiomycetes (syn. basidioma)
basidiomycetes (adj. basidiomycetous)—a group of fungi in the phylum Basidiomycota (kingdom Fungi) characterized by the formation of external basidiospores on basidia
basidiospore—a haploid (1*n*) sexual spore produced on a basidium
basidium (pl. basidia; adj. basidial)—a structure on which basidiospores are produced externally
basipetal—situated down from the apex toward the base of a shoot of a plant; developing in the direction of the base, so that the apical part is oldest (*see* acropetal)
beetle—any of numerous species of chewing insects in the order Coleoptera
bicellular—two celled
biennial—a plant that produces seed and dies at the end of its second year of growth (*see* annual; perennial)
binucleate—having two nuclei per cell (*see* multinucleate; uninucleate)
bioassay—any test (assay) using a living organism
biocide—a compound toxic to all forms of life
biological control—exploitation of the natural competition, parasitism, or antagonism of organisms for the management of pests and pathogens (syn. biocontrol)
biotic—pertaining to living organisms; with reference to disease, caused by a living organism
biotroph—an organism that can live and multiply only on another living organism (syn. obligate parasite) (*see* necrotroph)
biotype—a subdivision of a species, subspecies, or race based on some identifiable physiological trait, such as a particular virulence pattern
bitunicate—having two walls
blight—the sudden, severe, and extensive spotting, discoloration, wilting, or destruction of leaves, flowers, stems, or entire plants
blotch—an irregularly shaped, usually superficial spot or blot
botryose—shaped like a bunch of grapes
bract—a reduced leaf associated with a flower or inflorescence; a modified leaf from the axil of which a flower arises
breeding line—a plant strain used in a plant breeding program and usually containing one or more desirable agronomic or breeding characteristics
broadcast application—application of fertilizer by spreading or scattering it on the soil surface
bursa—an extension or flap of cuticle at the side of the male nematode sex organ; used for orienting the body during mating

calcareous—rich in calcium carbonate (lime)
callus—specialized tissue that forms over a wound or cut in a plant (e.g., cork cambium may form, and the cells produced will gradually seal the wound)
canker—a plant disease characterized (in woody plants) by the death of cambium tissue and malformation or loss of bark and (in nonwoody plants) by the formation of sharply delineated, dry, necrotic, localized lesions on the stem; a lesion caused by such a disease, particularly in woody plants
canopy—the expanded leafy top of a plant or plants
capsid—the protective layer of protein surrounding the nucleic acid core of a virus; the protein molecules that make up this layer (syn. coat protein)
carbohydrate—any of various chemical compounds composed of carbon, hydrogen, and oxygen, such as sugars, starches, and cellulose
carpel—the ovule-bearing structure of a flower in angiosperms
catenulate—with reference to the spores of certain fungi, formed in a chain
caterpillar—the larva of a butterfly
causal agent—the organism or agent that produces a given disease
chimera, chimaera—a plant or organ consisting of two or more genetically different tissues
chlamydospore—a thick-walled or double-walled, asexual resting spore formed from hyphal cells (terminal or intercalary) or by transformation of conidial cells; functions as an overwintering stage
chlorophyll (adj. chlorophyllous)—any of a group of green pigments found in chloroplasts and critical to photosynthesis
chloroplast—a disklike structure containing chlorophyll in which photosynthesis occurs in green plants
chlorosis (adj. chlorotic)—the failure of chlorophyll development, caused by disease or a nutritional disturbance; the fading of green plant parts to light green, yellow, or white
chromosome—the structure that contains the genes of an organism (in eukaryotes, chromosomes are in the nucleus and can be visualized with an optical microscope as threads or rods during meiosis and mitosis; in bacteria, the chromosome is usually a single circle of DNA and cannot be visualized with an optical microscope)
chronic—with reference to symptoms, slow-developing, persistent, or recurring (*see* acute)
circulative—with reference to viruses, passing through the gut and circulating in the body of an insect vector before being transmitted to a host
circulative transmission—virus transmission in which the virus must accumulate within or pass through the lymphatic system of an insect vector before it can be transmitted to a plant
cirrhus (pl. cirrhi), **cirrus** (pl. cirri)—a curled, tendril-like mass of exuded spores held together by a slimy matrix
clamp connection—a bridge- or buckle-like hyphal protrusion in basidiomycetous fungi, formed at cell division and connecting the newly divided cells
clavate—club shaped (syn. claviform)
cleistothecium (pl. cleistothecia)—a spherical ascocarp that is closed at maturity
coalesce—to grow together into one body, as enlarging lesions may join together to form a single spot
coenocytic—having multiple nuclei embedded in cytoplasm without cross-walls; nonseptate
colonize—with reference to plant pathogens, to infect and ramify through plant tissue
colony—a microorganism growing in mass, especially as a pure culture
conidiogenous—producing and bearing conidia
conidiophore—a simple or branched hypha on which conidia are produced
conidium (pl. conidia)—an asexual, nonmotile fungal spore that develops externally or is liberated from the cell that formed it
cortex (adj. cortical)—the region of parenchyma tissue between the epidermis and phloem in stems and roots; the region beneath the rind of a sclerotium
cotyledon—the primary embryonic leaf within the seed, in which nutrients for the new plant are stored (monocots have one cotyledon; dicots have two)

cross-protection—protection in a plant given by one virus strain over another virus strain
crown—the upper dome of a tree, bearing leaves, flowers, and fruits; the junction of the root and stem of a plant, usually at the soil line; in grafted woody plants, the rootstock portion of the plant near the soil surface
culm—the stem of grasses, cereals, and bamboos
cultivar (abbr. cv.)—a cultivated variety of a plant species, resulting from deliberate genetic manipulation and having recognizable characteristics, such as color; shape of flowers, fruits, and seeds; height; and form (syn. variety)
cultural practices—the methods by which plants are grown (e.g., application of nutrients, irrigation, cultivation, and so forth)
culture—the growth and propagation of microorganisms on nutrient media; the growth and propagation of living plants
cuticle (adj. cuticular)—the noncellular outer layer of an insect or nematode; the water-repellent, waxy layer of epidermal cells of plant parts, such as leaves, stems, and fruits
cyst—in bacteria, a specialized cell enclosed in a thick wall; in fungi, a resting structure in a protective membrane or shell-like enclosure; in nematodes, the egg-laden carcass of a female nematode; often dormant and resistant to environmental conditions
cytoplasm—the living protoplasm in a cell, not including the nucleus

damping-off—the death of a seedling before or shortly after emergence caused by decomposition of the root or lower stem (it is common to distinguish between preemergence and postemergence damping-off)
defoliation—the loss of leaves from a plant, whether normal or premature
defoliator—an insect that strips the leaves from plants
dehiscent—opening by breaking into parts
demicyclic—pertaining to the life cycle of rust fungi that lack the urediniospore (repeating) stage (*see* macrocyclic; microcyclic)
desiccate—to dry out
diagnostic (n. diagnosis)—pertaining to a distinguishing characteristic important for the identification of a disease or disorder
dichotomous—branching, often successively, into two more or less equal arms
dicot—a dicotyledon, or plant having two cotyledons (*see* monocot)
dieback (v. die back)—the progressive death of shoots, leaves, or roots, beginning at the tips
differential host—a plant host that develops disease symptoms that distinguish strains or races of a plant pathogen (syn. differential cultivar)
dikaryon (adj. dikaryotic)—an organism having two sexually compatible haploid nuclei per cell that divide simultaneously in the reproductive phase called dikaryophase
dilution end point—the stage of a serial dilution of cells or preparations at which growth or infection from a standard sample of the suspension no longer occurs
diploid—having two complete sets of chromosomes ($2n$ chromosomes) (*see* haploid; polyploid)
disease—the abnormal functioning of an organism
disease cycle—the succession of events and interactions among a host, a parasite, and the environment, from initial infection through pathogenesis to overseasoning, until another infection occurs
disinfest—to kill pathogens that have not yet initiated disease or other contaminating microorganisms that occur in or on inanimate objects (e.g., soil or tools) or on the surface of plant parts (e.g., seed)
dispersal—the spread of infectious material (inoculum) from diseased to healthy plants (syn. dissemination)
dissemination—*see* dispersal
distal—situated away from the point of attachment or origin or away from the main body (*see* proximal)
DNA—deoxyribonucleic acid; the double-stranded, helical molecule that contains genetic code information; consists of repeating units, or nucleotides, each of which is composed of deoxyribose (a sugar), a phosphate group, and a purine (adenine or guanine) or a pyrimidine (thymine or cytosine) base
dolipore septum—a type of cross-wall in fungi in the group basidiomycetes characterized by distinctive swellings and membranes in association with the septal pore
domesticate—a cultivated plant
dominant—pertaining to a phenotypic trait that is expressed in hybrid progeny of diploid organisms even if contributed by only one of the parents (*see* recessive)
dormancy (adj. dormant)—a condition of suspended growth and reduced metabolism of an organism; generally induced by internal factors or environmental conditions as a mechanism of survival
dormant—resting; living in a state of reduced physiological activity
dorsal—situated toward the back or top of the body (ant. ventral)
dwarfing—underdevelopment of a plant or plant organs caused by disease, inadequate nutrition, or unfavorable environmental conditions

echinulate—having small spines projecting from the cell wall
ecology—the study of the interactions between individual organisms, groups of organisms, and organisms and their environments
ectoparasite—a parasite that feeds from the exterior of its host (*see* endoparasite)
ectotrophic—pertaining to fungal development primarily over the root surface
edamame—a vegetable-type soybean harvested when pods are green
effuse (adj.)—stretched out, especially with reference to a filmlike growth
egg mass—a group of eggs held together by a gelatinous matrix
elicitor—a compound that when introduced into a living organism signal the synthesis or activation of another compound
ellipsoidal—elliptical in the plane section
embryo—an organism in the early stages of development, such as a young plant in the seed or a nematode before hatching from the egg
emergence—the growth of a seedling shoot through the surface of the soil
enation—an abnormal outgrowth from the surface of a stem or leaf
encapsidate—to enclose viral nucleic acid in a protein coat
encyst—to form a cyst or protective covering
endemic—native to a particular place; pertaining to a low and steady level of natural disease occurrence
endodermis—the layer of cells between the vascular tissue and the cortex of roots
endogenous—arising from inside (*see* exogenous)
endoparasite—a parasite that lives and feeds inside its host (*see* ectoparasite)
endophyte—a plant developing inside another organism; any of the various endoparasitic fungi associated with grass species
endosperm—the nutritive tissue formed within the embryo sac of seed plants
enzyme—a protein that catalyzes a specific biochemical reaction
enzyme-linked immunosorbent assay (abbr. ELISA)—a serological test in which the sensitivity of the reaction is increased by attaching an enzyme that produces a colored product in response to one of the reactants
epicotyl—the portion of the stem of a plant embryo or seedling above the node where the cotyledons are attached (*see* hypocotyl)
epidemic—an increase of disease in a population; a general and serious outbreak of disease (*see* epiphytotic)
epidemiology (adj. epidemiological)—the study of factors influencing the initiation, development, and spread of infectious disease; the study of disease in populations
epidermis (adj. epidermal)—the surface layer of cells of leaves and other plant parts
epinasty—the abnormal downward curling of a leaf, leaf part, or stem
epiphyllous—located on the upper surface of a leaf
epiphyte—an organism growing on a plant surface but not as a parasite
epiphytotic—pertaining to the widespread and destructive outbreak of a plant disease
eradicant—a chemical used to eliminate a pathogen from a host or environment
eradication—the control of plant disease by eliminating the pathogen after it has become established or by eliminating the plants that carry the pathogen
erumpent—bursting or erupting through the surface of a substrate
ethylene—a plant hormone influencing vegetative growth, fruit ripening, abscission of plant parts, and senescence of flowers
etiolation—the elongation of stems caused by low levels of light intensity
etiology—the study of the causes of diseases

eukaryote (adj. eukaryotic)—any of various organisms whose cells contain a membrane-bound nucleus and other organelles, including all higher plants, animals, fungi, and protists (*see* prokaryote)
exclusion—the act of keeping a pathogen from moving to a new location
exogenous—originating from the outside (*see* endogenous)
extracellular—not occurring or located within a cell
exudate—a liquid excreted by or discharged from diseased tissues, roots, leaves, and fungi

f. sp. (pl. ff. spp.)—*forma specialis* (pl. *formae speciales*); a taxonomic group within a pathogenic species, defined by its host range (members of different *formae speciales* infect different groups of plants)
facultative—capable of changing lifestyle (e.g., from saprophytic to parasitic or the reverse)
facultative parasite—an organism that is normally saprophytic but capable of living as a parasite
facultative saprophyte—an organism that is normally parasitic but capable of living as a saprophyte
fallow—pertaining to cultivated land kept free from a crop or weeds during the normal growing season
fasciation—the malformation of shoots or floral organs manifested as enlargement and flattening, as if several parts were fused
fascicle (adj. fasciculate)—a small group, bundle, or cluster
fastidious—with reference to prokaryotic organisms, having special requirements for growth and nutrition
feeder root—a fine root that absorbs water and dissolved nutrients
filament (adj. filamentous)—a thin, flexible, threadlike structure
filamentous—threadlike (syn. filiform)
filiform—long, needlelike (syn. filamentous)
flaccid—wilted; lacking turgor
flagellum (pl. flagella)—a hairlike, whiplike, or tinsel-like appendage of a motile cell, bacterium, or zoospore that provides locomotion
fleck—a minute, discolored spot in green tissue
flexuous—having turns or windings; capable of bending
focus (pl. foci)—a small area of diseased plants within a larger plot or field
foliar—pertaining to leaves
forma specialis (pl. *formae speciales*)—a taxonomic group within a pathogenic species, defined by its host range (members of different *formae speciales* infect different groups of plants); abbr. f. sp. (pl. ff. spp.)
fructification—any of the various spore-bearing organs formed by macro- and microfungi
fruiting body—any of the various complex, spore-bearing structures of fungi
fumigant (v. fumigate)—a gas or volatile substance used to kill or inhibit the growth of microorganisms or other pests
Fungi Imperfecti—a group of fungi lacking a sexual stage; a group comprising the asexual stages of fungi in the groups ascomycetes and basidiomycetes (syn. deuteromycetes)
fungicide (adj. fungicidal)—a chemical or physical agent that kills or inhibits the growth of fungi
fungistat (adj. fungistatic)—a compound that inhibits fungal growth or sporulation but does not kill fungi
fusiform—spindle shaped; tapering at each end

gall—an abnormal swelling or localized outgrowth, often roughly spherical, produced by a plant as a result of attack by a fungus, bacterium, nematode, insect, or other organism (syn. tumor)
gametangium (pl. gametangia)—a cell containing gametes or nuclei that act as gametes
gamete—sex cell
gelatinous—resembling gelatin or jelly
gene—a unit located on a chromosome and controlling the transmission of a heritable characteristic
gene silencing—the interruption of suppression of the expression of a gene at the transcriptional or translational level
gene stacking—the insertion of two more genes into the genome
genetic—pertaining to heredity or heritable characteristics
genetically modified organism (abbr. GMO)—an organism possessing a gene from another species; having been the subject of genetic engineering
geniculate—bent, like a knee
genome—the complete genetic information of an organism or virus
genotype—the genetic constitution of an individual or group; a class or group of individuals sharing a specific genetic makeup (*see* phenotype)
genus (pl. genera)—a taxonomic category that consists of species that are closely related structurally or phylogenetically (the genus or generic name is the first name in a Latin binomial)
germ tube—a hypha resulting from an outgrowth of the spore wall and cytoplasm after germination
germinate (n. germination)—to begin growth, as of a seed, spore, sclerotium, or other reproductive body
germplasm—the bearer of heredity material (often loosely applied to cultivars and breeding lines)
giant cell—an enlarged, multinucleate cell formed in a root by repeated nuclear division without cell division; induced by secretions of certain sedentary plant-parasitic nematodes
girdle—to circle and cut through a stem or the bark and outer few rings of wood, disrupting the phloem and xylem
globose—nearly spherical
graft transmission—the transmission of a pathogen from one host plant to another through fusion of living tissue from the diseased host with living tissue of a healthy host
gram-negative bacteria—bacteria that are stained red or pink after treatment with Gram's stain
gram-positive bacteria—bacteria that are stained violet or purple after treatment with Gram's stain
grasshopper—a plant-eating insect with long hind legs that are used for jumping and for producing a chirping sound
guttation—the exudation of watery, sticky liquid from hydathodes, especially along leaf margins

haploid—having a single complete set of chromosomes (*see* diploid; polyploid)
hardiness—the ability to withstand stress
haustorium (pl. haustoria)—a specialized branch of a parasite formed inside a host cell to absorb nutrients
herbaceous—pertaining to primary, soft, nonwoody tissue of a plant or plant part; having the characteristics of an herb
herbicide—any of various chemicals that kill plants or inhibit their growth (e.g., a weed or grass killer)
hermaphrodite (adj. hermaphroditic)—having both male and female reproductive organs in a single individual
heteroecious—with reference to rust fungi, passing different stages of the life cycle on alternate hosts (*see* autoecious)
heterokaryon (adj. heterokaryotic)—a cell with genetically different nuclei
heterothallism (adj. heterothallic)—a fungal life cycle in which sexual reproduction can occur only in the presence of genetically different mycelia (*see* homothallism)
heterozygous—having the same form (allele) of a gene on different chromosomes (*see* homozygous)
hilum—a scar on a seed that marks the point of attachment of the ovule
histopathology—the microscopic examination of diseased tissue
holomorph—a fungus having sexual and asexual forms in its life cycle, considered in all its forms (*see* anamorph; teleomorph)
homothallism (adj. homothallic)—a fungal life cycle in which sexual reproduction occurs in a single, self-fertile thallus (*see* heterothallism)
homozygous—having the same form (allele) of a gene on homologous chromosomes (*see* heterozygous)
honeydew—a sugary ooze or exudate often secreted by aphids; a characteristic symptom of ergot
host plant—a living plant attacked by or harboring a parasite or pathogen, from which the invader obtains part or all of its nourishment
host range—the range of plants on which an organism (particularly a parasite) feeds
hull—the outer coat of a seed
hyaline—transparent or nearly so; translucent; colorless
hybrid (v. hybridize)—the offspring of two individuals of different genotypes
hydathode—an epidermal leaf structure specialized for the secretion or exudation of water; the opening at the terminus of a leaf vein
hymenium—the continuous, spore-bearing layer of a fungal fruiting body
hyperplasia (adj. hyperplastic)—an abnormal increase in the number of cells, often resulting in the formation of galls or tumors

hypersensitive—extremely or excessively sensitive, often with reference to an extreme reaction to a pathogen

hypertrophy (adj. hypertrophic)—an abnormal increase in the size of cells in a tissue or organ, often resulting in the formation of galls or tumors

hypha (pl. hyphae; adj. hyphal)—a tubular filament of a fungal thallus or mycelium; the basic structural unit of a fungus

hypocotyl—the portion of the stem below the cotyledons and above the root (*see* epicotyl)

icosahedral (n. icosahedron)—having 20 faces, like polyhedral virus particles (*see* isometric)

immune—not capable of being infected by a given pathogen

immunogenic—able to induce the production of antibodies

imperfect state—the anamorph, or asexual form in the life cycle of certain fungi, in which asexual spores (e.g., conidia) or no spores are produced (*see* perfect state)

in vitro—in glass, on an artificial medium, or in an artificial environment; not within a living host

in vivo—within a living organism

inclusion body—a structure developed within a plant cell as a result of infection by a virus; often useful in identifying the virus

indicator plant—a plant in which specific or distinctive symptoms develop in reaction to a pathogen or certain environmental conditions; used to detect or identify the pathogen or determine the effects of the environmental conditions

infection—the process in which an organism enters, invades, or penetrates and establishes a parasitic relationship with a host plant

infection court—a site in or on a host plant where infection can occur

infection cushion—an organized mass of hyphae formed on the surface of a plant and producing numerous infective hyphae

infection focus—the initial site of infection, generally with reference to a population of plants

infection peg—the specialized, narrow hyphal strand located on the underside of an appressorium and penetrating host cells (syn. penetration peg)

infectious—with reference to disease, capable of spreading from plant to plant

infective—able to attack a host and cause infection; with reference to vectors, carrying or containing a pathogen and able to transfer it to a host plant

infest (n. infestation)—to attack as a pest (used especially for insects and nematodes); to contaminate (as with microorganisms); to be present in large numbers

inflorescence—a flower or flower cluster

initial inoculum—inoculum (usually from an overwintering source) that initiates disease in the field, as opposed to inoculum that spreads disease during the growing season (syn. primary inoculum)

injury—damage caused by transitory interaction with an agent such as an insect, chemical, or unfavorable environmental condition

inoculate (n. inoculation)—to place inoculum in an infection court; to insert a pathogen into healthy tissue

inoculum (pl. inocula)—a pathogen or its parts capable of causing infection when transferred to a favorable location

inoculum density—a measure of the number of propagules of a pathogenic organism per unit of area or volume

insecticide—a chemical used to kill insects

intercalary—inserted within (e.g., located along a hypha, as opposed to located at the end of a hypha)

intercellular—between or among cells

intercrop—to grow two or more crops simultaneously on the same area of land

internode (adj. internodal)—the portion of a stem between two successive nodes

interveinal—between veins (of a leaf)

intracellular—through or within cells

isolate—(n.) a culture or subpopulation of a microorganism separated from its parent population and maintained under controlled conditions; (v.) to remove from soil or host material and grow in pure culture

isometric—with reference to virus particles, having an icosahedral structure and thus appearing approximately round

juvenile—an immature form that appears similar to the adult stage but is usually smaller and is not sexually mature, as in nematodes or insect species that undergo gradual metamorphosis

knot—a gall or localized abnormal swelling

Koch's postulates—a set of criteria used to establish a causative relationship between a microbe and a disease

lamina—the expanded part of a leaf (*see* petiole)

land race—a locally developed strain of a plant

larva (pl. larvae)—the immature wormlike or caterpillar stage in the life cycle of an insect that undergoes complete metamorphosis

latent—present but not manifested or visible, as a symptomless infection

latent infection—an infection that does not have visible symptoms

latent period—the time elapsed between infection and the appearance of symptoms or production of new inoculum (sometimes synonymous with "incubation period"); the time elapsed after the acquisition of a pathogen by a vector but before the pathogen can be transmitted

Latin binomial—the convention established by the formal system for naming organisms in which the first word in a name identifies the genus and the second word identifies the specific epithet or species within the genus

leaf spot—a lesion typically restricted in development after reaching a characteristic size

leafhopper—any of various species of mobile insects with sucking mouthparts in the suborder Homoptera

leaflet—one of the separate blades or divisions of a compound leaf

legume—a simple, dry, dehiscent fruit developing from a simple pistil and splitting at maturity along two seams; any plant in the family Fabaceae (formerly Leguminosae)

lenticel—a natural opening in the surface of a stem, tuber, fruit, or root, permitting gas exchange

lenticular—lens shaped (convex on both faces)

lesion—a localized diseased area or wound

life cycle—the cyclical stages in the growth and development of an organism

lignin—a complex organic substance or group of substances that impregnate the cell walls of xylem vessels and certain other plant cells; constituting wood

local lesion—a small, restricted lesion; often the characteristic reaction of a differential cultivar to a specific pathogen, especially in response to mechanical inoculation with a virus

lodge—with reference to hay or grain crops, to fall over

lumen (pl. lumina)—the central cavity of a cell or other structure

macerate—to cause disintegration of tissues by separation of cells; to soften by soaking

macroconidium (pl. macroconidia)—the larger of two kinds of conidia formed by certain fungi (*see* microconidium)

macrocyclic—pertaining to the life cycle of rust fungi that typically exhibit all five spore stages (pycniospores, aeciospores, urediniospores, teliospores, and basidiospores) in the course of their development (*see* demicyclic; microcyclic)

macronematous—pertaining to a conidiophore that is morphologically different from vegetative hypha

manual transmission—the spread or introduction of inoculum to an infection court by hand

mating types—compatible strains (usually designated + and – or A and B), both of which are necessary for sexual reproduction in heterothallic fungi

maturity group—a group of soybean genotypes defined by adaptation within certain latitudes

mechanical injury—injury of a plant part by abrasion, mutilation, or wounding

mechanical transmission—the spread or introduction of inoculum to an infection court by human manipulation, accompanied by physical disruption of host tissues (wounding)

medium (pl. media)—a mixture of organic or inorganic chemical compounds and water providing the nutrients needed for the growth of a microorganism in vitro; with reference to higher plants, a mixture of fertilizers and other components in which plants are grown

meiosis—nuclear division in which the number of chromosomes per nucleus is halved (i.e., the diploid state is converted to the haploid state) (*see* mitosis)
melanin (adj. melanoid)—a brown-black pigment
meristem (adj. meristematic)—plant tissue characterized by frequent cell division, producing cells that become differentiated into specialized tissues
meristem culture—an aseptic culture of a plant or plant part from a portion of the meristem
mesophyll—the central, internal, nonvascular tissue of leaves, consisting of the palisade and spongy mesophyll
messenger RNA (abbr. mRNA)—a form of RNA that carries information to direct the synthesis of protein
metabolite—any chemical participating in metabolism; a nutrient
microclimate—weather conditions on a small scale, as at the surface of a plant or within a field
microconidium (pl. microconidia)—the smaller of two kinds of conidia formed by certain fungi (*see* macroconidium)
microcyclic—pertaining to the life cycle of rust fungi that produce only teliospores and basidiospores (*see* demicyclic; macrocyclic)
microflora—the combination of all of the microorganisms in a particular environment
microorganism—an organism small enough that it can be seen only with the aid of a microscope (syn. microbe)
microsclerotium (pl. microsclerotia)—a microscopic, dense aggregate of darkly pigmented, thick-walled hyphal cells
middle lamella—the layer between the walls of adjacent plant cells; consists largely of pectic substances
midrib—the central, thickened vein of a leaf
migratory—moving from place to place on a plant or from plant to plant when feeding (*see* sedentary)
mildew—a thin coating of mycelial growth and spores on the surfaces of infected plant parts
mite—a minute arachnid that has four pairs of legs
mitochondrion (pl. mitochondria)—a cellular organelle outside the nucleus; functions in respiration
mitosis—nuclear division in which the number of chromosomes per nucleus remains the same (*see* meiosis)
MLO—mycoplasmalike organism (*see* phytoplasma)
mold—any microfungus with conspicuous, profuse, or woolly superficial growth (mycelium or spore masses) on various substrates; commonly grow on damp or decaying matter and on the surfaces of plant tissues, especially with reference to economically important saprobes
mollicute—any of a group of prokaryotic organisms bounded by a flexuous membrane and lacking a cell wall (*see* phytoplasma; spiroplasma)
molt—to shed a cuticle or body encasement during a phase of growth
monocot—a monocotyledon, or a plant having only one cotyledon, such as species of grasses, including grain crops and corn (*see* dicot)
monoculture—the cultivation of plants of the same species in close proximity and with few or no other plant species present
monoecious—having male and female reproductive organs on a single individual
monogenic—determined by a single gene (*see* polygenic)
morphology (adj. morphological)—the study of the form of organisms; the form and structure of organisms
mosaic—a disease symptom characterized by nonuniform coloration, with intermingled normal, light-green and yellowish patches; usually caused by a virus (often used interchangeably with "mottle")
motile—capable of self-propulsion by means of flagella, cilia, or amoeboid movement
mottle—a disease symptom characterized by light and dark areas in an irregular pattern; usually caused by a virus (often used interchangeably with "mosaic")
mucilaginous—viscous; slimy
mulch—a layer of material, such as organic matter or plastic film, applied to the surface of the soil for purposes such as retention of water and inhibition of weeds
multinucleate—having more than one nucleus per cell (*see* binucleate; uninucleate)
multiseptate—having many septa (cross-walls)
muriform—having transverse and longitudinal septa
mutation—an abrupt heritable or genetic change in a gene or individual as a result of an alteration in a gene or chromosome or an increase in chromosome number
mycelium (pl. mycelia; adj. mycelial)—the mass of hyphae constituting the body (thallus) of a fungus
mycoparasite—a fungus that attacks another fungus
mycoplasmalike organism (abbr. MLO)—*see* phytoplasma
mycorrhiza (pl. mycorrhizae; adj. mycorrhizal)—a symbiotic association between a nonpathogenic or weakly pathogenic fungus and the roots of a plant
mycotoxin—any poisonous compound produced by a fungus

necrosis (adj. necrotic)—the death of cells or tissue; usually accompanied by blackening or browning
necrotroph—a parasite that typically kills and obtains its energy from dead host cells (*see* biotroph)
nectary—pertaining to a gland that secretes nectar (sugar-rich liquid)
nematicide—an agent that kills nematodes; usually a chemical
nematode—any of various unsegmented roundworms (animals) that is parasitic on plants or animals or free living in soil or water
node (adj. nodal)—the enlarged portion of a shoot at which leaves, branches, or recemes are attached
nodule—a small knot or irregular, rounded lump; with reference to leguminous plants, a structure located on the root and containing nitrogen-fixing bacteria
noninfectious disease—a disease caused by an abiotic agent and not capable of being transmitted from plant to plant
nonpersistent transmission—virus transmission in which a virus is acquired and transmitted by a vector after a short feeding time and retained by the vector for only a short time (syn. stylet-borne transmission)
nonseptate—lacking cross-walls; coenocytic
nymph—the juvenile stage in the life cycle of an insect with incomplete metamorphosis; superficially resembles the adult

obclavate—shaped like an upside-down club
obligate parasite—an organism that can grow only as a parasite in association with its host plant and cannot be grown in artificial culture media (syn. biotroph)
obovoid—shaped like an upside-down egg
obtuse—rounded or blunt; pertaining to an angle greater than 90°
oogonium (pl. oogonia)—a female gametangium of fungi in the group oomycetes containing one or more gametes
oomycetes (adj. oomycetous)—a group of funguslike organisms in the phylum Oomycota (kingdom Chromista); typically has nonseptate mycelium, asexual sporangia and zoospores, and sexual oospores
oospore—the thick-walled, sexually derived resting spore of organisms in the group oomycetes
ooze—a mass of bacterial cells mixed with host fluids
organelle—any of various membrane-bound structures contained within cells and having a specialized function (e.g., mitochondria and chloroplasts)
ostiole (adj. ostiolate)—a pore; the opening in the papilla or neck of a perithecium, pseudothecium, or pycnidium through which spores are released
ovary—the female reproductive structure of an organism; in plants, the enlarged basal portion of the pistil, containing the ovules and developing into the fruit
overseason—to survive or persist from one planting season to the next
oversummer—to survive or persist through the summer
overwinter—to survive or persist through the winter
oviposit—to deposit or lay eggs with an ovipositor
ovule—an enclosed structure that after fertilization becomes a seed; an egg contained within an ovary
ozone—a highly reactive form of oxygen (O_3) injurious to plants

palisade parenchyma—the tissue located beneath the upper epidermis of leaves; composed of elongate, tubular cells arranged upright in the manner of posts in a palisade fortification
papilla (pl. papillae; adj. papillate)—a small, blunt projection
paraphysis (pl. paraphyses)—a hairlike cell within a fungal fruiting structure
parasexual—pertaining to the recombination of genetic characters without a sexual process

parasite (adj. parasitic)—an organism that lives in intimate association with another organism, on which it depends for its nutrition; not necessarily a pathogen
parasitic plant—a plant that derives some or all of its nutritional requirements from another living plant
parenchyma (adj. parenchymatous)—the soft tissue of living plant cells with undifferentiated, thin, cellulose walls
parthenogenesis (adj. parthenogenetic)—reproduction by the development of an unfertilized egg
pasteurization—the process by which a material (usually a liquid) is treated with heat to eliminate selected harmful microorganisms
pathogen (adj. pathogenic)—a disease-producing organism or agent
pathogenesis—the initiation and development of disease
pathogenicity—the ability to cause disease
pathology—the study of diseases
pathotoxin—a chemical of biological origin (not an enzyme) that plays an important role in a plant disease
pathotype—a subdivision of a pathogenic species characterized by its pattern of virulence or avirulence to differential host varieties
pathovar (abbr. pv.)—a subdivision of a plant-pathogenic bacterial species, defined by its host range (a pathovar of a bacterial species is analogous to a *forma specialis* of a fungal species)
pedicel—a small, slender stalk; a stalk bearing an individual flower, inflorescence, or spore
peduncle—the stalk or main stem of an inflorescence; part of an inflorescence or fructification
penetration—the initial invasion of a host by a pathogen
penetration peg—the specialized, narrow hyphal strand located on the underside of an appressorium and penetrating host cells (syn. infection peg)
perennial—a plant that survives for several to many years (*see* annual; biennial)
perfect—sexual; capable of sexual reproduction
perfect state—the teleomorph, or sexual form in the life cycle of certain fungi (*see* imperfect state)
pericarp—the outer layer of a seed or fruit
perithecium (pl. perithecia)—the flask-shaped or subglobose, thin-walled fruiting body (ascocarp) of an ascomycetous fungus, containing asci and ascospores and having a pore (ostiole) at its apex, through which ascospores are expelled or released
peritrichous—having hairs or flagella distributed over the whole surface
persistent transmission—virus transmission in which a virus is acquired and transmitted by a vector after a relatively long feeding time and remains transmissible for a prolonged period while in association with the vector (syn. circulative transmission)
pest—any organism that damages plants or plant products
pesticide—any of various chemicals used to control pests
petiole—the part of a leaf by which it attaches to the stem (*see* lamina)
pH—the negative logarithm of the effective hydrogen ion concentration; a measure of acidity (pH 7 is neutral; values less than pH 7 are acidic, and values greater than pH 7 are alkaline)
phenotype—the composite of observable physical qualities of an organism produced by the interaction of its genotype with the environment
phialide—the end cell of a conidiophore with one or more open ends, through which a basipetal succession of conidia develops
phialospore—a conidium produced on a phialide
phloem—the food-conducting, food-storing tissue in the vascular system of roots, stems, and leaves
photochemical oxidant—any of various highly reactive compounds formed by the action of sunlight on less toxic precursors
photoperiod—the relative length of the periods of light and dark often associated with floral induction
photosynthate—a chemical product of photosynthesis
photosynthesis—the manufacture of carbohydrates from carbon dioxide and water in the presence of chlorophylls; uses light energy and releases oxygen
phycomycete—an obsolete term for any member of a group of fungi that lacks cross-walls (septa) in its mycelium
phyllody—a disorder in which floral organs are transformed into leaflike structures
phyllotaxy—the arrangement of leaves on a stem in relation to one another
physiological race—a group of individuals belonging to the same species and differing from other members of the species in their behavior or other characteristics but not in morphology; a physiological form
phytoalexin—any of various substances produced by higher plants in response to chemical, physical, or biological stimuli and inhibiting the growth of certain microorganisms
phytopathology—the study of plant diseases (syn. plant pathology)
phytoplasma—any of various plant-parasitic pleomorphic mollicutes (prokaryotes lacking cell walls) found in phloem tissue and not capable of growth on artificial nutrient media (previously called "mycoplasmalike organisms" (MLOs))
phytotoxic—harmful to plants; usually with reference to chemicals
pigment—a colored compound, such as chlorophyll, in the cells of plants or fungi
pionnotes—a spore mass with a fat- or grease-like appearance
pistil—the ovule-bearing organ of a plant, consisting of the ovary and its appendages (e.g., the style and stigma)
pith—parenchymatous tissue occupying the center of a stem
plant introduction—a germplasm accession; often referred to with the abbreviation "PI"
plant pathology—the study of plant diseases (syn. phytopathology)
plasmodium (pl. plasmodia)—a naked multinucleate mass of protoplasm moving and feeding in amoeboid fashion
pleomorphic—able to assume various shapes and perhaps sizes; having a life cycle characterized by a succession of distinctly different forms
plumule—a rudimentary shoot of the plant embryo
pod—a developing ovary after fertilization
polar—located at one end (pole) of the cell
pollen—male sex cells produced by the anthers of flowering plants or the cones of seed plants
pollination—the transfer of pollen from anther to stigma or from a staminate cone to an ovulate cone
polygenic—pertaining to or governed by many genes (*see* monogenic)
polymerase chain reaction (abbr. PCR)—a technique used to amplify DNA several orders of magnitude
polyploid—having three or more complete sets of chromosomes (*see* haploid; diploid)
positive-sense RNA—RNA that can serve directly as messenger RNA
posterior—situated toward the back or tail (*see* anterior)
potato dextrose agar (abbr. PDA)—a common medium used to culture mostly fungi
predispose (n. predisposition)—to make prone to infection and disease
primary inoculum—inoculum (usually from an overwintering source) that initiates disease in the field, as opposed to inoculum that spreads disease during the season (syn. initial inoculum)
primary leaf—the first true leaf that emerges from a plant following the cotyledons
primary root—a root that develops directly from the radicle of an embryo rather than from a crown or node
prokaryote (adj. prokaryotic)—any of various organisms lacking internal membrane-bound organelles and a distinct nucleus, such as bacteria and mollicutes (*see* eukaryote)
promycelium (pl. promycelia)—in rust and smut fungi, a germ tube issuing from a teliospore and bearing basidiospores
propagative virus—a virus that multiplies within its arthropod vector
propagule—any part of an organism capable of independent growth
protectant—an agent (usually a chemical) applied to a plant surface in advance of exposure to a pathogen for the purpose of preventing infection
protein—any of numerous nitrogen-containing organic compounds containing combinations of amino acids
protoplasm—the living contents of a cell
protoplast—the living contents of a cell, exclusive of the cell wall
proximal—situated toward or near the point of attachment or main body (*see* distal)
pseudothecium (pl. pseudothecia)—a perithecium-like fruiting body containing asci and ascospores dispersed rather than in an organized hymenium; an ascostroma with a single locule (cavity) containing bitunicate asci
pupa (pl. pupae; v. pupate)—the quiescent stage between the larval and adult stages of certain insects

pustule—a small, blisterlike elevation of the plant epidermis that forms as spores emerge
pv.—*see* pathovar
pycnidiospore—a spore (conidium) produced in a pycnidium
pycnidium (pl. pycnidia)—an asexual, globose or flask-shaped fruiting body of certain imperfect fungi, producing conidia
pycniospore—a haploid, sexually derived spore formed in the pycnium of a rust fungus (syn. spermatium)
pycnium (pl. pycnia)—the globose or flask-shaped haploid fruiting body of rust fungi, bearing receptive hyphae and pycniospores (syn. spermagonium)
pyriform—pear shaped

quarantine—an enforced isolation imposed to prevent the spread of disease; pertaining to plant disease, control of the transport of plants or plant parts to prevent the spread of pests or pathogens
quiescent—dormant; inactive

race—a subgroup or biotype within a species or variety that is distinguished from other races by virulence, symptom expression, or host range but not by morphology
raceme—the reproductive unit or inflorescence without branches and flowers attached to a central stalk with pedicels
rachis—the elongated main axis of an inflorescence
radicle—the part of the plant embryo that develops into the primary root
receptacle—the structure of a flower that bears the reproductive organs
receptive hypha—in rust fungi, the part of a pycnium (spermagonium) that receives the nucleus of a pycniospore (spermatium)
recessive—pertaining to a phenotypic trait that is expressed in the progeny of diploid organisms only if contributed by both parents (*see* dominant)
reniform—kidney shaped
resistant (n. resistance)—possessing properties that prevent or impede disease development (*see* susceptible)
respiration—a series of chemical reactions that make energy available through oxidation of carbohydrates and fat
resting spore—a spore (often thick walled) that can remain alive in a dormant state for some time, germinating later and capable of initiating infection
reticulate—having netlike markings
Rhizobium—a genus of bacteria that live symbiotically with roots of leguminous plants, converting atmospheric nitrogen into a form useable by the plants
***Rhizobium* nodules**—root galls caused by *Rhizobium* spp.
rhizome—a mostly horizontal, jointed, fleshy, often elongated, usually underground stem
rhizomorph—a macroscopic ropelike strand of compacted tissue formed by certain fungi
rhizosphere—the microenvironment in the soil immediately around a plant's roots
ribosome—a subcellular protoplasmic particle made up of one or more RNA molecules and several proteins; involved in protein synthesis
ring spot—a disease symptom characterized by yellowish or necrotic rings enclosing green tissue, as in some plant diseases caused by viruses
RNA—ribonucleic acid; any of several nucleic acids composed of repeating units of ribose (a sugar), a phosphate group, and a purine (adenine or guanine) or a pyrimidine (uracil or cytosine) base, transcribed from DNA and involved in translation to proteins
RNA interference (abbr. RNAi)—the inhibition of gene expression by RNA molecules
rogue—to remove and destroy individual plants that are diseased, infested by insects, or otherwise undesirable
root cap—a group of cells that protects the growing tip of a root
root hair—a threadlike, single-celled outgrowth from a root epidermal cell
rosette—a disease symptom characterized by short, bunchy growth habit caused by a shortening of internodes with no comparable reduction in leaf size
rot—softening, discoloration, and often disintegration of plant tissue as a result of fungal or bacterial infection
rotation—the practice of growing different kinds of crops in succession in the same field
rugose—wrinkled; roughened
runner—a slender, horizontal stem that grows close to the soil surface (syn. stolon)
runner plants—new plants produced asexually on a runner (stolon)
russet—a brownish, roughened area on the skin of fruit, resulting from cork formation
rust—any of several diseases caused by specialized fungi in the group basidiomycetes, some of the spores of which are a rusty color

sanitation—the destruction or removal of infected or infested plants or plant parts; the decontamination of tools, equipment, containers, work space, hands, and so on
sap feeder—an insect that feeds on vascular fluids
sap transmissible—pertaining to a virus that spreads via the direct transfer of plant sap
saprobe—*see* saprophyte
saprophyte (adj. saprophytic)—an organism that obtains nourishment from nonliving organic matter (syn. saprobe)
satellite virus—a virus that accompanies another virus and depends on it for its multiplication
scab—a roughened, crustlike diseased area on the surface of a plant organ
scald—a necrotic condition in which tissue is usually bleached and has the appearance of having been exposed to high temperature
sclerenchyma (adj. sclerenchymatous)—tissue made up of thick-walled plant cells
sclerotium (pl. sclerotia)—a vegetative, resting body of a fungus composed of a compact mass of hyphae with or without included host tissue; usually has a darkened rind
scorch—a symptom that suggests the action of flame or fire on the affected tissue, often at the margins of leaves
secondary infection—an infection resulting from the spread of infectious material from a primary or secondary infection without an intervening inactive period
secondary inoculum—inoculum produced by an infection established earlier in the same growing season
secondary organism—an organism that multiplies in already diseased tissue but is not the primary pathogen
secondary root—a branch of a primary root
sedentary—remaining in a fixed location (*see* migratory)
seed—a ripened ovule consisting of an embryo and stored food enclosed by a seed coat
seed transmission—primarily, the virus infection of plants from one generation to the next generation
seed treatment—the application of a biological agent, chemical substance, or physical treatment to seed to protect the seed or resulting plant from pathogens or to stimulate germination or plant growth
seedborne—carried on or in seed
selective medium—a culture medium containing substances that inhibit or prevent the growth of certain species of microorganisms
senesce (adj. senescent; n. senescence)—to decline, as with maturation, age, or disease stress
sepal—one of the modified leaves composing a calyx
septate—having septa (cross-walls)
septum (pl. septa; adj. septate)—a dividing wall; in fungi, a cross-wall
serology (adj. serological)—a method using the specificity of the antigen–antibody reaction for the detection and identification of antigenic substances and the organisms that carry them
sessile—with reference to leaves, leaflets, flowers, florets, fruits, ascocarps, basidiocarps, and so on, lacking a stalk, petiole, pedicel, stipe, or stem; with reference to nematodes, permanently attached and not capable of moving about
seta (pl. setae)—a bristle- or hairlike structure, usually deep yellow or brown and thick walled
sexual spore—a spore produced during the sexual cycle of a fungus
sheath—the lower part of a grass leaf, which clasps the culm; a membranous cover
shot hole—a symptom in which small lesions on a leaf fall out, giving the leaf the appearance of being hit by buckshot
sieve element—a tube-shaped living cell in the phloem, functioning in the transport of dissolved organic substances in the plant (syn. sieve tube element)
sign—an indication of disease from direct observation of a pathogen or its parts on a host plant (*see* symptom)

slime molds—saprophytic organisms in the class Myxomycetes forming vegetative amoeboid plasmodia and spores

smut—any of several diseases caused by fungi in the group basidiomycetes that typically release masses of black, dusty teliospores at maturity

soft rot—a mushy disintegration of fleshy fruit or vegetables as a result of bacterial or fungal infection

soil drench—the application of a solution or suspension of a chemical to the soil, especially a pesticide to control soilborne pathogens

soilborne—carried on or beneath the soil surface

solarization—a disease management practice in which soil is covered with polyethylene sheeting to increase heating by sunlight and thus control soilborne plant pathogens

sooty mold—black, nonparasitic, superficial fungal growth; often develops on honeydew produced by aphids and other phloem-feeding insects

sorus (pl. sori)—a compact fruiting structure, especially with reference to spore masses of rust and smut fungi

sp. (pl. spp.)—abbreviations of "species"; used only after genus names; the singular form is used to refer to an undetermined species, and the plural form is used to refer to two or more species without naming them individually (*see* species)

species—a group of living organisms that consists of similar individuals capable of exchanging genes or interbreeding

spermagonium (pl. spermagonia)—a structure in which male reproductive cells are produced; in rust fungi, the pycnium, a globose or flask-shaped haploid fruiting body composed of receptive hyphae and spermatia (pycniospores)

spermatium (pl. spermatia)—a male sex cell; a nonmotile male gamete; a haploid male gamete; in rust fungi, a pycniospore

spicule—the copulatory organ of a male nematode

spiroplasma—any of various spiral-shaped, plant-pathogenic mollicutes (prokaryotes lacking cell walls)

sporangiophore—the sporangium-bearing body of a fungus

sporangiospore—a nonmotile, asexual spore borne in a sporangium

sporangium (pl. sporangia)—a saclike fungal structure, the entire contents of which are converted into an indefinite number of asexual spores

spore—the reproductive structure of fungi and some other organisms, containing one or more cells; a bacterial cell modified to survive an adverse environment

sporidium (pl. sporidia)—the basidiospore of rust fungi, smut fungi, and other members of the group basidiomycetes

sporocarp—a spore-bearing fruiting body

sporodochium (pl. sporodochia)—a superficial, cushion-shaped, asexual fruiting body consisting of a cluster of conidiophores

sporulate—to produce spores

sporulating—producing and (often) liberating spores

spot—a small necrotic area occurring as a disease symptom on a leaf, flower, or stem

stamen (adj. staminal)—the male structure of a flower, composed of a pollen-bearing anther and a filament, or stalk

staminate flower—a male flower

stele—the central cylinder of vascular tissue, especially with reference to roots

stem feeder—an insect that prefers feeding on plant stems

stem pitting—a viral disease symptom characterized by depressions on the stem

sterigma (pl. sterigmata)—a small, usually pointed projection that supports a spore

sterile—unable to reproduce sexually; free of living microorganisms

sterile fungus—a fungus that is not known to produce any kind of spores

sterilization (adj. sterilized)—the total destruction of living organisms by various means, including heat, chemicals, and irradiation

stem borer—an insect larva that bores in plant stems

stigma—the portion of a flower that receives pollen and on which pollen germinates

stinkbug—a broad, shield-shaped insect (suborder Heteroptera) that feeds on the sap of plants and may be brightly colored

stipe—a stalk

stippling—a pattern of small dots or speckles on leaves where chlorophyll is absent

stipule—a small, leaflike appendage at the base of a leaf petiole; usually occurs as one of a pair

stolon—a slender, horizontal stem that grows close to the soil surface (syn. runner); in fungi, a hypha that grows horizontally along a surface

stoma, stomate (pl. stomata; adj. stomatal)—a structure composed of two guard cells and the opening between them in the epidermis of a leaf or stem; functions in gas exchange

strain—a distinct form of an organism within a species or virus, differing from other forms of the species biologically, physically, or chemically

striate (n. striations)—to mark with delicate lines, grooves, or ridges

stroma (pl. stromata)—a compact mass of mycelium (with or without host tissue) that supports fruiting bodies or in which fruiting bodies are embedded

stunting—a reduction in the height of plants resulting from a progressive reduction in the lengths of successive internodes or a decrease in their number

style—the slender part of many pistils, located between the stigma and the ovary and through which the pollen tube grows

stylet—the stiff, slender, hollow feeding organ of plant-parasitic nematodes or sap-sucking insects, such as aphids and leafhoppers

stylet-borne transmission—virus transmission in which a virus is acquired and transmitted by a vector after a short feeding time and is retained by the vector for only a short time (syn. nonpersistent transmission)

subepidermal—located or occurring just below the epidermis

subgenomic RNA—a piece of viral RNA, shorter than the entire genome of the virus, found in cells infected by the virus and sometimes encapsidated

substrate—the substance on which an organism lives or from which it obtains nutrients; a chemical substance acted upon, often by an enzyme

sunscald—the injury of plant tissues burned or scorched by direct sunlight

suscept—a susceptible plant

susceptible (n. susceptibility)—prone to develop disease when infected by a pathogen (*see* resistant)

symbiosis (adj. symbiotic)—a mutually beneficial association of two different kinds of organisms

symbiont—either of the organisms in a symbiotic association

symptom—an indication of disease by a reaction of the host (e.g., canker, leaf spot, wilt) (*see* sign)

symptomatology—the study of disease symptoms

symptomless carrier—a plant that is infected with a pathogen (usually a virus) but has no obvious symptoms

syncytium (pl. syncytia)—a multinucleate structure in root tissue formed by the dissolution of common cell walls induced by secretions of certain sedentary plant-parasitic nematodes (e.g., cyst nematodes)

synergism (adj. synergistic)—an interaction in which the total effect of the interacting factors is greater than the additive effects of the factors acting individually

systemic—pertaining to a disease in which the pathogen (or a single infection) spreads generally throughout a plant; pertaining to chemicals that spread internally through a plant

taproot—a primary root that grows vertically downward and from which smaller lateral roots branch

taxonomy (adj. taxonomic)—the science of naming and classifying organisms

teleomorph—the perfect state, or sexual form in the life cycle of certain fungi (*see* anamorph; holomorph)

teliospore—a thick-walled resting spore produced by some fungi (notably rust and smut fungi) from which a basidium is produced

telium (pl. telia)—the fruiting body (sorus) that produces teliospores of a rust fungus

testa (pl. testae)—seed coat

thallus—the vegetative body of a fungus

thermal inactivation point—the lowest temperature at which heating for a limited period (usually 10 min) is sufficient to cause a virus to lose its infectivity or an enzyme to lose its activity

thrip—a small, winged insect (order Thysanoptera) that sucks plant sap

tissue—a group of cells, usually of similar structure, that perform the same or related functions

tissue culture—an in vitro method of propagating healthy cells from plant tissues

titer—a measure of the concentration of a substance in a solution
tolerance (adj. tolerant)—the ability of a plant to endure an infectious or noninfectious disease, adverse conditions, or chemical injury without serious damage or yield loss; with reference to pesticides, the amount of chemical residue legally permitted on an agricultural product entering commercial channels, usually measured in parts per million (ppm)
tomentose—covered with a dense mat of hair
toxicity—the capacity of a substance to interfere with the vital processes of an organism
toxin—any poisonous substance of biological origin
transgenic—possessing a gene from another species; having been the subject of genetic engineering
translocation—the movement of water, nutrients, chemicals, or food materials within a plant
translucent—clear enough to allow the transmission of light
transmit (n. transmission)—to spread or transfer, as in spreading an infectious pathogen from plant to plant or from one generation of plants to another
transovarial passage—the passage of a virus through the eggs or offspring of its vector and then to the next generation of host plants
transpiration—the loss of water by evaporation from leaf surfaces and through stomata
trichome—a plant epidermal hair, of which several types exist
trifoliolate—a leaf that contains three leaflets
triturate—to grind, as with a mortar and pestle
truncate—to end abruptly, as though an end has been cut off
tuber—an underground stem adapted for storage; typically produced at the end of a stolon
tumor—an abnormal swelling or localized outgrowth, often roughly spherical, produced by a plant as a result of attack by a fungus, bacterium, nematode, insect, or other organism (syn. gall)
turgid—swollen or inflated; plump or swollen as a result of internal water pressure
turgidity—the state of being rigid or swollen as a result of internal water pressure
tylosis (pl. tyloses)—a balloonlike extrusion of parenchyma cells into the lumina of contiguous vessels, partially or completely blocking them

ultrastructure—the submicroscopic structure of a macromolecule, cell, or tissue
unicellular—one celled (*see* multicellular)
uniflagellate—having one flagellum
unifoliolate—a leaf that contains one leaflet
uninucleate—having one nucleus per cell (*see* binucleate; multinucleate)
urediniospore, **urediospore**, **uredospore**—the asexual, dikaryotic, often rust-colored spore of a rust fungus, produced in a uredinium; the repeating stage of a heteroecious rust fungus, in which it is capable of infecting the host plant on which it was produced
uredinium (pl. uredinia), **uredium** (pl. uredia)—the fruiting body (sorus) in which urediniospores of a rust fungus are produced

vacuole—a generally spherical organelle within a plant cell, bound by a membrane and containing dissolved materials, such as metabolic precursors, storage materials, and waste products
variegation—a pattern of two or more colors in a plant part, as in a green and white leaf
variety (adj. varietal)—a plant type within a species, resulting from deliberate genetic manipulation and having recognizable characteristics, such as color; shapes of flowers, fruit, and seeds; height; and form (syn. cultivar)
vascular—pertaining to fluid-conducting tissues (xylem and phloem) in plants
vascular bundle—a strand of conductive tissue; usually composed of xylem and phloem (in leaves, small bundles are called "veins")
vascular cylinder—the cylinder of vascular tissue in stems and roots (syn. stele)
vascular wilt disease—a xylem disease that disrupts the normal uptake of water and minerals, resulting in wilting and yellowing of foliage
vector—a living organism (e.g., insect, mite, bird, higher animal, nematode, parasitic plant, human) that is able to carry and transmit a pathogen; in genetic engineering, a cloning vehicle, or self-replicating DNA molecule, such as a plasmid or virus, used to introduce a fragment of foreign DNA into a host cell
vegetative—pertaining to the somatic or asexual parts of a plant, which are not involved in sexual reproduction
vegetative propagation—asexual reproduction; in plants, the use of cuttings, bulbs, tubers, or other vegetative parts to grow new plants
vein—a small vascular bundle in a leaf
vein banding—a symptom of virus disease in which the regions along veins are darker green than the tissue between veins
vein clearing—disappearance of green color in or around leaf veins
vermiform—worm shaped
vertical resistance—resistance thought to be conferred by a single gene or a few genes and completely protecting a host but only against specific races or strains (genetic variants) of a pathogen
vesicle—the thin sac in which zoospores are differentiated and from which they are released; the bulbous head terminating the conidiophores of *Aspergillus* spp.; a structure formed by an endomycorrhizal fungus within living cells of the root
vessel—the water-conducting structure of xylem tissue, with pit openings in end walls
viable (n. viability)—alive; capable of germination (e.g., of seeds, fungus spores, sclerotia, etc.); capable of growth
virescence—a state or condition in which normally white or colored tissues (e.g., flower petals) become green
virion—a complete virus particle
viroid—any of various small, unencapsidated (naked), circular, single-stranded RNA molecules capable of causing disease in plants
virulence—the degree or measure of pathogenicity; the relative capacity to cause disease
virulent—pathogenic; having the capacity to cause disease (*see* avirulent)
viruliferous—virus laden, usually with reference to insects or nematodes as vectors
virus—any of numerous submicroscopic, intracellular, obligate parasites consisting of a core of infectious nucleic acid (either RNA or DNA) usually surrounded by a protein coat
volunteer—a self-set plant; seeded by chance

water soaking—a disease symptom in which plant tissues or lesions appear wet, dark, and usually sunken and translucent
whitefly—a minute, winged insect (suborder Homoptera) that feeds on the sap of plants
whorl—a circular arrangement of like parts
wild type—the phenotype characteristic of most individuals of a species under natural conditions
wilt—the drooping of leaves and stems caused by a lack of water (inadequate water supply or excessive transpiration); a vascular disease in which normal water uptake is interrupted
witches'-broom—a disease symptom in which many weak shoots arise at or close to the same point, producing abnormal massed, brushlike growth

xylem—the water- and mineral-conducting, food-storing, supporting tissue of a plant

yeast extract agar—a medium often used to culture plant pathogens
yellows—a plant disease characterized by chlorosis and stunting

zonate—pertaining to the targetlike development of a tree canker, characterized by successive perennial rings of callus; pertaining to any symptom appearing in concentric rings
zoosporangium—a sporangium (spore case) bearing zoospores
zoospore—a fungal spore with flagella; capable of locomotion in water
zygomycetes (adj. zygomycetous)—a group of fungi in the phylum Zygomycota (kingdom Fungi) that produce resistant, spherical sexual spores following fusion of hyphal gametangia
zygospore—the sexual resting spore formed from the union of gametangia of fungi in the group zygomycetes

Index